# Fiber Optic Chemical Sensors and Biosensors

## Volume I

Editor

**Otto S. Wolfbeis, Ph.D.**
Professor
Institute of Organic Chemistry
Karl-Franzens University
Graz, Austria

**CRC Press**
**Boca Raton  Ann Arbor  Boston  London**

**Library of Congress Cataloging-in-Publication Data**

Fiber optic chemical sensors and biosensors / Wolfbeis, Otto S.

    p.   cm.
    Includes bibliographical references
    ISBN 0-8493-5508-7
    ISBN 0-8493-5509-5
    1. Fiber optics. 2. Sensors--chemical. 3. Spectroscopy.
    I. Otto S. Wolfbeis
    QD972.P962P65  1991       91-19435
    612'.10444--dc21

International Standard Book Number 0-8493-5508-7
International Standard Book Number 0-8493-5509-5

Library of Congress Card Number 91-19435
Printed in the United States

# PREFACE

The development of optical and fiber optic sensors for chemical and clinical parameters is a tremendously fast growing area. These devices have been named "optrodes" and, later, "optodes" by Lübbers in 1975. Particularly in combination with fiber optic waveguides (an offshot of the communication industry), optrodes offer quite new possibilities for remote and *in vivo* sensing, and for inexpensive disposable probes. After many years of worldwide activities it appeared to be timely to publish a book on the subject. It is intended to cover the various aspects of optical chemical sensors including spectroscopy, waveguide theory, physical and analytical chemistry, biochemistry and biophysics, medicine, opto-electronic components, and material sciences.

In view of the interdisciplinary character of the optrode technology, it is obviously necessary that experts from various fields contribute to such a monograph. Because I am aware of the fact that, in these days, nobody wants to read a two-volume book, it is built up in a "modular structure" so that the reader can select certain chapters when interested in a particular subject. Many chapters are self consistent. To get information, for instance, on the state of the art of fluorescence-based pH sensors and their configurations, it would suffice, at first, to read the chapters on Spectroscopic Methods, Sensing Schemes, and pH Sensors. Inevitably, however, there will be some overlap.

The book is organized in the following manner; in the first sections, fundamental aspects are treated (Chapters 1–7), followed by a description of specific sensors for important chemical species (Chapters 8–11) and their applications (Chapters 12–14). Optical methods for temperature measurement, which is an essential part of all analytical procedures requiring some level of precision, is treated in an Chapter 15. The final sections cover bioanalytical methods (Chapters 16–17) and their applications (Chapter 18). Luminescence-based probes for chemoreception which, in the editors opinion, hold great promise for the future, are treated in Chapter 21.

The editor would like to express his sincere appreciation to the authors for their effort and diligence in preparing their manuscripts. I'd like to thank my family which has allowed this book to become part of it for quite a while.

I'm aware of the fact that the present book is neither complete nor perfect. It could just as well have been organized in quite a different way. However, if we would have gone for perfection, it would not have been completed within the next decade.

Otto S. Wolfbeis
May 1991

# THE EDITOR

**Otto S. Wolfbeis, Ph.D.,** is Professor of Chemistry and Head of the Analytical Division of the Institute of Organic Chemistry at the Karl Franzens University in Graz, Austria.

Dr. Wolfbeis obtained his Ph. D. degree in 1972 from KFU Graz. He served as a post-doctoral fellow at the Max-Planck Institute for Radiation Chemistry in Mülheim, Germany from 1972–1974 and at the Technical University of Berlin in 1977. He has been at KFU since 1975. It was in 1987 that he assumed his present position.

Dr. Wolfbeis is a member of the Austrian, Swiss, and German Chemical Societies, the European Photochemistry Association (EPA), and the Society of Photoinstrumentation Engineers (SPIE). He heads the Austrian Working Group on Chemical Sensors, acts on the advisory boards of several analytical journals, and is active in several European commissions. He has organized or co-organized several conferences on topics related to optical spectroscopy and optical methods of analysis.

He received the Sandoz Prize in 1981, the Feigl Prize for microanalysis in 1986, and was awarded the highly reputed Heinrich-Emanuel Merck Prize for Analytics in 1988.

Dr. Wolfbeis has authored more than 200 papers on optical sensors, fluorescence spectroscopy of plant natural products such as coumarins, flavons, and alkaloids, and has authored a monograph on the fluorescence of organic natural products. Other fields of research include three-dimensional fluorescence spectroscopy of biological liquids and synthetic and spectroscopic work on fluorescent probes and indicators. He holds a number of patents related to fluorescence technology. He has given approximately 20 invited lectures at international meetings and numerous quest lectures at Universities and Institutes. His current major research interests are in optical sensors and biosensors.

# DEDICATION

To Tina, Claudia, Gudrun, and Barbara, and my many friends in the marvelous land between California and the New York Island.

# CONTRIBUTORS

**Mark A. Arnold**
Department of Chemistry
The University of Iowa
Iowa City, Iowa

**Floreal Blanc, CEA**
Centre d'Etudes Nucleaires
Boite Postale 6
Fontenay, France

**Loic J. Blum**
Laboratoire de Genie Enzymatique
CNRS-Universite Lyon 1
Batement 308
Villeurbanne, France

**Gilbert Boisde, CEA**
Centre d'Etudes Nucleaires
Boite Postale 6
Fontenay, France

**R. Stephan Brown**
Department of Chemistry
Erindale College
University of Toronto
Mississauga, Ontario, Canada

**Pierre R. Coulet**
Laboratoire de Genie Enzymatique
 CNRS-Universite Lyon 1
Batiment 308
Villeurbanne, France

**Dileep K. Dandge**
ST & E, Inc.
Pleasanton, California

**DeLyle Eastwood**
Lockheed Enginering & Sciences Co.
Las Vegas, Nevada

**Lawrence A. Eccles**
United States Environmental
 Protection Agency
Environmental Monitoring Systems
 Laboratory
Las Vegas, Nevada

**Kisholoy Goswami**
ST & E, Inc.
Pleasanton, California

**Kenneth T. V. Grattan**
Measurement and Instrumentation Centre
School of Electrical Engineering
The City University
London, England

**G. D. Griffin**
Health and Safety Research Division
Oak Ridge National Laboratory
Oak Ridge, Tennessee

**Nelson R. Herron**
Lockheed Engineering & Sciences Co.
Las Vegas, Nevada

**Stanley M. Klainer**
FiberChem, Inc.
Las Vegas, Nevada

**Ernst Koller**
Lambda Fluoreszenzechnologie GmbH
Grottenhof-Str.3
Graz, Austria

**Ulrich J. Krull**
Department of Chemistry
Erindate College
University of Toronto
3359 Mississauga Road N.
Mississauga, Ontario, Canada

**Mark J. P. Leiner**
AVL-List GmbH
Kleist-Str. 48
Graz, Austria

**Robert A. Lieberman**
AT & T Bell Laboratory
Murray Hill, New Jersey

**Patrick Mauchien, CEA**
Centre d'Entudes Nucleaires
Boite Postale 6
Fontenay, France

**Fred P. Milanovich**
Optical Sensor Consultants
Livermore, California

**Douglas Modlin**
Optical Sensor Consultants
Livermore, California

**Olivier Parriaux**
Centre Suisse d'Electroonique et de
 Microtechnologie S.A.
Maladiere 71
CH-2007 Neuchatel, Switzerland

**Jean-Jacques Perez, CEA**
Centre d'Etudes Muchleaires
Boite Postale 6
Fontenay, France

**John I. Peterson**
National Institutes of Health
Biomedical Engineering and
Instrumentation Branch
Bethesda, Maryland

**W. Rudolf Seitz**
University of New Hampshire
Parsons Hall
Department of Chemistry
Durham, New Hampshire

**M. J. Sepaniak**
University of Tennessee
Department of Chemistry
Knoxville, Tennessee

**Stephen J. Simon**
Lockheed Engineering & Sciences
 Company
Las Vegas, Nevada

**Einar Stefansson**
Duke University Eye Center
Durham, North Carolina
and
University of Iceland
Reykuavik, Iceland

**Elaine T. Vandenberg**
Department of Chemistry
Erindale College
University of Toronto
Mississauga, Ontario, Canada

**Tuan Vo-Dinh**
Bldg. 4005 S
MS 5258
Oak Ridge National Laboratory
Oak Ridge, Tennessee

**Julie Wangsa**
Department of Chemistry
The University of Iowa
Iowa City, Iowa

**Otto S. Wolfbeis**
Analytical Division
Institute of Organic Chemistry
Karl-Franzens University
Graz, Austria

# TABLE OF CONTENTS

## Volume I

## Volume II

Chapter 1

# INTRODUCTION

**Otto S. Wolfbeis**

## TABLE OF CONTENTS

## I. ANALYTICAL ASPECTS OF SENSORS

A sensor is a device capable of continuously and reversibly recording a physical parameter or the concentration of a chemical or biochemical species. Typical examples for a mechanical, electrochemical, and optical sensor, respectively, are the mercury thermometer, the pH glass electrode, and a nonbleeding pH paper strip. Ideally, such a sensor can be stuck directly into the sample, and the result of the measurement is displayed within a few seconds. No sampling, addition of reagent, or dilution is required. This is a most distinct feature of sensors, since these operations are known to introduce errors into all kinds of analytical assays.

Continuous sensing of chemical analytes is a matter of growing interest by virtue of the real-time nature of most sensors. Increasing concern about environmental quality, and — in a more and more cost-conscious world — considerable personnel savings in comparison to manual off-line methods contribute to the desirability of sensors. Hence, tremendous effort has been devoted to the development of various sensing devices which, partially in combination with lab robotic systems, will enable analyses to be performed fully automatically or on-line, e.g., in product and food control, in the clinical lab, and in many other fields of routine analysis.

Certain devices, which do not act fully reversibly but are suitable for single use, have also been called sensors. It is suggested that instead the word probe be used for these in order to make a rough differentiation, although the borderline between sensors and probes is vanishing. Another differentiation may be made between sensors (or probes) and so-called dosimeters, which are designed for cumulative assay. Dosimeters, in contrast to sensors and probes, do not give real-time data and are not reversible.

In view of the different requirements imposed on sensors for various purposes, their complexity can vary considerably. An optical sensor for groundwater studies, for example, can be a simple fiber that guides a laser beam to a fluorescent analyte such as the uranyl ion, and guides fluorescence back to a detector. The incoming light beam may be considered as the question, and the returning beam as the encoded answer. However, most molecules of analytical interest require more sophisticated instrumentation, particularly when they are not colored or fluorescent by themselves. In most of these cases, the sensor consists of a light source, a monochromator (except in case of laser light sources), a light coupler, the waveguide (a planar waveguide, a fiber, or a fiber bundle), an analyte-sensitive coating, a transducer (if necessary), a secondary filter (if necessary), an optical detection unit, an amplifier, a data acquisition and processing unit, and a data output or display.

A great variety of transducers sensitive to heat, light, anions and cations, gases, electron transfer, mass changes, conductance, and light depolarization are known and compete successfully with each other (Table 1). Among these, the electrochemical methods are most advanced and widely distributed. A frequently used electrochemical sensing method is potentiometry, and pH-sensing glass electrodes have been used over decades for the determination of pH as such, but also for carbon dioxide which reversibly changes the pH of a bicarbonate buffer, and ammonia, which affects the pH of an ammonium ion buffer. More recently, pH electrodes have been applied as transducers for bioprocesses in which protons are produced or consumed.[1]

Ion-selective electrodes (ISEs) have tremendously widened the field of application of potentiometry, although some seem to be of limited long-term stability. The ion-sensitive field effect transistor (ISFET) is often regarded as the second-generation ISE and presents a logical merger between the latter and solid-state integrated circuits. Devices are now in use for the direct *in vitro* analysis for blood electrolytes, mainly hydrogen ion, sodium, potassium, calcium, and chloride. ISEs are available for various other analytes and also act as transducers for enzyme substrate, inhibitor, coenzyme, or even bacterial electrodes.[1] The palladium gate metal oxide semiconductor FET (e.g., the Pd-MOSFET), on the other hand, is sensitive to hydrogen and other gases, such as ammonia, in the parts-per-million range.

Amperometric sensors have been applied mainly for oxygen determination and as transducers for oxygen-producing or oxygen-consuming enzymatic reactions,[1] but also for hydrogen peroxide. Major progress was achieved in the past when molecular oxygen (the physiological electron acceptor for oxidases) was replaced by an electron transfer mediator such as ferrocene. It shuttles electrons from the enzyme redox center to the surface of the electrode. Other promising nonoptical sensing techniques include mechanical and acoustic impedance, the piezo-electric effect, and calorimetry. They all possess their respective merits and potential fields of applications.

## II. OPTICAL SENSORS

Optical methods have always played the dominant role in various fields of analytical sciences. Colorimetry, photometry, and spot tests have been used to qualitatively determine chemical and biochemical species. A recent breakthrough resulted from the introduction of

**TABLE 1**
**Sensor and Transducer Types, Other than Optical, Exploited**
**in Analytical and Clinical Chemistry**

| Sensor type | Principle | Typical sensing application | Typical transducer application |
|---|---|---|---|
| Glass electrode | Potentiometry | pH | $CO_2$, $NH_3$, enzyme electrodes, titrimetry |
| Platin electrode | Potentiometry | Redox status | — |
| Field effect transistors | Potentiometry | Ions and gases | Enzymes, substrates, antibiotics, immunosystems |
| Ion selective electrodes | Potentiometry | Anions and cations | Titrimetry, enzyme assay |
| Polarographic electrodes | Amperometry | Heavy metal cations; as detectors in HPLC | — |
| Clark electrode | Amperometry | Oxygen, $H_2O_2$ halothane | In enzyme electrodes |
| Piezo-electric crystal | Mass determination | Gases, volatile liquids | — |
| Conductance sensor | Conductivity | — | Enzyme catalyzed reactions |
| Thermistors | Measurement of reaction enthalpy via its highly negative temperative coefficient | — | In chemical and biochemical reactions accompanied by consumption or release of heat |

strip tests based on dry reagent chemistries which allowed visual or instrumental evaluation of analytical results, mainly in the clinical chemistry field. Surprisingly enough, there are only a few examples known for fully reversible analysis ("sensing") by optical means. The pH paper strip containing cellulose-immobilized pH color indicators is probably the best known among these.

The history of optical sensors can be traced back to when pH indicator strips were developed by immobilizing pH-sensitive indicators on cellulose. In the 1930s, a method for continuously recording low levels of oxygen was reported by Kautsky and Hirsch,[2] who found that triplet oxygen quenches the phosphorescence of silica-gel-adsorbed trypaflavin and fluorescein. 0.0005 torr strongly quenched the phosphorescence, which after removal of oxygen regenerates within 1 to 2 s. The effect was utilized for on-line detection of low levels of oxygen produced during photosynthesis.

Bergman[3] was the first to exploit the quenching effect of oxygen on the fluorescence of polycyclic aromatic hydrocarbons. He devised an instrument for continuously recording oxygen at levels of above 1 torr. It consisted of a UV light source, a sensing layer (Vycor glass or a polyethylene film soaked with a solution of fluoranthene), a flow-through chamber, and a photocell at the opposite wall of the chamber. A typical configuration is shown in Figure 1. which is a schematic of a flow-through cell for continuous oxygen determination by fluorescence quenching. The poor spatial flexibility is obvious when compared with Figure 2.

Lübbers and Opitz[4] used pyrenebutyric acid to sense oxygen in an arrangement similar to the one of Bergman, and introduced the first optical sensing scheme for $CO_2$: the pH indicator

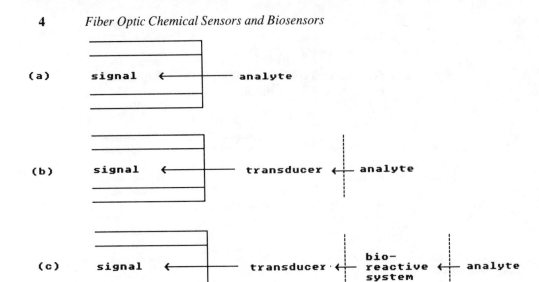

FIGURE 1. Schematic representation of the mode of gathering optical information in (a) a first generation, (b) second generation, and (c) third generation extrinsic fiber optic chemical sensor. In (a), the optical information is obtained directly from the absorbing or fluorescing analyte. In (b), a transducer element such as an indicator layer is attached to the fiber end. In (c), a bioreactive or bioreceptive element is added, for instance an enzyme layer, which converts the otherwise unmeasurable parameter into a measurable one. Evidently, the response times increase with increasing complexity of the sensor.

4-methylumbelliferone was dissolved in a 1 m$M$ bicarbonate buffer, and the solution covered with a 6 μm teflon membrane. $CO_2$ changes the pH of the internal buffer and this is recorded via the fluorescence of the indicator. The authors created two words for their device: optode[4] (from the greek οπτικος οδος, the "optical way" or "optical path") or optrode[5] (from optical electrode). Both expressions stress the fact that the primary information is optical rather than electrical. They also claimed, in a patent,[6] the use of optical fibers in combination with such sensors, but this idea does seem to have been performed in practice at that time.

## III. FIBER OPTIC SENSORS

A major breakthrough was achieved when conventional sensing techniques were coupled to fiber optics. Fibers are an outgrowth of the communication industry and allow the transmission of light over large distances. For chemical sensing, 1 to 100 m is a realistic range, although there may be exceptions. Notwithstanding the advent of high quality fibers, with their unique potential for the optical spectroscopist, several other new technologies are essential for current fiber optic sensing, and the considerable progress in these technologies over the past 10 years is indispensible: lasers with their unique properties are now available for routine application, the range of light-emitting diodes (LEDs) covers the 450 to >1000 nm range, and photodetectors and amplifiers have become available at low prices without compromising sensitivity and dark current. Finally, new methods in chemometrics, along with powerful, small-sized calculators and microprocessors which allow storage of large data volumes and rapid data processing, even in case of complex signal-to-concentration relations, have significantly contributed to the state of the art.

Optical fibers are based on the phenomenon of total internal reflection (Figure 2). Step index fibers consist of a core of refractive index $n_1$, surrounded by a cladding of lower refractive index. Most fibers are covered with a protective jacket which has no effect on the

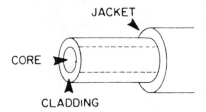

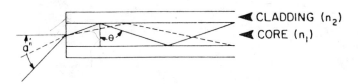

FIGURE 2. Top: Schematic of an optical fiber. Bottom: Path of light in a waveguide. $\alpha$ is the acceptance angle; $n_1$ and $n_2$ are the refractive indices for core and cladding; $\theta$ is the critical angle for total internal reflection to occur. Note that the ray entering at a large angle relative to the fiber axis (——) has to travel further to get the same distance down the fiber as the ray entering at a small angle (- - - -). (From Seitz, W. R., *Crit. Rev. Anal. Chem.*, 19, 135, 1988. With permission.)

waveguiding properties. Incident light is transmitted through the fiber if it strikes the cladding at an angle greater than the critical angle, so that it is totally internally reflected at the core/cladding interface. Light entering the fiber within an acceptance cone is transmitted. The acceptance cone half angle, $\alpha$, depends on the indices of refraction of core and clad as well as the refractive index of the outer medium, $n_0$:

$$\sin \alpha = \frac{\left(n_1^2 - n_2^2\right)^{1/2}}{n_0} \tag{1}$$

In case of air, $n_0$ is zero. More commonly, the range of angles is described in terms of the numerical aperture (NA), defined as

$$NA = n_0 \cdot \sin \alpha \tag{2}$$

or the f number, defined as

$$f/\# = \left(2 \tan \alpha\right)^{-1} \tag{3}$$

Basically, there are three kinds of fibers in use, namely the multimode step-index and multimode graded-index fibers, and the single mode step-index fibers. Their construction, refractive index profiles, and geometries are given in Figure 3.

The ray optic description of light propagation through a fiber (Figure 2) is an approximation that does not account for interference effects. A more rigorous description (Chapter 4) requires solution of Maxwell's equations. In essence, this leads to the conclusion that light propagates in discrete modes through the fiber. Each mode corresponds to a unique incidence angle and has a distribution of electric and magnetic field vectors such that a given ray of light does not interfere with itself.

Not all modes propagate through the fiber at the same rate. Instead, modes corresponding

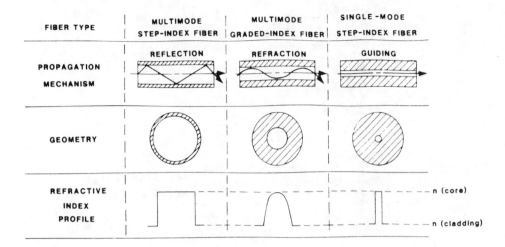

FIGURE 3. Overview on the most common types of optical fibers. (From Daly, J. C., Ed., *Fiber Optics*, CRC Press, Boca Raton, FL, 1984. With permission.)

to a larger angle between the entering ray and the fiber axis propagate more slowly because they have to travel a greater distance to get to the end of the fiber. This effect is called modal dispersion. The number of possible modes ($N_m$) is related to the NA and diameter d of the fiber by

$$N_m = 0.5(\pi \cdot d \cdot NA/\lambda)^2 \tag{4}$$

Aside from modal dispersion, there is also chromatic dispersion to be observed, which is due to the decrease in refractive index with increasing wavelength. This causes light of longer wavelength to travel slightly faster than shorter ones.

The NA, and thus modal dispersion, can be reduced by minimizing both the difference between $n_1$ and $n_2$ and the fiber diameter. A single mode fiber has a diameter so small (typically 3 to 5 µm) that only a single mode can propagate. An alternative way to reduce modal dispersion is to use graded-index fibers (Figure 3). The more off-axis the ray is, the further it penetrates into the region of lower refractive index. This accelerates the ray and compensates for the longer distance that off-ray axis rays have to travel.

In addition to the fibers themselves, the needs of the communication industry have led to the development of a variety of accessories including (1) couplers which act as beam splitters, (2) connectors for splicing together two pieces of a fiber with minimal light loss, (3) connectors for interfacing fibers to light sources of the LED or laser diode type, (4) tools for cutting and polishing fiber terminations and removing claddings, and (5) wavelength demultiplexers which resolve multiwavelength light into component wavelengths, thus acting as monochromators. A more detailed description of components is given in Chapter 6.

A typical instrumental arrangement for performing fiber optic sensing is shown in Figure 4. In essence, all the optical sensor types consist of a light source, coupler and decoupler, fiber light guide(s), sensor chemistries (as far as necessary), and light detector and signal processor.

It should be realized that for many applications a fiber is not required. Total internal reflection will occur at any interface between a transparent solid or liquid and air, as long as their refractive indices are different. Glass capillary tubes have successfully been used for chemical sensors. Furthermore, there is no requirement that the "fiber" be round. Rather, there are several advantages in using flat glass slides as the medium for light propagation. A flat

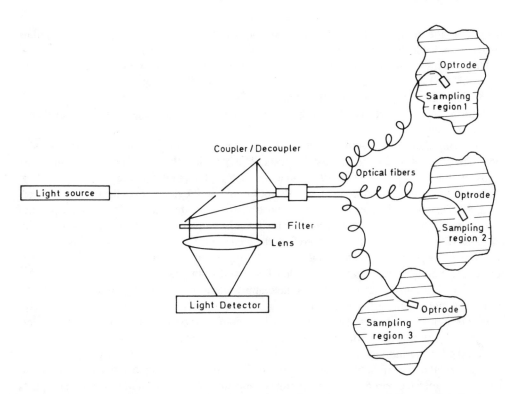

FIGURE 4. Typical experimental arrangement for performing fiber optic remote analysis at multiple locations.

surface is easier to coat with a reagent layer. Also, the angle of incidence of source radiation is more easily controlled.

Fiber optics, from soon after their availability, have been used in invasive sensing, but their function has been confined initially to a light pipe, much as a wire serves to conduct electrical energy to and from an electrical sensing element. Thus, blood oxygenation[7] was monitored by measuring the reflectance of hemoglobin and oxyhemoglobin, and cardiac output was determined from the speed of distribution of dyes injected into blood.[8] At the same time, Chance et al.[9] reported their work on the use of fiber optics for monitoring the intrinsic fluorescence of tissue which is mainly caused by reduced nicotine adenine dinucleotide.

It is very likely that Hesse[10] was the first to describe a fiber sensor for a chemical species. Specifically, sensors for oxygen and iodide were described in his patent, and both the fiber bundle and single fiber approach are discussed. To compensate for dye bleaching and temperature effects, the use of a reference cell is proposed, and both the reduction in fluorescence intensity as a result of quenching, and the reduction in lifetime (as measured by phase fluorimetry) are suggested methods for sensing oxygen and related quenchers.

Peterson et al.[11] were the first to describe, in great detail, the construction and performance of an invasive fiber catheter for measuring pH, and its application to blood pH monitoring. The device consisted of a pair of plastic fibers, both ending in a cellulosic dialysis tubing filled with a mixture of polystyrene particles and polyacrylamide beads dyed with the pH indicator phenol red. It was successfully employed for measuring blood pH *in vivo* and *in vitro*.

Hirschfeld was another researcher who was among the first to recognize the potential of fibers for chemical and biomedical sensing. His early work is difficult to trace back in the literature because it is hidden in various conference abstracts and technical papers,[12-14] but a lot of information is contained in his reviews[15] and patents (see the Appendix). At about the same time, the group of Seitz started the development of optical sensors based on chemilu-minescence,[16] which later was continued with other types of sensors.

Our activities in the field began in 1978 when we initiated a program for the development of sensors for blood gases and pH, all based on fluorescence. We became interested in sensing pH because of the well-expressed pH dependence of many natural products[17] and indicators,[18] and sensing oxygen[19] became obvious because it is a notorious quencher of fluorescence and phosphorescence.

Following these initial activities, a tremendous expansion in fiber optic chemical sensor technology has been seen in the last 10 years. This parallels the development of numerous kinds of fiber sensors for *physical* parameters such as temperature, pressure, magnetic or electric field, rotation, displacement, etc. Various books and reviews cover the physical sensor field.[20,21] This book, in contrast, focuses on chemical and biochemical sensors, with the exception of Chapter 15 on temperature. This is an important parameter to be considered in all kinds of chemical sensing and therefore deemed useful in a chemical sensor book. The literature surveyed is thought to be more or less complete up to 1990, with a few more recent references included. Particular attention has been paid to the patent literature.

Although various reviews and books have appeared, a book covering all aspects of fiber optic chemical sensing has not appeared so far. Several conference proceedings and a variety of useful reviews cover the subject from various points of view. While some treat a wider spectrum,[22,23] some are confined to measurement of chemical[24-34] and clinical[35-38] analytes, while others treat immunological[39] and optical[39,40] techniques, remote sensing aspects,[41,42] environmental monitoring,[14,23] or process control.[42,43] This book is intended to cover the whole aspect of chemical, environmental, biochemical, and clinical sensing including biosensing and immunosensing, but electronic signal amplification, signal processing, and data storage and processing are excluded. In each instance emphasis is given to practical applications rather than to purely theoretical treatment of the subject.

## IV. SENSOR CLASSIFICATION

The first sensor designed to collect information via fiber optics relied on the fact that alterations in a specific physical property of a medium being sensed would cause a predictable change in the light *transmission* characteristics of a fiber. Temperature, acoustic waves, acceleration, strain, position, and magnetic field are some of the physical properties measured with these initial sensors. They are frequently referred to as intrinsic fiber sensors and are treated in more detail in Chapters 3 and 5.

The field of application of fiber sensors in analytical chemistry greatly increased when other kinds of optical spectroscopy were coupled with the fiber optic technique. As a result, sensing is no longer restricted to measuring parameters that change the transmission of a fiber, but can be extended to numerous organic, inorganic, clinical, and biomedical analytes which have an intrinsic color or fluorescence, for which indicators are known, or which give rise to any kind of measurable change in the optical system. These are referred to as extrinsic fiber sensors and may be subdivided into first, second, and third generation sensors, respectively. In the first group, the fiber simply acts as a light guide. It allows remote spectrometric analysis of any analyte having an intrinsic optical property (such as light absorption or emission) that can be discerned from the background. Thus, the analyte provides the analytical information directly (Figure 1a). These sensors are also called bare-ended fiber sensors, or plain fiber sensors, or passive optrodes. A typical example is provided by a method for continuously monitoring copper(II)ion in electroplating baths[44] by measuring its absorption at approxi-mately 800 nm in a small flow-through cell within the light path between two fibers, or continuously monitoring groundwater contaminants such as uranyl ion[41] via their intrinsic fluorescence.

However, a variety of chemical species does not have an analytically useful intrinsic absorbance or fluorescence (examples: oxygen, $H^+$) and therefore cannot be probed directly. Moreover, various colored species may be contained in a matrix with optical properties similar to the species of interest so that they cannot be recognized easily. This situation gave rise to the development of fiber sensors in which the analytical information is mediated by some sort of indicator chemistry (Figure 1b). They are sometimes referred to as second generation sensors or active optrodes. Typical examples are pH sensors with immobilized pH indicators at their distal end. A subgroup of second generation sensors is called "reservoir sensors": since a variety of chemical species can be detected optically only by continuous addition of reagent, fiber sensors were constructed with a reaction volume at the end which is fully accessible to the analyte and continuously is supplied with a chemical reagent contained in a reservoir near the sensing site. The reaction between reagent and analyte produces a measurable chromophore or fluorophore there (see Chapters 3 and 12).

Unfortunately, for a variety of analytes there are no indicators known which (1) give fully reversible color changes at room temperature, (2) react without addition of aggressive reagents at near neutral pH, and (3) fulfill the condition of being highly specific. This is particularly true for biomolecules such as glucose and lactate. In order to detect these species, a biocatalytic process is usually coupled to either plain fiber sensors or an indicator-mediated sensor such as an oxygen optrode. The result is a third generation sensor as shown in Figure 1c. Typical examples are the fiber optic glucose sensors which function on the basis of the enzymatic action of glucose oxidase (GOD) according to

$$\text{glucose} + O_2 \rightarrow \text{gluconolactone} + H_2O_2 \rightarrow \text{gluconate} + H^+ + H_2O_2$$

The production of protons or $H_2O_2$, or the consumption of oxygen, may be monitored optically. It is clear from Figure 1 that response times become increasingly long with the complexity of the sensors and the thickness of catalytic layer and transducer layer. A complete classification scheme for chemical and biochemical optical sensors is given in Table 2.

Another kind of classification is based on the nature of optical information that is utilized. Thus, one may differentiate between absorbance, reflectance, luminescence intensity, luminescence lifetime, refractive index, surface plasmon resonance, or ellipsometric sensors. The optical methods underlying this classification are treated in Chapter 2.

A final way of classification is according to the field of application and one may discern between chemical sensors, enzyme-based biosensors, and immunosensors. The term biosensors is occasionally applied to all kinds of sensors suitable for sensing biological systems, for example to biomedical pH and oxygen optrodes. In this book, the term is confined to sensors where a classical biomolecule is involved in the process of signal production, for example a coenzyme, an enzyme, or another protein. Immunosensors form another group since the signal is the result of a binding or unbinding process between antibody and antigen, with no chemical transformation usually being involved. Biosensors and immunosensors are treated in Chapters 17 and 18, respectively.

There are two fundamental technologies known for the construction of optical sensors. In the first, the sensor chemistry is manufactured first and then attached to the fiber or fiber bundle. The sensor layers are usually produced on planar supports such as glass, cellulose, or plastic and then glued or mechanically fixed at the fiber tip. In the second type, the chemistry is manufactured directly on the fiber, after coating and clad have been removed from the end. It looks as if planar sensors can be produced more reproducibly and in larger quantities and that they are more easily subject to quality control.

**TABLE 2**
**Classification Scheme for Optical Chemical Sensors and Biosensors, and Typical Examples**

| Sensor type | Chemical sensors | Biosensors | Immunosensors |
|---|---|---|---|
| 1st Generation | Remote spectrometers for on-line monitoring of chemical species[a] | Remote monitoring of coenzyme fluorescence[b] | Detection of Ag/Ab binding via the intrinsic Trp fluorescence of proteins[c] |
| 2nd Generation | Fiber sensors based on immobilized indicators[d] | Fiber sensors based on immobilized bioprobes[e] | — |
| 3rd Generation | 2nd Generation sensor coupled to enzymatic reaction[f] | 2nd Generation sensor coupled to enzymatic reaction[g] | Fiber optic ELISA |

[a]  Example: remote detection of methane via its NIR absorption.
[b]  Example: determination of glucose via the intrinsic fluorescence of GOD.
[c]  Example: monitoring the binding of anti-HSA to HSA via the intrinsic fluorescence of HSA.
[d]  Example: monitoring pH via color changes of a pH indicator.
[e]  Example: determination of thiamine via its quenching effect on the fluorescence of 1-alkoxy-pyrenetrisulfonate.
[f]  Example: organic chloride sensor based on the enzymatic degradation of alkylhalides by certain enzymes. The consumption of oxygen is monitored with an oxygen optrode.
[g]  Glucose biosensor.

# V. ADVANTAGES AND DISADVANTAGES OF OPTICAL SENSING

Depending on the field of application, optical fiber sensing can offer one or more of the following advantages over other sensor types:

1.   Optrodes do not require a reference signal as in potentiometry, where the difference of two absolute potentials is measured. The need for reference electrodes makes potentiometric sensors, particularly those intended for use in disposable probes, relatively costly. Moreover, the liquid-liquid junction between the two electrolytes is prone to perturbations and can be considered to be weakest "link" (the Achilles' heel)[45] in all potentiometers.
2.   The ease of miniaturization allows the development of very small, light, and flexible fiber sensors. This is of great utility in the case of minute sample volumes and in designing small catheters for invasive sensing in clinical chemistry and medicine. By now, mechanically stable optical sensor heads much smaller than any electrochemical sensor (including field-effect transistors) can be manufactured.
3.   Low-loss optical fibers allow transmittance of optical signals over wide distances, typically 10 to 1000 m, and even larger distances seem feasible when use is made of amplifiers currently used in optical telecommunication. Remote sensing makes it possible to perform analyses in ultraclean rooms, when samples are hard to reach, dangerous, too hot or too cold, in harsh environments, or radioactive. One may state that in fiber optic sensors the instrument comes to the sample rather than the sample to the meter. Even when no remote sensing is required, the use of fibers provides much more spatial flexibility for arrangement of other optical components.

4.  Because the primary signal is optical, it is not subject to electrical interferences by static electricity of the body, strong magnetic fields, or surface potentials of the sensor head. On the other hand, fibers do not present a risk to patients since there are no electrical connections to the body.

5.  Analyses can be performed in almost real-time since no sampling with its inherent drawbacks is necessary.

6.  Since several fiber sensors placed in different sites can be coupled to one fluorimeter via a chopper, the method allows multiple analyses with a single control instrument. Placement of the spectrometer at a central location remote from the sensor head (which experiences varying experimental conditions) makes routine maintenance checkout possible, assures that calibration will be preserved, and consequently renders the instrument more reliable.

7.  Coupling of small sensors for different analytes to produce a sensor bundle of small size allows simultaneous monitoring of various analytes by hybrid sensors without crosstalk of the single strands.

8.  In many cases the sensor head does not consume the analyte in a measurable rate as, for instance, in the case of polarographic electrodes. This fact is of particular advantage in case of extremely small sample volumes.

9.  Fiber optical sensing is a nondestructive analytical method (except for some of the reservoir optrodes).

10. Fibers are manufactured from nonrusting materials, so that they have excellent stability when in permanent contact with electrolyte solutions. Plastic and glass easily withstand agressive reductants in the anlyte solutions including $SO_2$, $H_2S$, and $H_2O_2$. They also appear to be resistant to radiation doses in the order of $10^7$ rad or more.

11. Optical sensors have been developed which respond to chemical analytes or physical parameters for which electrodes are not available.

12. A fiber optic can transmit much more information than can an electrical lead. High information density can be achieved since the optical signals can differ with respect to wavelength, phase, decay profile, polarization, or intensity modulation. Thus, one single fiber may guide green and red light in one direction, and blue and yellow light into the other. As a result, a single fiber can, in principle, guide a huge number of signals simultaneously. In practice, a single fiber may be used to assay several analytes at the same time because different analytes or indicators can respond to different analytical wavelengths. This has been demonstrated in a single sensor for both oxygen and carbon dioxide.[46] Time resolution along with spectral selection offers a particularly fascinating new technique in fiber optical sensing and can make superfluous the need for hybrid sensors.

13. Sensors with immobilized indicator or reagent layers are not subject to inner filter effects caused by interfering analytes having similar absorption. This is due to the usually high optical density of the indicator layer which (with certain limitations) may also serve as a filter to the underground emission from the sample.

14. While most pH optrodes have a much smaller dynamic range as compared to electrodes, sensors based on dynamic fluorescence quenching have a useful dynamic range often larger than that of electrochemical sensors. For instance, the fiber optic oxygen sensor shows much better precision than does the Clark electrode in the pressure range above 200 torr.

15. Many sensors are simple in design and can easily be replaced by substitute parts, even if manufacturing the sensor head requires relatively complex chemistry. They therefore

can offer cost advantages over other sensor types. Consequently, it is likely that single use and disposable optical sensors will experience a bright future.

16. Most fiber sensors can be employed over a wider temperature range than electrodes and some have a smaller temperature dependence.
17. Many oxygen and pH sensors are steam-sterilizable.
18. The feasibility of fluorescence immunoassay (FIA) using fiber optics has already been demonstrated, although such sensors have poor, if any, reversibility, and therefore may rather be called probes. This provides a promising principle for invasive nonradioactive determination of antigens and, eventually, antibodies and seems to present the only alternative to the few electrochemical sensing methods. RIA, in contrast, is not applicable for sensing purposes.
19. Optical fibers are a quasi-one-dimensional medium. This offers the unique advantage for spatially distributed measurement over a tailorable path.

Notwithstanding a number of advantages over other sensor types, fiber sensors exhibit the following disadvantages:

1. Ambient light can interfere. Therefore, optical sensors must either be used in dim or dark surroundings, or the optical signal must be encoded so that it can be resolved from ambient background light, or it must be covered with an optically tight layer ("optical isolation").
2. Sensors with indicator phases are likely to have limited long-term stability because of photobleaching or wash out. Signal drifts can be compensated for by relating the signals obtained under two different excitation wavelengths, or ratioing the signal to an intrinsic standard, or by measuring parameters such as lifetime, which are independent of indicator concentration. Photobleaching efficiency increases with increasing irradiation intensity. Consequently, a powerful laser should be used only when necessary, for instance, when long optical cables with their considerable attenuation make lasers indispensible.
3. Since in indicator phase sensor analyte and indicator are in different phases, there is a mass transfer necessary before a steady-state equilibrium — and consequently a constant response — is established. This, in turn, limits response time for analytes with small diffusion coefficients. This situation makes it desirable to keep the volume of the analyte-accessible indicator phase much smaller than that of the sample in order not to dilute the sample volume.
   The required mass transfer may, however, contribute to improved selectivity. Thus, when oxygen *electrodes* are used as transducers for enzymatic reactions, interferences are observed with easily reducible species including ascorbic acid, hypoxanthine, uric acid, or cystin. When an oxygen *optrode* is employed, interference is negligible because none of these species is transferred into the hydrophobic oxygen-sensitive polymer layer.
4. Sensors with immobilized pH indicators as well as chelating reagents have limited dynamic ranges as compared to electrodes since the respective association equilibria obey the mass action law. The corresponding plots of optical signal vs. log of analyte concentration are sigmoidal rather than linear as in the case of the Nernst relation. As a result, the total signal change occurs over a much smaller analytical range.
5. The fiber optics used at present have impurities of a spectral nature that can give background absorption, fluorescence, and Raman scatter. Low-priced (plastic) fibers are confined to wavelengths between about 420 and 800 nm, whereas UV light is efficiently

transmitted by rather expensive quartz fibers only. The intensity losses in very long fibers are further complicated by spectral attenuation and change in the numerical aperture as a function of fiber length.

6. Commercial accessories of the optical system are not optimal yet. Stable and long-lived light sources. better connectors, terminations, optical fibers, inexpensive lasers, and, in particular, blue LEDs and semiconductor lasers for the whole visible range are needed.

7. More selective indicators have to be found for various important analytes, and the immobilization chemistry has to be improved so as to achieve both better selectivity and sensitivity.

8. Many indicators suffer a reduction in sensitivity after immobilization or when dissolved in a polymer resulting in smaller slopes of the response curves. In particular, dynamic quenching efficiency is frequently drastically diminished. Consequently, the respective conventional method will be much more sensitive than the opto-sensor method.

9. In indicator phase sensors for pH and electrolytes it is the concentration of the *dye* that is quantitatively determined, rather than the analyte itself. The analyte concentration can be calculated from the indicator concentration because there is a defined relation. This relation can vary with ionic strength, solvent composition, and matrix composition ("protein error") and may give rise to considerable bias in precision. Moreover, the sensor registers the concentration of a species rather than its activity, as do electrodes.

In summary, optical sensors offer a variety of new aspects. Despite several limitations, they have the potential of becoming an attractive alternative to other sensing methods and to perform diagnostic, environmental, or clinical functions better, faster, more accurately, or less expensively than existing approaches.

# VI. FIELDS OF APPLICATION

One may differentiate between applications where a plain fiber is used as a light pipe, and true optrode applications. Plain fibers will have application in all situations where classical optical analysis is to be performed over a certain distance. Instrumentation for the remote acquisition of UV/VIS data is commercially available, e.g., from Oriel (Stamford, CT), Photonetics (Marly, France), or Guided Wave, Inc. (Helsingborg, Sweden, and El Dorado Hills, CA). They are intended for use in radioactive areas, fermentation systems, high-voltage areas, explosive and dangerous areas, jet and rocket machines undergoing testing, biological hazard stations, tracer studies in geology, and in marine river and reservoir locations. An instrument specifically designed for the detection of the fluorescence of NADH in bioreactors has been developed by the Ingold AG (Urdorf, Switzerland). Representative plain fiber sensing schemes are described in Chapter 3.

Optrodes (in the sense of indicator phase sensors) offer a much wider field of application and are of potential utility in all kinds of analytical sciences including those covered by electrodes at present. Typical areas are pollution and process control, biotechnology, defense, seawater analysis, clinical chemistry, and invasive biomedical techniques. Their ruggedness and capability of transmitting signals over wide distances makes possible utilization in locations too harsh and inaccessible to the instrument, since the meter can be located in the benign environment of the lab whereas the fiber comes to the sample.

## A. GROUNDWATER MONITORING

In the face of increasing public concern about the quality of drinking water, continuous monitoring of groundwater has become a major aspect of modern analytical chemistry. Rather than digging a well field with numerous boreholes large enough to admit sample collectors which are brought to a laboratory for analysis, it has been proposed[13,41] to introduce long-distance communication-grade fibers down to the groundwater level and to monitor pH, chloride, uranium, organic pollutants, and tracer substances using the corresponding optrodes. Several fibers may then be coupled to one spectrometer at a central location up to 1 km remote. The ability to make up to 50 unattended *in situ* measurements using a reasonably priced centralized fluorimeter system has been discussed[14] and should result in acceptable economy.

The much smaller diameter of the fiber hold (typically 1 to 2 cm) allows the use of small boreholes. It has been stated[32] that at typical sites such as chemical or nuclear plants, savings in drilling costs can be as much as $500,000. Moreover, the sample can be studied *in situ* and in real time, with almost no opportunity for contamination of a container or the well itself from outside.

For nuclear waste repositories, environmental monitoring of nuclear installations, or study of underground nuclear tests, the hazards associated with the samples can be avoided by leaving them safely underground. Even if the fiber is damaged by radiation, it is easier and cheaper to replace than the whole monitor.

The principles of optical groundwater monitoring do, of course, also hold for geological tracer studies using highly fluorescent markers such as rhodamine 6G, or pH gradient studies using pH indicators. They are detectable in boreholes in picomol quantities when lasers are used as excitation light sources.[32,41]

## B. POLLUTION MONITORING

There are two fascinating new approaches for continuous control of water and air pollution: the possibility of remote sensing of environmental parameters without the use of fibers has been studied by several groups.[47,48] Increasing efforts were directed to the detection of polycyclic aromatic hydrocarbons in the environment which are by-products of many new coal-processing factories, including coal-liquidification plants. Similarly, mineral oil spills released by ships can be distinguished from seawater background by this technique and can even be classified into subgroups.

On the other hand, airborne laser fluorosensor measurement of chlorophyll concentration (phytoplankton population) has been applied to water-mapping studies both in inland waters and in the ocean. Abundance, distribution, and biological activity in natural waters can easily be monitored over wide areas and even at various depths.

A variety of optical sensors for primary pollutants is known to work in principle. Thus, air pollution sensing for industrial applications and environmental research may be performed with a network of specific optrodes hooked to a central measurement station by optical fibers. The sensors may be instantaneously concentration-sensitive or be designed for cumulative measurements of the total integrated exposure (fiber dosimeters). Sensors have been developed for formaldehyde, ammonia, nitrogen oxides, chloroform, hydrogen sulfide, sulfur dioxide, and reactive hydrocarbons. Finally, seawater quality and pollution control appears to become a major future field of application that is particularly compatible with fiber sensing methods.

## C. PROCESS CONTROL

Production efficiency and product quality are critically dependent on process control. More and more, on-line sensors are replacing classical sampling techniques, and analytical chemistry is emigrating from the laboratory to the factory. In view of the costs resulting from a failing chemical reaction and its environmental consequences, there are extreme reliability requirements on process control instrumentation. Another aspect is fire hazard. Fibers are likely to present no risk of sparks, which turns out as a decisive advantage in explosive environment. In an evaluation of the possibility that radiant energy emerging from a fiber optic system could ignite a flammable vapor, it was demonstrated[49] that ignition indeed can occur. When light was focussed onto a 50-μm dust particle, ether vapor started to explode when approximately 6 mW of energy was absorbed within a very short period of time. This, however, is an energy density which is unrealistic in most sensor applications.

On the other side, most sensors for continuous process control are faced with a harsh environment. Typically, a reliability of better than 99% is required under extreme and rapidly varying temperatures, high noise and vibration, and substantial chemical exposures, but with a minimum of servicing and maintenance at rather long intervals. Furthermore, the instrument is expected to stay accurate over a prolonged period, despite the absence of recalibration or even checkup.

Clearly, bioprocess control is becoming another major field of application for optosensing. The development of long-range, low-cost, and high-performance fibers along with the design of sterilizable optrodes for the most important chemical and physical parameters has provided a new dimension of continuous bioprocess management[50] (see Section VI.G. herein).

## D. BIOMEDICAL APPLICATIONS

Possibly the greatest field of application is sensing clinically and biochemically important analytes such as blood gases and electrolytes, metabolites, enzymes, coenzymes, immunoproteins, bacteria and inhibitors. Optrodes for blood analytes are being used and will be used *in vivo* as sensors for the continuous monitoring of the critically ill and as devices for testing blood samples *in vitro*. Various kinds of methods for continuous measurements of critical parameters which give warning of life-threatening trends such as pH, oxygen, carbon dioxide, and blood pressure are well established in principle and in practice. However, one may expect from fiber optics to see continued improvements in biocompatibility, signal stability, ease of calibration, and sterilization. A selection of potential biomedical sensor applications is compiled in Table 3.

Within the last decade, clinical practitioners have gradually moved from the diagnosis of established disease toward presymptomatic prognosis and preventive measures. Continuous tests for electrolytes, total protein, urea, glucose, creatinine, cholesterol, triglycerides, and others in blood or urine are the subject of intense research and development to produce devices capable of automatically recording results by using computerized data logging and output systems.

Aside from sensors for these analytes which are present in relatively high concentrations, there is a substantial demand forseeable for biosensors and immunosensors for substrates being present in considerably lower concentrations such as hormones, steroids, thyroid function constituents, and pregnancy markers. It is assumed that in these cases fluorosensors will be of particular utility by virtue of the sensitivity and versatility of fluorimetry.

The area of therapeutic drug monitoring is another one of rapidly increasing commercial significance, along with early detection of infection diseases and of the various forms of cancer. Table 3 demonstrates the enormous increase in the number of clinical tests performed in one year.

A highly significant trend is now visible which will lead to the decentralization of clinical testing away from hospitals, ideally into the patient's home. This will produce a huge market for simple and inexpensive but reliable equipment in the next 10 years. According to various estimations, the home/self-care product market in the U.S. will increase from $5.5 billion in 1985 to $15.0 billion in 1992. Most of the techniques utilized in these tests will be based on enzymatic and immunological reactions, combined with sensitive optical methods such as fluorescence.

Biosensors based on enzymatic reactions coupled to conventional pH or oxygen sensors will mainly cover the large field of low-molecular weight analytes, while immunosensors are likely to be most useful for larger analytes. The interaction of antibodies with antigens, a

**TABLE 3**
**Expansion of Clinical Testing Markets in the U.S.**
**(Nonisotopic Tests Only)[a]**

| Test discipline | 1983[b] | 1984[b] |
|---|---|---|
| Infectious diseases | 51 | 112 |
| Therapeutic drug monitoring | 27 | 56 |
| Autoimmune disease | 14 | 30 |
| Plasma proteins | 14 | 28 |
| Cancer tests | 2 | 5 |
| Others | 11 | 27 |
| TOTAL | 119[c] | 258[c] |

[a] Source: BBI/Information Resources Int., Inc.
[b] Millions of tests.
[c] Minimum values.

process known to be of outstanding selectivity, can be followed by various techniques. Evanescent wave sensors with immobilized antibodies on the waveguide surface, fluorescence polarization studies of labeled binding partners, fiber optic decay measurements, surface plasmon resonance, or combinations thereof, offer numerous possibilities. Bilirubin, steroids, albumin, and enzymes have all been assayed by antibody-binding methods which, however, have been transferred to the fiber optical sensing technology only recently.

In summary, the opportunity exists for fiber and waveguide sensors to participate in a rapidly growing segment of the market for clinical analytes, although there may be tough competition from other sensor instrumentation including electrodes, FETs, piezo-electric devices, Fourier-transform-IR spectroscopy, calorimetry, and dry-reagent chemistry. Probably, an interdisciplinary approach to the solution of existing hurdles will be essential.

## E. DEFENSE

Given the option of performing optical tests over large distances, the optical waveguide is an ideal means for early detection of biological or chemical agents, some of which are not easily detectable by other means. Assuming most B- or C-type warfare agents to be transported with the wind, one sees that the time advantage provided by a fiber sensor is simply given by dividing the operative length of the fiber through the speed of wind (in the fiber direction). Realistic lengths for present day fibers are 5 km, and the time advantage can therefore be as large as of 10 to 60 min.

The detection of chemical species will mostly rely on biosensors and bioreceptors, and on the inhibition of enzymatic processes into which certain enzymes such as acetylcholinesterase are involved. Sensors responsive to the six to ten important types of the mode of action of chemical agents may be coupled to a sensor bundle to give a universal warning system that can respond to all present and future agents. Biological species, on the other hand, will be detectable most appropriately by immuno methods or by detection of the various exhibitions of bacterial activity.

In that respect, the one-dimensional nature of the fiber medium is another attractive and important feature since it offers the advantage of spatially distributed measurement over a chosen path. This opens the way to cover large areas with a network of parallel or radially arranged fibers.

## F. BIOTECHNOLOGY

Proper management of biotechnological processes is a prerequisite for successfully running bioreactors. Various efforts have been undertaken to sense parameters such as oxygen, pH, $pCO_2$, glucose, and glutamate in fermentation plants using classical electrodes or electrode-based biosensors. The need for sterilization as well as the fact that all sensors are covered with protein or biomass within a short time imposes, however, severe constraints on sensors for use in this field. It looks, at present, that most kinds of fiber optic oxygen sensors can meet the requirement of steam sterilization at 115 to 130°C without severely compromising their performance. Therefore it is likely that they will be applied for sensing oxygen in both the fluid phase and in exhaust gases.

Process control in biotechnology involves two analytical procedures: (1) the measurement of physical and chemical parameters in the reactor, using sterilized sensors, and (2) measurement of chemical parameters in an external loop or during down stream processing. The need for sterilization is a difficult hurdle for practically all kinds of biosensors with incorporated biocatalysts, and for immunosensors. Therefore, it is likely that these probes initially will be used in an external loop or in combination with flow injection analysis rather than in the bioreactor itself. No such problem exists with plain fiber sensors which measure the intrinsic fluorescence of the broth, for instance that of NADH.[50,51] In this case, the fiber acts as a light pipe, having no sensor chemistry at its end. It can easily be sterilized and can give useful information on biomass, oxygen saturation, and even substrate supply. Flow-injection analysis (FIA), which has had such an outstanding success in analytical biotechnology, will be a major field of application for optical sensors anyway.

## G. TITRIMETRY

Although not a continuous method, titrimetry is still one of the most widespread tools in analytical chemistry. Recent work demonstrates that fibers or optrodes can replace electrodes in various titrimetric procedures, including acid-base titrations, argentometry, bromometry and iodometry, complexometry, or redox titrations. There are two ways to perform this. In the first, an indicator is added to the solution to be titrated, and the relative change in fluorescence during titration is followed using a (bifurcated) fiber. This method can offer considerable cost reductions since optical fibers along with indicator solutions are much less expensive than electrodes. The optical and electrochemical detection systems are of comparable price.

In the second method, the analyte-selective indicator is immobilized at the end of the fiber to give an analyte-selective optrode, which is dipped into the solution during titration. It may be predicted that the endpoints of the majority of standard titration methods can be determined by fiber optical methods with the same precision as with electrochemical methods, although the full potential of fiber optical titrimetry is not yet fully exploited. The state of the art in this field is presented in Chapter 13.

# APPENDIX: A SELECTION OF RELEVANT PATENTS

Owing to the practical significance and obvious commercial importance of fiber optic sensors, a selection of useful and relevant patents is given in the following, along with brief statements as to its contents. The list is not considered to be complete, and only the first application that has come to the author's attention is cited. It should therefore be borne in mind that most patents are filed in a variety of countries. The following abbreviations are used: US, United States Patent; EP, European Patent; DE, West German Patent; UK, United Kingdom Patent; JP, Japanese Patent; DD, East German Patent; AT, Austrian Patent; PCT, International Patent within the Patent Cooperation Treaty.

## A. OXYGEN

UK 1,190,583 (Bergman): method for determination of oxygen based on fluorescence quenching of, e.g., Vycor-adsorbed fluoranthene. US 3,725,658 (Stanley and Kropp): apparatus and method for sensing oxygen via fluorescence quenching; sensor also detects $SO_2$ and $NO_x$; electroluminescence also suggested for excitation. US 3,612,866 (Stevens): oxygen sensor based on quenching of fluorescence, reference chamber used, radioactive source for luminescence excitation suggested. US 4,003,707 (Lübbers): gas sensing device with a diffusion membrane and immobilized indicators; mainly for oxygen and $CO_2$. US 4,476,870 (Peterson): fiber optic oxygen probe for implantation in body. US 3,709,663 (Hendricks): oxygen detection by exposing a film of poly(ethylene naphthalenecarboxylate) to air and light; subsequent heating results in oxygen-dependent thermoluminescence. PCT WO 87/00023 (Kahlil et al.): oxygen sensor comprising a plastic film containing a luminescent substance quenched by oxygen; substances include metallo-derivatives of (fluorinated) porphyrins. US 4,557,900 (Heitzman and Kroneis): Optical gas sensor with selectively permeable matrix of hydrophobic material and hydrophilic (water-soaked) beads dispersed in the matrix; useful for oxygen, $CO_2$, and other diffusible gases.

DE Offen. 3,346,810 (Bacon): oxygen sensor based on quenching of luminescence intensity or lifetime of metalorganic complexes; visual detection posible with reference cell that has variable dye concentration; mixture of fluorophors with different oxygen sensitivities and emission colors also suggested. EP Appl. 190,829 and 190,830 (Lefkowitz): plastic optical waveguide with a portion of the core exposed to a fluorescent oxygen-sensitive dye; evanescent wave excitation. US 4,578,101 (Marsoner and Kroneis): oxygen sensor with plasticizer and optical isolation layer filled with iron oxide particles. US 4,657,736 (Marsoner et al.): solubilization of oxygen indicators in silicone by butylation. DE Offen. 3,148,830 (Barnikol and Burkhard): oxygen-sensitive layer composed of a monolayer of beads with surface-adsorbed indicator and glued onto a tape; response time extremely short (5 ms) and no cross-sensitivity to humidity. US 4,321,057 (Buckles): oxygen sensor with an active length devoid of a clad and a core that is penetrated by the analyte; alternatively, the clad is the oxygen-sensitive material and fluorescence is reflected into the core; claims also extended to intrinsically colored species that can penetrate into the clad and to glucose assay via measurement of oxygen consumption by glucose oxidase catalyzed oxidation; also comprises a method for immuno assay; see also US 4,399,099.

US 4,712,865 (Hsu and Heitzmann): covalent binding of oxygen indicators to a polymer matrix, especially silicone. DE 2,346,792 (Berthold): oxygen sensor similar to the one of Bergman (GB), but with fluorescence collected in remission mode and sensitive layer being covered with gas-permeable membrane. EP Appl. 243,116 (Yafuso et al.): solubilization of decacyclene for use in oxygen optrodes. US 4,764,343 (Nyberg): oxygen sensor for exhaust gases comprising a reflective surface reversibly reacting with oxygen. US 4,775,514 (Barnikol): luminescent oxygen-sensitive silicone layer with dyed hydrophobic particles incorporated.

## B. OTHER CHEMICAL SPECIES

US 4,557,900 (Heitzmann and Kroneis): $CO_2$ sensor consisting of particles soaked with dye solution and buffer and embedded in silicone polymer. US Appl. 470,920 (1983, Vurek): fiber optic $CO_2$ sensor comprising a solution of phenol red in bicarbonate covered with silicone membrane; absorbance changes measured with two fibers. DE 2,508,637 (Lübbers): optical isolation of pH, $CO_2$ and oxygen sensors by covering sensing layer with black or reflective analyte-permeable membrane in order to avoid sample fluorescence to enter the optical system of the optrode. EP Appl. 137,157 and US 4,548,907 (Seitz and Zhang): pH sensor based on a membrane-immobilized fluorescent pH indicator at the end of a fiber; measurement at two analytical wavelengths to account for drift and leach problems. US 4,200,110 and 4,476,870

(Peterson et al.): pH-sensitive indicator in a permeable envelope at the end of a fiber optic. US 4,716,118 (Wolfbeis and Offenbacher): determination of ionic strength by optical measurement of pH with two sensors of different responses to ionic strength. US 4,376,827 (Stiso et al.): optical dip stick determination of ionic strength using bromthymol blue. US 4,473,650 (Wang): optical test means for determining ionic strength or specific gravity; see also U.S. 4,318,709 and U.S. 4,532,216. US 4,166,804 (Bleha et al.): immobilized pH indicators for continuous optical determination of pH. US 4,568,518 (Wolfbeis et al.): pH-sensitive fluorescent cellulose membrane with indicator linked to a cellulose with an interpenetrating polyethylene-imine network. AT 377,364 (Kroneis and Wolfbeis): measurement of oxygen and pH (or $CO_2$) with one indicator: pH (or $CO_2$) via fluorescence intensity as measured under two excitation wavelengths, oxygen via dynamic quenching.

US 4,272,485 (Lübbers): use of carrier molecules to selectively transport the analyte to the sensing layer. EP Appl. 247,261 (Hirschfeld and Wong): fiber optic fluorosensor for invasive pH measurement based on glass-immobilized fluorescein. EP 105,870 (Marsoner et al.): optical $CO_2$ sensor consisting of a homogeneous dispersion in the form of droplets of an aqueous buffer solution containing a fluorescent pH indicator in a $CO_2$-permeable polymer. US 4,272,484 (Lübbers): enzyme optrode with enzymes in the sensing layer together with optical indicator. US 3,112,999 (Grosskopf): reagent card for optical detection of carbon monoxide.

JP 60,209,149 (Nishizawa and Yamazaki): hydrogen sensor comprising an optical waveguide covered with a reactive thin film (Pd); $H_2$ detected by changes in light transmission through the waveguide resulting from a change in the optical absorption coefficient of the reactive thin film. UK Appl. 2,192,710 (Mosely et al.): sensor for hydrogen, CO or $H_2S$ comprising an optical fiber with coating (e.g. Pd) that catalyzes oxidation of gas; resulting temperature change detected by various optical means. US Appl. 709,251 (1985, Giuliani): optical waveguide sensor for methane by depositing a very thin film of water on the waveguide surface and measuring the IR absorption of methane (and higher alkanes) by the evanescent wave technique.

US 4,513,087 (Giuliani and Wohltjen): optical waveguide sensor for ammonia, hydrazine, pyridine, and the like using a fiber coated with a pH-sensitive dye; measurement by evanescent wave absorption. JP 88/18251 (Igarashi et al.): optical waveguide ammonia sensor. US 4,560,881 (Briggs): fiber optic fluorescent particle detector. EP Appl. 125,554 and 125,555 (Charlton et al.): optical test means for detection of ions (such as $K^+$) applying ion-pair extraction into a plastic carrier matrix; though not applied to fibers, it easily could.

AT 384,891 (Urbano et al.): halide sensor based on fluorescence quenching of immobilized quaternized heterocyclic dyes. DE 3,343,636 (Wolfbeis et al.): pH sensor with porous support of large specific surface. DE 3,627,876 (Federmann et al.): fiber optic gas absorption cell; methane determined by comparison of beat frequency of light with the Zeeman effect using a laser resonator.

EP Appl. 257,955 (Webb and Daniels): surface plasmon resonance device with chemically sensitive layer on a metal layer; suitable for detection of molecular interactions by measuring changes in angle of incidence at which maximum light reflectance occurs; suitable for, e.g., interaction of valinomycin with $K^+$, biotin with avidin, halothane with oil film, or enzymes with enzyme substrate; see also UK Appl. 2,197,065 (Batchelder and Willson). Eur. Appl. 126,600 (Welti and Kirkbright): optical fiber sensor based on spectral changes of an immobilized indicator.

US 4,502,937 (Yagi): electrical signals of an ion-selective electrode transformed into optical signals of an LED which are transported via fibers. EP Appl. 198,815 (Wolfbeis and Hochmuth): electrolyte optrode (specifically for potassium) based on optical measurement of a potential created at a lipid-water interface. EP 105,870 (Marsoner et al.): $CO_2$ sensor with the indicator solution finely dispersed in a gas-permeable polymer such as silicone rubber (which may contain an oxygen sensitive dye, thus giving a sensor for $CO_2$ and $O_2$). US 4,509,522 (Manuccia and Eden): fiber optic device to measure hemoglobin-bound or dissolved oxygen, $CO_2$, or CO in blood by infrared absorptiometry at 9.0, 4.3, and 5.13 µm, respectively.

JP 58/215,547 (Horiba Ltd.): pH optrode comprising two fibers using a liquid crystal as the optical pulse width modulator. JP 60/85,365 (Showa Co.): optical fiber with liquid crystal whose transmittance changes when exposed to gas sample. JP 60/100,743 (Showa et al.): $NH_3$ sensor based on measurement of absorption at 1.18 to 1.22 µm; see also JP Kokai 59/183,348. JP 61/66,949 (Watanabe): hydrogen optrode based on the measurement of IR transmittance of fiber at 1.4 µm. US 4,718,747 (Bianchi et al.): optical fiber cable with hydrogen-sensitive layer. US 4,608,345 (Feldman and Panzer): colorimetric detection of alcohols in gasoline. E.P. Appl. 231,086 (Hirschfeld): optrode for remotely monitoring the concentration of dissolved gases or volatile components in solution. US 4,352,983 (Silvus et al.): determination of oil in water using a refractive index-based fiber sensor with lipophilic coating.

## C. BIOSENSORS AND IMMUNOSENSORS

DE Offen. 2,948,904 (Lübbers et al.): optrode with additional reaction volume with immobilized enzyme, resulting in biosensors that use a pH, oxygen, or $CO_2$ optrode as a transducer; suitable for determination of glucose, ethanol, cholesterol, starch. DE Offen. 2,856,251 (Lübbers): optical biosensor comprising an indicator-based optrode and an immobilized enzyme or an imobilized antibody. DE Offen. 3,001,669 (Lübbers et al.): immobilization of dyes or enzymes to membranes. EP Appl. 249,957 (Meserol et al.): prism containing a gel with immobilized enzyme and chromogenic reagent and measurement of absorbance changes after multiple internal reflection; also applicable to immunoassay. US 4,041,932 (Fostick): monitoring of blood constituents by fluorimetric measurement of the concentration of gases collected in a chamber sealingly attached to a skin "window" formed by removing the *stratum corneum* over a small area of the patient's skin.

UK Appl. 2,197,065 (Batchelder and Willson): surface plasmon resonance sensor for detection of antigens in blood. US 4,447,546 (Hirschfeld): disposable fiber optic immunosensor based on competitive binding of labeled and unlabeled antibody to antigen bound to the fiber core; see also US 4,582,809 (Block and Hirschfeld). DE 3,701,833 (Wolfbeis): fiberoptic determination of enzyme activities using chromogenic enzyme substrates. US 4,755,684 (Leiner and Schaur): fiberoptic method for differentiation between normal sera and cancerous sera by measuring fluorescence of proteins at two or three excitation wavelengths and ratioing the two signals. DD 243,351 (Knorre et al.): fiber optic sensor for proteins and nucleic acids; response is based on optical effects occurring at the surface of the bent (fiber) waveguide; transmission of light measured via changing the bending radius. US 4,368,047 (Andrade and VanWagenem): use of the intrinsic fluorescence of tryptophane in an evanescent wave planar immunosensor.

US 4,344,438 (Schultz): fiber sensor for blood plasma constituents such as glucose based on competitive binding of analyte to specific receptor sites; fluorescently labeled ligand is displaced simultaneously and enters the cone of the fiber numerical aperture and thereby is "seen" by the optical system. US 3,979,184 (Giaever): immunoassay on planar sputtered plates by interferometry. US 4,368,047 (Andrade and VanWagenem): total internal reflection immunoassay based on measurement of intrinsic (Trp) fluoresence of proteins. US 3,939,350 (Kronick and Little): evanescent wave immunoassay on planar waveguide based on competitive binding of free hapten to a labeled hapten-albumin conjugate. US 4,509,522 (Manuccia and Eden): fiber optic method for determination of CO and $CO_2$ in blood via IR absorption. JP 60/77,740 (Sumimoto): optical fiber sensor for *in situ* artery blood analysis via reflectometry and light scattering. US Appl. 275,142 (1988, Ares-Serono): surface plasmon resonance biosensor. Eur. Appl. 59,032 (Elings and Briggs): measurement of dye concentration in bloodstream. US 4,451,149 (Noeller): fiber optic fluorescence polarization immunoassay.

## D. MISCELLANEOUS PATENTS

US 4,306,877 (Lübbers): use of reference substance in optical sensors for calibration by measurement at two wavelengths. US 4,580,059 (Wolfbeis and Urbano): determination of two interfering species using two sensors with different response to each species. US 4,573,761 (McLachlan et al.): fiber optic probe for Raman analysis with at least one fiber carrying exciting light and at least two fibers for collecting scattered light. EP Appl. 214,768 (Hirschfeld): chemical sensor based on energy transfer from fluorescent indicator to a nonfluorescent absorber, or from a fluorescent absorber to a nonfluorescent indicator. UK Appl. 2,103,786 (Cramp et al.): fiber sensor with indicator covering the core of the fiber end and whose absorbance or fluorescence is measured as a function of the parameter to be determined; see also US 4,600,310 (Cramp and Reid). PCT WO 86/07,149 (Tiefenthaler et al.): optical chemical sensors for gases, liquids, and biomolecules based on measurement of changes in refractive index in an opto-coupler.

US 4,321,057 (Buckles): sensor for oxygen, glucose, antigen/antibody reactions based on evanescent wave technique and analyte-sensitive chemistry on fiber core. PCT WO 87/00920 (Costello): fiber optic invasive catheter with absorption cell. EP Appl. 244,394 (Marsoner et al.): multiplicity of optical sensors, light sources, and optical filters arranged in series. AT 383,684 (Kroneis and Offenbacher): improved method for stray light suppression in planar optical sensors or fiber sensors with planar sensing membranes exploiting effects of total internal reflection. US 4,710,623 (Lipson et al.): elongated fiber with a hole drilled into the end containing an analyte-sensitive chemistry which thereby is protected from physical damage. EPA 263,805 (Marsoner et al.): optical arrangement for efficiently coupling excitation light into a sensing layer and collecting emission by a photosensor beneath the sensing layer. US 4,443,700 (Macedo et al.): optical sensing apparatus and method. US 4,270,049 (Tanaka et al.): optical liquid leakage detection system. EPA 261,642 (Kawaguchi and Shiro): multilayer optical waveguide biodetector kit. US 4,608,344 (Carter et al.): refractive index-based waveguide sensor with analyte-specific coating and evanescent wave detection.

# REFERENCES

1. **Turner, A. P. F., Karube, I., and Wilson, G. S. Eds.,** *Biosensors,* Oxford University Press, Oxford, 1987.
2. **Kautsky, H. and Hirsch, A.,** Detection of minutest amounts of oxygen via phosphorescence quenching, *Z. Anorg. Allg. Chem.,* 222, 126, 1935.
3. **Bergman, I.,** Rapid response atmospheric oxygen monitor based on fluorescence quenching, *Nature (London),* 218, 396, 1968.
4. **Lübbers, D. W. and Opitz, N.,** The pCO$_2$/pO$_2$-Optode: a new probe for measurement of pCO$_2$ and pO$_2$ of gases and liquids, *Z. Naturforsch. Teil C,* 30C, 532, 1975.
5. **Opitz, N. and Lübbers, D. W.,** A new fast-responding optical method to measure pCO$_2$ in gases and solutions, *Pfluegers Arch. (Eur. J. Physiol.),* 355, R120, 1975.
6. **Lübbers, D. W. and Opitz, N.,** Optical sensor for gases, U.S. Patent 4,003,707 (claim 14).
7. **Johnson, C. C.,** Fiber optic probe for oxygen saturation and dye concentration monitoring, *Biomed. Sci. Instrum.,* 10, 45, 1974.
8. **Polanyi, M. L. and Hehir, R. M.,** *Rev. Sci. Instrum.,* 33, 1050, 1962.
9. **Chance B., Mayevsky, A., Goodwin, C., and Mela, L.,** Factors in oxygen delivery to tissue, *Microvasc. Res.,* 8, 276, 1974.
10. **Hesse, H. C.,** East German Patent 106,086, 1974.
11. **Peterson, I. J., Goldstein, S. R., Fitzgerald, R. V., and Buckhold, D. K.,** *Anal. Chem.,* 52, 864, 1980.
12. **Hirschfeld, T., Johnson, D., Haugen, G., and Hrubesh, L.,** Remote spectrophotometric techniques based on long distance fiber optics, paper presented at the 180th Natl. Meet. American Chem. Society, Las Vegas, August 1980.
13. **Hirschfeld, T., Johnson, D., Miller, S., and Klainer, S.,** Long range analysis via fiber optics, paper presented at the Pittsburgh Conf., Atlantic City, March 1981.
14. **Hirschfeld, T., Deaton, T., Milanovich, F., and Klainer, S.,** Feasibility of using fiber optics for monitoring groundwater contaminants, *Opt. Eng.,* 22, 27, 1983.
15. **Milanovich, F. P. and Hirschfeld, T.,** Process, product, and waste stream monitoring with fiber optics, *Adv. Instrum.,* 38, 407, 1983.
16. **Freeman, T. M. and Seitz, W. R.,** Oxygen probe based on chemiluminescence, *Anal. Chem.,* 53, 98, 1981; Chemiluminescence fiber optic probe for hydrogen peroxide, *Anal. Chem.,* 50, 1242, 1977.
17. **Wolfbeis, O. S.,** The fluorescence of organic natural products, in *Molecular Luminescence Spectrosopy: Methods & Applications,* Vol. 1, Schulman, S. G., Ed., John Wiley & Sons, New York, 1985, chap. 3.
18. **Wolfbeis, O. S. and Lippert, E.,** Multiple fluorescence of 7-hydroxylepidone, *Z. Naturforsch. Teil A,* 33A, 238, 1978.
19. **Marsoner, H., Kroneis, H., and Wolfbeis, O. S.,** European Patent 109.959 (22. 4. 1980); U.S. Patent 4,587,101.
20. **Giallorenzi, T. G., Bucaro, J. A., and Dandridge, A.,** Optical fiber sensor technology, *IEEE J. Quantum Electron,* 18, 626, 1982.
21. **Chester, A. N., Martellucci, S., and Scheggi, A. M., Eds.,** *Optical Fiber Sensors,* Martinus Nijhoff, Dordrecht, 1987.
22. **Arnold, M. A., Ed.,** *Talanta,* special issue 35, number 2, 1988.
23. **Wolfbeis, O. S.,** Optic and fiber optic fluorosensors in analytical and clinical chemistry, in *Molecular Luminescence Spectroscopy, Methods & Applications,* Vol. 2, Schulman, S. G., Ed., John Wiley & Sons, New York, 1988, chap. 3.
24. **Walt, D. R.,** Design, Preparation, and applications of fiber optic chemical sensors, *Chemical Sensors and Microinstrumentation,* Murray, R. W., Ed., ACS Symposium Ser. Vol. 403, American Chemical Society, Washington, D.C., 1989.
25. **Narayanaswamy, R.,** Optical fibre sensors for chemical species, *J. Phys. E.,* 21, 10, 1988.
26. **Alder, J. F.,** Optical fibre chemical sensors, *Fresenius Z. Anal. Chem.,* 324, 372, 1986.
27. **Edmonds, T. E. and Ross, I. D.,** Low cost fibre optical chemical sensors, *Anal. Proc.,* 22, 206, 1985.
28. **Saari, L. A.,** Trends in fiber optic sensor development, *Trends Anal. Chem.,* 6, 85, 1987.
29. **Wolfbeis, O. S.,** Fluorescence optical sensors in analytical chemistry, *Trends Anal. Chem.,* 4, 184, 1985.
30. **Lübbers, D. W. and Opitz, N.,** Optical fluorescence sensors for continuous measurement of chemical concentrations in biological systems, *Sensors Actuators,* 4, 641, 1983.
31. **Seitz, W. R.,** Chemical sensors based on immobilized indicators and fiber optics, *Crit. Rev. Anal. Chem.,* 19, 135, 1988.
32. **Maugh, T. H.,** Remote spectrometry with fiber optics, *Science,* 218, 875, 1982.
33. **Borman, S. A.,** Optrodes, *Anal. Chem.,* 53, 1618A, 1981.

34. **Wolfbeis, O. S.,** Analytical chemistry with optical sensors, *Fresenius Z. Anal. Chem.,* 325, 387, 1986.
35. **Peterson, J. I. and Vurek, G. G.,** Fiber-optic sensors for biomedical applications, *Science,* 224, 123, 1984.
36. **Wise, D. L. and Wingard, L. B., Eds.,** *Biosensors with Fiber Optics, Humana Press,* Clifton, NJ, 1991.
37. **Martin, M. J., Wickramasinghe, Y. A. B. D., Newson, T. P., and Crowe, J. A.,** Fibre optics and optical sensors in medicine, *Med. Biol. Eng. Comp.,* 1987, 597.
38. **Wolfbeis, O. S.,** Fiber-optic sensors in biomedical sciences, *Pure Appl. Chem.,* 59, 663, 1987.
39. **Harmer, A. L.,** Guided wave chemical sensors, *Proc. Electrochem. Soc.,* 87, 409, 1987.
40. **Dakin, J. and Culshaw, B., Eds.,** *Optical Fibre Sensors,* Vols. 1 and 2, Artech House, Boston — London, 1988.
41. **Hirschfeld, T., Deaton, T., Milanovich, F., and Klainer, S. M.,** The feasibility of using fiber optics for monitoring groundwater contaminants, EPA Rep. AD-89-F-2A074, 1983.
42. **Hirschfeld, T., Callis, J. B., and Kowalski, B. R.,** Chemical sensing in process analysis, *Science,* 226, 312, 1984.
43. **Schaffar, B. P. H. and Wolfbeis, O. S.,** Chemically mediated fiber optic biosensors, in *Biosensors: Principles and Applications,* Blum, L. J. and Coulet, P. R., Eds., M. Dekker, New York, 1991, chap. 8.
44. **Freeman, J. E., Childers, A. G., Steele, A. W., and Hieftje, G. M.,** A fiber optic absorption cell for remote determination of copper in industrial electroplating baths, *Anal. Chim. Acta,* 177, 121, 1985.
45. **Christiansen, T. F.,** The Achilles heel of potentiometric measurements, the liquid junction potential, *IEEE Trans. Biomed. Eng.,* 33, 79, 1986.
46. **Wolfbeis, O. S., Weis, L. J., Leiner, M. J. P., and Ziegler, W. E.,** Fiber optic fluorosensor for oxygen and carbon dioxide, *Anal. Chem.,* 60, 2028, 1988.
47. **Zirino, A., Ed.,** Mapping strategies in chemical oceanography, *Adv. Chem. Ser.* 209, American Chemical Society, Washington, D.C. 1985.
48. **O'Neil, R. A., Buja-Bijunas, L., and Rayner, D. N.,** Field performance of a laser fluorosensor for the detection of oil spills, *Appl. Opt.,* 19, 86, 1980, and Referencess. 1 to 8 cited therein.
49. **Tortoishell, G.,** Safety of fiber optic systems in flammable atmospheres, *Proc. SPIE,* 522, 132, 1985.
50. **Wolfbeis, O. S.,** Fiber optic sensors in bioprocess control, in *Sensors in Bioprocess Control,* Twork, J. V. and Yacynych, A. M., Eds., Marcel Dekker, New York, 1990.
51. **Beyeler, W., Einsele, A., and Fiechter, A.,** On-line measurements of culture fluorescence: method and application, *Eur. J. Appl. Microbiol. Biotechnol.,* 13, 10, 1981.

Chapter 2

# SPECTROSCOPIC TECHNIQUES

### Otto S. Wolfbeis

## TABLE OF CONTENTS

# I. INTRODUCTION

All optical methods of chemical analysis rely on the interaction of electromagnetic radiation with matter. As a result, optical spectrometry can be a fast and nondestructive analytical tool. The wavelength range covered by optical methods applied in fiber optic sensors can be divided into the UV region (200 to 400 nm, 6.2 to 3.1 eV), the visible (400 to 780 nm, 3.1 to 1.6 eV), the near infrared (0.78 to 3 μm, 1.6 to 0.4 eV), and infrared (3 to 50 μm, 0.4 to 0.025 eV).

The main types of interaction of light with matter are dispersion, absorption, diffraction, interference, and reflectance, all of which can be explained by the dual nature of light, i.e., treating light as a particle (photon) and/or a wave. The wave character of light is usually described in terms of an electric and a magnetic field vector perpendicular to each other. Changes in the amplitude of these vectors can induce a change in charge distribution of a molecule, resulting in a change in the polarization of the molecule. In gases and liquids, this causes molecules to rotate because of their reorientation (orientation polarization) with respect to their position at a frequency in the radiowave range. Because of the change of the charge distribution within the molecules, the relative position of the nuclei and of the electrons will also be changed. Vibronic excitation of a molecule occurs in the infrared frequency domain, while changes in electronic distribution are caused by UV and visible light.

A polarized molecule can act as a new dipole and is capable of emitting a second wave of a frequency identical with that of the first wave. In liquids and gases, molecules are statistically distributed with respect to orientation, so that the secondary waves will be emitted in all directions. Since polarization efficiency also depends on the interaction cross-radius between wave and molecule (usually taken as the square of the radius of the molecule), the efficiency of second wave emission ("scattering") is small in case of molecules of a diameter smaller than $\lambda/100$. When the diameter comes to lie in the order of $\lambda/10$ or larger, elastic scattering will become more intense and manifests itself in various forms including Rayleigh scattering. When the diameter of a molecule exceeds the wavelength of interacting light, there may be more than one center of scattering, and the resulting secondary waves may interfere. This results in so-called Mie scatter. Its use is widespread in both static light scattering methods (such as nephelometry and turbidimetry) and dynamic scattering methods such as photon correlation spectroscopy and laser velocimetry.

As a general rule it can be said that the particle radius R and wavelength $\lambda$ are the most critical parameters for scatter intensity $I_s$, because if

$$R < \lambda, \text{ then} \qquad I_s = K \cdot R^6/\lambda^4$$
$$R \text{ ca. } \lambda, \text{ then} \qquad I_s = K \cdot R^4/\lambda^2, \text{ and}$$
$$R > \lambda, \text{ then} \qquad I_s = K \cdot R^2$$

It is for this reason that, in fluorimetry, shortwave excitation results in stronger background scatter than longwave excitation.

Aside from the polarization effects discussed so far, there will be resonant interaction of radiation with molecules. The most popular method for describing the energy levels of a molecules is based on the Jablonski diagram. The ground state (in the case of organic molecules mostly a singlet state) is designed as the $S_o$ state which, at room temperature, is vibrationally almost fully relaxed. Rotational energy levels are usually neglected. Absorption of light creates an excited state species (the $S_1$ state) which, after vibrational relaxation, can return to the $S_o$ state in various ways including radiative deactivation (fluorescence) or radiationless deactivation. Another path is via the triplet state by either phosphorescence or

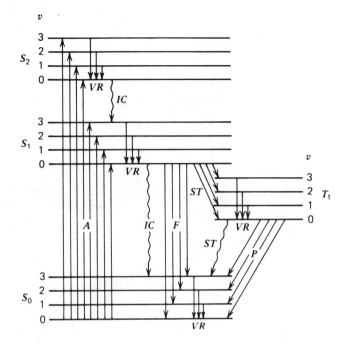

FIGURE 1. Jablonski diagram showing the relative energies of ground state ($S_0$) and first excited singlet ($S_1$) and triplet states ($T_1$) and their vibrational sub-niveaus ($v = 0, 1, 2, 3$). Rotational energy levels are omitted for clarity. Electronic absorption (*A*) from $S_0$ to $S_1$ is followed by rapid, radiationless internal conversion (*IC*) and vibrational relaxation (*VR*) to the lowest vibrational level ($v = 0$) of $S_1$. Competing for deactivation of $S_1$ are the radiationless internal conversion (*IC*) and singlet-triplet intersystem crossing (*ST*) as well as fluorescence (*F*). The latter is followed by vibrational relaxation (*VR*) in the ground state. Intersystem crossing (*ST*) is followed by vibrational relaxation (*VR*) in the triplet state ($T_1$). Both phosphorescence (*P*) and nonradiative $T_1$–$S_0$ intersystem crossing (*ST*) return the molecule from the $T_1$ to the $S_0$ state. Vibrational relaxation in $S_0$ then thermalizes the hot ground state molecule.

radiationless relaxation. A final path (and very undesirable in optical sensing) away from the $S_1$ state is an irreversible photochemical reaction. A popular method to describe the energy levels of a molecule is the so-called Jablonski diagram shown in Figure 1. It can be used to explain the absorption and emission processes that will be discussed in the following, with particular emphasis given to the fiber aspect[1] and their respective merits for sensing purposes.

## II. ABSORPTIOMETRY

UV/visible absorptiometry in its various forms is probably the most popular method in conventional analytical chemistry. It is based on the absorption of light by the analyte or an indicator to give the respective $S_1$ state. According to Beer and Lambert, the following linear relation exists between absorbance A and the concentration C of an absorbing analyte

$$A = \log I_0 / I = \varepsilon \cdot c \cdot d \tag{1}$$

where $I_0$ is the intensity of incoming light, and I the intensity of light after passing the sample. Absorbance is sometimes also referred to as optical density (OD) or extinction (E). $\varepsilon$ is the decadic molar absorption coefficient having typical values of $10^4$ to $10^5$ $M^{-1}cm^{-1}$ (with $M$ being moles per liter), and d is the path length of the light in the cuvette or sample. When $\varepsilon$ is high and d is large, absorptiometry can be very sensitive.

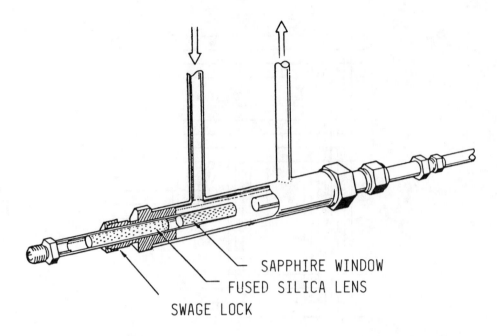

SAPPHIRE WINDOW
FUSED SILICA LENS
SWAGE LOCK

FIGURE 2.  Fiber optic absorption flow-through cell.

Photometry has a relatively weak temperature coefficient and is an intrinsically self-reference method in that it always is a ratio of intensities that is measured rather than an absolute intensity (as in fluorometry, Raman spectrometry, and reflectometry). Thus, it cannot be made more sensitive by raising the intensity of the light source ($I_o$). This can turn out as an advantage because strong illumination of indicators favors photodecomposition.

Numerous textbooks cover the subject[2] and there seems to be no need to go into further detail. Absorptiometry has been applied over a wide spectral range to determine species having an intrinsic UV, VIS, NIR, or IR absorption. A selection of fiber optic absorptiometric sensing methods using plain fibers is presented in Chapter 3, Section III. Analytes lacking such an absorption can be assayed indirectly through a chemical transduction system. This is described in Chapter 3, Section IV on indicator-mediated sensing.

Potential pitfalls in absorptiometry result from the tendency of larger molecules to self-aggregate, thereby forming dimers and higher aggregates which cause deviations from linearity between A and C. In fact, most longwave absorbing molecules, including chlorophyll, acridine orange, and rhodamines, form aggregates at concentrations higher than $10^{-4}$ $M$. In addition to changes in $\varepsilon$, the dimers give rise to new absorption bands. Also, fiber photometry suffers from changes in the refractive index of the sample solution. Finally, dyes tend to deposit on the surface of fibers.

In fiber optic absorptiometry, the intrinsic absorption of fibers has to be compensated for. Typical absorption spectra of commercial fibers are shown in Chapters 6 and 14. The intrinsic absorption of fibers (as obtained in a blank run) can be substracted from the actual spectrum by electronic means, or measurements are made with two identical fibers, one of which acts as a reference.

It is surprising to see that, when compared to either reflectance, fluorescence, or NIR techniques, absorptiometry in the visible has rather seldom been used in combination with plain fiber optic chemical sensing. A few examples are given in Chapters 3 and 14. One reason may be the poor selectivity of absorptiometry. Various types of flow-through cells for absorption measurements are known, and a few examples are given in Figure 2. Others are found in Chapter 3.

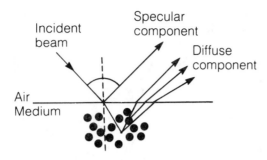

FIGURE 3. The difference between regular (specular) and diffuse reflection of light from a solid surface.

A typical practical application is continuous measurement of the absorption of the copper(II) ion in electroplating baths.[1] Copper (II) ion has an absorption that perfectly matches the emission of some IREDs. It is obvious, though, that these arrangements are prone to interference by all kinds of particles passing the beam, and by any other analyte absorbing at this wavelength. Other ions with NIR absorption include nickel(II) and chromium(III). Also, changes in the refractive index of the sample change the numerical aperture of the fiber and, consequently, the amount of light coupled into the second fiber. Also based on direct absorption measurement is the first sensor for nitrate in drinking water.[4] Using UV transmitting fibers and quartz optics, nitrate was measured via its absorption at 220 nm.

## III. REFLECTOMETRY

Two types of reflectance are known to occur at interfaces between media of different optical density.[5] On a polished surface, such as on metal (mirrors), reflectance is of the guided type ("regular" reflectance). This kind of reflectance is not used in chemical sensors. On a rough surface, on the other hand, reflectance will be of the eiffusive type provided the size of the scattering center is in the order of the wavelength of light or larger (Figure 3). Diffuse reflectance spectrometry is a frequently employed optical sensing technique and also is the method by which we view most of our environment. Other fields of applications are blood oximetry (see Chapter 19), the measurement of dyes and pigments, papers and textiles, as well as dry reagent chemistry, partially in combination with fibers.

There are several reasons for the popularity of reflectometry. It can be applied to optically dense materials being even inhomogeneous, and to opaque materials which in absorbance would give a high background. Reflectometry allows large spatial flexibility, and various parameters such as intensity, spectral variation, angular distribution, and polarization may be exploited. On the other hand, the relation between optical signal and analyte concentration is more complex, instrumentation is somewhat more expensive, and reproducibility strongly depends on the reproducible fabrication of the sensing matrix.

The processes that can occur when light hits the interface of two media having different dielectric constants are shown in Figure 3. If the surface is plane, part of the wave will be reflected at the same angle $\alpha$ as the incident light (regular reflection). Another fraction will be refracted and protrude in phase 2 at an angle $\beta$ (Figure 3). In the presence of scattering particles, however, diffuse reflection is observed, and practically all light that enters phase 2 reappears on the surface. Part of the light will be absorbed by the (dyed) particles, another part will be reflected into all directions. The distribution of diffusively reflected light is rather homogeneous and largely independent of size and shape of the particles.

Kubelka and Munk[6] have developed a theory for the quantitation of reflectance spectra by introducing an absorption coefficient K and a scattering coefficient S. K is proportional to the

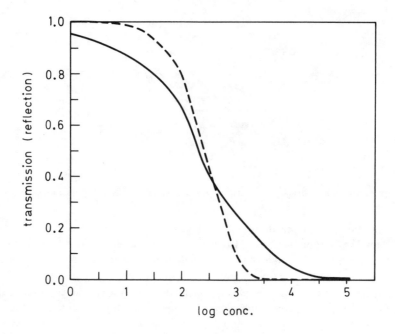

FIGURE 4. Comparison of the analytical ranges as measured by reflection ( —— ) and transmission (- - - -).

absorption coefficient in the Lambert-Beer law. Since, in practice, light is scattered and absorbed even within a sensing layer of defined thickness, the relation between the measured reflectance intensity R and K and S is rather complex. For an infinite layer thickness

$$K/S = \frac{(1-R)^2}{2R} \qquad (2)$$

Since $K = \varepsilon_R \cdot C$ (where $\varepsilon_R$ is comparable to, but *not* identical with the molar absorption in the Lambert-Beer law), it can be derived that the relation between analyte concentration C and signal R is given by

$$C = \frac{S(1-R)^2}{2} \varepsilon \cdot R \qquad (3)$$

Because the total intensity of reflected light cannot be measured easily in optical sensors, a factor has to be introduced in practice to account for incomplete light gathering. The situation is similar to fluorimetry, where light is emitted into all directions too and no absolute concentrations can be determined (which contrasts absorptiometry) unless the system is precalibrated. Another significant difference when compared to absorptiometry is the larger dynamic range (Figure 4) which is highly desired in sensor application in order to avoid sample dilution.

Kubelka-Munk functions are widely used in dry reagent chemistry instrumentation[7,8] such as those of Boehringer and Ames-Miles, and of reflectometry-based optical probes and sensors for chemical species.[9-11] Reagent layers having a reflecting zone beneath the reaction zone require a different mathematical treatment. These multilayer chemistries are applied in instrumentation available from Kodak and Fuji and are equipped with a more sophisticated optical system is shown in Figure 5. A cross-section through a typical multilayer system. The sample is in contact with the permeable and reflective layer RL. The analyte diffuses into the

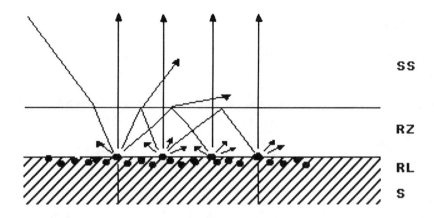

FIGURE 5. Williams-Clapper model of a multilayer reflector. RL, reflective layer; RZ, reagent zone; SS, solid support; S, sample. Arrows show the incoming and reflected light.

reagent zone RZ to cause a color reaction. Light enters the reagent zone through an optically transparent solid and impermeable support SS and partially also reaches the RL where it will be scattered back by diffuse reflection. After further multiple reflection in the RZ, light leaves the layers on the surface and re-enters the fiber on top.

Because this situation is not compatible with the Kubelka-Munk model, Williams and Clapper[12] have developed an equation that describes this situation. It is rather complex and is treated in detail elsewhere.[13] Essentially, the intensity of reflected light depends on the transmissivity of the reagent layer, the reflectivity of the RL, the inner Fresnel reflection, and the index of refraction of the RL. Integrating the Williams-Clapper equation can result in a linear relation between reflection density and analyte concentration. The reader is refered to a more specialized treatment of the subject.[14]

Aside from these two theories, other models for quantitative reflectometry have been proposed. Thus, the Rozenberg model[15] is applicable to mixtures of absorbing and nonabsorbing powders, while the Pitts-Giovanelli theory[16] has been shown to be valid for scattering particles suspended in an absorbing aqueous medium. These methods have been compared for several model systems, and the conclusion was that for an adsorber-absorber system such as that of a dyed resin particle the Kubelka-Munk theory was best.[17] Particularly for dyed pH-sensitive particles[9] and a fluoride-sensitive material,[18] more detailed treatments have been presented, and Goldman[19] has discussed the application of diffuse reflectance fluorimetry in TLC.

Diffuse reflectance spectrometry is not confined to the visible. It can be used in the NIR and IR as well, preferentially in combination with Fourier transform methods (FTIR). The sample is irradiated with IR light which penetrates the sample and then is scattered. The technique is applicable to most solids, involves little or no sample preparation, and avoids the unnecessary complications of absorptions due to solvents or mulling agents. By far its greatest advantage, however, is its ability to handle a wide range of sample, including powders, crystals, solids with rough surfaces, plastics, etc.

The absorption properties of a matrix surrounding an analyte influence the band intensity in near-infrared diffuse reflectance spectra. If the matrix does not absorb radiation at the same wavelength as the analytical band, then use of the Kubelka-Munk equation provides a linear relationship between band intensity and concentration over a major portion of the concentration range for the analyte.

If the matrix surrounding the analyte absorbs radiation at the same wavelength as the analytical band, then deviations from linearity of plots of the Kubelka-Munk function vs. concentration occur. In this case, the use of log 1/R values instead of the Kubelka-Munk function has been shown empirically[20] to provide a more linear relationship between reflectance and concentration. It has also been shown that the concentration range over which linearity holds is dependent upon particle size and on the strength of the absorption by the matrix. The reason for this behavior is explained by the effective penetration depth of the beam, which is shown to be only one or two particle diameters when absorption by the matrix is strong.

## IV. LUMINESCENCE

Luminescence is observed when the energy of an electronically excited state species is released in the form of light. Depending on whether the excited state is singlet or triplet, the emission is called fluorescence or phosphorescence. Other classifications are according to how the excited state is produced, e.g., by photoexcitation (photoluminescence), chemical reaction (chemi- and bioluminescence), electricity (electroluminescence), stress (triboluminescence), and others. Photoluminescence is the most distributed method.[21-24]

Since both fluorescence and phosphorescence are intrinsically at least tri-parametric (e.g. excitation wavelength, emission wavelength, emission intensity), a full description of photoluminescence emission requires an excitation-emission data matrix.[25] Because fluorescence and phosphorescence are performed at two different analytical wavelengths, and only a fraction of the molecules present in a matrix exhibits luminescence, these methods are generally more selective than photometry. Chemiluminescence has the attractive feature of not requiring a light source. A variety of useful books is available.[26-28]

Quantitative luminometry is based on the measurement of luminescence intensity, lifetime, or polarization. The former two are more readily applicable to fiber optics, which can maintain polarization with difficulty only. One has to differentiate between situations in which the luminophore is the analyte, i.e., in plain fiber sensors, and where it is an indicator whose properties vary as a result of ground state interactions (such as pH shifts) or excited state interactions (quenching) with the analyte.

Plain fiber luminescence sensors utilize the intrinsic luminescence of analytes. As in conventional fluorimetry, luminescence sensing can be performed under either right-angle illumination or front-face illumination. The first situation is found in fiber optical fluorescence detection when the ends of the two fibers guiding the exciting and fluorescent light, respectively, are directed toward the sample at an angle of typically 90° to each other. In fluorosensing, however, front-face fluorimetry is the more common situation since both single fibers and bifurcated fibers are usually immersed in the sample solution and remitted light is collected.

Parker has shown[21] that the relation between luminescence intensity I and analyte concentration C is given by:

$$I = 2.3 \cdot I_0 \Phi_f \varepsilon \cdot C \cdot l \left[ 1 - \frac{2.3\, \varepsilon \cdot C \cdot l}{2!} + \frac{(2.3\, \varepsilon\, C \cdot l)^2}{3!} + \cdots \right] k \qquad (4)$$

where $I_o$ is the intensity of the exciting light, $\varepsilon$ being the molar extinction coefficient at the excitation wavelength, $\Phi_f$ the luminescence quantum yield, and l the optical depth. A factor k that takes into account the geometrical arrangement of the instrument has to be included.

For weakly absorbing solutions, for which $\varepsilon \cdot C \cdot l$ is small, Equation 4 simplifies to Parker's law:

$$F = 2.3I_0\Phi_f \cdot \varepsilon \cdot C \cdot l \cdot k \tag{5}$$

Again, the constant k accounts for the fact that only a fraction of the total emission is observed. This equation holds with sufficient precision for most fluorosensors applied in optically thin media such as groundwater and air. Unless l can be made very small, it is not applicable to analysis of strongly absorbing samples such as blood.

## A. FLUORESCENCE INTENSITY-BASED SENSORS

Fluorescence is the light emitted by a molecule in its first excited singlet state. Fluorimetry is sensitive, fairly selective, and versatile. Numerous organic or bioorganic,[29] pharmaceutical,[30] and inorganic analytes[31] are known to be fluorescent or to undergo fluorogenic reactions (see Chapter 7). The principles of fluorescence spectroscopy have been described in numerous books.[21-24] With respect to fiber sensing, the advantage of fluorosensing[32] is its spatial flexibility, the various parameters that can be exploited for the purpose, and the inherent sensitivity even when the smallest fibers are used. An excellent treatise of luminescence methods on surfaces, including theory, practical considerations, and scattering models, has been presented by Hurtubise.[33] Fluorescence methods are applicable over a wide concentration range and therefore are ideally suited for fiber fluorimetry, preferentially with laser excitation.[34]

Luminescence methods are easily implemented with single fibers, and the functional dependence of luminescence intensity on analyte concentration is more easily defined than is the case in reflectance. The mathematical laws underlying plain fiber fluorimetry and indicator-mediated fluorimetry will be presented in Chapter 3.

On the other hand, fluorimetry is prone to interferences by external quenchers, inner filter effects, and misinterpretation by unskilled users, partially because of instrumental artifacts and effects of viscosity, temperature, oxygen quenching, energy transfer, and solvents. Raman scatter can interfere in case of weak fluorescence intensity. In addition, most semiconductor light sources cannot be applied to fluorosensors which mostly require shortwave excitation. Finally, because intense light sources may be needed to obtain measurable fluorescence, photobleaching becomes more serious in fluorimetry than in reflectometry.

## B. FLUORESCENCE LIFETIME-BASED SENSORS

Lifetime is defined as the average time a molecule remains in the excited state. Typical fluorophore lifetimes range from 2 to 20 ns, while phosphorescence lifetimes are much longer (1 μs to 10 s). Metal-organic complexes are in between. It should be kept in mind that lifetime is an average value of a statistically significant number of molecules. In case of a single-exponential decay, the lifetime $\tau$ is defined as the time after which only 1/e of the initial population still exists in the excited state (i.e., approximately 37%).

There are two widely used methods for lifetime determination, namely the pulse method and the phase modulation method.[22,35] In the former, the sample is excited with a brief pulse of light and the time-dependent decay of luminescence intensity is measured. In the phase method, the sample is excited with sinusoidally modulated light, and phase shift between the sine function of luminescence to that of exciting light is determined. Each method possesses its unique advantages and disadvantages which have been discussed in detail.[22]

Lifetime-based sensors are useful in optical sensing if the analyte affects the lifetime of an indicator, or the lifetime of one partner in an energy-transfer (ET) system is affected by the analyte. Therefore, all dynamic quenchers of the luminescence of suitable indicators may be assayed by this technique, and all ET- based sensors have the potential of being so.

Since, according to Stern and Volmer, the relation between lifetimes in the absence ($\tau_0$) and presence ($\tau$) of a quencher is given by

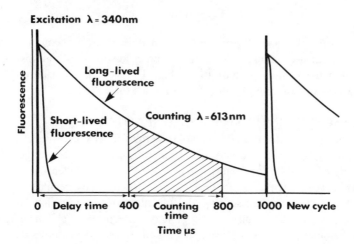

FIGURE 6.   Fluorescence decay profile of an europium chelate as used in a time-resolved immunoassay. Background fluorescence disappears after a few nanoseconds, whereas the chelate decays in the millisecond time regime.

$$\tau_0/\tau = 1 + k_0 \cdot \tau_0[Q] = 1 + k_{sv}[Q] \qquad (6)$$

there is a linear relation between quencher concentration and $(\tau_0/\tau - 1)$. Because decay time is affected by the dynamic quenching process only, lifetime plots are usually linear. Static fluorescence quenching has no effect on lifetimes.

A typical example for a sensor based on lifetime measurement is the oxygen optrode based on luminescence quenching of a long-lived ruthenium complex.[36] Lifetime methods are ideally suited for straylight and background suppression because the fluorescence of some chelates lasts much longer than elastic scatter and background fluorescence (Figure 6). Therefore, time-delayed fluorimetry holds great potential for fiber optic time-resolved immunosensing with its scatter and background problems.[37]

While lifetime-based sensing requires more sophisticated optoelectronic instrumentation, it has some decisive advantages over intensity-based sensing: (1) It has negligible drift arising from leaching and photobleaching because the decay time is independent of fluorophore concentration. (2) It displays excellent long-term stability because it is an internally referenced system in that the emission phase is compared with the excitation phase. (3) Drifts in both light source intensity and photodetector sensitivity play no role because lifetime is independent of light source and photodetector intensity. (4) It is likely that time-resolved fluorometry can overcome problems associated with fiber bending. It has been calculated that fiber lengths approaching 1 km can be used without significant loss in accuracy or precision due to dispersion for lifetime determination in the >1 ns time domain.[38]

## C. ENERGY TRANSFER-BASED SENSORS

An excited molecule (a "donor") can transfer its electronic energy to another species (an "acceptor").[22,39] This process, called energy transfer (ET), occurs without the appearance of a photon and results from a dipole-dipole interaction between donor and acceptor. The rate depends on the fluorescence quantum yield of the donor, the overlap of the emission spectrum of the donor with the absorption spectrum of the acceptor, and their relative orientation and distance. It is the dependence on the overlap of the spectra and the distance that have been exploited in optrodes. Typical critical distances over which energy can be transfered are 0.5 to 10 nm. It should be kept in mind that ET is not a diffusional process. Aside from dipole-

dipole interaction, there is another type of ET which involves emission of light and its reabsorption by the acceptor. This type ("trivial ET") depends mainly on the optical density of the sample and geometrical factors and is less useful for sensing purposes. According to Förster, the efficiency of intermolecular dipole-dipole ET is expressed in terms of a critical separation distance $R_o$ at which the probability of ET is equal to that of spontaneous transfer. The energy transfer efficiency E can be calculated from both the relative fluorescence yield and the lifetimes, in the presence ($F_{da}$) and absence ($F_d$) of an acceptor according to

$$E = 1 - \left( \frac{F_{da}}{F_a} \right) \tag{7}$$

and

$$E = 1 - \left( \frac{\tau_{da}}{\tau_d} \right) \tag{8}$$

The transfer efficiency is related to the distance r between donor A and acceptor B by

$$E = \frac{R_o^{\,6}}{\left( R_o^{\,6} + r^6 \right)} \tag{9}$$

provided donor and acceptor have a fixed distance such as in labeled proteins and immobilized indicators. In solution and fluid membranes, there is random distribution and more complex expressions are required. A fixed D-A distance also results in a single exponential decay.

Another popular parameter to express the efficiency of ET is the rate constant $k_{ET}$, which is related to the lifetime of the donor ($\tau_D$) by

$$k_{ET} = \frac{R_0^{\,6}}{r^6 \tau_D} \tag{10}$$

The book of Lakowicz[22] is an excellent source for a more detailed description of ET and how to extract physical parameters from experimental data. One consequence for optical sensors is that the D-A distance in indicator-mediated sensors should be kept constant, e.g., by covalently linking D and A via spacer groups.

$R_o$ depends on the refractive index n of the medium, the spectral overlap J of the absorption of A and emission of D, the quantum yield $\Phi_f$ of the donor, and a geometric factor $\kappa^2$:

$$R_o = 9.79 \cdot 10^3 \left( J \cdot n^{-4} \cdot \Phi_f \kappa^2 \right)^{\frac{1}{6}} \tag{11}$$

The geometric factor $\kappa$ is the angular component of the dipole-dipole interaction tensor averaged over all orientations of D and A. In case of a freely rotating D and A, $\kappa^2 = 2/3$.

The relationship between r and acceptor concentration [A] is

$$r^3 = \frac{750}{\pi N[A]} \tag{12}$$

where N is Avogadro's number. Therefore, the efficiency of ET depends on the concentration of A. When $r = R_o$, [A] is the concentration of A that reduces the fluorescence of D by one half. This relation is the basis for a simple graphic evaluation of $R_o$.

Two ET situations may be discerned. In the one, the acceptor is fluorescent, in the other it is not. The first is realized in a fiber optic pH sensor composed of immobilized eosin as a donor (fluorescer), and immobilized phenol red acting as a pH-dependent *acceptor*.[40-42] The second situation is found in ET-based sensors for oxygen[43] and sulfur dioxide[44] based on excited state ET quenching of a fluorescent acceptor whose fluorescence is modulated by the direct quenching of the *donor* by the analyte.

The dramatic dependence of E on r has been utilized in a glucose biosensor that employs fluorescein as the donor and rhodamine as the fluorescent acceptor.[45] In this case it is not a pH-dependent change in acceptor concentration or the dynamic quenching of the fluorescence by varying concentrations of a quencher, but rather the distance between two fluorophors that varies in a closed system as the glucose concentration varies.

Similarly, the distance between two fluorophors can vary with ionic strength.[46] The Hirschfeld patent application[42] describes, in some detail, ET-based sensors for hydrogen sulfide (based on the formation of black PbS which acts as a filter to excite fluorescein), nitrous oxides (by formation of a purple absorber that quenches the acridine orange emission), pH (bromothymolblue quenches the luminescence of uranium glass), and alkali metal ions (using chromogenic crown ethers that quench the fluorescein emission).

The potential of ET-based sensors relies on the fact that it allows, in principle, all absorbance-based indicators to be applied in combination with fluorometric methods. The donor may be chosen from the most stable and longwave excitable fluorophors known, and the very stable and longwave absorbing indicators (for pH, metal ions, etc.) may be coupled to these fluorophors provided there is a sufficiently large spectral overlap.

A further extension that is forseeable is the combination of lifetime measurement with ET. This should result in a further improvement of the sensing system with respect to long-term stability. Thus, a lifetime-based ET system for pH has been presented[47] which is composed of a nonfluorescent pH-sensitive dye which quenches the fluorescence of a pH-independent fluorophore in accordance with its actual concentration which is governed by pH. The advantage is obvious: lifetime, rather than intensity, is modulated by pH. Other potential applications of lifetime ET are in double-labeling immunoassay, one label being a fluorophore, the other a quencher, and the distance between antigen and antibody being the analytical information obtained via lifetime. Finally, all radiative ET sensor systems (e.g., those for oxygen or $SO_2$) may be adapted to lifetime measurements.

## D. PHOSPHORESCENCE

Phosphorescence is mostly studied at temperatures of 77K or lower in rigid matrixes where diffusion of the ubiquitous quencher oxygen and solvent molecules is slow enough to prevent radiationless deactivation of the triplet state.[21] Phosphorescence can, however, be observed at room temperature when the indicator is placed on a rigid support such as silica gel or cellulose, sometimes even in the presence of oxygen.[33,48] While phosphorescent indicators so far have not found application for analytes involved in ground-state reactions, some quenching analytes can be assayed by this method. Mention should be made of phosphorescence-based oxygen sensors. However, the frequently observed sensitivity of room temperature phosphorescence towards humidity can be a major obstacle for chemical applications.

The fundamental relations between analyte concentration and phosphorescence intensity and lifetime are the same as in fluorimetry (see Equations 4 and 5, and Chapter 3). Because of the longer lifetimes ($\tau_p$) of the triplet state, much smaller quencher concentrations are sufficient to cause the same quenching efficiency (lifetime reduction) as in fluorimetry. Roughly, it can be said that the concentration of a phosphorescence quencher has to be only $\tau_F/\tau_p$ of the concentration of a fluorescence quencher for achieving the same quenching effect. Since $\tau_F/\tau_p$ is in the order of 1000 to 100,000, phosphorescence quenching is inherently more sensitive. Again, time discrimination can separate between indicator phosphorescence and Rayleigh, Raman, or Mie scatter. Methods for suppression of straylight are presented in Chapter 3.

## E. CHEMILUMINESCENCE

Some chemical reactions, in particular oxidations, are accompanied by the emission of light. This phenomenon is called chemiluminescence (CL) or bioluminescence (BL), depending on the system that emits light. From the photophysical point of view, there is no principal difference between CL and BL. Various books treat the subject from various points of view.[26-28] Typical processes during which CL is emitted include the oxidation of white phosphor by oxygen, the luminous emission of burning fire, and various oxidations involving hydrogen peroxide or ozone as reagent. CL methods are therefore particularly suited to probe oxidase-catalyzed reactions which proceed under formation of hydrogen peroxide. It should be pointed out, however, that all CL procedures involve a reaction with a finite reaction time. Reagent will be consumed after a while unless continuously supplied or present in large excess. For the generalized chemiluminescent reaction

$$\text{(catalyst)}$$
$$A + B \rightarrow C^*$$
$$C^* \rightarrow C + h\upsilon$$

any of the components, including the catalyst, can be measured. Because the signal usually is transient, measurement of CL intensity is time-dependent. The signal is often recorded at a specific time after interaction of reagent and analyte, or by integration of light over a fixed time. In chemical probes, the steady state intensity is usually measured, using immobilized reagents (in excess) and enzymes.

The main advantages of CL methods are low detection limits, wide dynamic ranges, and the fact that there is no need for a light source with its power requirements since the chemically sensitive material itself is the emitter. Instrumentation for CL sensing therefore has the potential of being quite simple.

These features already define the field of application for CL and BL based sensors which include disposable one-way probes, all kinds of portable or peripheral instrumentation, bedside care, and applications where high sensitivity is required, for instance in immunoassay. Other applications are in flow injection opto-sensing and in ultra-sensitive ATP and NADH assay.

Freeman and Seitz[49] have developed a hydrogen peroxide (HP) sensor based on the peroxidase-mediated oxidation of luminol by HP at alkaline pH. As long as luminol is present in excess, the CL intensity reflects the amount of peroxide. The authors have established a model that describes the relation between signal intensity and layer thickness, enzyme parameters ($V_{max}$ and $K_m$), and diffusion coefficient and bulk concentration of analyte. The most interesting aspect is the fact that the intensity is proportional to the square root of the enzyme activity. Thus, variations in activity will affect CL intensity not to the extent that they would in solution where there is a linear relation.

When compared to electrodes, the optrode approach to sense HP has a few different features. The enzyme electrode not only requires the substrate to diffuse into the enzyme phase but also that the product diffuses to the surface of the electrode. In the optical approach, the reaction is detected directly, and there is no need for the product to diffuse to the surface of the fiber optic. This has a beneficial effect as regards response time. On the other hand, the reagent layer must be composed of a material tranparent to light. This limits the number of possible enzyme phases. Also, the products of the chemical reaction can quench the emission.

Enzyme-mediated CL reactions usually require alkaline pH where other enzymes that may be coupled to the system already may display low activity. Also, if the chemical reaction involves the production of protons, the pH (and thus the response function) can be changed. However, if such side affects can be compensated for, CL and BL probing holds great promise for the reasons given above and in selected fields of application.

Aside from probing hydrogen peroxide, CL methods have been applied to monitor oxygen[50] and chlorine dioxide,[51] and it was stated that the optical probe, unlike the oxygen electrode, does not suffer from interferences by chlorine and hydrogen sulfide because they do not affect the CL reaction. Another CL probe was developed for glucose.[52] Horseradish peroxidase and glucose oxidase were immobilized on a photodiode which is immersed in a basic solution of luminol and analyte. The glucose bioprobe detects glucose in the 0.1 to 1.5 *M* range via the hydrogen peroxide produced by the enzymatic reaction. A more detailed treatise on CL probes is given in Chapter 20.

# V. INFRARED SPECTROMETRY

So far, infrared (IR) and Raman spectrometry exclusively have been exploited in plain fiber sensors, and not in indicator-mediated sensors. Infrared spectrometry is an absorption technique and covers the near-infrared (NIR) and infrared (IR) regions, i.e., from 800 to about 2500 nm for NIR and about 2.5 to 20 μm for IR. Glass fibers have good transmittance to VIS and NIR light. Telecommunication grade fibers have the least attenuation between 1050 and 1600 nm. IR signals can only be transmitted with special fiber such as those having a zirconium fluoride or silver halide core. They have been used, for instance, to acquire IR spectra via optical fibers.[53] In view of this situation, and because excellent laser and LED light sources are available for the NIR range, NIR and Raman techniques are most frequently employed in chemical sensors.

The wavelengths of the NIR absorption of typical chemical bonds are summarized in Table 1. It is mainly C–H, O–H, and N–H bonds that absorb in the NIR and therefore can be detected by this technique. The relation between analyte concentration and absorption obeys the Lambert-Beer law. However, the molar absorption is very low[54] as compared to VIS absorption spectra so that large "cell lengths" are required to achieve good sensitivity. The absorption coefficient for the $CH_4$ harmonics is in the order of $9 \times 10^{-6}$ ppm$^{-1}$·m$^{-1}$. Because solvents frequently interfere, a major field of NIR optrodes is in the detection of analytes in air.

There are two types of sensing methods. In the nondispersive one, a single (laser) line is used to gain analytical information. This is the preferred method in case the molecule has a unique absorption band, and a proper laser line is available. In dispersive NIR spectrometry, a whole spectrum is acquired and then deconvoluted, usually by the Fourier transform technique.

Typical applications of NIR fiber sensing are in the field of mixed solvent composition studies,[55] in gas sensing[56a] (see Chapter 3, Section II, and Chapter 11), and the determination of water in the 0.1 to 1% range in other liquids via the absorption of water at 5181 cm$^{-1}$ and 6802 cm$^{-1}$. Improved sensitivity is achieved in methane gas sensing by making use of frequency-modulated laser spectroscopy in the near infrared.[56b] A useful application of IR spectrometry is the determination of oil in water via the C–H absorption at about 2950 cm$^{-1}$. Oil is usually extracted first into a solvent or polymer such as poly(chlorotrifluoroethylene) which has no C–H bond. The ratio of aromatic and aliphatic hydrocarbons (an important parameter in oil industry) can be determined via NIR fiber spectrometry by virtue of their different C–H absorptions. A final application is the determination of hydroxyl number (HN) at 2055 nm. HN is frequently used as a measure of the degree of polymerization in reactions

**TABLE 1**
**Near Infrared Absorption Bands of C–H, O–H, N–H, and C=O Bonds**

| Vibration | Group | Wavelength (nm) |
|-----------|-------|-----------------|
| C–H | $-CH_3$ | 1194—1196 |
|  | $-CH_2-$ | 1211—1217 |
|  | Ar–H | 1144—1153 |
|  | R–CHO | 2100—2210 |
|  | Ar–CHO | 2210, 2250, 1250 |
|  | Epoxy | 1650—2200 |
|  | Cyclopropane | 1620—1650 |
|  |  | 2220—2270 |
|  | $-C–C=CH_2$ | 1613—2110 |
|  | $-O–CH=CH_2$ | 1619 |
| O–H | $-CH_2-OH$ | 2751 |
|  | $-CH–OH$ | 2759 |
|  | $-O–OH$ | 1460—2080 |
| C=O |  | 2780—3000 |
| N–H | $-NH_2$ | 2860—3030 |
|  |  | ca. 2000 |
|  |  | 1430—1540 |
|  |  | ca. 1000 |
|  | >N–H | 2860—3030 |
|  |  | 1430—1540 |
|  |  | ca. 1000 |
|  | $-NH_3^+$ | 2170—2220 |

where the OH group is the reacting entity in polymer chain development, such as in polyurethanes. NIR fiber photometers are commercially available.

The measurement of low levels of water in solvents is of considerable industrial significance and shall be discussed, as an example of NIR fiber absorptiometry, in some detail. Water has a distinct absorption at around 1900 nm (Figure 7) which can be measured remotely in a flow-through cell, provided the solvent has no, or little, background absorption. The absorption is highly temperature-dependent (Figure 8). Generally, nonpolar solvents have less background than polar solvents. Fiber NIR absorptiometry is usually performed by a single beam technique, i.e., a blank run is required to determine background caused by the cuvette and water-free solvent. A calibration curve is usually established with standards. A typical analytical range is from 100 to 1000 ppm water in apolar solvents, with lower levels also being determinable using very accurate standards and computer-aided data analysis.

An interesting feature of some ion and semiconductor lasers results from the fact that their emission lines match the vibrational overtones of certain analytically important gases. For example, an erbium laser has been applied as a light source for sensing methane in the atmosphere,[57] owing to the coincidence of the $^2v_3$ R(6) line of methane and the laser emission centered at 1644.9 nm. This is a striking example for a nondispersive NIR fiber sensor.

Tenge et al.[58] have conducted a feasibility study to see how well calibration techniques perform in the infrared region of fiber optics. Calibration in the 1131 to 1531 nm region was compared to calibration in the 1131 to 2531 nm region, using mixtures of three organic solvents. The latter region is considered more information rich. Fiber IR absorptiometry has also been applied in a fiber optic catheter for detection of $D_2O$ in blood by virtue of the 4-μm absorption of $D_2O$ which is not interfered by water or blood.[59a] An interesting biosensing scheme based on the measurement of the infrared absorption of species produced by enzymes

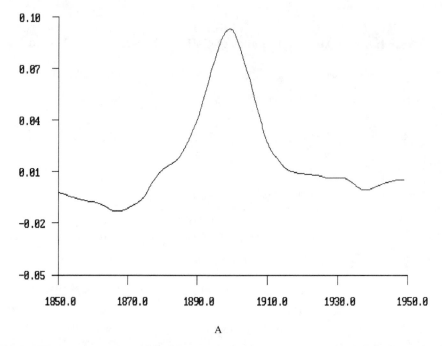

A

FIGURE 7.  NIR absorption spectrum of (a) 600 ppm water in toluene and (b) water in ethylene glycol. (Reprinted from Fitch, P. and Gargus, A. G., *Am. Lab.,* 17(12), 64, 1985.)

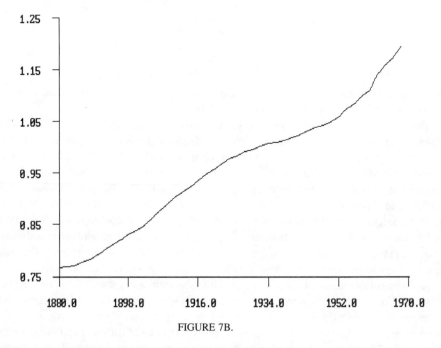

FIGURE 7B.

was introduced by Weigl and Kellner[59b]. When immobilized glucose oxidase is brought into contact with glucose, gluconic acid will be formed as a result of enzymatic action. An ATR crystal was covered with oxidase and the increase in the IR absorption around 1120 cm$^{-1}$ due to formation of gluconic acid was found to be a specific and quantitative parameter for glucose concentration.

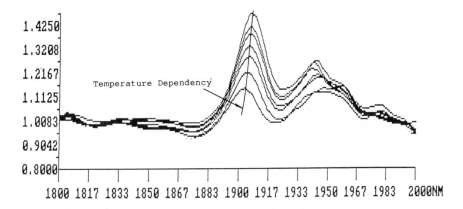

FIGURE 8.  Temperature dependence of the NIR absorption of water in toluene. (From Foulk, S. and Gargus, A. G., *Am. Lab.*, 19(12), 32, 1987. With permission.)

## VI. RAMAN SPECTROMETRY

Raman spectrometry is an emission technique involving inelastic scatter of absorbed (laser) light. Raman bands arise from changes in polarizability in a molecule during a vibration, while infrared bands reflect a change in polarization during such a vibration. The usual infrared spectrum represents absorptions of radiation that do not reach a detector, while Raman spectra are seen as scatter or emissions which are detected. Thus the instrumentations for the two spectroscopies are quite different. Sample preparation is generally easier in Raman because scatter rather than transmission is to be measured.

Although most vibrations give signal in both IR and Raman, the selection rules often provide very different relative intensities and band shapes. Symmetric vibrations and nonpolar groups are most readily seen in Raman, asymmetric and polar groups in IR. Water, for example, gives very little signal in Raman, but absorbs strongly in infrared, giving Raman an inside advantage in studying the components of aqueous media in biological and pharmaceutical chemistry. The high signal-to-noise ratios available in FTIR and recent sample handling techniques based on attentuated total reflection (ATR) are closing this gap, however.

Raman is often the method of choice when attention is focused on the highly polarizable bonds of a sample, such as $C=C$, $C\equiv C$, $C\equiv N$, or S–S. These bonds are often invisible in IR, but always produce strong Raman bands. An IR spectrum by itself can often be misleading when considering the presence or number of such bonds within a molecule, for example, in proteins. The S–S frequency is measurable by Raman, but transparent to IR. Even the SH bond is more easily seen in Raman than in IR, although SH information obtained by IR has proven useful in studying simple molecules and proteins.

Because Raman most often uses either visible or NIR light, the fibers can be constructed entirely of glass. IR demands more exotic materials to still transmit the light. IR fibers and elements in general cannot withstand many highly corrosive solutions, as can glass and quartz.

Modern Raman instruments consist generally of six parts: (1) a monochromatic light source, usually an argon or krypton ion laser; (2) (fiber) optics for transmitting the light to and from the sample; (3) a sample chamber (the sensing area); (4) a monochromator for light dispersion; (5) a light detector, either photomultiplier or multichannel; (6) a microprocessor or computer for instrument control, data acquisition, manipulation, and analysis.

The technique has been used exclusively in direct (i.e., not indicator-mediated) sensing. Fiber optic Raman spectroscopy detects signal intensity in proportion to the number of emission sites at the fiber tip or along the path over which there is interaction with the analyte.

Since Raman emission is observed with virtually all organic molecules, Raman sensors are not limited to a specific class of molecules, as in the case of fluorescence sensors. The disadvantage of Raman scattering is that it is a very weak effect, typically 4 to 6 orders of magnitude weaker than fluorescence. Therefore, Raman sensors require rather sophisticated instrumentation to obtain the desired signal levels at low concentrations.

Since Raman spectrometry normally uses visible or NIR excitation light which is efficiently transmitted by optical fibers, it can provide vibrational information about the sample yet still be easily coupled to fibers. Lasers are usually applied as light sources, and the Raman emission bands are found at emission frequencies smaller by 1500 to 4500 $cm^{-1}$ than the exciting light. It has been stated[60] that the closest one can go to the laser line is about 1255 cm in order to still be able to distinguish Raman emission from elastically scattered light. Because of the poor Raman signal, it is also strongly subject to interferences by traces of fluorescent impurities, present especially in biological and colored samples. Fluorescence is very broad and often so intense that it overwhelms the Raman signal.

Quite a variety of methods has been proposed to enhance Raman signal by separating it from luminescence.[61,62] Also, the light being transmitted down a fiber excites a number of molecules in the fiber itself and thus generates a large background. For this reason, multifiber probes with the strands performing independent functions of transmitting and collecting should be used, usually in a ratio of 1:1 to 1:10.

A very promising recent development for overcoming fluorescence involves the use of a Fourier transform (FT) near-infrared spectrometer as a detection system, and a Nd/YAG laser, operating at 1.06 μm, as a source. Though there is considerable intensity loss at this wavelength in the Raman signals, fluorescence is reduced to essentially zero. The multiplex advantage of the FT interferometer compensates for the decrease in Raman signal. Excellent Raman spectra of highly fluorescent compounds that were previously unobtainable are now available.[63]

Schwab and McCreery[64] have performed calculations as to the collection efficiency and sampling depth of the probe. Depending on configuration, the signal from the fiber probe (which does not employ a focused beam) was from one to nine times as large as that from a conventional liquid sampling system. An argon laser served as a light source, with 1 input fiber and 18 collecting fibers in the lightguide bundle. The device has been applied to Raman studies on liquids, solids, t-RNA, solution, low temperature samples, and electrochemically generated radical cations of the drug chlorpromazine. Various other types of Raman based optrodes have been recorded,[65,66] including devices for continuous analysis of glucose/water mixtures, solvent mixtures, and gas mixtures. The most likely applications of fiber optic Raman sensors will be for routine or remote analysis of liquids and gases. A valuable aspect is their simplicity. The data (spectra or single-line data) may be obtained by an unskilled operator with no special cuvettes and little attention paid to sample alignment. The low specificity in complex samples, however, can turn out as a disadvantage. A number of physical fiber sensors relies on this particular technique.[67]

Quantitative measurements of Raman scattering from three-layer asymmetric slab dielectric waveguides made of organic polymer have been obtained[68] from homogeneous distributions of scatterers as a function of mode by direct measurement of all mode-dependent quantities. For low-order modes, precision of the Raman intensity is limited to approximately 5% by uncertainty in the measurement of the coupling efficiency. Observation of Raman instead of Rayleigh scatter vs. distance in the waveguide allowed measurement of the optical loss function to be made down to 0.2 dB/cm with less than 1% uncertainty, so it never is the limiting factor.

Hollow fibers have been used to hold samples for both spontaneous Raman[69] and coherent anti-Stokes Raman spectrometry[70] (CARS). In the CARS experiment, two beams must cross at a certain angle known as the phase-matching angle, which matches the momenta of the photons in the two beams along a direction. A third beam is generated at another angle to the incident pair. In a 50-μm capillary, for instance, one beam may be directed along the axis of the core, and the second incident beam may cross the first at an angle of 2°. Thus, the two beams cross at many points throughout the length of the path and generate CARS. CARS generated in a flame has been transmitted to a spectrometer using fiber optics,[71] and a fiber optic Raman probe was reported[72] with fibers both carrying the laser light to the sample, and scattered signal back.

Intense spontaneous Raman radiation has been obtained from benzene and tetrachloroethylene by passing the beam of an argon laser (5 to 250 mW) through filled hollow fused quartz fibers having 10 to 25 m length.[71] Spectral intensifications by factors of 100 to 1000 compared to conventional sample techniques have been obtained. Notwithstanding the length of the fiber, only small amount of sample is required.[73] A 20-m fiber having a 75-μm core requires a sample volume of about 0.1 cm$^3$. In other words, the core of a rather long fiber cable can act as a sample volume in a continuous sensor.

A relatively new technique is surface-enhanced Raman spectrometry (SERS) which has an enhancement in signal 3 to 6 orders of magnitude.[74] The effect is observed with species absorbed on surfaces such as roughened silver, metal-coated microspheres, silver and indium island films, and colloids. The SERS technique has the potential of resolving a mixture into its different Raman active components[75a] and recently has been adapted to fiber optic techniques[75b] and $TiO_2$ substrates.[75c]

## VII. EVANESCENT WAVE SPECTROMETRY

All methods described so far can be used in the conventional way or in the evanescent wave mode. In an optical waveguide, light is totally reflected at the interface between an optical dense medium and an optically rare one, when the angle of refraction is larger than a critical angle β. The relation between the two refraction indices and β is given by

$$\sin\beta = \frac{n_1}{n_2} \tag{13}$$

Interestingly enough, light is not instantaneously reflected when it reaches the interface. Rather, light penetrates to some extent into the optically rare phase. More precisely, the amplitude of the electric field does not drop abruptly to zero at the boundary but has a tail that decreases exponentially in the direction of an outward normal to the boundary. This phenomenon is refered to as an *evanescent wave*.[76]

How light evanesces from the phase with $n_1$ into the phase with $n_2$ is shown in Figure 9. The displacement of incoming and reflected ray (D) is called the Goos-Hänchen shift. Provided the reflection angle is close to the critical angle β this shift can be calculated by the following equation:

$$D = d_p \cos\beta \tag{14}$$

The depth of penetration ($d_p$) is defined as the distance within which the electric field of the wave falls to $1/e$ (Equation 3):

$$E = \frac{E_0 e^{-z}}{d_p} \tag{15}$$

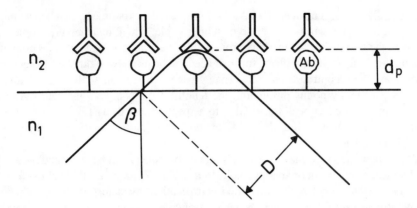

**FIGURE 9.** Total internal reflection of light at the interface between optical media having refraction indices of $n_1$ and $n_2$, demonstrating that light is evanescing into the phase with $n_2$. $d_p$ is the penetration depth, D the Goos-Hänchen shift. Because light penetrates the outer phase it can be used to probe immobilized molecules there, for instance an antibody (Ab).

with E being the amplitude of the electric field at depth z. $d_p$ depends on the wavelength of the light and the refraction indices of the two media:

$$\frac{\lambda}{d_p} = 2\pi n_1 \left[ \sin^2\beta - \left(\frac{n_2}{n_1}\right)^2 \right]^{\frac{1}{2}} \tag{16}$$

Typically, $d_p$ ranges from 50 to 1200 nm for visible light, which is more than the thickness of a reagent or protein layer immobilized on a surface, as shown in Figure 9. Light protruding into this phase is able to induce the fluorescence of the fluorophore, or to be absorbed, or to be scattered.

According to the Maxwell equations, a standing sinusoidal wave perpendicular to the reflecting surface is established in the sensor medium, whereas in the second medium the electric field amplitude shows an exponential decay (Equation 15). This is shown schematically in Figure 10. In contrast to the electromagnetic field of normal light, the evanescent field comprises a longitudinal component, too.

The field amplitude depends on the polarization of the incident light. Hence, the effective layer thickness $d_e$ is different for an incident beam being polarized parallel from that of an incident beam being polarized perpendicular to the plane of incidence. For the two directions of polarization of the incident wave the effective thickness $d_p$(parallel) and $d_p$(perpendicular), respectively, can be calculated from the refraction indices. Obviously, the geometry of the optical arrangement in evanescent wave sensors is rather sensitive to wavelength and polarization effects.

Light penetrating the second medium will be attenuated according to the Beer-Lambert law. This is the basis of absorption sensors (ATR sensors) utilizing evanescent waves.[76-85] Sensitivity can be enhanced by combining the evanescent wave principle with multiple internal reflections. The number of reflections (N) is a function of the length (L) and thickness (T) of waveguide and the reflection angle $\beta$:

$$N = L \cot \beta / T \tag{17}$$

This is of particular utility in case of weakly absorbing analytes such as methane with its

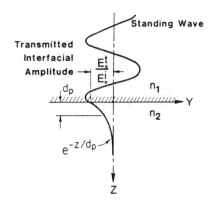

FIGURE 10.   Schematic of the surface wave at the interface of two optical media showing the standing-wave pattern in the medium having $n_1$, and the exponential decay of the interfacial electric field amplitude (evanescent wave) into medium having index $n_2$.

poor NIR absorption.[78] Instead of the intrinsic absorption of the analyte, the color changes of an indicator on the core may be monitored.[77,79] In another example,[80] the binding of methothrexate (a cancer drug) to anti-methothrexate has been monitored by evanescent wave absorptiometry at 320 nm.

The situation is considerably more complex for evanescent wave-induced fluorescence,[81-84] partly because of the different refractive indices of exciting and emitted light, differences in critical reflection angles, and polarization of emitted light. Evanescent wave sensors have been applied successfully to measure the fluorescence of indicators in solution[82] and on fibers (e.g., in an oxygen sensor[84]), and of proteins and labeled antigen/antibody couples. The major field of evanescent wave fluorimetry appears to be in immunodetection.[83,85,86]

Evanescent wave-type sensors offer the typical advantages of attenuated total reflection spectroscopy. Interface layers that are too thin to support optical propagation as well as monodispersed microspheres can easily be probed. The interface can be between the surface of a fiber, the face of a prism, a simple glass slide, the surface of a thin-film waveguide, and the material of interest. In these cases a sample film which can be considerably thinner than 1 μm is formed on the surface of a fiber or a slab of glass that acts as waveguide. Both the cylindrical and slab waveguide with evanescent coupling to the sample offer the advantage that the effective pathlength can be made relatively long. Evanescent wave spectrometry along with time resolution techniques[87] will considerably improve the selectivity of the method. The sensitivity of evanescent wave sensors to changes in the refractive index of the analyte solution may turn out as a severe limitation for some chemical applications.

By interposing a 50-nm layer of silver between the glass surface and the sample, the intensity of the evanescent field can be increased by one to two orders of magnitude. The incident field couples to the electrons in the metal layer, an effect that is called surface plasmon resonance (see later). The depth of penetration into the sample is a function of the incidence angle only and is not affected by the presence or absence of a metal layer.

When the cores of two fibers are nearly adjacent over some distance, as shown in Figure 11, light is coupled from one core to the other. This holds for both single- and multi-mode fibers. These sensor configurations have found application in a variety of promising sensing schemes, but mainly for physical parameters. Some chemical applications will be presented in the following chapters.

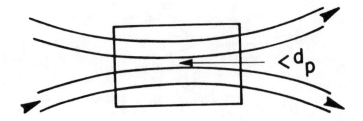

FIGURE 11.   Schematic of a fiber optic evanescent wave coupler. When the distance between the fibers is less than the penetration depth, light couples from one fiber into the other.

## VIII. FIBER REFRACTOMETRY

Since the feasibility of transporting light over large distances is based on total reflection of light at an interface between two dielectric phases, modulation of $n_2$ (the index of the clad) by chemical means is an obvious and very direct way to transduce a chemical parameter via Snell's law into optical information. Most refractive index sensors rely on modulation of light intensity. Refractive index is a sensitive but nonspecific way to measure the concentration or composition of solutions and solvent mixtures. A variety of probe geometries are possible:

1. Prisms (generally 45° prism), and other flat surfaces are attached to the fiber end. This requires machining of the fiber and mounting, and attenuation losses in air are usually high, typically 10 dB or more, due to alignment of the fiber ends or the prism.[88]
2. A declad fiber, in which the optical cladding is removed and replaced by the liquid, is equivalent to a variable numerical aperture measurement.[89] This technique has been used in a liquid chromatograph in which a He–Ne laser beam is injected through aspheric optics down a 1 mm diameter glass rod. The rod is built into a chromatography cell of 7 μl volume, and the sensitivity is $1 \cdot 10^{-5}$ refractive index units.[90]
3. A fiber can be bent in a single continuous curve, either as a U bend or a spiral of multiple turns. The fiber may also have the cladding removed.[88,91] An improvement is to use a series of bends in alternate directions,[92] or a half ball at the end of two fibers.

Kapany and Pike,[89] have described a photorefractometer in which light loss from glass rods was used to monitor the refractive index of a surrounding liquid. The design involved a "turret" of rods of different refractive index covering the range of interest, immersed in the liquid under test. The loss of light transmitted along the rods can be related to the degree of mismatch between the refractive indices of the rod and the liquid, and transmission is reduced to zero when the refractive index of the liquid matches that of the fiber.

A photorefractometer based on optical fibers using the same fiber as both sensor and transmission line was described by Takao and Hattori.[91] The sensor was formed by stripping the cladding from part of the fiber and bending the exposed part to a uniform radius. When this part of the fiber was immersed in a liquid, the loss of light transmission allowed the calculation of the refractive index of the liquid. The effects of fiber material, fiber parameters, and bending radius on useful refractive index range, light transmission, and sensitivity have been studied experimentally and theoretically.[93] A variety of other papers deal with the coupling of light between media of different refractive index, and it has been demonstrated[94] that through control of the refractive index surrounding a fused biconical-taper single-mode coupler the coupling coefficient can be varied from 0 to 96%.

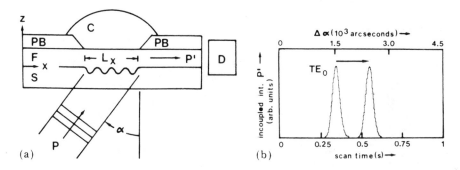

FIGURE 12. (a) Schematic of an input grating coupler sensor. P, power of incident laser beam; $\alpha$, angle of incidence; P′, power of incoupled guided mode; D, detector; C, sample cell; F, waveguiding film; S, solid support; PB, protective buffer layer on the waveguide outside the grating region $L_x$. (b) Measured incoupling power P′ of the $TE_0$ mode vs. angle $\alpha$ before (left) and after (right) adsorption of protein molecules from the aqueous solution in C.

Lukosz et al.[95-98] have utilized the effect of adsorption and desorption of chemical species on the effective refractive index of the surface of a planar waveguide to devise various kinds of gas sensors and integrated biosensors. The transducer (Figure 12) consists of a planar 180-μm $SiO_2$-$TiO_2$ waveguide film equipped with a surface relief grating and attached to a glass microscope glass as solid support (Bragg reflector). The beam of a He–Ne laser is directed onto the grating which is covered by a flow-through cell. There are three possible reasons for a change in the $n_D$ of the guided mode (F): (1) A chemical reaction causes a change in the refractive index of the liquid or gas covering the waveguide. (2) In waveguides showing microporosity, the cover liquid penetrates the pores of F, resulting in a change in the $n_D$ of the waveguiding film. (3) Adsorption of molecules contained in the liquid cover on the waveguide surface leads to the formation of an adlayer.

The effect of refractive index on the coupling of light between two crossed or twisted fibers (Figure 11; Section VII herein) has been exploited in many refractive systems. In these, light is focused into one fiber and detected in the other. Because the two fibers are tightly twisted together, some of the guided modes of the first fiber are converted into radiating modes. After interaction with the medium surrounding the fibers, the radiating modes of the first fiber that have entered the second are then converted into the guided modes of the second, since the two have, to zero*th* order, the same radius of curvature. Because the environment modulates the interaction, the system can be used as a sensor.

The medium can influence the interaction in various ways: (1) Through changes in the index of refraction; (2) through absorption characteristics such as resonant absorption of dyes and scattering by suspending particles; (3) through physical interaction with the fiber such as absorption of the liquid by the fiber. The intensity of the light from the second fiber ranges from approximately 0.5 to 10% compared with that of the first depending on the fibers environment. A typical experimental arrangement is shown in Figure 13.

Smela and Santiago-Aviles[99] have applied such an arrangement to sense the refractive index of pure solvents, mixtures of used and fresh motor oil, and mixtures of trichloroethane and oil. The effect of varying the number of twists as well as the length of the twists was also examined. In addition, by surrounding the fibers with a thin film of solution and observing the changes in output over time, additional information can be obtained.

Halliday and Alder[100] observed that where two fiber waveguides are in contact but with their axes skewed, light in the one fiber is transferred to the other when a droplet of liquid having a refractive index very close to that of the fibers is allowed to fall on the crossing point. The amount of light transferred decreases at all other refractive indices depending on the angle

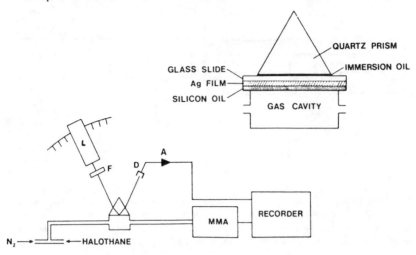

FIGURE 13.   (a) Typical surface plasmon resonance curve from a plot of reflectivity as a function of θ. (a) With an immobilized biotin layer; (b) after exposing the layer to biotin.

at which the fibers are inclined and on the angle of propagation of the light guided in the excited fiber. This provides a sensitive on-line means of detecting liquids of specific refractive index and discriminating against those of a different index, for instance in quality assurance.

Various fields of applications have been reported for fiber optic refractive index-based sensors which include liquid level detectors,[101] remote detection of refractive index using a handheld meter,[102] glucose sensors,[103] and detection of oil in water by using an unclad fiber coated with an lipophilic onterphase.[104] The normal total internal reflection is degraded when the outer phase absorbs oil products from water. Diesel oil (17 mg l$^{-1}$) and crude oil (3 mg l$^{-1}$) were detectable using the 633-nm line of a He-Ne laser.

As shown by Attridge et al.,[105] even pH sensors can be based on this technique. Rather than using customary absorption or fluorescence, the change in the index of refraction of a phase with immobilized pH indicator is detected using a He-Ne laser. A fiber directional coupler is used, and the change in $n_D$ caused by pH changes results in a change in net transfer of light from one guide to the other.

Problems associated with refractive index measurements are the change in refractive index with temperature, and contamination of the probe with dirty liquids. Automatic temperature compensation can be achieved by using the variation of the light output from an LED with temperature, or by having two probes of different material with different refractive index temperature sensitivities.[92]

## IX. SURFACE PLASMON RESONANCE

A surface plasmon resonance (SPR) experiment is performed by recording the reflectance of a metal-coated glass surface as a function of the angle of incidence.[106] At a certain angle $\theta_r$, a sharp minimum in the reflectance occurs, indicating the excitation of a so-called SPR which occurs when photons are reflected at a di-electric/metal interface under conditions such that the momentum of the photons becomes coupled to the electron gas in the metal, thereby exciting collective longitudinal oscillations known as plasmons. A schematic of the frequently used Kretschmann configuration is shown in Figure 13. The angle $\theta_r$ strongly depends on the dielectric profile in the immediate vicinity of the metal layer which can be coated with a chemically sensitive material. Changing the material properties shifts the resonance angle, providing a highly sensitive means of monitoring the surface reaction. The angular resolution of a typical set-up is better than $5 \cdot 10^{-2}$ degree.

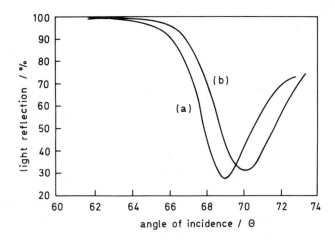

FIGURE 14. Top: Schematic view of an SPR sensor in the Kretschmann configuration. Bottom: Experimental arrangement for on-line halothane monitoring. L, laser; F, filter; D, photodetector; A, amplifier.

The SPR condition is fulfilled when the component of the evanescent field wave vector parallel to the interface between metal and dielectric ($K_x$) is equal to the surface plasmon wave vector ($K_{SP}$) as given by

$$K_x \cdot c = w\left(\varepsilon_1 \cdot \sin\theta\right)^{\frac{1}{2}} = w\left(\varepsilon_2^{-1} + \varepsilon_m^{-1}\right)^{-1/2} \qquad (18)$$

where w is the optical frequency, c the free space velocity of light, and $\varepsilon_m$ is the real part of the dielectric constant of the metal. $\varepsilon_1$ is the dielectric constant of the prism, and $\varepsilon_2$ that of the sample $\theta$ is the angle of incidence of the optical beam at the metal/dielectric interface. Thus, the value of the wave vector at resonance is a function of both dielectric constants, the wavelength, and of the metal.

As illustrated in Figure 14, the arrangement comprises a glass prism with a thin metal layer attached to one face of the prism. Typical metal layer thicknesses are between 20 and 100 nm. An oil of refractive index matching that of the glass can be placed between glass support and prism during the experimental phase in order to avoid the delicate procedure of cleaning and coating the prism during successive experiments. The polarized laser beam is directed onto the prism placed on a rotation stage. Reflectance is measured as a function of angle. The sample solution is contained in a cuvette having a metal-coated glass wall, usually a glass slide.

A major field of application of SPR is in the immunoglogical field, because the binding of, for example, an antigen on the metal surface to an antibody in solution causes the critical angle $\theta$ to change significantly. Moreover, the method is rather sensitive. Kooyman et al.[107] found that maximum sensitivity is obtained for a silver layer about 55 nm thick and in direct contact with the species to be quantified. Application of an intermediate layer with high permittivity for the analyte can be useful in suppressing background responses.

In experiments described by a Swedish group, Kretschmann's arrangement (Figure 14) was used to detect halothane[108a] (an inhalation narcotic), other gases,[108b] and as an immunosensor.[109] A thin film of silicone/glycol copolymer oil which reversibly absorbs halothane was applied to the exposed surface of the metal layer. Absorption of the narcotic produced a change in the dielectric properties of the oil which, in turn, gave rise to a detectable shift in $\theta$ which is in order of $3 \times 10^{-5}$ degree per ppm halothane. The extent of shift is linearly related to the concentration of gas over a limited range of concentrations. Others have reported basic

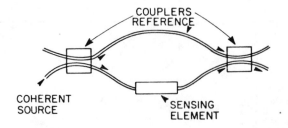

FIGURE 15.   Schematic of a fiber optic Mach-Zehnder interferometer.

schemes for immunosensors[110] when it was shown that the adsorption angle θ of HSA on a silver film after binding to anti-HSA is shifted by 3.2°. Ethanol/water mixtures were detected using a Fourier transform optical set up along with a diode array.[111] As little as 0.01% ethanol are detectable in water. The temperature coefficient of the angular shift for pure water was about 0.02° per degree Celsius. Because of the size of the device, the use of optical fibers and microoptic components was suggested. However, a fiber optic SPR sensor has not been reported so far.

# X. INTERFEROMETRY

Interferometry is based on the interferences of two waves being in phase to each other. Its principles are outlined in Chapter 15, along with the principles of the Mach-Zehnder, Michelson, and Fabry-Perot arrangements. Monomode fibers are required throughout. In all configurations the light beam is first divided and then combined again. In the Mach-Zehnder configurations (Figure 15) coherent monochromatic light is launched into one fiber, and half of the light is coupled, in a first coupler, into the other fiber. One fiber passes a sensing element before it enters coupler two where the two fiber branches are combined and light recombines. If there is a phase shift caused by the sensing element, the detector records changes in the fringe pattern (a) series of maxima and minima as illustrated in Figure 16 (see also Chapters 4, 5, and 15).

The method is extremely sensitive (this, in fact, can be a source of difficulty) and has so far been exploited mainly for sensing physical parameters.[112] It also is very temperature-dependent. Interferometric sensors have been emerging in the past years as a significant new class of measuring instrument for a range of physical quantities,[112,113] most of which can be transduced into mechanical strain which is measured, in terms of displacement, by interferometers. The reference fiber compensates for unspecific interferences, provided it is exposed to the identical conditions as the sensing fiber. One disadvantage of interferometric sensors is their inherently lower information content than devices based on amplitude measurements at two or more wavelengths and the resulting ease of self-referencing.

In a typical instrument where conventional beam splitters and mirrors are used to control the beam division and recombination, the output will fluctuate due to random noise perturbing the alignment of the optical components. To a large extent, these problems are nullified if the interferometer is constructed entirely from fibers. But although the fiber approach solves these problems, it introduces a new and potentially major source of error, namely the variability of the optical constants of the fiber itself. There are several other fundamental properties of all interferometers, whether all-fiber or not, which tend to restrict their use in sensor applications. They have been discussed by Jackson.[113]

Though the configuration of the Mach-Zehnder interferometer is slightly more complex than the Michelson, it does offer two significant advantages: (1) optical feedback to the light source is minimum; this is important when laser sources are used, as any out-of-phase light

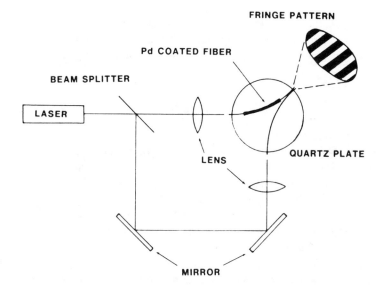

FIGURE 16.   Schematic of the Mach-Zehnder interferometer used to detect hydrogen gas. A 0.5-mW helium-neon laser was used as a light source. (From Butler, M. A., *Appl. Phys. Lett.,* 45, 1007, 1984. With permission.)

fed back into the laser tends to induce random changes in its optical output frequency; (2) there are two antiphase outputs from the interferometer which can be used to reduce the effects of intensity noise in the laser output.

In the all-fiber Mach-Zehnder, the directional couplers replace both the conventional beam splitters and mirrors. Indeed, it is this component which make these interferometers unique, as it virtually eliminates all the effects of random mechanical disturbances which normally induce misalignment. A more detailed discussion of the various kinds of fiber interferometers including differential interferometers, the Sagnac arrangement, and laser Doppler anemometers has been presented.[113,114]

There are a few examples of interferometric chemical sensors: Butler[115] has measured hydrogen gas using a palladium-coated sensing fiber which expands on exposure to hydrogen because the metal very specifically absorbs the gas. This changes the effective path length of the fiber which is detected interferometrically in an arrangement shown in Figure 16. The experiments suggest a high sensitivity and a wide analytical range. In a somewhat different arrangment, Farahi et al.[116] demonstrated a dynamic range of 1 to 200 Pa hydrogen easily being possible. The authors also demonstrated[117] the feasibility of sensing flammable gases. The sensing element is a 100-mm monomode fiber coated with platinum metal which catalyzes the exothermic reaction of hydrocarbon gases with oxygen. The resultant heat of the chemical reaction is transduced to a phase retardance in the fiber-guided beam and thereby can be detected interferometrically.

## XI. MISCELLANEOUS SCATTERING TECHNIQUES

Aside from Raman scatter (Section VI herein), there are other kinds of light scatter which can be utilized in optical sensing.[67] Rayleigh and Mie scatter are of the elastic scatter type and occur at the wavelength of incident radiation. Rayleigh scatter is caused by all molecules of size at around or below the wavelength of light. Mie scatter is observed when the particles are large as compared to the wavelength of radiation. Both wavelength $\lambda$ and particle diameter R influence the intensity of scattered light ($I_s$). If $R < \lambda$, then $I_s = k \cdot R^6 / \lambda^4$; if R is about $\lambda$, then

$I_s = k \cdot R^4/\lambda^2$; if $R > \lambda$, then $I_s = k \cdot R^2$. The latter condition is found in case of Mie scatter (i.e., in turbidimetry and nephelometry). Logarithmation and rearrangement yields the familiar relation between intensity of incident light ($I_o$), scattered light ($I$), and particle concentration (C):

$$\log I_0/I = k_s \cdot C \qquad (19)$$

where $k_s$ is a constant. In both Rayleigh and Mie scatter, the polarization of the particle remains constant. Nephelometry (which is based on Mie scattering) is frequently used in analytical chemistry and process control. It requires, however, the particles to be homogeneously distributed in the sample and not to change their size and shape upon dilution. One source of error in nephelometry is the occurrence of gas bubbles.

Nephelometry frequently makes use of standard suspensions, for instance barium sulfate or the more recent formazin standard suspension made up from hydrazine sulfate and hexamethylene tetramine in water. This solution gets turbid and forms a fairly stable standard suspension having 4 FTU (formazine turbidity units) which may be adjusted to any lower value by dilution with water. Nephelometry plays an important role in beer quality assurance.

A useful example for a light-scatter-based fiber device is a marine oil-in-water monitor which is based on measurement of scattered light from a laser diode at 850 nm connected to a fiber optic lead placed in the water flow.[118] Straight-through and sideways scattered light is collected. The angular dependence allows distinction between large and small particles. The device shows some temperature sensitivity: for light crude oils the change is less than 0.2% per degree Celsius: for heavy crudes there is a larger change. The assembly includes an automatic window cleaning system to counter dirt accumulation. The device measures in the up to 1000 ppm range.

High pressure cells, up to 250 bars pressure, for turbidity measurement, have been built with up to 50 cm optical path length. The construction requires accurate alignment of the optics and high pressure seals. These have been used for differential measurement of particle contamination in cooling water. Alternatively, dual fiber optic probes measure back-scattered light in front face configuration.[118] For other areas such as electric power stations, oil content is measured[1] turbidimetrically as 0 to 2 ppm (+/- 0.1 ppm).

Evanescent-wave nephelometry was applied in a no-label, homogeneous optical immunoassay for IgG, using a planar quartz waveguide.[119] Following adsorption of antiserum to the surface of an optical waveguide, the immobilized antibody was then reacted with a solution containing antigen. The reaction was detected utilizing the evanescent wave component of a light beam totally internally reflected within the waveguide. The growing antigen-antibody layer resulted in an increase of scattered light which was monitored kinetically (Figure 17).

## XII. PHOTOACOUSTIC SPECTROMETRY

Following the absorption of electromagnetic radiation, the excited state can undergo radiationless deactivation (see the Jablonski scheme, Figure 1) which results in a warming up. In a closed system, this leads to a pressure increase. If the exciting light is modulated (typical frequency 20 to 12,000 Hz), the variation in pressure results in an acoustic wave which can be detected using a microphone. According to the theory established by Rosencwaig and Gersho,[120-122] there are several interesting parameters that influence the signal: (1) signal intensity depends on intensity of exciting light; (2) the analytical depth in a sample is inversely proportional to the modulation frequency; this offers the possibility of acquiring depth profiles via PAS; (3) PAS is hardly affected by the nature of the sample or its surface. It is therefore very suitable for studies on rough surfaces, powders, lubricants, and grease. This capability is based on the fact that only the absorbed light is converted to sound. Scattered light, which

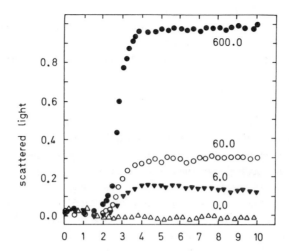

FIGURE 17. Generation of light-scattering signal at the surface of an optical waveguide bearing immobilized antibody. IgG was added after 2 min in concentrations ranging from 0.0 to 60.0 μg·mL⁻¹, and scattering monitored at 440 nm.

presents such a serious problem when dealing with many solid materials by convential spectroscopic techniques, presents no difficulties in PAS. On the other hand, PAS requires rather large samples.

The only chemical sensor-type arrangements based on PAS that have come to the attention of the author are a gas sensor using an enzyme-photoacoustic detector[123] and a disposable glucose probe.[124] Fibers have been used to transport light into sample cells.[125] When coupled to a Mach-Zehnder type interferometer, photo-acoustic spectroscopy with fibers can be used to detect $NO_2$ in the 0.5 ppm range.[126]

# XIII. MISCELLANEOUS TECHNIQUES

### A. ELLIPSOMETRY

Ellipsometry[127] is an optical technique that exploits the reflection of polarized light from surfaces in order to measure thin surface layers or monolayers.[128] It is receiving increasing attention as a tool for investigating biological interactions at liquid-solid interfaces.[129] It could thus prove valuable for the detection of binding processes between biomolecules which otherwise are difficult to follow. As a matter of fact, ellipsometry requires large, complex apparatus and does not lend itself to implementation methods unless polarization-maintaining fiber optics are available.

### B. POLARIMETRY

For the latter reason, fiber polarimetry has found no practical application so far. Mermelstein[130] has constructed a compensated and stabilized all-fiber polarimetric sensor from about 50 m of bow-tie fiber with interferometric detection. A signal/noise ratio analysis was presented to weigh the relative merits of the all-fiber polarimetric and Mach-Zehnder interferometers. It was concluded that the polarimeter, because of the simplicity in design, improved environmental immunity, and high SNR, is a worthy candidate for many fiber sensor applications. Previously, an all-fiber ellipsometer had been demonstrated by Yoshino and Kurosawa.[131] Several polarimetric fiber configurations and their respective merits have been described in a review by Jackson.[113] Noeller[132a] has described a polarization fluoroimmunoassay apparatus for directing a polarized beam of light which is guided through a fiber optical element onto a sample containing fluorescent tagged species. The intensity of the emission is measured simultaneously with crossed polarizers to give information on the amount of bound vs. unbound immunopartner. A fiber optic polarimetric immunosensor has been presented using polarization-maintaining fibers with elliptical core.[132b] Differential fiber polarimeters have been described as well.[132c] Polarization-maintaining fibers became commercially available in the late 1980s.

## C. DIFFRACTION

Diffraction effects may be used in at least four different ways in combination with optical fibers: first, where light is delivered to and from a diffraction grating via optical fibers for analysis of, say, Raman scattering or changes in the diffraction grating structure due to perhaps antigen-antibody binding occurring on the grating. Second, the grating may be embossed onto the optical fiber itself[133] and now the fiber itself plays an intrinsic role in the scattering process to be observed. This method of operation has been suggested as a probe technique for biotechnological application, perhaps sensing immunoassay reactions.[134] Third, Hill et al.[135] have reported the fabrication of diffraction gratings along the length of fibers through a photorefractive effect. Signal processing elements such as filters constructed from such gratings will be highly sensitive to internal perturbation such as stress. Fourth, Hutley[136] has suggested the use of diffraction gratings with optical fibers to allow sensing of wavelength encoded information from a transducer.

## D. TIME-OF-FLIGHT FLUOROMETRY

It is possible to determine the spectrum of an optical pulse from measurements of the photon time-of-flight distribution through a fiber, provided that the time resolution of the experimental system, including the duration of the original pulse and the time broadening due to multimode dispersion, is sufficiently high. Because of the wavelength dependence of the group velocity of light, different frequency components are temporally dispersed after an optical pulse passes through a fiber. The time-domain distribution of the transmitted photons is measured by a fast-response detector, and this distribution is transformed into the ordinary frequency-domain spectrum using the wavelength dependence of the photon time-of-flight and the transmission spectrum of the fiber.

The feasibility of time-domain luminescence spectroscopy using an optical fiber as a dispersive element has been demonstrated[137] for the fluorescence of hematoporphyrin in dioxane. A cw mode-locked argon laser was used for the excitation, and a time-correlated single-photon counting technique for detection. The effects of several time-broadening factors and possible applications of this spectroscopic method are discussed.

## E. OTHER TOTAL INTERNAL REFLECTION TECHNIQUES

The frustrated-total-internal-reflection (FTIR) sensor[136] consists of two fibers with their ends polished at an angle to the fiber axis which produces total internal reflection (TIR) for all modes propagating in the fiber (Figure 18). With the fiber ends sufficiently close, a large fraction of the light power can be coupled between the fibers. If one fiber is stationary and the second fiber experiences a vertical displacement, the light power coupled between the fibers varies due to this displacement and the light transmitted by the output fiber is modulated. The disadvantage of the sensor is that it requires tight mechanical tolerances.

The critical angle sensor (Figure 19) is similar to the FTIR sensor. It employs a single-mode fiber cut at an angle just below the critical angle. If the critical angle $\theta$, given by $\sin^{-1}(n_2/n_1)$, is not near 45° (45° requires medium 2 to be a gas), then an additional angular cut is needed to allow the reflected beam to travel back through the fiber. This back-reflected beam is monitored after it exits the fiber. While this sensor has not yet been demonstrated in a fiber configuration, the technique is included since it operates on a fundamental optical mechanism which does not require mechanical designs typical of other sensors.

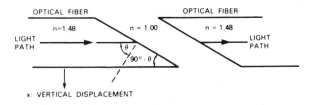

FIGURE 18. Schematic of a frustrated-total-internal reflection sensor.

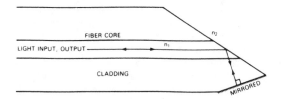

FIGURE 19. Schematic of a critical angle fiber sensor.

## F. PHOTON CORRELATION SPECTROSCOPY

When an assembly of small particles in a suspension is illuminated by a beam of laser light, the scattered electric field from each particle contributes to a speckle pattern observed in the far-field. The particles move naturally under Brownian motion, thus the observed intensity in a single speckle (or coherence area) fluctuates with time. In photon correlation spectroscopy (PCS), this intensity is measured with a sensitive detector. The intensity fluctuations are analyzed by autocorrelation in a digital (photon) correlator. The decay time of the correlation function is related to the translational diffusion coefficient of the particles. Subsequent analysis relates the diffusion coefficient to their mean hydrodynamic radius. The design of a fiber optic on-line PCS-based meter for the measurement of the size of proteins in the eluant of a gel-permeation liquid chromatography column has been described.[138] The arrangement involves two fibers, one acting as a launch and the other collecting scattered light.

A different situation is found in case of absorbing species. The basic concept of absorbance correlation spectrometry involves the passage of light sequentially through two absorbance cells, a reference cell containing a known quantity of the species to be detected and a sampling cell where the presence of the species is to be determined. Usually, the method is applied for sensing gases such as methane. An optical signal passing sequentially through the cells will suffer absorption in each of the cells. If the absorption in the reference cell is periodically modulated, then the total absorption depends on whether the gas absorption lines in the sampling cell correlate with those in the reference cell gas. If the absorption lines of the gases do not correlate strongly, then the modulation index of the optical signal is essentially unaffected by the presence of the different gas in the sampling cell. Therefore, the concentration of a specific gas in the sampling cell can be quantified by this means. Hence, this method provides unique specificity for *direct* optical detection of NIR absorbing species.[139]

Pressure modulation of the gas within the reference cell has been achieved by the use of a novel acoustic resonator.[140] This device provides a reasonable pressure ratio whilst being compact and easily driven. Unlike previous piston-compressors, the high modulation frequency of the resonator improves the tolerance to transients in the optical signal produced by the passage of dust through the optical beam. The system can be tailored to detect a desired gas by changing the reference gas and a broadband section filter. Results of methane are extremely promising. A novel method for detection of ammonia gas has been reported as well.[140] This involves Stark modulation of the gas in the reference cell by the application of an electric field. This method is applicable to gases with strong electric dipole moments, such as $NH_3$, CO, $NO_x$ and HCl.

# REFERENCES

1. **Chester, A. N., Martellucci, S., and Scheggi, A. M., Eds.,** *Optical Fiber Sensors*, Martinus Nijhoff, Dordrecht, 1987.
2. **Lange, B. and Vejdelek, Z. J.,** *Photometric Analysis*, Verlag Chemie, Weinheim, 1980.
3. **Freeman, J. E., Childers, A. G., Steele, A. W., and Hieftje, G. M.,** A fiber optic absorption cell for remote determination of copper in industrial electroplating baths, *Anal. Chim. Acta*, 177, 121, 1985.
4. **MacCraith, B. D. and Maxwell,** Fiber optic sensor for nitrates in water, *Proc. SPIE*, 1510, xxx, 1991; in press.
5. **Kortüm, G.,** *Reflection Spectrometry*, Springer Verlag, Berlin, 1969.
6. **Kubelka, P. and Munk, F.,** A contribution to the optics of pigments, *Z. Tech. Phys.*, 12, 593, 1931.
7. **Sonntag, O.,** *Dry Reagent Chemistry*, Georg Thieme Verlag, Stuttgart, 1988.
8. **Curme, H. G.,** Multilayer film elements for clinical analysis: general concepts, *Clin. Chem.*, 24, 1335, 1978.
9. **Guthrie, A. J., Narayanaswamy, R., and Russell, D. A.,** Application of Kubelka-Munk diffuse reflectance theory to optical fibre sensors, *Analyst*, 113, 457, 1988.
10. **Kealey, D.,** Quantitative reflectometry: principles and scope, *Talanta*, 19, 1563, 1972.
11. **Kealey, D.,** Quantitative reflectometry: precision and interference, *Talanta*, 21, 475, 1974.
12. **Williams, F. C. and Clapper, F. R.,** Multiple internal reflections in photographic colorprints, *J. Opt. Soc. Am.*, 43, 595, 1953.
13. **Neeley, W. E.,** Reflectance photometer for multilayer dry film slides, *Anal. Chem.*, 56, 742, 1984.
14. **Werner, W. and Rittersdorf, W.,** Reflectance Photometry, in *Methods of Enzymatic Analysis*, Bergmeyer, H. U., Ed., Verlag Chemie, Weinheim, 1983, 305.
15. **Hecht, H. G.,** Quantitative analysis of powder mixtures by diffuse reflectance, *Appl. Spectrosc.*, 34, 16, 1980.
16. **Hecht, H. G.,** Comparision of continuum models in quantitative diffuse reflectance spectrometry, *Anal. Chem.*, 48, 1775, 1976.
17. **Hecht, H. G.,** A comparision of the Kubelka-Munk, Rozenberg and Pitts-Giovanelli methods of diffuse reflectance, *Appl. Spectrosc.*, 37, 348, 1983.
18. **Narayanaswamy, R., Russell, D. A., and Sevilla, F.,** Optical fibre sensing of fluoride ions in a flow-stream, *Talanta*, 35, 83, 1988.
19. **Goldman, J.,** Quantitative analysis on thin layer chromatograms. Theory of absorption and fluorescent densitometry, *Anal. Chim. Acta*, 78, 7, 1973.
20. **Olinger, J. M. and Griffiths, P. R.,** Quantitative effects of an absorbing matrix on NIR diffuse reflectance spectra, *Anal. Chem.*, 60, 2427, 1988.
21. **Parker, C. A.,** *Photoluminescence of Solutions*, Elsevier, Amsterdam, 1968.
22. **Lakowicz, J. R.,** *Principles of Fluorescence Spectroscopy*, Plenum Press, New York, 1983.
23. **Korkidis, K., Baeyens, W. R. G., and DeKeukelaire, D.,** Marcel Dekker, New York, 1990.
24. **Schulman, S. G., Ed.,** *Molecular Luminescence Spectroscopy: Methods & Applications*, John Wiley & Sons, New York, Vol. 1, 1985, and Vol. 2, 1988.
25. **Leiner, M. J. P., Hubmann, M. E., and Wolfbeis, O. S.,** Total luminescence spectroscopy and its application in biomedical sciences, in *Handbook of Chemical and Biochemical Analysis*, Marcel Dekker, New York, 1990, chap. 15.
26. **Burr, J. G., Ed.,** *Chemiluminescence and Bioluminescence*, Marcel Dekker, New York, 1985.
27. **DeLuca, M. and McElroy, W. D., Eds.,** *Bioluminescence and Chemiluminescence: Basic Chemistry and Analytical Applications*, Academic Press, New York, 1981.
28. **Van Dyke, K., Ed.,** *Bioluminescence and Chemiluminescence: Instruments and Applications*, Vols. 1 and 2, CRC Press, Boca Raton, FL, 1985.
29. **Wolfbeis, O. S.,** The fluorescence of organic natural products, in *Molecular Luminescence Spectroscopy: Methods & Applications*, Vol 1, Schulman, S. G., Ed., John Wiley & Sons, New York, 1985, chap 3.
30. **Baeyens, W. R. G.,** Fluorescence and phosphorescence of pharmaceuticals, in *Molecular Luminescence Spectroscopy: Methods & Applications*, Vol 1, Schulman, S. G., Ed., John Wiley & Sons, New York, 1985, chap 2.
31. **Fernandez-Guttierrez, and Munoz, A.,** Determination of inorganic substances by luminescence methods, in *Molecular Luminescence Spectroscopy: Methods & Applications*, Vol 1, Schulman, S. G., Ed., John Wiley & Sons, New York, 1985, chap 4.
32. **Wolfbeis, O. S.,** Fiber optical fluorosensors in analytical and clinical chemistry, in *Molecular Luminescence Spectroscopy: Methods & Applications*, Vol 2, Schulman, S. G., Ed., John Wiley & Sons, New York, 1988, chap 3.
33. **Hurtubise, R. J.,** Luminescence from solid surfaces, in *Molecular Luminescence Spectroscopy: Methods & Applications*, Vol 2, Schulman, S. G., Ed., John Wiley & Sons, New York, 1988, chap 1.

34. **Sepaniak, M. J.,** The clinical use of laser-excited fluorometry, *Clin. Chem.*, 31, 671, 1985.
35. **Demas, J. N.,** Time-resolved and phase-resolved emission spectroscopy, in *Molecular Luminescence Spectroscopy: Methods & Applications*, Vol 2, Schulman, S. G., Ed., John Wiley & Sons, New York, 1988, chap 2.
36. **Lippitsch, M. E., Pusterhofer, J., Leiner, M. J. P., and Wolfbeis, O. S.,** Fibre optic oxygen sensor with the decay time as the information carrrier, *Anal. Chim. Acta*, 205, 1, 1988.
37. **Petrea, R. D., Sepaniak, M. J., and Vo-Dinh, T.,** Fiber optic time resolved fluorimetry for immunoassays, *Talanta*, 35, 139, 1988.
38. **Vickers, G. H., Miller, R. M., and Hieftje, G. M.,** Time-resolved fluorescence with an optical fiber probe, *Anal. Chim. Acta*, 192, 145, 1987.
39. **Stryer, L. and Haugland, R. P.,** Energy transfer: a spectroscopic ruler, *Proc. Natl. Acad. Sci.*, 58, 719, 1967.
40. **Jordan, D. M., Walt, D. R., and Milanovich, F. P.,** Physiological pH fiber optic chemical sensor based on energy transfer, *Anal. Chem.*, 59, 437, 1987.
41. **Yvan, P. and Walt, D. R.,** Calculation for fluorescence modulation by absorbing species and its application to measurements using optical fibers, *Anal. Chem.*, 59, 2391, 1987.
42. **Hirschfeld, T. B.,** European Patent Appl. 214,768, 1987.
43. **Sharma, A. and Wolfbeis, O. S.,** Fiberoptic oxygen sensor based on fluorescence quenching and energy transfer, *Appl. Spectrosc.*, 42, 1009, 1988.
44. **Sharma, A. and Wolfbeis, O. S.,** Fiber optic fluorosensor for sulfur dioxide based on energy transfer and exciplex quenching, *Proc. SPIE*, 990, 116, 1989.
45. **Meadows, D. and Shultz, J. S.,** Fiber optic biosensors based on fluorescence energy transfer, *Talanta*, 35, 145, 1988.
46. **Christian, L. M. and Seitz, W. R.,** Optical ionic strength sensor based on polyelectrolyte association and fluorescence energy transfer, *Talanta*, 35, 119, 1988.
47. **Wolfbeis, O. S. and Schulman, S. G.,** Optical measurement of pH via fluorescence lifetime, unpublished results, 1990.
48. **Vo-Dinh, T.,** *Room Temperature Phosphorimetry for Chemical Analysis*, John Wiley & Sons, New York, 1984.
49. **Freeman, T. M. and Seitz, W. R.,** Chemiluminescence fiber optic probe for hydrogen peroxide based on the luminol reaction, *Anal. Chem.*, 50, 1242, 1978.
50. **Freeman, T. M. and Seitz, W. R.,** Oxygen probe based on chemiluminescence, *Anal. Chem.*, 53, 98, 1981.
51. **Smart, R. B.,** Measurement of chlorine dioxide with a membrane chemiluminescence cell, *Anal. Lett.*, 14A, 189, 1981.
52a. **Aizawa, M., Yoshishito, I. and Kuno, H.,** Photovoltaic determination of hydrogen peroxide with a biophotodiode, *Anal. Lett.*, 17, 555, 1984.
53. **Simhony, S., Katzir, A. and Kosower, E. M.,** FTIR spectra of organic compounds in solution and as thin layers using an attenuated total internal reflectance fiber optic cell, *Anal. Chem.*, 60, 1908, 1988.
54. **Kaye, W.,** Near-infrared spectroscopy — a review, *Spectrochim. Acta*, 6, 257, 1954.
55. **Weyer, L. G., Becker, K. J., and Leach, H. B.,** Remote sensing fiber optic probe NIR spectroscopy coupled with chemometric data treatment, *Appl. Spectrosc.*, 41, 786, 1987.
56a. **Inaba, H.,** Optical remote sensing of environmental pollution and danger using low-loss fiber network system, *Springer Ser. Opt. Sci.*, 39, 288, 1983; and references cited therein.
56b. **Mohebati, A. and King, T. A.,** Fibre optic remote gas sensor with diode laser FM spectroscopy, *Proc. SPIE*, 1172, 186, 1990.
57. **White, K. O. and Watkins, W. R.,** Erbium laser as a remote sensor for methane, *Appl. Opt.*, 14, 2812, 1975.
58. **Tenge, B., Buchanan, B. R., and Honigs, D. E.,** Calibration in the fiber optic region of the near-infrared, *Appl. Spectrosc.*, 41, 779, 1987.
59a. **Böck, J., Gersing, E., Sundmacher, F., and Hellige, G.,** Intravascular fiberoptic detection of $D_2O$ concentration in blood, *Proc. SPIE*, 906, 169, 1988.
59b. **Weigl, R. and Kellner, R.,** Development and performance of a novel IR-ATR-based glucose sensor system, *Proc. SPIE*, 1145, 134, 1989.
60. **Reichert, W. M., Ives, J. T., and Suci, P. A.,** A scaled down laser spectroscopy configuration for solution-phase fiber optic sensing, *Appl. Spectrosc.*, 41, 1347, 1987.
61. **Wirth, M. J. and Chon, S. H.,** Comparison of time and frequency domain methods for rejecting fluorescence from Raman spectra, *Anal. Chem.*, 60, 1882, 1988.
62. **Bright, F. V.,** Multicomponent suppression of fluorescent interferants using phase-resolved Raman spectroscopy, *Anal. Chem.*, 60, 1622, 1988, and references cited therein.
63. **Schrader, B., Hoffmann, A., Simon, A., and Sawatzki, J.,** Can a Raman renaissance be expected via the near-infrared Fourier transform technique?, *Vibrational Spectrosc. ,*1, 239, 1991.

64. **Schwab, S. D. and McCreery, R. L.,** Versatile, efficient Raman sampling with fiber optics, *Anal. Chem.*, 56, 2199, 1984.

65. **Yamada, H. and Yamamoto,** Illumination of flat or unstable samples for Raman measurements using optical fibers, *J. Raman Spectrosc.*, 9, 401, 1980.

66. **Nguyen, Q. D. and Plaza, P.,** Possibilities of remote and multisite analysis by laser Raman spectroscopy with optical fibers, *Analusis*, 14, 119, 1986.

67. **Brown, R. G. W.,** Optical fiber sensing using light scattering techniques, *J. Phys. E*, 20, 1312, 1987.

68. **Miller, D. R., Han, O. H., and Bohn, P. W.,** Quantitative Raman spectroscopy of homogeneous molecular profiles in optical waveguides via direct measurement, *Appl. Spectrosc.*, 41, 245, 1987.

69. **Walrafen, G. E. and Stone, J.,** Intensification of spontaneous Raman spectra by use of liquid core optical fibers, *Appl. Spectrosc.*, 26, 585, 1972.

70. **Schaefer, J. C. and Chabay, I.,** Generation of enhanced CARS signals in liquid-filled waveguides, *Opt. Lett.*, 4, 227, 1979.

71. **Eckbreth, A.,** Remote detection of CARS employing fiber optic guides, *Appl. Opt.*, 18, 3215, 1980.

72. **McCreery, R. L., Fleischmann, M., and Hendra, P.,** Fiber optic probe for remote Raman spectrometry, *Anal. Chem.*, 55, 146, 1983.

73. **Walrafen, G. E.,** New slitless optical fiber laser Raman spectrometer, *Appl. Spectrosc.*, 29, 179, 1975.

74. **Chang, R. and Furtak, T., Eds.,** *Surface-Enhanced Raman Scattering*, Plenum Press, New York, 1982.

75a. **Vo-Dinh, T., Meyer, M., and Wokaum, A.,** Surface-enhanced Raman spectrometry with silver particles, *Anal. Chim. Acta*, 181, 139, 1986.

75b. **Myrick, M. L. and Angel, S. M.,** Normal and surface-enhanced Raman scattering with optical fibers, *Proc. SPIE*, 1172, 38, 1989.

75c. **Belle, J. M., Stokes, D. L., and Vo-Dinh, T.,** Titanium dioxide based substrate for optical monitors in surface-enhanced Raman scattering analysis, *Anal. Chem.*, 61, 1779, 1989.

76. **Müller, G. J.,** Spectroscopy with the evanescent wave in the visible region of the spectrum, *Am. Chem. Soc. Symp. Ser.*, 102, 239, 1979.

77. **Hirschfeld, T.,** U.S. Patent 3,604,927, 1971.

78. **Tai, H., Tanaka, H., and Yoshino, T.,** Fiber optic evanescent wave methane sensor using optical absorption for the 3.392-μm line of a He–Ne laser, *Opt. Lett.*, 12, 437, 1987.

79. **Ruddy, V., MacCraith, B., and Murphy, J. A.,** Spectroscopy of fluids using eranescent absorption on multimode fiber, *Proc. SPIE*, 1172, 83, 1989.

80. **Sutherland, R. M., Dähne, C., Place, J. F., and Ringrose, A. S.,** Optical detection of antibody-antigen reactions at a glass-liquid interface, *Clin. Chem.*, 30, 1533, 1984.

81. **Ives, J. T., Reichert, W. M., Lin, J. N., Hlady, V., Suci, P. A., VanWagenem, R. A., Newby, K., and Andrade, J. D.,** Total internal reflection fluorescence surface sensors, in *Optical Fiber Sensors*, Chester, A. N., Martellucci, S., and Scheggi, A. M., Eds., Martinus Nijhoff, Dordrecht, 1987, 391, and references cited therein.

82. **Harrick, N. J. and Loeb, G. I.,** Internal reflection techniques in fluorescence spectroscopy, in *Modern Fluorescence Spectroscopy*, Wehry, E. L., Ed., Heyden, London, 1976, 211.

83. **Place, J. F., Sutherland, R. M., and Dähne, C.,** Opto-electronic immunosensors: a review of optical immunoassays at continuous surfaces, *Biosensors*, 1, 321, 1985.

84. **Murray, R. C. and Lefkowitz, S. M.,** European Patent Appl., 190,830, 1986.

85. **Andrade, J. D., VanWagenem, J. D., Gregonis, D. E., Newby, K., and Lin, J. N.,** Remote fiber optic biosensors based on evanescent excited fluoro-immunoassay: concepts and progress, *IEEE Trans. Electron Devices*, 32, 1175, 1985.

86. **Reichert, W. M., Ives, J. T., Suci, P. A., and Hlady, V.,** Excitation of fluorescent emission from solutions at the surface of polymer thin-film waveguides, *Appl. Spectrosc.*, 41, 636, 1987.

87. **Dodson, B. W.,** Time-resolved spectroscopy of evanescent-wave excited fluorescence, *Proc. SPIE*, 497, 91, 1984.

88. **Kapany, N. S. and Pontarelli, D. A.,** A photorefractometer: extension of sensitivity and range, *Appl. Opt.*, 2, 425, 1963.

89. **Kapany, N. S. and Pike, J. N.,** Fiber optics, IV. A photorefractometer, *J. Opt. Soc. Am.*, 47, 1109, 1957.

90. **David, D. J., Shaw, D., Tucker, H., and Unterleitner, F. C.,** Design, development, and performance of a fiber optic refractometer: application to HPLC, *Rev. Sci. Instrum.*, 47, 989, 1976.

91. **Takao, T. and Hattori, H.,** Fluid observation with an optical fiber refractometer, *Jpn. J. Appl. Phys.*, 21, 1509, 1982.

92. **Harmer, A. L.,** Optical fiber refractometer using attention of cladding modes, *IEE Conf. Proc.*, 221, 104, 1983.

93. **Haubenreisser, W., Lehmann, J., Perthel, R., and Willsch, R.,** Studies on fiber optic refractometry, *Exp. Tech. Phys.*, 32, 519, 1984.

94. **Lamont, T. G., Johnson, D. C., and Hill, K. O.,** Power transfer in fused single-mode fiber couplers: dependence on external refractive index, *Appl. Opt.*, 24, 327, 1985, and references cited therein.

95. **Tiefenthaler, K. and Lukosz, W.,** Grating couplers as integrated humidity and gas sensors, *Thin Solid Films*, 126, 205, 1985.

96. **Seifert, M., Tiefenthaler, K., Heuberger, K., Lukosz, W., and Mosbach, K.,** An integrated optical biosensor, *Anal. Lett.*, 19, 205, 1986.

97. **Lukosz, W. and Tiefenthaler, K.,** Sensitivity of integrated optical grating and prism couplers as (bio)chemical sensors, *Sensors Actuat.*, 15, 273, 1988.

98. **Nellen, P. M., Tiefenthaler, K., and Lukosz, W.,** Integrated optical input grating couplers as biochemical sensors, *Sensors Actuat.*, 15, 285, 1988.

99. **Smela, E. and Santiago-Aviles, J. J.,** A versatile twisted optical fiber sensor, *Sensors Actuat.*, 13, 117, 1988.

100. **Halliday, N. and Alder, J. F.,** Refractive index specific optical fibre detector and spectrophotometer for liquid droplets, *Analyst*, 112, 1751, 1988.

101. **Ravishankar, M. K. and Pappu, S. V.,** Fiber optic sensor-based refractometer-cum-liquid level indicator, *Appl. Opt.*, 25, 480, 1986.

102. **Kuchejda, M.,** Fiber optic refractometer with inherent turbidity measurement, *Proc. SPIE*, 798, 246, 1987.

103. **Ross, I. N. and Mbanu, A.,** Optical monitoring of glucose concentration, *Opt. Laser Technol.*, 17, 31, 1985.

104. **Kawahara, F. K., Fiutem, R. A., Silvus, H. S., Newman, F. M., and Frazar, J. H.,** Development of a novel method for monitoring oils in water, *Anal. Chim. Acta*, 151, 315, 1983.

105. **Attridge, J. W., Learer, K. D., and Cozens, J. R.,** Design of a fibre optic pH sensor with rapid response, *J. Phys. E*, 20, 548, 1987.

106. **Räther, H.,** Surface plasma oscillations and their applications, *Phys. Thin Films*, 9, 145, 1977 and references cited therein.

107. **Kooyman, R. P. H., Kolkman, H., Van Gent, J., and Greve, J.,** Surface plasmon resonance immunosensors: sensitivity considerations, *Anal. Chim. Acta*, 213, 35, 1988.

108a. **Nylander, C., Liedberg, B., and Lind, T.,** Gas detection by means of surface plasmon resonance, *Sensors Actuat.*, 3, 79, 1983.

108b. **Zhu, D. G., Petty, M. C., and Harris, M.,** An optical sensor for $NO_2$ based on a copper-phythalecyanine Langmuir-Blodgett film, *Sensors & Actuators*, 2B, 265, 1989.

109. **Liedberg, B., Nylander, C., and Lundström, I.,** SPR for gas detection and biosensing, *Sensors Actuat.*, 4, 299, 1983.

110. **Flanagan, M. F. and Pantell,** SPR and immunosensors, *Electron. Lett.*, 20, 968, 1984.

111. **Matsubara, K., Kawata, S., and Minami, S.,** Optical chemical sensor based on SPR, *Appl. Opt.*, 27, 1160, 1988.

112. **Giallorenzi, T. G.,** Optical fiber sensor technology, *IEEE J. Quantum Electron.*, 18, 626, 1982.

113. **Jackson, D. A.,** Monomode optical fibers interferometers for precision measurement, *J. Phys. E*, 18, 981, 1985.

114. **Jackson, D. A,** Monomode fiber optic interferometers and their application in sensing systems, in *Optical Fiber Sensors*, Chester, A. N., Martellucci, S., and Scheggi, A. M., Eds., Martinus Nijhoff, Dordrecht, 1987, 1.

115. **Butler, M. A.,** Optical fiber hydrogen sensor, *Appl. Phys. Lett.*, 45, 1007, 1984.

116. **Farahi, F., Akhavan-Leilabady, P., Jones, J. D. C., and Jackson, D. A.,** Interferometric fiber optic hydrogen sensor, *J. Phys. E*, 20, 432, 1987.

117. **Farahi, F., Akharan-Leilabady, P., Jones, J. D. C., and Jackson, D. A.,** Optical fiber flammable gas sensor, *J. Phys. E*, 20, 435, 1987.

118. **Papa, L., Piano, E., and Pontiggia,** Turbidity monitoring by fiber optics instrumentation, *Appl. Opt.*, 22, 375, 1983.

119. **Sutherland, R., Dähne, C., and Place, J.,** Preliminary results obtained with a no-label homogeneous optical immunoassay for human IgG, *Anal. Lett.*, 17, 43, 1984.

120. **Rosencwaig, A. and Gersho, A.,** Theory of the photo-acoustic effect with solids, *J. Appl. Phys.*, 47, 64, 1976.

121. **Burggraf, L. W. and Leyden, D. E.,** Quantitative photo-acoustic spectroscopy of intensely light-scattering thermally thick samples, *Anal. Chem.*, 53, 759, 1981.

122. **Perkampus, H. H.,** Photo-acoustic spectroscopy with solids, *Naturwissenschaften*, 69, 162, 1982.

123. **Öhler, O., Seifert, M., Cliffe, S., and Mosbach, K.,** Detection of gases produced by biological systems with an enzyme-photoacoustic sensor, *Infrared Phys.*, 25, 319, 1985.

124. **Yaegachi, M., Itago, F., and Sugitani, Y.,** Photo-acoustic determination of glucose using a multilayered film, *Bunseki Kagaku*, 34, 772, 1985.

125. **McQueen, D. H.,** A simplified open photo-acoustic cell and its applications, *J. Phys E*, 16, 738, 1983.

126. **Munir, Q., Weber, H. P., and Bättig, R.,** Resonant photoacoustic gas spectrometer fiber sensor, *Proc. SPIE*, 514, 81, 1984.

127. **Azzam, R. M. A. and Bashara, N. M.,** *Ellipsometry and Polarized Light,* Elsevier, Amsterdam, 1977.
128. **Debe, M. K.,** Optical probes of thin films, *Prog. Surf. Sci.*, 24, 1, 1987.
129. **Stenberg, M. and Nygren, H.,** A receptor-ligand reaction studied by a novel analytical tool — the isoscope ellipsometer, *Anal. Biochem.*, 127, 183, 1982.
130. **Mermelstein, M.,** All-fiber polarimetric sensor, *Appl. Opt.*, 25, 1256, 1986.
131. **Yoshino, T. and Kurosawa, I.,** All fiber ellipsometry, *Appl. Opt.*, 23, 1100, 1984.
132a. **Noeller, H. G.,** U.S. Patent 4,451,149, 1984.
132b. **Heideman, R. G., Blikman, A., Koster, R., Kooyman, R. P. H., and Greve, J.,** Polarimetric optical fiber sensor for biochemical measurements, *Proc. SPIE,* 1510, xxx, 1991; in press.
132c. **Kersey, A. D., Davis, M. A., and Marrone, M. J.,** Differential fiber optic sensor configuration with dual wavelength operation, *Appl. Opt.,* 28, 204, 1989.
133. **Russell, P. S. T. and Ulrich, R.,** Grating fiber coupler as a high resolution spectrometer, *Opt. Lett.*, 10, 291, 1985.
134. **Brown, R. G. W.,** Laser fiber optics in biotechnology, *Trends Biotechnol.*, 3, 200, 1985.
135. **Hill, K. O., Fujii, Y., Johnson, D. C., and Kawasaki, B. S.,** Photosensitivity in optical fiber waveguides: application to reflection filter fabrication, *Appl. Phys. Lett.*, 32, 647, 1978.
136. **Hutley, M. C.,** Wavelength-encoded optical fiber sensors, *Proc. SPIE*, 514, 111, 1984.
137. **Watanabe, J., Kinoshita, S., and Kushida, T.,** Photon time of flight fluorescence spectroscopy through an optical fiber, *Jpn. J. Appl. Phys.*, 24, 761, 1985.
138. **Chow, K. M., Stansfield, A. G., Carr, R. J. G., Rarity, J. G., and Brown, R. G. W.,** On-line photon correlation spectroscopy using fiber optic probes, *J. Phys. E*, 21, 1186, 1988.
139. **Goody, R.,** Cross-correlating spectrometer, *J. Opt. Soc. Am.,* 58, 900, 1968.
140. **Dakin, J. P. and Edwards, K. O.,** Progress in fibre remoted gas correlation spectrometry, *Proc. SPIE,* 1510, xxx, 1991, in press.

Chapter 3

# SENSING SCHEMES

## Otto S. Wolfbeis

# TABLE OF CONTENTS

# I. INTRODUCTION

Following the discussion of the most common spectroscopic techniques in the preceding chapter, a presentation of the various sensing schemes will be given in this chapter. One may differentiate between three situations: (1) the analyte directly affects the light-guiding properties of the waveguide (Section II), (2) the fiber is used to guide light from the source to the chemical species of interest whose optical properties are sensed by one of the spectroscopic methods (Section IV), (3) an indicator or chemical reagent is required to produce an optically detectable signal change that can be monitored. These so-called indicator-mediated sensors will be described in some detail in Section V. A variety of bio- and immunosensing schemes have been presented that will be compared and discussed separately in Section VI. Finally, schemes for internal referencing and suppression of background light and false light will be treated in the last two sections.

# II. INTRINSIC PLAIN FIBER SENSING

These sensor types are based on the fact that a chemical species can affect the waveguide properties. Hence, it is not the absorption or emission properties of an analyte that are measured, but rather the effect of the analyte upon the optical properties of the fiber, in particular the refractive index of the clad (or the outer phase). When the analyte itself is the outer medium of the waveguide, its refractive index can modulate the waveguide properties as well, in particular via its effect on the evanescent wave. Not surprisingly, all fibers for use in telecommunication are more or less insensitive towards most chemical species of interest, and special instrumental techniques or coatings are required to effect the desired signal changes.[1]

None of the common equations that describe the signal-to-analyte relations in optical analysis (Lambert-Beer, Kubelka-Munk, Parker, Stern-Volmer, etc.) are applicable to these situations. Rather, such sensors are based on one or more of the following effects of the analyte: (a) An increase in the strain of the coating; (b) a change of the refractive index of the clad through absorption, or adsorption into the clad; (c) a change of the refractive index through swelling.

One prerequisite for these sensors is that the fiber be in full contact with the sample, which is in contrast to indicator-mediated sensors where the sample must be in contact with the sensing chemistry, but not with the waveguide. However, since most waveguide materials are highly resistant to aggressive chemicals, this does not present a real limitation.

Unfortunately, only a few types of intrinsic plain fiber sensors meet the requirements of fair specificity for a given species. A notable example is the interferometric hydrogen sensor presented in Chapter 2.X. in which the sensing element is a piece of single-mode optical fiber (with the plastic jacket stripped off) which has a 10-$\mu$m palladium coating about 3 cm long. Exposure of the coated fiber to hydrogen gas results in the formation of palladium hydrides which have a lattice constant that is larger than that of pure palladium. The process is fully reversible. The expansion of the hydride will stretch the fiber in both axial and radial direction (Chapter 2, Figure 16) and change its effective optical pathlength. This is detected by inclusion of the sensing fiber in one arm of an Mach-Zehnder interferometer by gluing coated and uncoated fibers to a fused quartz plate with an adhesive that matches the index of refraction of the fiber to make it an effective mode stripper. The technique is said to be capable of detecting 1 to 30,000 ppm hydrogen in a gas, but the device has a rather large signal noise because of its sensitivity to vibrations and sound.

The effect of refractive index on the efficiency of light coupling can be employed to monitor changes in the $n_D$ of liquids and of coatings on fiber cores (Chapter 2.VIII). Various configurations have been proposed. A configuration in which changes in refractive index $n_2$

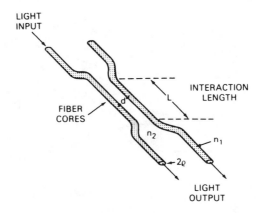

FIGURE 1. Sensor configuration in which light intensity in one fiber is modulated by varying the evanescent coupling between two fibers.

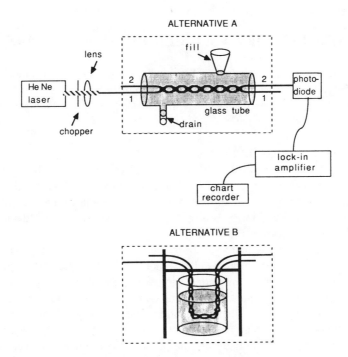

FIGURE 2. (A) Evanescent wave coupling between two twisted fibers in a flow-through cell. (B) Same technique applied to a liquid in a beaker.

change the coupling efficiency between the two fibers is shown in Figure 1. By twisting the fibers by each other as shown in Figure 2, a much larger interaction length is obtained.[2] The intensity of the light in the second fiber can be as much as 10% that in the first fiber.

Although the demonstrated laboratory performance of the evanescent wave sensors is encouraging, a better understanding of the complex transduction mechanism is required. Optimization involves both optical and mechanical considerations. Any sensor utilizing a transduction mechanism which requires mechanical alignment with tolerances in the order of micrometers may be difficult to implement in practical devices.

A third class of "waveguide" sensors in which the optical properties are directly modulated by the analyte are the surface plasmon resonance-based devices described in chapter 2. A

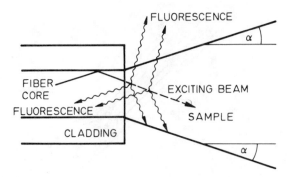

FIGURE 3.  Schematic of the fluorescence produced at a single fiber termination.

typical experimental arrangement is shown in Chapter 2, Figure 13. Except for SPR sensors coupled to antigen/antibody reactions, none is specific for a given analyte and therefore can be applied to well-defined analytical problems only, for instance the detection of narcotic gases in breath gas.

## III. SINGLE FIBER VS. FIBER BUNDLE SENSORS

While the sensing schemes described in Section II require the use of single fibers, the methods described in the next sections, in contrast, can mostly be applied to both single fiber and fiber bundle sensing. It appears appropriate at this stage to discuss the respective merits and limitations.

### A. SINGLE FIBER CONFIGURATIONS

Because of the capability of optical fibers to transmit light of various wavelengths simultaneously and in both directions, a single fiber is sufficient to guide both the light from the light source (the "question") to the sampling region, and the analyte-modulated light (the "encoded answer") back to a detector. This is in striking contrast to the poor and unidirectional data transmission capability of electrical wires.

Single fibers are widely used for both absorption and fluorescence measurements, for instance in various kinds of flow-through cells, where the intrinsic spectral properties of an absorbing or fluorescent analyte are measured. However, precise absorbance measurements require a correction for reflected and scattered source radiation.[3] It is shown in Figure 3 how fluorescence is produced at a single fiber termination. The advantage of this configuration is the almost perfect overlap of the sample cone that is excited and the accepting cone. Differences between the two cones only arise from the fact that the refraction indices of core and clad are smaller for the longwave fluorescence. A major limitation can be the intrinsic fluorescence of the fiber which fully interferes with the sample fluorescence. A typical detection limit is $3\cdot10^{-6}$ $M$ only for rhodamine B using a 600-μm PCS fiber.[4]

The situation is somewhat better in case of indicator-mediated optrodes (Section V herein) if one manages to immobilize a sufficient amount of strongly fluorescent indicator at the fiber end so that fluorescence mainly originates from the indicator. In this case it is not the weak fluorescence of an analyte being present in low concentration that limits sensitivity, but rather a poor sensitivity of the indicator (in terms of signal change with analyte concentration).

The intrinsic fluorescence and Raman scatter of the fiber also hamper fiber sensing based on Raman and Mie scatter. Wherever possible, the multifiber approach is favored in these

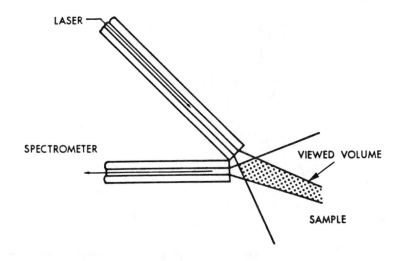

FIGURE 4. Double fiber terminal showing the limited cone overlap when compared with Figure 3.

cases. Single fibers are not a good choice either in reflectometry because the fiber efficiently collects specularly reflected light in addition to diffusely scattered light.

Aside from intrinsic fluorescence, Raman bands at 490 and 600 cm[1] are present in silica quartz glass.[5] Another source of serious background in single fibers is stimulated Raman scatter which becomes very intense when the input power of laser light into fibers exceeds certain limits, typically 100 mW. The major sources of absorption and scattering losses in fluorozirconate fibers are transition element impurities and hydroxide contamination. Following irradiation with alpha and gamma rays, absorption bands with maxima centered at 440 and 630 nm have been found in silica fibers.[5] They are less significant in high-OH fibers. Gamma ($\gamma$) irradiation can also result in increased fluorescence in the region of 640 to 720 nm, with a maximum at around 670 nm. A more detailed description of the effects of radiation is presented in Chapter 14.

Single fibers are attractive because of their minute space requirements. Therefore, they are ideally suited for invasive methods. By coupling several single fibers with different analyte-sensitive chemistries at the end, multiple sensor bundles can be obtained that are good for sensing several parameters simultaneously. These bundles should not be confused with the bifurcated fiber bundles discussed in the next chapter. Single fibers, in particular single-mode fibers, are the only choice in certain spectroscopic techniques such as interferometry.

## B. FIBER BUNDLE CONFIGURATIONS

In double fiber arrangements (Figure 4) one fiber guides the excitation radiation to the sample, and a second fiber collects scattered light or fluorescence and guides it back to the photodetector. Bifurcated fiber optic bundles (Figure 5) function in a similar manner, with part of the fibers carrying the excitation light, and the rest collecting fluorescence and/or scattered light and returning it to the detector.

In sensors based on scatter or fluorescence the efficiency of light collection is limited when fiber bundles are used. This is particularly true in case of parallel fibers with their limited overlap of the cones of exciting and collecting fibers (Figure 5). Moreover, the fiber bundle approach requires the sensing material to be placed a certain critical distance away from the fiber end. An obvious means to increase collection efficiency is to enlarge the numerical aperture of the fiber.

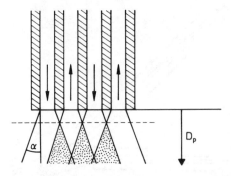

FIGURE 5. Fiber bundle terminal demonstrating the overlapping cones of input and output bundles, and the minimum penetration depth $D_p$ (marked with a dashed line) required in fiber bundle sensing.

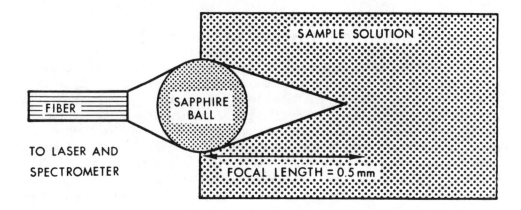

FIGURE 6. Sapphire ball optrode.

The limiting background absorption, fluorescence, and Raman emission of single fiber sensors is tremendously reduced in multiple fiber sensors. Hence, sensitivity can be distinctly improved, particularly in case of plain fiber sensors which detect the often weak optical signals of the analyte. In indicator-mediated sensors, the signal is mostly strong enough because it comes from an immobilized indicator. The disadvantages of the multifiber arrangements include a larger space requirement, poor collection efficiencies, the difficulty to optimize the angle between excitation and emission fibers, and the distinctly higher costs because of the larger fiber amounts required.

### C. IMPROVING COLLECTION EFFICIENCIES

While guiding light from the source to the sample usually is not a serious problem, the collection of light modulated by the analyte (i.e., the analytical information) is more difficult and has resulted in various ingenious configurations. There are two approaches: (1) the collection efficiency is improved, and (2) the amount of interfering primary light is minimized.

The already good collection efficiency of single fiber terminations can be improved upon by placing a lens or, less costly, a sapphire ball at the fiber end.[6] It has the double function of focusing the incoming beam into the sample, as well as helping to better collect fluorescence or Raman light from the sample (Figure 6). Another possibility[6] is to apply the so-called

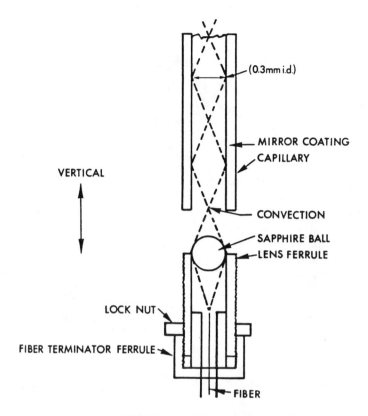

FIGURE 7. Capillary optrode.

capillary technique in which an analyte-filled capillary is attached to the fiber end. Because of total reflection inside the silver capillary, both the absorbance and fluorescence are enhanced (Figure 7). When measuring in the absorbance mode, the end of the capillary is covered with a reflecting cap.

In an experimental comparison of single- and double-fiber configurations,[7] an approximate angle of 20° between fibers gave the optimal signal-to-background ratio for the double fiber configuration. However, with the fiber optics tested there is only a 50% difference in the fluorescence collection efficiency of single- and double-fiber arrangements. As the signals obtained with these two basic configurations differ by no more than a factor 2 or 3, other factors such as size, ruggedness, or signal-to-noise ratio may be important in choosing the configuration for a given application.

The efficient collection of Raman light is particularly important in view of its poor intensity. This also holds for weakly fluorescent analytes. Modifications to enhance the signal obtained with double fiber configurations[8,9] have been studied experimentally and theoretically. A popular fiber arrangement[10] is the one shown in Figure 8, in which the central fiber guides incoming light and the other ring of fibers collects the signal. The two-fiber probe shown in Figure 9, with a 45° reflecting mirror in front of the incoming fiber, is commercially available. Alternatively, the two fibers may be positioned to each other at a fixed angle. These configurations are intended for use in fluorimetry and scattering techniques.

Methods to improve the interaction between light and analyte or indicator (including multiple reflection, evanescent wave spectrometry, and methods utilizing differences in the index of refraction of guiding and sensing phase) will be included in the next sections.

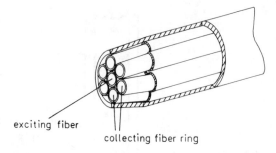

exciting fiber

collecting fiber ring

FIGURE 8. Fiber bundle configuration for fluorescence and Raman measurements, with the central fiber acting as the exciting fiber, and the outer fibers collecting the signal.

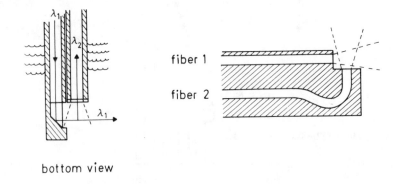

bottom view

FIGURE 9. Right-angle fiber configurations for measurement of light scattering and fluorescence.

## IV. EXTRINSIC PLAIN FIBER SENSING

In these sensors, the fiber (or the fiber bundle) acts as a light conduit to remotely probe the spectral properties of the analyte. The respective laws of Lambert-Beer, Parker, or Kubelka-Munk (as described in the previous chapter) are applicable. In contrast to the intrinsic fiber sensors, the analyte should *not* affect the lightguiding properties of the fiber. Fiber and analyte do not necessarily have to be in full contact.

Numerous sensing configurations have been proposed. A few typical examples of absorption cells are shown in Figure 10. The simplest case (A) is a flow-through cell whose diameter defines the optical pathlength, with two fibers attached to the cell at both sides. The fiber optic peers through the cell, and both liquids and gases can be investigated. Another single fiber approach, but with two sample volumes and more versatile geometry with respect to sensor insertion into a sample is shown in Figure 10 (B). While 0.1 to 10 cm pathlengths are sufficient for UV/VIS absorptiometry, the small molar absorbances of NIR absorption have necessitated to increase the pathlength l in order to improve sensitivity (according to the Lambert-Beer law). One way is to use a multiple reflection chamber as shown in Figure 10 (C). The increase in sensitivity is furnished by a considerably prolonged light path.

An insertable plain fiber absorbance probe has been described by Sepaniak et al.[6,11] A single fiber served to guide light to a sample chamber of well-defined pathlength at the tip of a 19-gauge hypothermic needle. The needle was drilled with holes at the side, and the piece of aluminium foil (acting as a reflector) fixed at the fiber tip. This needle was inserted into a 15-gauge needle to form a cannula. Fluids can be drawn into the irradiation cavity with a syringe. The device has been used to monitor bilirubin concentrations in spiked human serum.

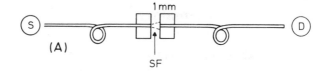

FIGURE 10.  Fiber optic absorption cell types. S, light source; D, detector; M, mirror; SF, sample flow. In all these configurations, the second fiber collects only part of the light escaping the first fiber. Losses increase in going from configurations (A) to (C), but the pathlength and, hence, sensitivity also increases.

The Lambert-Beer law (Equation 1) relates analyte concentration and path length, with absorbance. The highest precision in obtained when the absorbance is at around $e^{-1}$. Because of the constant background of fibers at certain wavelengths, or the sample solvent, usually an additional term (B) is added to the Lambert-Beer law which now reads

$$\log I_0/I = A = \varepsilon \cdot C \cdot 1 + B \tag{1}$$

The background (B) can be ascertained from a blank run.

In samples with varying solvent absorption, in binary mixtures of UV or NIR absorbing solvents, or with two absorbing dyes, the total absorption at a given wavelength is the sum of the individual absorptions of analyte and solvent. The total absorbance A therefore is

$$A = 1\left(\varepsilon_1 \cdot C_1 + \varepsilon_2 \cdot C_2\right) \tag{2}$$

where $\varepsilon$, $C_{1\,Cl}$, $\varepsilon_2$, $C_2$ are the absorptivity and concentration of solute and solvent, respectively. In binary mixtures, where the composition is frequently assayed by NIR absorptiometry, $C_2 = 1-C_1$, so that

$$A = 1\left[C_1\left(\varepsilon_1 - \varepsilon_2\right) - \varepsilon_2\right] \tag{3}$$

Hence, for a given pathlength l the absorbance should be a linear function of concentration $C_1$. To compensate for variations in the light source, variable connector losses, dirt on windows, etc., a reference wavelength can be introduced that is not absorbed by either the solute or the solvent. The absorbance is then defined as

$$A = ^{-10}\log\left(\frac{T_1}{T_2}\right) \tag{4}$$

where $T_1$ is the transmittance ($I/I_0$) at the absorbing wavelength, and $T_2$ the transmittance at the reference wavelength. Multi-component liquids can be analyzed in the same way by a careful choice of more than one reference wavelength.

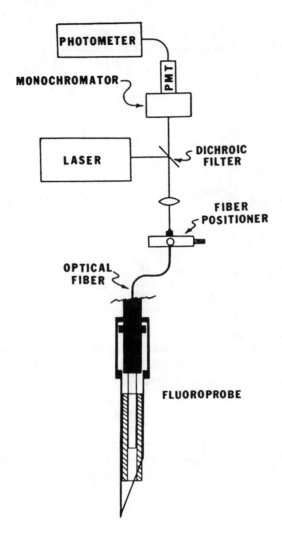

FIGURE 11. Single strand fiber optic sensor for invasive measurements using a 20-gauge needle.

Plain fiber fluorosensing is usually performed using one of the configurations shown in Figures 3 to 9. However, fiber fluorimetry is never as sensitive as conventional fluorimetry in cells because of considerable fluorescence and Raman background. Typical experimental arrangements for separating the strong excitation light from sample fluorescence in single fiber configurations are shown in Chapters 6 and 12. When bifurcated fibers are used, exciting light can easily be separated from scattered or reflected excitation using conventional filters. Sepaniak et al.[12] used a 20-gauge needle to insert a fiber into the body (Figure 11). After having stripped off protective coating and optical cladding of a 200-μm quartz fiber and placing it in a glass capillary, the fiber was inserted into the needle. The fiber could be withdrawn in the capillary and "sees" a volume of about 200 nl. The drug doxorubicin was detected this way at levels of $10^{-7}$ *M*.

Fluorescence is a rather selective technique that is particularly useful in case of complex mixtures. Although a variety of industrially and environmentally important analytes have a strong absorption and fluorescence in the UV and thus appear to be ideally detectable by one of these methods, working in the UV at below 300 nm is rather critical for several reasons: it requires expensive fused silica fibers (Chapter 2, Figure 2), there will be poor selectivity because of the number of molecules absorbing in this region, and the intensity of straylight is higher by a factor of 16 at 250 nm when compared to 500 nm. Finally, background fluorescence in the far UV is strong and limits sensitivity, and no inexpensive lasers are available for this wavelength range at present. Consequently, fiber sensing at below 300 nm will be confined to special situations.

Aside from the fact that plain fiber methods are rugged and simple, they do not require sophisticated calibration and are likely to have excellent long-term stability and lifetimes. On the other hand, direct absorbance, fluorescence, or scatter measurements are relatively non-selective, and only a limited number of analytes have an intrinsic absorbance of sufficient intensity in the UV, VIS or NIR. Even fewer molecules are known that have a useful intrinsic fluorescence. Fortunately, several low molecular weight pollutants have narrow overtone IR absorptions in the NIR that can serve as an alternative for quantitation.

Another disadvantage in plain fiber extrinsic sensors is due to the fact that the numerical aperture of light leaving the fiber at the distal end also depends on the refractive index of the medium ($n_0$) to be probed. The amount of light coupled into the second fiber therefore not only reflects changes in the absorbance in the sample cell, but also changes in the $n_D$ of the analyte medium. In fact, this is the basis for many liquid level sensors and fiber refractometers. One approach to overcome this limitation is to work at two wavelengths, one of which is the analytical, and referencing out interferences by $n_D$ changes with the second wavelength which is not modulated by the analyte. This problem is minimized in single-fiber configurations but is also true for evanescent wave sensors. Such effects play a minor role in indicator-mediated sensors.

Owing to the poor absorptivities and low concentration of some gaseous or fluid analytes, plain fiber sensing is not very sensitive. One way to improve sensitivity is to make the fiber a multiple-reflection cell (Figure 10) Alternatively, the clad of the fiber can be removed so that the light wave evanesces into the outer (sample) phase. This is treated in some detail in Chapter 2 Section VII; see also Chapter 2, Figure 9. As a result, the amplitude of the propagating line is decreased if the wavelength matches the absorption of the gas. The fiber transmittance P is given by

$$P = P_0 \exp(-\alpha rcL) \tag{5}$$

where $P_0$ is the transmittance in the absence of the absorbing gas, $\alpha$ is the absorption coefficient of the gas in free space (e.g., for methane 8.3 atm cm$^{-1}$), r is the ratio of the power of the evanescent wave to that of the total propagating wave, c is the gas concentration, and L is the sensor length. Log($P_0$/P) is, of course, the equivalent to the absorption in the Lambert-Beer law.

It is known that the evanescent-wave ratio r depends on the mode that is excited. The number of modes that can propagate is a function of the V value:

$$V = \left( \frac{\pi d}{\lambda} \right) \left( n^2 - 1 \right)^{\frac{1}{2}} \tag{6}$$

where d is the fiber diameter, $\lambda$ is the wavelength, and n is the refractive index of the fiber. For a typical case, n is 1.5 and $\lambda$ is 3.392 μm, so that Equation 6 becomes

$$V = 1.04d \tag{7}$$

The condition for the fiber to be single mode is, from V < 2.405,

$$d < 2.3 \ \mu m \tag{8}$$

In the multimode fiber various modes can exist at the same time, so r depends on the excitation of the modes. This is likely to cause instability of the measurement sensitivity. An effective method for reducing such instability is to give random bending to the sensing (cladless) part of the fiber, so all propagating modes are nearly equally excited.

In a single mode fiber there is no sensitivity instability associated with the excitation of the modes. Various kinds of evanescent wave sensing schemes have been presented. In all these, the evanescent wave probes the spectral properties of the analyte itself, for instance the UV fluorescence of a protein or the absorption of drugs,[13] or one of the analytes shown in Tables 1 to 3.

Commercial instrumentation for remote fiber absorbance, reflectance, or fluorescence measurements is available from American, British, and French companies. Plain fiber sensing has found widespread application in industrial process monitoring and environmental sciences. A main application in practice seems to be the determination of the composition of solvent mixtures and impurities by measuring the NIR overtone absorption of certain functional groups. A representative collection is given in Table 1. A dedicated sterilizable fluorosensor for specifically measuring the NADH fluorescence in fermenter broths is available from a Swiss company.

The spatial flexibility of fibers along with their minute size was probably the driving force that led to the many applications of plain fiber sensors in biomedical sciences. Main applications are the measurement of blood color which reflects oxygen saturation (see Chapter 1), and color changes of other body fluids and tissue.

Because fibers can probe tiny sample volumes and can be operated in rather hostile environments such as in radioactive areas and near flames, they have been employed in various chromatography and spectroscopic studies. An impression of the versatility of plain fiber applications is given in Table 3.

## V. INDICATOR-MEDIATED SENSING

Only a limited number of analytes has an intrinsic absorption or a related spectroscopic property that can be utilized for direct sensing without compromising selectivity. For several important species including pH, metal ions, and oxygen in water, no direct and sensitive methods are known. The remedy for this situation is the well-established indicator chemistry. By immobilizing a proper indicator on the waveguide, a device is obtained whose spectral properties reflect the analyte concentration. Practically all known indicator-mediated sensors rely on absorption or luminescence measurement. Unlike the case of intrinsic and most cases of extrinsic plain fiber sensors, there is no need for direct physical contact between the reagent and the fiber. This can be important in applications where frequent changes in the reagent phase may be required. An interesting classification scheme for reagent phase indicators according to the role of the indicator (reactant, adsorbant/extractant, catalyst, or substrate) has been given by Seitz.[14] In this book, indicator-mediated sensors are classified according to the spectroscopic techniques and experimental arrangements employed.

### A. GENERAL CONSIDERATIONS

A major advantage of indicator-mediated sensors over plain fiber sensors relies on the fact that, in a first approximation, they are not affected by the refractive index of the medium. Only

**TABLE 1**
**A Selection of Plain Fiber Extrinsic Chemical Sensors for Industrial and**
**Environmental Applications**

| Analyte/application | Method | Ref. |
|---|---|---|
| Methane and related hydrocarbons | Measurement of the NIR absorption at around 1.34-1.66 μm | 79—82 |
| Ammonia | Absorption by ammonia of the $CO_2$ laser line in an absorption cell via silver halide fibers | 83 |
|  | NIR absorption at between 1.18 and 1.67 μm | 86 |
| $UF_6$ | Measurement of laser-induced fluorescence of $UF_6$ following pulsed near-UV excitation via 50-m single fibers |  |
| $NO_2$ | Absorption by $NO_2$ at 496 nm measured via 20 and 500 m fibers, using an absorption cell at the end of double fibers | 87 |
| $N_2/CO_2$ | Analysis of gas mixtures via laser-induced remote Raman spectrometry | 88 |
| $Pu/F_6$, $UF_6$, $F_2$, $CIF_3$ | Measurement of analyte absorption in the UV (254 and 365 nm) via fiber optics | 89 |
| $CO_2$ | Fiber optic measurement of IR absorption at 4.9 μm | 90 |
| Uranium/plutonium | Multi-wavelength fiber absorptiometry | 91 |
| Ethanol, glucose, fructose | Laser Raman spectrometry and multivariate, least-square analysis | 92 |
| NADH in biocultures | Plain fiber fluorosensor for NADH whose concentration is related to various other parameters in fermentation | 93 |
| NADH in biocultures | Plain fiber fluorosensing of the NADH-dependent culture fluorescence of immobilized cells | 94 |
| Liquid mixtures | Analysis of $CHCl_3/CH_2 Cl_2$ mixtures by laser Raman spectroscopy using optical fibers | 88 |
| Monitoring polymerization | Measurement of polymer fluorescence which starts to increase upon polymerization | 95, 96 |
| $Cu^{2+}$ in electroplating baths | Remote measurement of the $Cu^{2+}$ absorption at 820 nm | 97 |
| Water in solvents | Fiber optic measurement of the water NIR absorption at 1900 nm | 98 |
| Alcohols in organic solvents | Fiber optic measurement of the alcohol NIR absorption between 1100 and 1600 nm | 99 |
| Hydroxyl number in polymers | Fiber optic measurement of the hydroxy absorption in the NIR | 100a |
| Polymer melt analysis | NIR absorptiometry | 100b |
| Fuel quality analysis | NIR absorptiometry | 100c |
| Hydrocarbon sensing | IR absorptiometry | 100d |
| Combustion monitoring | IR absorptiometry | 100e |
| Solvent mixture composition | Fiber optic measurement of NIR absorption spectra and multivariate analysis | 101 |
| Water tracer studies | Remote detection of tracer fluorophors | 6 |
| Phytoplankton monitoring | Measurement of chlorophyll fluoresence via plain fibers | 102—104 |
| Dust | Light attenuation by dust | 105 |
| Phenols in groundwater | Detection of the intrinsic fluorescence of phenols | 106 |

**TABLE 2**
**Biomedical Applications of Plain Fiber Extrinsic Chemical Sensors**

| Measurand | Method | Ref. |
|---|---|---|
| Hemoglobin-bound oxygen | Diffuse reflectance of blood cells as measured at >1 wavelength | See Chapter 19 |
| Colored body substances | Fiber optic absorption measurement in a sample cell at the end of an implantable catheter | 107 |
| Intracellular redox state | Measurement of mitochondrial redox states via fiber fluorimetry of NADH | 108 |
| Fluorescent cells | Optical fiber probe for measuring the fluorescence of labeled cells and antibodies | 109 |
| Muscle elastin | Fiber optic reflectance and fluorescence of tissue | 110 |
| Blood CO, $O_2$, and $CO_2$ | Measurement of CO at 5.13 μm, $CO_2$ at 4.3 μm, oxygen at 9.0 μm | 111 |
| Bilirubin | Fiber optic absorbance measurement at 458 nm in a sample cell at the end of a 19-gauge hypodermic needle | 11 |
| Doxorubicin, adriamycin | Single fiber in a 20-gauge needle used to excite the fluoresence of the drug | 11 |
| Hematoporphyrin derivative | *In vivo* detection of HpD in tissue via its longwave intrinsic fluoresence | 112 |
| $D_2O$ in blood | Fiber optic measurement of the $D_2O$ absorption at ≈ 4 μm where $H_2O$ does not absorb | 113 |

if the medium causes a change in the refractive index of the sensing layer, interferences will be observed. The disadvantages of indicator-mediated sensors are also obvious: the indicator dye must not leach out or bleach, the response times are usually longer, and swelling of the sensing material can cause artifacts. Finally, the sensor construction and fabrication is distinctly more complicated than in plain fiber sensors.

An important variable is the amount of indicator in the reagent phase. It can affect response in several ways. The amount of an indicator interacting with the analyte in the ground state must be significantly less than the amount of analyte in the sample. Otherwise, the amount of analyte combining with, or dissociating from, the sample may be large enough to significantly alter the amount of analyte in the sample. For example, an optical pH sensor with a large amount of indicator will be subject to error when measuring the pH of a sample with a low buffer capacity.

The amount of indicator will also affect the magnitude of the observed optical signals. In the case of reflectometry or absorptiometry, the relative change in reflected or transmitted intensity per unit change in analyte concentration will depend on the amount of indicator and will tend to be greater when the absorbance is low, i.e., between 0.1 and 0.5. The actual optimum will depend on the functional relationship between reflected or transmitted light and the amount of indicator chromophore.

In the case of luminescence-based indicators, luminescence increases with the amount of indicator up to a point. In the absence of self-quenching, intensity is governed by the inner filter effect (see Section V.E herein). No further increase in luminescence is observed once the amount of indicator is sufficient to absorb essentially all the excitation light. For a typical example see Chapter 8, Figure 11. Such effects may be observed even at a relatively low amount of indicators.

An important issue in chemical sensing is the response time. In indicator-mediated sensing it is generally much longer than in plain fiber sensors, because of the mass transfer involved and the kinetics of the interaction between analyte and indicator. Most devices developed so

**TABLE 3**
**Miscellaneous Applications of Plain Fiber Extrinsic Sensors in**
**Chemical Sciences**

| Application | Method | Ref. |
|---|---|---|
| Atomic emission spectrometry | Optical fibers used in a multi-element analyzer to guide an emission of defined wavelength to a detector | 114, 115 |
| | Analysis of alkali elements | 116, 117 |
| Electron beam diagnostics | Optical fibers used to observe luminescence produced by interraction of e-beam with air | 118 |
| Molecular beam luminescence studies | Detection of molecular species via laser induced luminescence | 119 |
| Combustion monitoring | Determination of Na and OH in flames | 120 |
| Fluorescence detection in GC | Fibers used to excite and collect fluorescence of PAHs in GC eluates | 121 |
| Micro-cell spectrometry | Absorbance and fluorescence in <0.2 $\mu$l flow-through cell measured via fibers | 122 |
| | Laser-induced fluorescence in 3—60 nl cell, dye detection limit 12—90 fg | 123 |
| Fluorescence detection in LC | Bile acids and steroids detected via fluorescence in capillary column LC using optical fibers | 124 |
| | PAHs detected in LC by fluorimetry via fibers | 125 |
| | High pressure cell | 126 |
| 2-wavelength LC detector | Amino acid assay with ninhydrin operating at an analytical and a reference wavelength | 127 |
| Spectroelectrochemistry | Remote acquisition of absorption of metal-organic ruthenium complexes as a function of applied potential | 128 |
| Thermal lens spectrophotometry | Fibers used for detection of the thermal lens effect in small sample cells | 129 |
| Remote acquisition of fluorescence spectra | Fiber bundle used to guide the 514-nm Ar laser line into rhodamine 6G and to collect fluorescence | 130 |
| Fiber optic ELISA | Comparison of fiber optic microliter readings with data of conventional spectrometer and ELISA reader | 131 |
| Dry reagent chemistry | Fibers used to measure fluorescence and reflectance from reagent strips | 132 |
| Review on various fields of application | Remote UV/VIS/NIR spectroscopy | 133 |
| Therapeutic drug monitoring | Remote measurement of the lifetime of a fumor drug | 134 |

far involve rapid reactions, so that mass transfer has been the dominant factor establishing response time. However, in enzyme-based sensors possessing chemical sensors as transducers the complex mass transfer and, occasionally, the slow enzyme kinetics can lead to considerably prolonged response times. A discussion of the definition of response times is given in chapter 8 on pH sensors.

## B. SENSORS VS. PROBES

In this book, a sensor is defined as a device capable of continuously and reversibly reporting the concentration of an analyte or a physical parameter. However, the reversibility of some so–called sensors is very poor or even nonexistent. Such devices are suitable for "single shot" assay in a fashion much like a dipstick test. It is suggested that these devices be referred to as probes, although it may be difficult in some cases to make a clear differentiation.

All fluorosensors based on excited state interaction of indicator and analyte (such as quenching) are, of course, fully reversible. In sensors based on ground state interactions, the reversibility depends on the binding constant (or the Stern–Volmer static quenching constant), the analytical range, and the relative concentrations (or amounts) of analyte (A) and reagent (R). Considering, for instance, the equilibrium constant for a reaction between A and R, we see that the equilibrium

$$A + R \rightleftharpoons AR$$

can be driven completely to the right side if [A] (the analyte concentration) is kept low and the equilibrium constant $K_e$ favors binding over dissociation. AR is the optically detected product. In such a "stochiometric" reaction,[14] the analyte is reacted with an excess of reagent and is completely converted into AR whose concentration is measured. The signal increases with time until all reagent is consumed. At this point, reaction has gone to completion and the probe has to be disposed or regenerated outside the sample by chemical means. Because it is the total amount of analyte that is determined this way, the volume of the sample needs to be known in order to determine the analyte concentration. Under conditions of unlimited supply of analyte, the relative signal change is proportional to the analyte concentration in case of sensors, while in probes it is the product of analyte concentration and time of exposure.

Typical probes include metal ion "sensors" with almost irreversible binding of metal ion by a ligand, all types of reservoir "sensors" (Section IV.F herein), all irreversible color reactions adapted to fiber technology, and practically all immuno-"sensors". Probes have serious disadvantages over sensors in that not only do they act irreversibly, but they also consume the analyte from a sample, report the integrated amount of analyte that has contacted the reagent since the sensor was first contacted with the sample, and are difficult to calibrate. Fully correctly,[14] it has been stated that "it is naive and incorrect to assume that all chemistries used for laboratory photometric analysis can be adapted for continuous *in situ* measurements using fiber optics."

The main field of application of probes will be in fairly reliable testing devices such as disposable kits which may be precalibrated or self-calibrating and can be evaluated with portable instrumentation. A major problem will result from the need to prepare probes with sufficient reproducibility. This imposes severe requirements on quality control.

Similar considerations apply to chemiluminescence probes, where the number of photons emitted per time depends on the number of analyte molecules, available reagent molecules, and the rate of reaction. Only if reagent is in excess, the steady-state intensity is a measure of analyte concentration. One way to maintain a fairly constant reagent supply is provided by so-called reservoir sensors (Section V.G herein).

In true sensor applications (as opposed to probes), the concentration of indicator is much lower than that of the analyte and the analytical range is near the equilibrium constant. For example, the pH range sensed by an acid-base indicator depends on the $pK_a$ of the indicator. In order to cover a reasonable analytical range, the binding constant should not be too large, i.e., binding should not be almost irreversible.

Indicators having such properties are strongly preferred in optical sensing by virtue of their continuous and reversible response. The resulting optrodes fulfill the basic requirements of a sensor, and do not perturb the sample. The equilibrium response of these indicators does not depend on mass transfer (although the time required to reach equilibrium is mass-transfer-dependent).

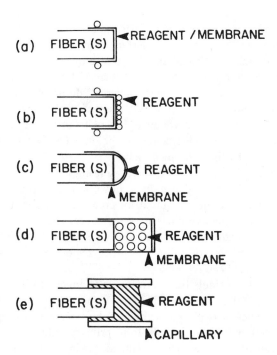

FIGURE 12. Typical extrinsic configurations with an immobilized reagent on the waveguide.

With the exception of pH, these indicator types have not generally been used for quantitative analysis because they are subject to error by small sample-to-sample variations which shift the equilibrium constant, and do not offer the same sensitivity as "irreversibly" binding indicators. In fact, it is the fully reversible formation of a complexed species in the presence of the uncomplexed ligand that provides the unique possibility of optically determining the ratio of the two species.

Since ratio methods are among the best methods for internal referencing (Section VIII herein), the resulting sensors are inherently more stable with respect to drifts. Unlike in probes, the ratio of the concentrations of analyte and the analyte-ligand complex is independent of the amount of ligand (because $[A] < [L]$), and the analyte concentration can be determined without knowledge of the sample volume.

Conceivably, this can become the basis for the design of precalibrated sensors based on intensity ratio measurements. Some of these concepts are currently pursued by companies active in the optical sensor field. Ideally, the calibration curve (and its variation with time) is provided by the manufacturer in the form of a bar code or by electronic means, and the analyst can proceed directly to analysis. This is not practical with currently available electrochemical sensors.

## C. SENSOR CONFIGURATIONS

There are two principally different ways to fabricate indicator-mediated fiber sensors. In the first approach, the chemistry is produced directly on the fiber. In the second, the sensor chemistry is first built up separately, and the material, in a final step, is placed on the fiber. The latter method appears to give a better reproducibility and has found much more widespread application.

Typical indicator-mediated sensor configurations are shown in Figure 12. In (a), the indicator is immobilized directly on a membrane positioned at the end of a fiber (bundle). Typical examples include indicators covalently immobilized on cellulose membranes (such as certain pH indicators) or electrostatically bound to ion exchange membranes. When the immobilized indicator is in form of a powder, it can be somehow attached to a membrane as shown in (b). Powder can be glued onto the surface of the membrane[14] or incorporated directly into the membrane during its fabrication.[15] The indicator may also be retained in position at the fiber end as shown in (c). It can be present there in homogeneous solution as long as it cannot pass the membrane. Thus, charged indicators[15] do not pass hydrophobic membranes and most dissolved enzymes[16] do not pass dialysis membranes. The membrane can also help to isolate the optical system from interferences by optical effects of the sample and to prevent certain (bio)chemical species to enter the reagent volume. On the other hand, the rate of mass transfer is slowed down by the outer membrane.

Another arrangement in which the powdered sensing material is physically retained in a tubular membrane permeable to the analyte is shown in Figure 12(d). When the diameter of the tube is small enough, the time required for complete mass transfer of analyte into the reagent phase is short and response times are satisfactory. The sensors described by Peterson in Chapter 16 are representative examples for this type. Despite of the obvious importance of the sensor geometry, modeling of the optical processes occurring in various reagent phase configurations has not yet been reported.

Arrangements for use in single fibers are shown in Figure 13. In (a), an indicator-loaded bead is cemented into a dip at the terminal end of the fiber. The bead is permeable to the analyte. In (b), the indicator is placed in front of the fiber which is spliced to allow better separation of reflected or fluorescent light from primary light, and to suppress fiber background. The two branches are usually not identical: the incoming arm is mostly smaller than the signal-collecting arm. In (c), the indicator is placed on the fiber whose clad was removed, and in (d) the chemistry is placed in a hole obtained by removing the core at the fiber end. This helps to protect a mechanically sensitive chemistry from physical damage.

Aside from positioning the sensing chemistry at the distal end of the fiber(s), it may be placed directly on the core after the clad has been removed. A particularly simple arrangement is shown in Figure 14(a). The source is an LED and the detector a phototransistor. No other optical component is required. Sensitivity is greatly enhanced by multiple internal reflections. Light attenuation by the colored reagent layer occurs because the wave evanescences from the core into the reagent. When the fiber end is mirrored as shown in Figure 13(b), detection can be at the same side as illumination. The clad is removed at the distal end only, which makes it easier to insert the sensor into a sample.

In all these configurations the indicator can be either immobilized directly on the fiber tip or core, or first onto a polymer which then is placed on the waveguide. Direct immobilization after chemical modification of the glass surface by silane coupling reagents gives very thin sensing layers with correspondingly short response times. The method has found application in pH and oxygen sensors. Polymer-linked indicators give thicker layers, longer response, and higher signal intensities. An interesting approach is to use an indicator-covered fiber whose emission excites the fluorescence of a doped fiber core. The working principle, shown in Figure 15, is that the sensing dye (such as an oxygen-sensitive fluorophore) is dissolved in the cladding and excited at $\lambda_1$. It emits fluorescence at $\lambda_2$ which is coupled into the core to excite a fluorescent-core fiber whose emission occurs at $\lambda_3$.

This two-stage approach has advantages over the direct evanescent field coupling method. First, coupling of light from the clad into the core is more efficient because light no longer just travels across the core and out the other side, but rather is reabsorbed and re-emitted by the core. Second, the emission of the $\lambda_3$ fluorescence is isotropically distributed. Finally, dyes can be attached to the core in very high concentrations. For a discussion of the potential of intrinsic fiber sensors, see Chapter 5.

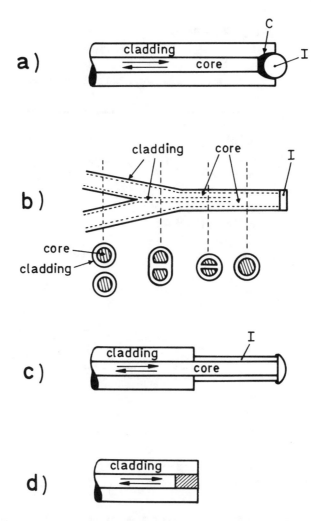

FIGURE 13. Single fiber configurations for indicator-based sensors. (a) An indicator-loaded bead I is cemented into the fiber end. (b) The indicator is placed at the fiber end, and the single fiber is spliced as shown in the cross-sections. (c) Indicator I is placed directly on the fiber whose cladding is partially removed. (d) The chemistry is placed in a hole drilled into the fiber end.

FIGURE 14. Schematic of an LED-based chemical sensor. C, capillary tube acting as a waveguide; R, thin reagent coating on the tube; PT, phototransistor.

Unlike the sensors treated so far, it is possible to prepare sensors where the *core* is the sensing material.[17] When placed on a suitable substrate, light can propagate directly through the reagent film which is effectively serving as the new "core" of a waveguide (Figure 16). This requires the sample index of refraction to be lower than that of the "old" core. The advantage of such a system is the high efficiency with which exciting light is used, the large sampling areas that can be achieved, and the possibility of working with sensing materials

**SENSOR DESIGN**

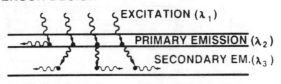

**PRINCIPLE:**

- **SENSING DYE IN CLADDING:**
  - **- EXCITATION AT** $\lambda_1$
  - **- EMISSION AT** $\lambda_2$
- **COUPLING DYE IN CORE**
  - **- EXCITATION AT** $\lambda_2$
  - **- EMISSION AT** $\lambda_3$

**ADVANTAGE:  INCREASED SIGNAL**

FIGURE 15.   Illustration of two-stage fluorescence coupling.

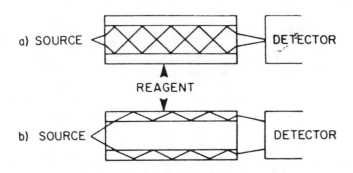

FIGURE 16.   Schematic of a chemical sensor in which the reagent is a thin coating on an optically transparent core. (a) The refractive index of the cladding is less than that of the core, and the reagent is probed by internal reflection spectroscopy. (b) The refractive index of the reagent is greater than the core, so the reagent layer acts as the waveguide.

containing very low amounts of indicators (some of which are poorly soluble). For a more detailed discussion, see Chapter 5.

Meserol et al.[18] have patented an interesting method for optical analysis of a sample in contact with a reagent gel on a solid support. The gel has an index of refraction larger than the support and the sample so that it can act as a planar waveguide. In the gel a colorimetric reagent is contained, e.g., an NADH-dependent enzyme, a chromogenic enzyme substrate, or an indicator. If the reagent is an antibody, its reaction with an antigen will result in an increase in light scatter.

Planar fluorosensors have been described[19] in which the sensing material acts as the planar waveguide and a series of photodetectors is placed beneath the waveguide. Light is effectively launched into the sensor layer with a grating coupler, and the emitted fluorescence is very efficiently detected by the diodes. In another configuration, a number of light sources and light detectors are placed on a plane support and covered with the analyte-sensitive fluorescent indicator layer. In one version, each light source is surrounded by a photodiode as the outer ring. Both sources and detectors are covered with appropriate interference filters and with analyte-sensitive indicator layers.

Kroneis and Offenbacher[20] have described an arrangement in which a light guide transports light from a source to a sensing material on the surface of the fiber. Because this layer has an index of refraction larger than that of the core, light couples into the sensing layer and excites the fluorescence of the indicator. Fluorescence intensity is measured using a photodetector placed on the other side of the fiber. Since fluorescence passes the fiber without being totally reflected, the photodetector sees practically no exciting light but rather only fluorescence.

An interesting sensing scheme was reported by Opitz and Lübbers.[21a] An electroluminescent foil (ELF) acts as a planar light source which shines onto a fluorescent sensing layer. An array of photodiodes, placed in a series of boreholes drilled into the ELF, detects re-emitted fluorescence of the sensing layer. ELFs are operated at about 115 V and 400 Hz and are available with blue, green, and yellow emissions. Signal-to-noise ratios of about 2000 were achieved.

## D. ABSORPTIOMETRIC DIRECT INDICATION

The absorption of a variety of indicators is modified by either pH changes, complexation by metal ions, or redox reaction. The resulting absorbance changes can be monitored. Most of these resulting sensors involve an equilibrium between analyte, A, and immobilized reagent, R. If the stochiometry of this reaction is 1:1, the reaction may by represented by A + R $\rightleftharpoons$ AR. The equilibrium is governed by the equilibrium constant, K, defined as

$$K = \frac{[AR]}{[A][R]} \tag{9}$$

which holds for homogeneous solutions and also for sensing layers with homogeneous distribution of indicator and analyte. It is assumed that activity coefficients in the immobilized phase are equivalent for R and AR and thus cancel out. AR and R vary with analyte concentration as follows:

$$[AR] = \frac{K[A]C_R}{(1 - K[A])} \tag{10}$$

$$[R] = \frac{C_R}{(1 + K[A])} \tag{11}$$

where $C_R$ is the sum of free and combined reagent molecules (AR + R). A plot of the response vs. analyte concentration for the case in which the measured optical parameter is proportional to [AR] is shown in Figure 17. At low concentrations, [A] is proportional to [AR] (i.e., absorbance of the complexed indicator), but at higher [A] there will be indicator saturation.

The alternative case in which the optical measurement is proportional to [R] is shown in Figure 17b. This situation is encountered with many phenolic pH indicators. In this case, increasing the analyte concentration (lowering the pH) leads to a decrease in measured parameter.

Equation 11 can be rearranged to give

$$\frac{C_R}{[R]} = 1 + K[A] \tag{12}$$

which is the equation for a colored dye whose absorbance is extinguished by the analyte. In the case of static fluorescence quenching, the equation is usually written as

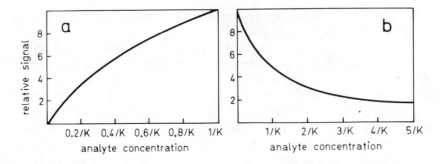

FIGURE 17.    Relation between relative fluorescence intensity, binding constant, and analyte concentration when (a) the reagent forms a fluorescent complex with the analyte and (b) the fluorescence of an indicator is statically quenched (e.g., by complexation with a transition metal ion).

$$\frac{I_0}{I} = 1 + K_s[A] \tag{13}$$

where $I_o$ reflects the concentration (intensity) of the fluorophore in the absence of the analyte, and I its intensity (concentration) in the presence of [A]. Both equations strikingly resemble the Stern-Volmer equation (Chapter 2, Equation 6). A plot of $C_R$/[R] vs. [A] will be linear with an intercept of one.

Many optical sensors are operated on a two-wavelength basis. If it is possible to selectively measure the optical properties of A and AR, the following equation

$$\frac{[AR]}{[R]} = K[A] \tag{14}$$

can be used to determine [A] directly. Since the ratio of AR to R is independent of $C_R$, the method is insensitive to slow loss of reagent. However, AR and R must be present in comparable concentrations to make measurement adequately precise. This limits the dynamic range. Aside from ionic species, the theory may also be applied to sense neutral species such as ethanol based on reversible binding to chromophores.[21b]

It should be noted that it is the equilibrium constant which governs the response function in these sensors and the analytical range of sensors. The conditions for optical sensing thus differ from the analytical situation where large equilibrium constants and reagent excesses are desired to drive a reaction to completion. If that were the case with a sensor, essentially all the analyte would be extracted into the solvent until the point where the reagent phase were saturated. The device would not function as a reversible sensor but rather as a probe. This not only holds for absorbance-based sensors, but also for fluorescent probes as discussed in the next section.

## E. FLUORIMETRIC DIRECT INDICATION

Fluorimetric indicators are probably the most popular ones in fiber sensing via indicators. This is due to the advantages of fluorimetry described in Chapter 2, where the relationship between analyte-concentration and fluorescence are given as well.

There are two kinds of fluorimetric direct indication. In the first, an indicator interacts with the analyte in a ground state process such as protonation or chelation. It is the *concentration* of one species of the complexed and uncomplexed forms that is detected. In the second kind, the indicator interacts with the analyte in the excited state by a so-called dynamic quenching process. It is the *quantum yield* or *lifetime* of the indicator that is monitored. The concentration

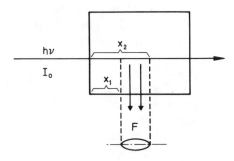

FIGURE 18. Demonstration of the inner filter effect in right angle fluorimetry.

of the indicator remains constant, and it does not undergo any spectral shifts or chemical modifications by complexation. These two situations will be discussed in Section V.E.1 and V.E.3 herein.

The selectivity of fluorosensing is known to be much better than absorptiometry and may further be improved by making use of time resolution, two-photon excitation,[22,23] or sequential excitation.

## 1. Indication Based on Ground State Interaction

Parker's law describes the relation between analyte concentration and luminescence intensity, but is an approximation (Chapter 2, Equation 5). In indicator-mediated fluorosensing, the high absorbances of indicator layers prevent the Parker law to be valid and require Equation 4 in Chapter 2 to be used instead. The mutual interaction between signal change, binding constants, and analytical range is very much like in the absorbance case (Section V.4 herein). It has been found, however, that these equations are valid only if the sensing membrane is not overloaded with indicator.[24]

Fluorescent indicators have been widely used in pH sensors (Chapter 8) and electrolyte sensors (Chapter 9). A selection of indicators is given in Chapter 7. Ground state indicators are more prone to interferences by the inner-filter effect than quenchable fluorophors (Section V.E.3 herein) because of the varying absorption of dye with varying analyte concentration.

## 2. The Inner Filter Effect

A major source of error in fluorimetry can result from the so-called inner filter effect, i.e., the absorption of part of the exciting light before it enters the zone viewed by the detector. When, for instance, exciting light of intensity $I_0$ leaves a fiber to excite the sample contained in a compartment as shown in Figure 18, and a second fiber views fluorescence F through a hole of width $(x_2 - x_1)$, the actual excitation energy $I_x$ at the center of the compartment will be smaller than $I_0$ because of partial light absorption over distance $x_1$. The correction factor to be applied to the observed fluorescence intensity F to give the value $F_0$ that would have been observed in the absence of an inner filter effect is now[25]

$$\frac{F_0}{F} = \frac{2.3 \cdot A\left(x_2 - x_1\right)}{\left[{}^{10}\exp\left(-Ax_1\right) - {}^{10}\exp\left(-Ax_2\right)\right]} \tag{15}$$

A is the absorbance per centimeter. A more detailed discussion of inner filter effects in fiber fluorimetry has been given recently,[26] and several useful mathematical treatments are available.[25,27-30]

The mathematics of a situation where bifurcated fiber optics are used for excitation and front-surface fluorescence collection in a 32-μl flow-through cell was described.[30] The inner filter effect correction function is formed by deconvoluting the fiber optic/cell light function with mathematical descriptions for primary and secondary absorbance effects. The absorption-corrected fluorescence of quinine sulfate was found to vary linear with concentration, despite primary self-absorption values extending to 3. The cell design of the fiber optic fluorimeter eliminates the collimation requirements of other techniques and thereby improves light throughput and signal-to-noise ratio.

Fluorosensing in combination with fiber light guides is usually performed under frontal illumination. Fortunately, inner filter effects are in some ways less serious with this method. Although the excitation spectra may become distorted at high absorbances, the emission spectra will only be scarcely so. A major advantage of front-face fluorimetry results from the fact that, unlike in right-angle fluorimetry, the exciting light can be totally absorbed by the sample, which in turn leads to higher emission intensities. There are no losses of excitation light as in right-angle fluorimetry.

Thus, when a catheter is inserted into a blood vessel, the exciting light will be totally absorbed within a few micrometers. Only a fraction of the absorbed light will reach the fluorophore, whereas most of the light will be lost to hemoglobin. The fraction of light that is exciting a fluorophore is given by the ratio between the absorbance of the fluorophore ($A_f$) and the total absorbance of the sample ($A_t$). A useful example for compensation of an inner filter effect in hemolyzed blood is given by Eisinger and Flores.[29]

### 3. Indication Based on Excited State Interaction

Aside from fluorogenic reagents which undergo a *ground state* reaction, a large group of fluorophors is known which undergoes fluorescence quenching by virtue of an *excited state* interaction with the analyte. The difference between the two processes is best illustrated by looking at the lifetime characteristics of the two processes (Figure 19). Ground-state interaction (static quenching) does not influence the lifetime of the fluorophore, but rather its emission intensity. Excited state interaction (dynamic quenching), in contrast, reduces the lifetime.

Dynamic quenching obeys the so-called Stern-Volmer law, provided that it is exclusively dynamic ("collisional"). The relation between quencher concentration [Q] and measured luminescence intensity I is given by

$$\frac{I_0}{I} = \frac{\tau_0}{\tau} = 1 + K_d[Q] = 1 + k_q \tau_0[Q] \tag{16}$$

with $I_0$ being the luminescence intensity of the indicator in the absence of any quencher and $K_d$ the overall ("Stern-Volmer") quenching constant. $\tau_0$ and $\tau$ are, respectively, the lifetimes in the absence and presence of quencher. $K_d$ can be shown to be the product of the lifetime $\tau_0$ of the fluorophore in the absence of quencher and the bimolecular quenching constant $k_q$:

$$K_d = k_q \cdot \tau_0 \tag{17}$$

Luminescence can consequently be expected to decrease with (1) increasing analyte (= quencher) concentration, (2) increasing indicator decay time, and (3) increasing $k_q$. Since $k_q$ is viscosity-dependent, quenching efficiency can be governed to some extent by proper choice of solvent viscosity.

The bimolecular quenching process proceeds via encounters between analyte and fluorophore, each encounter consisting of perhaps a hundred actual collisions, because the surround-

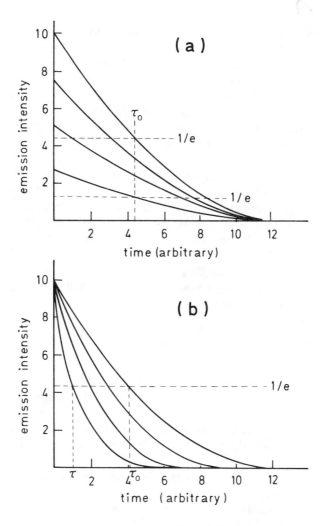

FIGURE 19. Fluorescence decay curves in case of (a) static and (b) dynamic fluorescence quenching. (a) The total intensity is diminished, but $\tau_0$ remains unchanged. (b) The initial intensity remains constant, but the decay is faster so that $\tau_0$ is reduced.

ing solvent molecules act as a kind of cage, preventing the separation of reactants after each collision. If the activation energy ($\Delta E$) for the reaction is high, the observed rate will be given by the product of the collision rate, a steric factor, and the Boltzmann factor $e^{(-\Delta E/RT)}$. If the activation energy is low, reaction will take place on each encounter, and the rate of reaction will be controlled solely by the rate at which reactants can diffuse together. This rate depends only on viscosity and temperature.[31]

Stern-Volmer constants can be obtained from plots of $I_0/I$ or $\tau_0/\tau$ vs. [Q], which should result in a straight line of slope $K_d$ and an intercept of 1. It is useful to note that $1/K_d$ is the quencher concentration at which half the maximal fluorescence can be observed (i.e., when $I_0/I = 2$). Linearity between [Q] and $I_0/I$ is occasionally lost when Q is present in high concentrations because of both static and dynamic quenching. This situation will be discussed later. Deviations from the linear Stern-Volmer relation can, in certain cases, be predicted.[32,33] When the fluorophores and quenchers are ions, the solution dielectric constant and ionic strength can also influence the deviation.

The most important application of dynamic quenching is in oxygen sensing and the determination of $I_o$ and $K_d$ shall be exemplified for this case.[34] A two-point calibration is required at least. For calibration at a zero oxygen concentration, water purged with pure nitrogen can be used together with another solution of defined oxygen concentration $[Q]_x$, e.g., water equilibrated with air. In practice, difficulties may be sometimes encountered in establishing a linear relationship between $I_o/I$ and the oxygen concentration $[Q]$. This is mostly due to the fact that a constant level of oxygen-independent light intensity is passing the filter system. Origins of, and methods for reducing, such "false light" are described in Section IX herein.

The intensity of this false light ($I_f$) can be determined by using a blank sensor containing no fluorescent indicator. However, because of the various thicknesses of the sensors, the sensitivity of the straylight intensity towards small positional changes and some stretching of the sensors in the flow-through cell, this approach is of little practical value. A more attractive approach is to determine the false light together with $K_d$ by means of a three-point calibration, as described below.

As the apparent value, $I_o'$, measured by the instrument, is composed of the "true" value $I_o$ and a certain amount of false light, we can set $I_o = (I_o' - I_f)$ and, by analogy, I equal to $(I' - I_f)$. Equation 16 can now be rewritten as:

$$\frac{\left(I_0' - I_f\right)}{\left(I' - I_f\right)} = 1 + K_d[Q] \tag{18}$$

$I_o'$ and $I'$ are the intensities measured by the detector in the absence and presence, respectively, of the quencher at a concentration $[Q]$; $I_f$ is considered to be independent of the concentration of the quencher.

Values for $I_f$ and $K_d$ were calculated from the intensity values $I_1'$, $I_2'$, and $I_3'$, obtained from three measurements at three different quencher concentrations $[Q]_1$, $[Q]_2$, and $[Q]_3$, respectively.[34]

In practice, for tonometered (gas-equilibrated) liquids, it is difficult to obtain a calibration point for an absolutely oxygen-free aqueous solution ($I_o$). The zero oxygen value can, therefore, be calculated preferentially from the following equation, which can be derived[34] from Equation 18:

$$I'_0 = I_1 + I'_1 K_d[Q]_1 - I_f K_d[Q]_1 \tag{19}$$

Static quenching, in contrast to dynamic quenching, results from complexation between analyte and indicator in the ground state. Typical static quenchers are oxygen at high pressure[31] and some ions which form ion pairs with fluorophores.[35] From a formal point of view, static quenching also includes the extinction of the fluorescence of a pH indicator by protons.

The effect of static quenching on luminescence decay curves is shown in Figure 19. The relation between quencher concentration $[Q]$ and fluorescence intensity is described by a Stern-Volmer type of equation

$$\frac{I_0}{I} = 1 + K_s[Q] \tag{20}$$

with $K_s$ (the state quenching or "binding" constant) replacing $K_d$ of Equation 16. $K_s$ is defined as the ratio of the concentrations of the complex between reagent or indicator R to the product of $[R]$ and $[Q]$ (see also Equation 12). It is useful to note that, in case of exclusive dynamic quenching, $\tau_o/\tau$ is $(1 + K_d[Q])$, but in case of exclusive static quenching it is 1.

Despite the formal analogy between the equations describing dynamic and static quenching, there are several aspects which demonstrate the entirely different nature of the two processes and may serve to differentiate between these:

1. The temperature coefficients are opposite. Dynamic quenching becomes more efficient with increasing temperature, whereas in static quenching the reverse is usually the case, since the stability of most complexes decreases with temperature. Since temperature also lowers the viscosity of the solvent, an additional increase in dynamic quenching efficiency with increasing temperature will be observed. The increase is proportional to $T/\eta$.

2. Since collisional quenching only affects the excited states of the fluorophores, it will have no effect on the absorption spectra. Static quenching, which is a ground state process, will lead to a perturbation of the absorption spectrum of the fluorophore. The changes are mostly small, but can be recognized by subtraction of the spectra obtained in the presence and absence of a quencher.

3. The lifetime $\tau$ of a fluorophore is not changed after addition of a static quencher. In the case of dynamic quenching, however, $I_o/I = \tau_o/\tau$. Lifetime measurements are therefore a most definite method to distinguish between the two kinds of quenching processes.

A fluorosensor for mercury ion, based on efficient static quenching of indole immobilized on quartz, may be considered as a typical representative for a sensor based on static quenching.[26] That the process is static is demonstrated by the practical invariance of $\tau_o$ of indoles with mercury(II) concentration.

### 4. Combined Static and Dynamic Quenching

It may be anticipated that, to a certain extent, all quenching processes proceed both statically and dynamically. Frequently, however, one of the two so much prevails over the other that one can speak of exclusive static or dynamic quenching. On the other hand, a number of indicators are known to be quenched via both mechanisms at a comparable rate, a fact that can result in positive deviations of otherwise linear Stern-Volmer plots. The following describes how to obtain the kinetic data for the two processes.[31]

The dynamic portion of the observed quenching can be determined by lifetime measurements, since

$$\frac{\tau_o}{\tau} = 1 + K_d[Q] \tag{21}$$

The quenching constant for the static process cannot be extracted from a single intensity measurement, since this will provide an apparent (composed) quenching constant $K_{app}$, defined as

$$K_{app} = K_d + K_s \left(1 + K_d[Q]\right) \tag{22}$$

where $K_d$ and $K_s$ are the quenching constants for the dynamic and static process, respectively. The relation between fluorescence intensity $I_o$, $K_{app}$, and analyte concentration $[Q]$ can be described by

$$\frac{I_0}{I} = 1 + K_{app}[A] = 1 + \left(K_d + K_s\right)[Q] + K_d K_s [Q]^2 \tag{23}$$

with

$$K_{app} = K_d + K_s + C_d K_s [Q] = \frac{I_o/I - 1}{[Q]} \tag{24}$$

Plotting $K_{app}$ as calculated at various quencher concentrations with the help of Equation 24 against [Q] results in a straight line with an intercept I of $(K_s + K_d)$ and a slope S of $K_s K_d$. Therefore, $K_s$ and $K_d$ can be calculated from

$$K_s^2 - K_s I + S = 0 \qquad (25)$$

$$K_d^2 - K_d I + S = 0 \qquad (26)$$

There are two solutions for each of these quadratic equations. The correct value for $K_d$ can be obtained independently from lifetime measurements.

Another modified Stern-Volmer equation that can be used to determine the dynamic quenching constants is given[31] by

$$\frac{I_0}{I} = \left(1 + K_d [Q]\right) e^{[Q] v N / 1000} \qquad (27)$$

where $v$ is volume of sphere within which the probability of immediate quenching is 100%, and N is Avogadro's number. For a more detailed discussion of the various mechanisms of quenching processes see Reference 26.

Numerous fluorescent indicators are known to be quenched by more than one analyte. This limits the selectivity of fluorosensors. It has been shown that interferences by other quenchers can be eliminated by simple mathematical methods in combination with a multiple-sensor technique.

If the fluorescence of an indicator is quenched by several quenching analytes present in concentrations $Q_1$, $Q_2$, ..., $Q_n$, their contributions to the overall quenching process has to be taken into account by adding further terms to the Stern-Volmer equation:

$$\frac{I_0}{I} = 1 + {}^1K[Q_1] + {}^2K[Q_2] + {}^3K[Q_3] + \cdots \qquad (28)$$

To calculate the concentration of n analytes, one needs n independent equations, obtained by measuring $I_0/I$ with n indicators. All K's are constants, which have to be determined once from plots of $I_0/I$ vs. [Q] for each indicator-quencher combination.

For the more simple and frequent case of two quenchers being present in concentrations $[Q_1]$ and $[Q_2]$, a simple mathematical expression can be given.[36,37] These are generally applicable to the quantitation of a variety of quenchers (ions as well as uncharged molecules), provided that they act independently. The advantage of this approach relies on the fact that specific indicators are no longer necessary. Among the analytically important quenchers that can be determined by this multiple sensor technique, mention should be made of oxygen, sulfur dioxide, dinitrogen monoxide, iodine, acrylic acid derivatives, halothane, and olefins, all of which unselectively quench the fluorescence of most polycyclic aromatic hydrocarbons.

In the two-sensor technique, it is not necessary that there be two different indicators with their different excitation and emission maxima which complicate the optical system of a sensor. It is sufficient that the fluorescent indicator has two different quenching constants for the same analyte. This situation is simply achieved by dissolving the indicator in different solvents, for instance, in two different polymers. Aside from a more simple optical system, the use of one indicator instead of two requires the determination of three (rather than four) quenching constants, which improves the accuracy of the corresponding sensor.

In the most refined version, the sensing polymer membrane is covered with a polymer layer that is permeable for one quencher only. The second sensing membrane is identical with the first, except that it is not covered with the partially impermeable layer. As a result the mathematical procedure can be greatly simplified. The following equation has been derived:[37]

$$[Q_1] = \frac{(\alpha - \beta)}{{}^2K_a} \qquad (29)$$

where $\alpha$ and $\beta$ are the signals ($I_o/I - 1$) of the two sensors (A and B). ${}^2K_a$ is the dynamic quenching constant describing quenching of the analyte being present in concentration $[Q_1]$.

The equation allows the determination of one quenching analyte in the presence of an interferent, using a system of two sensing membranes, one of which is accessible to one analyte only. From a technical point of view, the manufacturing of such a sensor combination is very simple, since they differ only by a thin cover. The optical system can be kept simple because it is the same indicator whose fluorescence is measured in two sensing layers, and precision can be expected to be better because two (instead of four) quenching constants with their inherent errors have to be determined. The method has successfully been applied to determine oxygen, halothane, or both, using a fiber catheter.[37] It requires, however, the two analytes to have similar diffusion constants.

## F. INDIRECT INDICATOR-MEDIATED SENSING SCHEMES

No direct indicators are known for optical detection of acidic or basic gases and vapors. However, because they are capable of passing hydrophobic membranes, they can be assayed indirectly via the pH changes they can induce in an aqueous solution. For the important case of $CO_2$, the following chemical equilibria hold:

(1)     $CO_2(aq) + H_2O \rightleftharpoons H_2CO_3$     (hydration)
(2)     $H_2CO_3 \rightleftharpoons H^+ + HCO_3^-$     (dissociation, step 1)
(3)     $HCO_3^- \rightleftharpoons H^+ + CO_3^{2-}$     (dissociation, step 2)

These are governed by the following equilibrium constants:

$$K_h = \frac{[H_2CO_3]}{[CO_2]_{aq}} = 0.0026 \qquad (30)$$

$$K_1 = \frac{[H^+][HCO_3^-]}{[H_2CO_3]} = 1.72 \times 10^{-4} \qquad (31)$$

$$K_2 = \frac{[H^+][CO_3^{2-}]}{[HCO_3^{2-}]} = 5.59 \times 10^{-11} \qquad (32)$$

In most studies on $CO_2$ sensors, the total analytical concentration of carbon dioxide (i.e., $[CO_2]_{aq} + [H_2CO_3]$) has been related to the response.

The effect of varying buffer concentration on the response of a $pCO_2$ sensor is shown in Figure 20. Obviously, sensitivity and dynamic range of sensors based on such a scheme nicely can be adjusted via the capacity of the internal buffer. In addition, the slopes of the graphs are governed by the $pK_a$ of the indicator and ionic strength. Unfortunately, high buffer concen-

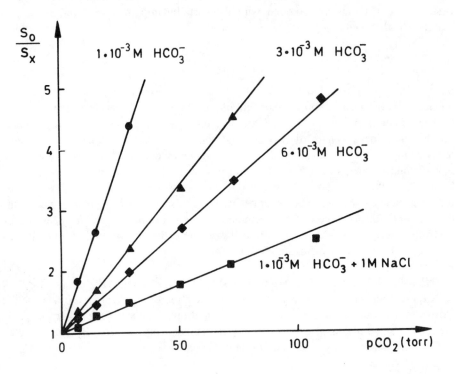

FIGURE 20.   Linearized calibration curves of a carbon dioxide sensor based on $CO_2$-induced pH changes of a buffer solution containing various amounts of hydrogen carbonate and NaCl. $S_o$ is the fluorescence intensity under zero $pCO_2$; $S_x$ the intensity under various $CO_2$ pressures.

trations result in very long response times. These are further prolonged with increasing thickness of the sensor layer (typically 20 µm thick), black poly(tetrafluoroethylene) covers acting as an optical isolation, and slow kinetics of the hydration of $CO_2$. Response is generally slower at low $pCO_2$ than at high $pCO_2$.

Zhujun and Seitz[38a] used HPTS in bicarbonate, covered with silicone, to sense $CO_2$. The equation that was used to relate the $CO_2$ partial pressure to hydrogen ion concentration is

$$\left[H^+\right]^3 + N\left[H^+\right]^2 - \left(K_1 C + K_w\right)\left[H^+\right] - K_1 K_2 C = 0 \tag{33}$$

where N is the internal bicarbonate concentration; $K_1$, $K_2$, and $K_w$ are the dissociation constants of carbonic acid and water, respectively; and C is the analytical $CO_2$ concentration, including both hydrated and unhydrated $CO_2$. It was shown that within a limited range, there is linearity between $CO_2$ pressure and [H⁺] according to

$$\left[H^+\right] = \frac{K_1}{CN} \tag{34}$$

In practice, the internal $HCO_3^-$ concentration should be such that the $CO_2$ concentrations of interest yield pH changes in a range where the internal pH sensor is most sensitive.

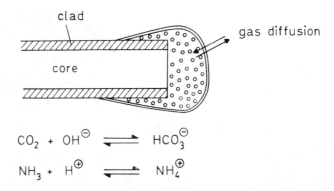

$$CO_2 + OH^{\ominus} \rightleftharpoons HCO_3^{\ominus}$$

$$NH_3 + H^{\oplus} \rightleftharpoons NH_4^{\oplus}$$

FIGURE 21. Schematic of a sensor for acidic or basic gases: a drop of silicone polymer at the fiber end contains an emulsion of a buffer whose internal pH is changed by the gas. The pH change is displayed by the color change of an optical indicator. The sensor is frequently covered with black or white material to prevent interence from extraneous light. White covers also act as reflecting layers in absorbance-based sensors.

Various configurations have been used. In one of those (Figure 21), the buffer is entrapped, as a fine emulsion, in a drop of silicone rubber that is placed at the fiber end. One limitation of these devices is the tendency to dry out, and the emulsion type usually needs to be stored under controlled and constant conditions. It tends to get destabilized and requires a very long time to produce a stable baseline because of the mechanical tension of the polymer on the water emulsion which makes diffusion of water into the droplets slow. Water can be a limiting species when excess $CO_2$ is supplied and a hydrophilic material is added to the buffer in order to prevent dry out.

In another configuration, a conventional pH sensor (see Chapter 10, Figure 21, and Chapter 19) is covered with a silicone film, rather than producing an emulsion in the silicone. The silicone is impermeable to protons, but permeable to the gas. The sensor has excellent response characteristics, but the main problem in covering pH sensors with hydrophobic membranes is to maintain long-term stability. This is treated in more detail in Chapter 11.

Aside from $CO_2$, various other acidic gases and vapors including $SO_2$, $NO_x$, HCN, $H_2S$, HF, acetic acid, and HCl can be assayed by this sensing scheme. They mutually interfere, but the cross-sensitivity towards the weakly acidic gases can be kept small in case of sensors for strongly acidic gases which require internal buffers of rather low pH. Thus, when the internal buffer of an $SO_2$ sensor is a 0.01 $M$ sodium hydrogen sulfite solution of pH 3.7, gases such as $CO_2$, HCN, and $H_2S$ practically would not interfere.

Similarly, various basic gases and amines can be assayed via measurement of pH in an internal buffer, with the pH shifting to higher values this time. Typical examples are sensors for ammonia, hydrazine, pyridines, and volatile organic amines. The bases mutually interfere, as do bases with sensors for acidic gases, and acidic gases with sensors for bases.

Another indirect sensing scheme has been introduced by the group of Simon.[38b] Since direct indicators for alkali and earth alkali ions exhibit both poor selectivity and sensitivity, the well-known high specificity of neutral ion carries (such as valinomycin) for metal ions was exploited to recognize such species. The transport of ions by such carriers into a pvc membrane was coupled to a proton transfer mechanism through ion exchange (ion in, proton out, and vice versa). This is a most useful transduction scheme with extremely broad applicability and can be extended to numerous other neutral ionophores, other spectroscopic techniques,[38c] and biosensing.[38d] Aside from ion exchange, co-extraction of an anion along with a cation is another promising sensing scheme.

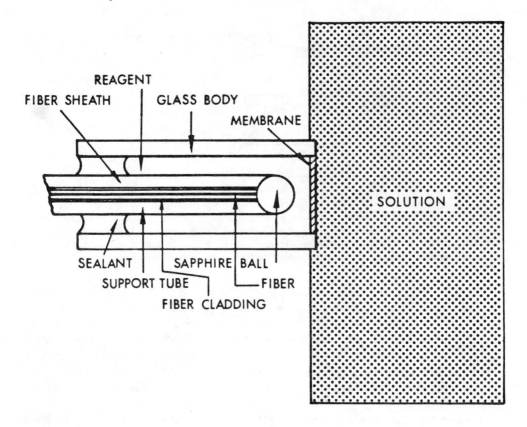

FIGURE 22.  Reservoir optrode in which the analyte in the solution penetrates through the membrane into the reagent compartment to form a colored or fluorescent species that is "seen" by the fiber.

## G. RESERVOIR SENSORS

In order to make a variety of irreversible chemical reactions applicable to continuous sensing, the so-called reservoir optrode has been developed. In this case, a chromogenic or fluorogenic reagent, or even acid or base, is continously added to the sensing solution seen by the waveguide. It is an extension of the methods used in dipstick and strip tests, with the product of reaction continuously being removed, and the reagent being supplied at a constant rate. In chemiluminescence (CL) "sensors" (which are of the reservoir type too), the reagent is mostly contained in excess in the sensing layer, or slowly allowed to diffuse into the reaction volume.

Typical configurations are shown in Figure 22 and in Chapter 12. The tip of the fiber is surrounded by a reagent solution which is in contact with the sample solution via a membrane. The chromogenic or CL reaction takes place at the membrane surface, or close to it. The optical signal is collected and guided back. Because of the minute size of the tip, the usage rate of reagent can be as small as 1 ml per month. However, any processes that may perturb the required careful control of mass transfer of both analyte and reagent are potential sources of error. Also, the product of reaction must be transported away from the sensor surface. Unless remote fiber sensing is essential in such an approach, the application of flow-injection analysis to such problems may be more effective and reliable.

Typical representatives of fiber reservoir optrodes are those for uranyl ion and polychlorinated organics (see Chapter 12), pH, certain gases,[38e-h] an ions,[38i] and some of the CL sensors mentioned in Chapter 2. In the former case, the weak intrinsic emission of uranyl ion is greatly enhanced by complexation with phosphate at low pH. This is achieved by slow and continuous addition of dilute phosphoric acid.

Generally, reservoir optrodes will always represent an interesting alternative to indicator phase sensors when the analyte does not exhibit native optical properties. Obviously, however, reactions that proceed very slowly or require drastic conditions, such as high temperature or concentrated acids or bases, are of limited value in reservoir sensors. Among the advantages over plain fiber sensors, improved selectivity and sensitivity are most noteworthy. In addition, the analytical wavelength (which is frequently determined by available laser lines) can be governed to some extent by a proper choice of reagents. The potential of the reservoir "sensor" principle for chemiluminescence "sensors" is evident.

# VI. SCHEMES FOR BIOSENSING

## A. ENZYME-MEDIATED BIOSENSORS

Most biomolecules of analytical interest do not have an intrinsic optical property that allows on-line monitoring with sufficient selectivity in a real sample. Thus, most sugars including glucose have no absorbance in the useful range, and their polarimetric determination is both unselective in the presence of other optically active species, and difficult because fibers do not easily maintain light polarization. The intrinsic fluorescence of biomolecules[3a9] such as proteins and nucleic acids is far in the UV and very poor for the latter. The same is true for a variety of clinically important metabolites. While oxidized nicotinamide dinucleotide ($NAD^+$) is nonfluorescent, the reduced form (NADH) displays strong fluorescence at around 450 nm when excited at 345 nm (see Chapter 17). In order to prevent leakage of NADH, has been immobilized on poly(ethylene glycol).[39b]

Similarly, the coenzyme flavine mononucleotide (FMN) and, less so, flavine adenine dinucleotide (FAD) have an intrinsic fluorescence at 520 nm when excited at around 450 nm. Both FMN and FAD change their fluorescence when reduced. This is the basis for a group of biosensors based on the action of an enzyme on a substrate (the analyte): the spectral properties of the coenzyme (NADH, FAD) can be monitored. NADH is produced from $NAD^+$ during the enzymatic action of dehydrogenases. This is discussed in detail in Chapter 17. The reduced form of FAD is produced as an intermediate only because it accepts oxygen as a substrate and thereby is oxidized to form FAD again. Hydrogen peroxide is released. In contrast to the NAD-associated enzymatic dehydrogenations, the reaction of FAD enzymes (oxidases) is therefore fully reversible in the presence of oxygen.[16] The typical response of an optical biosensor for lactate, based on the measurement of the intrinsic fluorescence of the enzyme, is shown in Figure 23.

A second sensing scheme for biomolecules is based on the consumption or production of easily detectable small molecules during enzymatic action. Thus, esterases produce acids, oxido-reductases consume oxygen, and decarboxylases and deaminases produce $CO_2$ and $NH_3$, respectively. Thus, by coupling one of the sensors for pH, oxygen, $pCO_2$ or $pNH_3$ to an enzyme with its usually outstanding selectivity, another type of biosensor is obtained. A cross section of a sensing layer of a glucose optrode with an oxygen optrode as the transducer is shown in Figure 24. The uppermost enzyme layer recognizes glucose and oxidizes it. The consumption of oxygen is recorded optically by the oxygen-sensitive material beneath the enzyme layer.

It is important to keep in mind that in enzyme based biosensors it is a kinetic (diffusion-controlled) equilibrium that is established, rather than a thermodynamic equilibrium. Consequently, the response of such sensors strongly depends on the diffusion rates of all species involved in the chemical reaction. More than in any other sensor type, the performance of such devices is affected by the choice of polymer materials, the kind of immobilization, analyte matrix effects, and temperature.

There are two kinds of limitations in the transducer approach, namely the varying background of the species tha measured (e.g., oxygen and the pH dependence of the relation

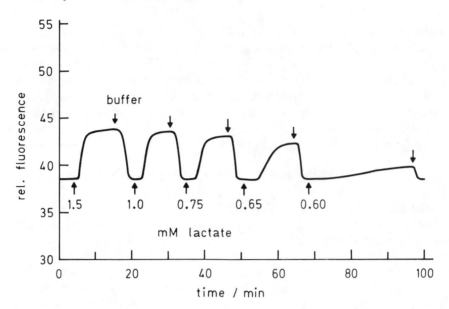

FIGURE 23.    Signal change, response time, reversibility, and dynamic range of an optical biosensor for lactate based on the changes in the intrinsic fluorescence of lactate mono-oxygenase. (From: Trettnak, W. and Wolfbeis, O. S., *Fresenius Z. Anal. Chem.*, 334, 427, 1989.

between $[CO_2]$ and $[HCO_3^-$ (and $[NH_3]$ and $NH_4^+$). Hence, when a pH sensor acts as the transducer in a glucose sensor incorporating glucose oxidase, the sample pH along with buffer capacity, oxygen supply, and ionic strength of the sample have to be kept constant in order to make the sensor specific for glucose. Even if an oxygen sensor is used as a transducer, the pH still must be kept constant because enzymes have a pH-dependent activity. This also holds for measuring $pCO_2$ or $pNH_3$. Ammonia, in particular, is a species that is not present at high levels in many biosamples and therefore is a most useful species to detect if the pH can be kept constant or its effect can be compensated for.

Sensors having oxygen optrodes as transducers do not suffer from the interferences by sample pH. However, variations in external oxygen level still can cause an apparent signal change. They can be taken into account using a second sensor having no enzyme and consequently responding to oxygen only. This has been shown for the specific case of an optical ethanol sensor where the oxygen consumption caused by the oxidation of ethanol by ethanol oxidase is measured[40] but is of general validity.

The Stern-Volmer equation predicts the following relations between optical signal and analyte concentrations: sensor A (containing no enzymes) responds as in Equation 35:

$$\frac{I_0^a}{I^a} - 1 = K_{sv} \cdot \left[O_2\right] \tag{35}$$

with $I_o^a$ and $I^a$ being the fluorescence intensities measured in the absence and presence of oxygen, respectively. $K_{sv}$ is the Stern-Volmer quenching constant, and $[O_2]$ is the concentration of oxygen.

The response of sensor B, i.e., the one containing the enzyme, can be described by a modified Stern-Volmer equation (Equation 36):

$$\frac{I_0^b}{I^b} - 1 = K_{sv} \cdot \left[O_2\right] - K_{sv} \cdot \left[O_2\right]' \tag{36}$$

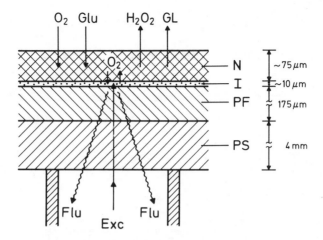

FIGURE 24. Cross-section through the sensing layer of a fiber optic glucose sensor having an oxygen-sensitive layer as a transducer. PS, plexiglass support; PF, polyester film (on which the layers are fabricated); I, indicator layer (decacyclene dissolved in silicone); N, nylon membrane with immobilized enzyme. The arrows indicate the diffusional processes involved: Glu, glucose; GL, glucono-lacton. Exciting light (Exc) hits the dyed silicone layer, and fluorescence (Flu) is emitted back into the fiber.

with $I_0^b$ being the fluorescence intensity of sensor B in the absence of both oxygen and ethanol, and $I^b$ being the intensity in the presence of certain levels of these species. $[O_2]$ is the respective oxygen concentration (the same as in Equation 35), and $[O_2]'$ is the difference in oxygen concentration that is caused by the oxidation of ethanol. The relation between $[O_2]'$ and [EtOH] can be described by:

$$\left[O_2\right]' = \frac{f[EtOH]}{2} \tag{37}$$

where factor f accounts for the unknown rate of conversion of ethanol into acetaldehyde and hydrogen peroxide, and factor 2 has to be introduced because 1 mol of oxygen is consumed when 2 mol of ethanol is oxidized. Equation 36 can therefore be written as

$$\frac{I_0^b}{I^b} - 1 = K_{sv} \cdot \left[O_2\right] - K_{sv} \cdot f \cdot \frac{[EtOH]}{2} \tag{38}$$

By setting $\alpha$ for $(I_0^a/I - 1)$ (which is the signal of sensor A) and $\beta$ for $(I_0^b/I^b - 1)$ (the signal of sensor B), and subtracting Equation 38 from Equation 35 we obtain

$$\alpha - \beta = \frac{K_{SV} \cdot f \cdot [EtOH]}{2} \tag{39}$$

Thus, the difference between signals $\alpha$ and $\beta$ is linearly related to the ethanol concentration via a constant c which is equal to $K_{sv} \cdot f/2$. A plot of $(\alpha - \beta)$ vs. [EtOH] should result in a straight line that goes through zero and has a slope K.

Using this calibration graph, the concentration of ethanol may be calculated from the two signals even under varying levels of oxygen. In fact, the two-sensor technique may be used for continuous recording of both oxygen and enzyme substrates such as glucose or ethanol,

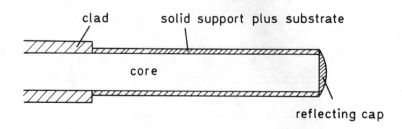

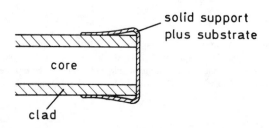

FIGURE 25. Fiber optic sensing of enzyme activities by monitoring the formation of a colored or fluorescent product by enzymatic action upon an enzyme substrate. The substrate is immobilized on a solid support which, in (A), forms the cladding of a waveguide, while in (B) it is fixed at the distal end of the fiber. (From Wolfbeis, O. S., in *Analytical Uses of Immobilized Biological Compounds for Detection, Medical, and Industrial Uses*, Guilbault, G. G. and Mascini, M., Eds., Reidel, Dordrecht, 1988, 219, which also shows further configurations.)

which is of considerable interest in biotechnology. Because of the mass transfer involved, very thin sensing layers are required to achieve fast response time. Langmuir-Blodgett films may be one solution[41] of the problem (see also Chapter 20). The most elegant approach to shuttle electrons from the substrate directly onto an electrode via a mediator,[42] thus circumventing the reduction of oxygen, cannot be applied to optrodes.

Aside from determination of the rate of formation of substrates (which is the basis for most present-day biosensors), enzymatic reactions have been used to quantify the activity of enzymes.[43] The rate of the formation of colored or fluorescent products from non-fluorescent enzyme substrates as a result of enzymic activity is used as the analytical information. The substrate can be immobilized on a membrane at the fiber tip, but other configurations have also been used,[44] some of which are illustrated in Figure 25.

In a method for the detection of enzyme inhibitors,[45] an enzymatic reaction is maintained by continously supplying a reaction coil with a red enzyme substrate. The enzyme hydrolyzes the substrate to give a blue product of hydrolysis. If the enzyme is inhibited, the blue product (whose absorption is followed using a yellow LED or a He–Ne laser) is no longer formed.

A promising sensing scheme was described by Burgess,[46] who covered one arm of the two fibers of a Mach-Zehnder interferometer with an enzyme over about 1 cm of its (uncladded) length. When the assembly is exposed to a flowing stream in which boluses of an enzyme substrate occur, the enthalpy of the reaction alters the propagation characteristic, producing a phase shift in the interferometric pattern. It has been used successfully in the peroxide-peroxidase and urea-urease systems.

Conceivably, enzymatic reactions may serve a fourth purpose in fiber sensing: if an enzymatic reaction is catalyzed by a metal ion, low concentrations of this ion may become rate-limiting. Usually, only catalytic amounts of metal ions such as Ca(II), Mg(II), or Zn(II) are required. This can form the basis for an extremely sensitive enzymatic assay of such ions.

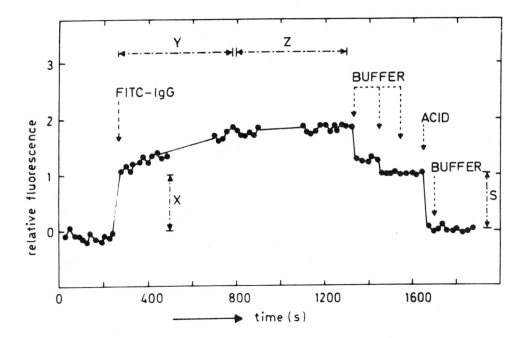

FIGURE 26.   Specific binding of FITC-labeled immunoglobulin G (IgG) by antiserum immobilized at a glass-liquid interface of a quartz slide. The binding process is monitored by the increase in fluorescence which is excited by the evanescent wave of the beam propagating in the slide. When labeled IgG is added, there is an increase in fluorescence (X) due to free molecules fluorescing close to the waveguide surface. Over distance Y, a plateau level Z is reached. After removing supernatant IgG with buffer, an intensity S remains which is a measure of specific binding. Treatment with strong acid (0.01 *M* hydrochloric acid) removes bound IgG and makes the probe reusable. (Redrawn from Sutherland, R. M., Dähne, C., Place J. F., and Ringrose, A. S., *Clin. Chem.*, 30, 1533, 1984.)

Strictly spoken, all these devices are probes rather than sensors because they require a continuous supply of reagent or can be operated only until reagent is exhausted. However, in case of some biosensors it is the ubiquitous oxygen which is a second substrate and whose supply practically never becomes limiting.

## B. SCHEMES FOR IMMUNOSENSING

Immunosensors, in contrast to enzyme-based sensors, rely on the establishment of a thermodynamic equilibrium between bound and free antibodies and antigens. The kinetic conditions of enzyme-based sensors (Chapter 16) thus strongly contrast the associative conditions of immunosensors treated in detail in Chapter 17.

The immunosensing schemes can be divided into two large groups, namely the no-label and the label immunoassays. No-label immunoassays utilize the intrinsic absorption or fluorescence of, or the scatter occurring at, the Ab/Ag complex. The intrinsic absorption can be due to tryptophane or an hapten such as benzopyrene or methothrexate. In a label immunoassay, the spectral properties of a label are monitored. Luminescent labels are used almost exclusively, and fluorescence intensity, lifetime, energy transfer, chemiluminescence, quenching, and electroluminescence have been applied. In Figure 26 it is shown how the fluorescence, as observed by the evanescent wave technique, increases when fluorescently labeled immunoglobulin binds to surface-adsorbed anti-IgG. Binding is very strong and can be reversed with strong acid only.

Other no-label immunoassays are based on light scattering (Chapter 2, Figure 18, and Chapter 18) or on bending effects of a fiber. Thus, specific macromolecules such as proteins and nucleic acids were detected[47] with a sensor comprising a fiber bent in the sensing region, a light source, and a detector. The radius of the curvature of the U-shaped fiber can be adjusted with a micrometer screw. The U-shaped part is covered with a reagent layer (such as an antibody), dipped into an analyte-free solution, and the amplifier adjusted to give a 100% transmission reading. When brought into contact with the analyte (resulting in a reaction in the reagent layer) the reading decreases in accordance with the analyte concentration.

The optical effect results from light leakage out of microbends (tight radius bends or crimps) in the fiber which may change the upper limit of total internal reflection. This immunosensor appears to be the only example of a microbend sensor.

In 1987, Badley et al.[48a] reported details of the fluorescence capillary-fill device (FCFD), a novel type of immunosensor. The devices resulted from an adaption of the technology used to mass-manufacture liquid-crystal display cells. On sample contact, a result is given in about 1 min. There is no need for operator attention, physical separation, and washing steps. FCFD consists of two pieces of glass separated by 100-$\mu$m gap. The lower plate is covered with an immobilized layer of specific antibody (Ab) and acts as an optical waveguide. the other plate has a soluble layer of antigen (Ag) labeled with fluorescent dye. The sample under test is drawn up by capillary action and dissolves the labeled reagent which then competes with the sample Ag for the limited number of Ag binding sites. The quantity of bound reagent is measured by evanescent wave fluorimetry. The technology lends itself to high-volume manufacture, the light system consisting of low cost components. First practical results have been published.[48b]

A related competitive binding assay has been proposed previously by Hirschfeld.[49] A fiber in the center of a cylindrical sample volume is coated with labeled Ag already bound to Ab. Labeled (Ag*) and unlabeled (sample) Ag compete for binding to Ab. The fluorescence of the Ab*/Ag complex is excited by the evanescent wave of the fiber and collected by it. The disadvantage of this approach is that the sample Ag must scavenge part of the preload Ag, a process that proceeds with rather low rate only.

Recently,[50] we have developed an optical immunoassay which can be applied to both planar and fiber waveguides. In the planar version, one piece of waveguide is covered with an Ab labeled with a fluorophore that is dynamically quenched by a quencher such as iodide deposited on the surface of the second plate in an arrangement not unlike the one of Badley et al. Formation of the Ab/Ag complex prevents the quencher from approaching the fluoro-phore and quenching it. Both the increase in fluorescence intensity and lifetime can be used as the analytical information.

One problem associated with optical immunoassays is the strong luminescence background of most samples of biological origin. This probably is the main reason why evanescent wave techniques are so popular: the evanescent wave "sees" only a small layer of an immobilized protein on the waveguide surface, but not the bulk solution above it. Because of the high specificity of the Ab/Ag interaction and the limited penetration depth of the lightwave, the assay can be very selective.

An even bigger problem at present is the strong binding between Ab and Ag, thus making the process almost irreversible and the device a one-shot probe rather than a sensor. It may be possible to reset the reagent phase to zero by removing it from the sample and exposing it to conditions where the binding reaction is reversed (e.g., in 0.1 N hydrochloric acid), but this is not a very elegant solution of the problem. Attempts have been made and are being made to reduce the strong binding of Ab to Ag, for example by introducing chemical or structural modifications at the binding sites.

## C. OTHER BIOSENSING SCHEMES

An interesting and potentially broadly applicable sensing scheme has been introduced by Schultz et al.[51] Fluorescein-labeled dextran competes with glucose for binding to concanavalin A which is immobilized on sepharose. A schematic of the device is shown in Figure 27. The end of a bifurcated fiber optic fits into a hollow fiber with a plug on the end. The substrate

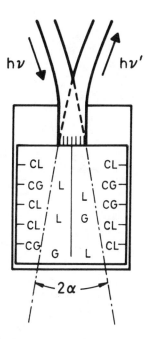

FIGURE 27.   Schematic of a glucose sensor based on competitive binding. The geometry of the cylinder at the fiber end (F) is chosen such that the fiber "sees", within its numerical aperture 2a, only the labeled dextrane (L) in solution, not the bound one. Concanavalin (C) is immobilized on the walls of the cylinder which, at the lower end, is permeable to glucose. Glucose (G) and L compete for binding to C. Increasing glucose levels displace more labeled dextran from C into solution, resulting in slow increase in fluorescence. The process is slow but reversible. (Redrawn from Schultz, J. S., Mansouri, S., and Goldstein, I. J., *Diabetes Care*, 5, 245, 1982.)

is fixed on the walls of the fiber and thus is in a position out of its numerical aperture. Therefore, it cannot be seen by the fiber. Glucose can freely diffuse through the hollow fiber or a dialysis membrane, which is impermeable to the large dextrane molecules. An increasing glucose concentration displaces the labeled dextrane, causing it to diffuse into the illuminated solution volume. Thus, fluorescence intensity as seen by the fiber follows the glucose concentration. This sensor principle is of particular interest because the design idea has broad potential for application to any analytical problem for which a specific competitive binding system can be devised. The method has been combined recently with energy-transfer techniques,[52a] but suffers from poor selectivity and slow response.

Provided receptors with sufficient selectivity are found and can be immobilized on waveguides, receptor-based sensing is one of the most promising detection methods. Although strongly dependent on its environment,[52b] such sensors provide unique possibilities. It is also known that some membrane-bound receptors can transport proteins through membranes, a process that can be coupled to a proton transport. Experiments have been performed in which lactose permease, incorporated in a lipid bilayer, was used to detect lactose. When contacted with lactose, the pH near the lipid bilayer increased and was monitored using a pH-densitive dye.[52c]

The first sensing schemes for differentiation between optical isomers of optically active compounds have been presented recently. Thus a sensing membrane was described that responds with some specificity to the (R)-form of the β-blocker propanolol.[53a] It is based on the use of an enantio-seletive lipophilic receptor (bis-tert.-butyl tartrate) contained in a pvc membrane along with a pH-sensitive lipophilic dye. When the receptor binds the β-blocker (which is present as a cation at physiological pH) and is transported into the pvc membrane, a proton is released from the pH-sensitive dye into the sample solution in order to maintain electro-neutrality. The dissociation of a proton from the dye causes its fluorescence to change.

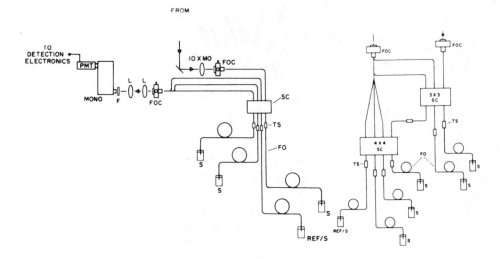

FIGURE 28.   Schematic diagram demonstrating the optical multiplexing of fiber waveguides by using a 4¥4 star coupler; PMT, photomultiplier tube; MONO, monochromator; FAC, fiber optic coupler; SC, star coupler; MO, microscope objective (10¥); FO, fiber optic; S, sample solutions; REF/S, reference solution.

This is the optical parameter that is detected. The sensor fully reversibly responds to propranolol over the 0.02 to 10 mM concentration range with selectivity factors ranging from 0.04 to 0.30. Lipophilic rhodamines were meanwhile found to be superior and more stable fluorophores. Similar results were obtained with other biogenic amines,[53b] and by using absorptiometric dyes along with other selectors.[53c]

# VII. MISCELLANEOUS SENSING SCHEMES

Optical wavelength division multiplexers/demultiplexers (WDM) have been used in telecommunications to increase the information capacity of single fiber optic systems. From a spectroscopic point of view, a WDM is a monochromator with fiber optic inputs and outputs. The stability, ruggedness, and compactness of these devices make them very attractive for use in on-line chemical process monitoring. Of the many WDM designs, the use of a dispersive element offers the advantage of a high channel capacity, while avoiding cumulative losses. A graded-index rod-prism grating approach has been used[54] to construct an inexpensive device with 11 input/output fibers covering a range of 90 nm. Coupled with a 35-element self-scanned photodiode array, the whole system is a compact spectrometer with multichannel analysis capability. The performance of this system operated in several spectroscopic modes using various fiber configurations was described.

The general concept of a multi-element optical waveguide sensor has been described.[55] The device consists of eight optical waveguides, each coated with a thin film known to react specifically with one or more components in a multicomponent system. An array of eight sequentially-activated light-emitting diodes is attached to the waveguide assembly in such a fashion as to activate each detection channel separately. Each waveguide is a fiber-optic coupled to a single high-gain, low-noise photomultiplier tube or photodiode/operational amplifier detector. The amplified signals can be displayed visually or input to a microprocessor pattern-recognition algorithm. Analog switches/multiplexers are used in feedback loops to control automatic gain-ranging, light-level adjustment, and channel-sequencing. Preliminary experiments involving the monitoring of redox/pH changes were discussed.

Optical multiplexing with an N × N fiber optic star coupler (with N = 3, 4, or 8) and two-photon excited fluorescence can be used to achieve multipoint measurements in highly absorbing environments. Differentiation of fluorescence signals from various sampling points can be attained by implementing the time-of-flight characteristics of fiber lightguides (see Chapter 2). Because the transit time of a light pulse through a fiber optic depends largely on the length of the waveguide, fibers of various lengths permit discrimination of the different sampling points in time. With the help of nanosecond time resolution it was possible[22] to determine the concentration information at several sensing locations simultaneously. Calibration graphs for a strong fluorophore were linear for all sizes of the star coupler, with submillimolar detection limits illustrated in the optical arrangement. The optical arrangement is illustrated in Figure 28.

Optical fibers have the advantage of providing what is essentially a one-dimensional measurement medium, allowing either line-integrations or line-differentiations to be performed over any chosen path. The attractions of these features are manifold. The line integrating property provides the means for attaining large sensitivities via long interaction paths. The line-differentiation property provides the means whereby the spatial distribution of the analyte (or temperature; see Chapter 15). Consequently, both the spatial and temporal variations may be determined. These sensors are referred to as distributed optical fiber sensors. Their applications in the physical sensor field are numerous, and distributed chemical sensing schemes are in the wake (Chapter 5). A review on present methods and physical applications has been given by Rogers.[56]

An effect not exploited so far for sensing purposes is the enhanced fluorescence from molecules deposited on the surface of an optical waveguide structure.[57] The enhancement is attributed to the near-field interaction between the molecules deposited on the waveguide and its modal fields. The process does not require the use of prism or grating couplers and can give rise to an enhancement factor in excess of 100.

Effects of gases on the transmissivity of liquid crystals are the basis for an optical gas detector.[58] The device consists of a sample cell containing a cholesteric liquid crystal (e.g., cholesteryl oleate) and a component for introducing a sample gas such as oxygen. A color change is effected by heating the liquid crystal; this is monitored via fiber cables.

The photoluminescence (PL) of semiconductor electrodes coated with palladium film is influenced by hydrogen gas. When the surface of an n-GaAs semiconductor is derivatized with a redox-active film such as ferrocene, its fluorescence (at 865 nm) is rapidly quenched by reductants such as iodine or bromine. Hydrogen, oxygen, CO, $CH_4$, and water vapor remain without effect. The gaseous reductant hydrazine, in contrast, rapidly reverses the effect of iodine. It was concluded that the quenching corresponds to oxidation of the ferrocene to give ferricenium iodide.[96] The effects can be used to optically sense hydrogen, iodine, bromine, or hydrazine (a rocket fuel).

The combination of optical spectroscopy and electrochemistry, often termed spectro-electrochemistry, has provided powerful tools for probing complex redox processes near the solution electrode interface. Optical fibers can be placed directly at the active surface of the working electrode so that the spectral properties of the compound under investigation can be monitored. The concept was demonstrated with molten salts, the ascorbate/dopamine model, and in isolated animal brain electrochemical studies[60] (see Table 3).

Fujiwara et al.[61] have studied the possibility of making the core of a fiber the sample cell of a photometer. When the core is filled with a solvent of high index of refraction (e.g., $CS_2$), waveguiding is observed. With a 5-m cell, 10 ng of iodine could be detected based on the absorption at 540 nm. Similarly, 0.4 pg perylene per milliliter could be detected via its intrinsic fluorescence in a 12-m fiber.

A new type of fiber optic chemical sensor implements either electrical spark or radiofrequency excitation of analyte species at the probe tip. The sensor head contains the excitation components and optics, an umbilical which houses electrical cables and gas lines, and fiber optics. The analyte emission is stimulated by either flame, radiofrequency, or spark. The probe is intended for use in monitoring vapors, aerosols, and groundwater contaminants such as polyhalogenated hydrocarbons.[62]

Rather than immobilizing chemically sensitive indicators on the fiber, it has been shown[63] that pH and $pO_2$ can be monitored in fermentation broths by simply adding the respective indicators to the sample. Effects of varying sample volume are compensated for by making a two-wavelength measurement. In addition, an integrated scatter scanning technique can be used to monitor all diameter *in situ*. Results have been obtained during baker's yeast fermentation and compare very favorably with electrodes.

Ion pair extraction analysis is based on the finding that certain organic cations or anions are soluble in water only, as long as they have small counterions. If, however, a large cation and a large anion are present simultaneously, the pair can be extracted into an organic phase. If one partner is a dye, and the extraction stochiometry is well defined (for instance cation:anion 1:1), the photometric determination of the dye allows quantitation of the uncolored counterion. This is the basis of most assays for charged detergents. It has also been applied to sense alkali ions as described in Chapter 9.

Charlton[64] has disclosed test means for determining ions (such as potassium) comprising a hydrophilic carrier matrix incorporated with finely divided globules of a hydrophobic vehicle. The latter contains an ionophore of the crown ether type, and a reporter substance which forms a colored complex with the ionophore/cation complex. Though not applied in combination with optical fibers, it easily could. Alternatively, the crown ether may be linked directly to a dye.[65] This is the basis for the commercially available Chromolyte® spectrophotometric Na and K assay.

Various nucleophilic gases including ammonia form strongly colored complexes with transition metal ions. Typically, an alkaline solution of copper(II) tartrate, contained in a $NH_3$-permeable hydrophobic polymer, when exposed to $NH_3$ in gases or liquids, results in a shift of the $Cu^{2+}$ absorption maximum from 705 nm to 640 nm. By monitoring the transmitted intensity at 580 nm, or the ratio of the intensities at 580 and 780 nm, using the respective LEDs, a $pNH_3$-dependent signal is obtained.[66]

## VIII. INTERNAL REFERENCING

The primary purpose for internal referencing is to compensate for changes in any of the variables other than analyte concentration that affect the value of the analytical signal. These include fluctuations in source intensity and color temperature, changes in either the amount of indicator (e.g., by leaching or bleaching) or the optical properties of the indicator phase (e.g., by swelling), changes in the "bending loss" of the fiber due to bends which change the angle at which the transmitted beam strikes the core/cladding interface, connector losses, and drifts in the photodetection and electronic amplification system. Other interferences may result from the system's sensitivity to physical parameters such as temperature, pressure, and vibration. In interferometric sensors, periodically repeating transfer functions are another source of error.

One of the advantages of the optical approach to sensing is the possibility and ease of relating analyte concentration to the ration of intensities at two different wavelengths or to a phase shift. A common method is to derive the analytical output from the ratio of the two signals, both of which are equally sensitive to the spurious interferences and, at the same time, unequally sensitive to the analyte. In most cases, at least one of the signals changes with analyte concentration, while the other either is constant or varies with analyte concentration in a different way than the analytical intensity. The two intensities are referred to as "analytical" and "reference" intensities, respectively.

Measurement of analyte-dependent lifetime is an intrinsically self-referenced method, while interferometry is only partly self-referenced. The simplest but least efficient reference method is to measure source intensity to compensate for source fluctuations. Rather than measuring the total lamp spectrum (which may vary, although its intensity remains constant at the analytical wavelength) it is obviously better to measure at the latter. A much more attractive approach is to use a reference intensity which follows the same optical path as the analytical intensity, since this will compensate for changes in the complete system.

Given the importance of internal referencing for precise measurements, a number of approaches has been presented. Jones and Spooncer[67] have described a useful method of referencing out adventitious intensity variations in an optical fiber intensity-modulated sensor ("shutter sensor") by measuring the transmitted light intensity at two wavelengths; one carries the signal information and the other is used to normalize the intensity. This can be utilized in reflectance-based sensors as well.

Back-scattered excitation radiation has been used as a reference in an oxygen sensor based on fluorescence (Chapter 16). Alternatively, a fluorophor that is insensitive to analyte concentration has been incorporated into the indicator phase to provide a reference signal,[68] or the intrinsic Raman or fluorescence intensity of the fiber may be exploited. In a single fiber probe for remote fluorescence measurements, the ratio of fluorescence intensity to Raman signal from the aqueous sample was found to be a useful internal standard.[69] Alternatively, fluorescently doped fibers may serve as an internal reference.[70] In sensors based on absorption, the reference signal can be obtained at a wavelength where no absorption occurs. Triple wavelength referencing is performed in some fiber sensors for measurement of blood oxygenation (see Chapter 19). However, these methods have their limitations when applied to indicator-mediated sensors because they do not account for loss of indicator due to leaching or decomposition, and are subject to error associated with differences in the wavelength dependence of any variable that changes with time. This plays, of course, no role in plain fiber sensors.

Where feasible, the best approach of internal referencing in indicator-mediated sensors is to incorporate the reference signal into the immobilized reagent itself. In pH indicators with their different absorptions (and, sometimes, emissions) for the acid and conjugate base form, luminescence intensity can be measured at two absorption, excitation, or emission wavelengths.[24,71] Measurement of reflectivity or fluorescence at the isosbestic wavelength of the reflectivity spectrum or excitation spectrum gives a pH-independent reference signal that perfectly reports any indicator concentration changes. This type of reference signal compensates not only for instrumental fluctuations, but also for any loss of indicator and variations in the optical properties of the reagent layer. In energy transfer-based sensors, where a pH-independent fluorescer (donor) transfers its energy to a pH-dependent fluorophor (acceptor), measurement of the ratio of emission intensities of donor and acceptor provides another means of self referencing. Because of the $R^{-6}$ dependence of the energy transfer (Chapter 2), it is necessary, however, that donor and acceptor concentrations vary in the same way.

Another possibility for two-wavelength referencing we found most useful[72] is provided by a new type of photodiodes ("color sensors") where two pn-junctions are incorporated in one chip. Shortwave light has access to the upper photodiode only, while the lower is reached by longwave light only. A typical spectral response is shown in Figure 29. The current ratio between the two photodiodes can serve as a signal that is independent of various sources of sensor signal fluctuations. The method is particularly suited for working at wavelengths above 500 nm.

However, in all these referencing schemes there is no compensation for changes in inner filter effects that are different at the two wavelengths. Moreover, because the index of refraction (and hence numerical aperture) is a function of wavelength, it stands to reason that effects such as fiber bending losses are not equal at all wavelengths. The result is that one can only hope to approach a perfect ratioing scheme by picking the two wavelengths as close together as possible.

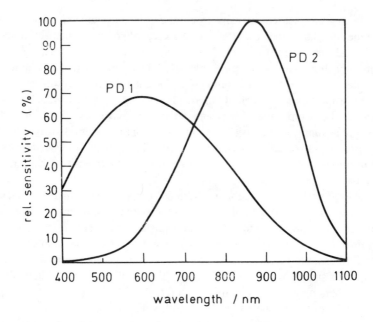

FIGURE 29.   Spectral sensitivity of the Sharp PD 150 color sensor at 25°C.

Lifetime-based sensors are intrinsically self-referenced in that (a) the phase of the exciting pulse is related to the phase of luminescence emission, (b) lifetimes are independent of indicator concentrations, (c) inner filter effects are minimized, and (d) lamp source and detector sensitivity fluctuations do not affect the measured phase shift. Both oxygen and pH (or $CO_2$) can be measured by lifetime as outlined in the respective chapters. Fiber optic lifetime measurements over large distances may, however, be distorted by modal and chromatic dispersion.

Optical waveguide chemical sensors with the cladding stripped off and replaced by a transducing or recognizing element are prone to interferences by changes in the refractive index of the medium to be sensed. This holds for both chemical sensors based on measurement of $n_D$ of an indicator layer and to evanescent wave-based sensors. To overcome this cross-sensitivity, changes in $n_D$ have been measured at a wavelength outside the analytical wavelength for the chemical transducer system, e.g., in the NIR using a IRED or, even better, a cw diode laser.[72]

Other methods of internal referencing rely on the coupling of light from one fiber into the other when the core distance comes to lie below the penetration depth of the evanescent wave ($d_p$). Various configurations of cross-talk two-fiber referencing have been proposed.[73,74]

## IX. REJECTION OF INTERFERING LIGHT

There are several sources of light that can interfere with the optical signal that is used to calculate a chemical concentration. If the background signal is constant, it may be tolerated provided it is not too high (i.e., less than about 10%). Unfortunately, even so-called "constant" background signals frequently turn out to be highly temperature-sensitive. In all cases where the background is variable, one has to provide means for compensating it.

The sources for background light are numerous: ambient light, light from the light source passing the blocking region of optical filters; cross-talk between fibers in fiber bundles; mirror-type reflection at the interface between lightguide and connectors and sensing layers; diffuse reflection; Rayleigh, Mie, and Raman scatter; and intrinsic fluorescence of fibers and sensor materials. This light, to which I refer as "false light", can be much stronger than, e.g.,

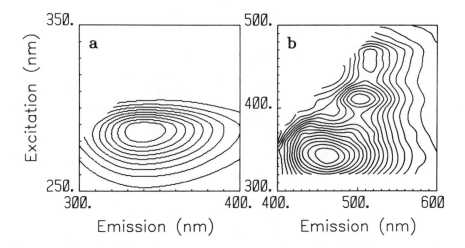

FIGURE 30. Total fluorescence of (a) the UV part and (b) the visible part of human serum at pH 8.11. Strong visible fluorescences at 460, 510, and 525 nm make fluorescent indicators having the same emission wavelengths prone to interferences by the intrinsic fluorescence of blood. (From Wolfbeis, O. S. and Leiner, M., *Anal. Chim. Acta*, 167, 203, 1985.)

fluorescence or Raman light. Raman light, on the other hand, can be a serious source of error because it occurs at wavelengths similar to those of fluorescence and passes emission filters without any losses being incurred. A method for the determination of the intensity of false light (which interferes in the determination of quenching constants) is described in Section V.E herein. Finally, the intrinsic fluorescence (or similar optical properties) of the sample may interfere in indicator-mediated sensors.

There are three main methods for suppression of false light, namely electronic subtraction, spectroscopic differentiation between light of different origin, and mechanical means. Electronic subtraction is the method of choice for eliminating interference by ambient light. When the light source is pulsed, light intensity can be measured with the source on and off, and the difference can be taken as the optical signal. This works well as long as the level of ambient light is not too high. It also requires the light source to be rapidly pulsed. This is possible with, e.g., LEDs, xenon lamps, and most lasers. Continuously burning lamps can be pulsed using shutter wheels.

Among the spectroscopic means for false-light separation, mention should be made of fluorescence lifetime measurements. They can elegantly suppress all sorts of scattered light and short-lived fluorescence but can be prone to interference by continuous light background. It also helps to reduce luminescence background in Raman spectroscopy. Phase resolution methods appear to be particularly suited.[75] Demas and Keller[76] have compared phase resolution with other background suppression techniques in terms of (1) being able to readily achieve both the suppression of Raman in luminescence spectra and of luminescence in Raman spectra and (2) ease of operation and found it superior to all other methods. Particular attention has to be paid to fluorescence suppression in Raman spectrometry[77] (see also Chapter 2, Section VI).

A simple mechanical means for eliminating optical interferences from the sample is to cover the sensing layer with an optical isolation, i.e., an analyte-permeable but nontransparent membrane. An optically isolated glucose sensing layer is shown in Figure 24. While this slows down the response of the sensor, it prevents both the exciting light passing the sensor layer to hit the sample, and reflected or fluorescent light to enter the fiber optic. The nontransparent (usually white, red, or black) optical isolation also helps to reduce the amount of scattered light and to keep it at a constant level. The total fluorescence of human serum is shown in Figure 30 with its strong fluorescence of bilirubin at about the same wavelength (460/515 nm) as that of the popular but poor pH indicator fluorescein. In the absence of an optical isolation, high bilirubin levels can cause additional fluorescence at the same wavelength as the popular pH indicators fluorescein or HPTS and thus feign too high a pH.

Single fiber probes are most susceptible to straylight. A considerable reduction in straylight was accomplished[78] via a coupling cell which matches the refractive index of the optical fiber, hence minimizing reflections, and by polishing the terminal ends. An optical delay line based on high-frequency modulation further reduced straylight.

# REFERENCES

1. **Giallorenzi, T. G.,** Optical fiber sensor technology, *IEEE J. Quantum Electron.*, QE-18, 626, 1982.
2. **Smela, E. and Santiago-Aviles, J. J.,** A versatile twisted optical fiber sensor, *Sensors Actuators*, 13, 117, 1988.
3. **Coleman, J. T., Eastham, J. F., and Sepaniak, M. J.,** Fiber optic based sensor for bioanalytical absorbance measurements, *Anal. Chem.*, 56, 2246, 1984.
4. **Dakin, J. P. and King, A. J.,** Limitations of a single optical fiber fluorimeter due to background fluorescence, *IEE Conf. Publ.*, 221, 195, 1983.
5. **Miller, S. E. and Chynoweth, A. G., Eds.,** *Optical Fiber Telecommunications*, Academic Press, New York, 1979.
6. **Hirschfeld, T., Deaton, T., Milanovich, F., and Klainer, S. M.,** The feasibility of using fiber optics for monitoring groundwater contaminants, EPA Rep. AD-89-F-2 A074, 1983.
7. **Louch, J. and Ingle, J. D.,** Experimental comparison of single- and double-fiber configurations for remote fiber optic fluorescence sensing, *Anal. Chem.*, 60, 2537, 1988.
8. **Schwab, P., and McCreery, R. L.,** Versatile efficient Raman sampling with fiber optics, *Anal. Chem.*, 56, 2199, 1984.
9. **Plaza, P., Dao, N. Q., Jouan, M., Feurier, H. and Saisse, H.,** Simulation and optimization of adjacent fiber optic chemical sensors, *Appl. Opt.*, 25, 3448, 1988.
10. **McLachlan, R. D., Jewett, G. L., and Evans, J. C.,** U.S. Patent 4,573,761, 1986.
11. **Tromberg, B. J., Eastham, J. F., and Sepaniak, M. J.,** Optical fiber fluoroprobes for biological measurements, *Appl. Spectrosc.*, 38, 38, 1984.
12. **Sepaniak, M. J., Tromberg, B. J., and Eastham, J. F.,** Optical fiber fluoroprobes in clinical analysis, *Clin. Chem.*, 29, 1678, 1983.
13. **Place, J. F., Sutherland, R. M., and Dähne, C.,** Opto-electronic immunosensors: a review of optical immunoassays at continuous surfaces, *Biosensors*, 1, 321, 1985.
14. **Seitz, W. R.,** Chemical sensors based on immobilized indicators and fiber optics, *Crit. Rev. Anal. Chem.*, 19, 135, 1988.
15. **Wolfbeis, O. S., Weis, L. J., Leiner, M. J. P., and Ziegler, W. E.,** Fiber optic fluorosensor for oxygen and carbon dioxide, *Anal. Chem.*, 60, 2028, 1988, and references cited therein.
16. **Trettnak, W. and Wolfbeis, O. S.,** A fully reversible fiber optic glucose sensor based on the intrinsic fluorescence of glucose oxidase, *Anal. Chim. Acta*, 221, 195, 1989.
17. **Hardy, E. E., David, D. J., Kapany, N. S., and Unterleitner, F. C.,** Coated optical waveguides for spectrophotometry of chemical reactions, *Nature (London)*, 257, 666, 1975.
18. **Meserol, P. M., Prodell, R. C., Bernstein, P., and Gupta, G.,** European Patent Appl. 249,957, 1987.
19. **Marsoner, H., Kroneis, H., Karpf, H., Wolfbeis, O. S., List, H., and Leitner, A.,** European Patent Appls. 244,394, 1987 and 263,805, 1988.
20. **Kroneis, H. K. and Offenbacher, H.,** U.S. Patent 4,703,182, 1987.
21a. **Opitz, N. and Lübbers, D. W.,** Towards intelligent fluorescence optical sensors: optodes with integrated excitation and fluorescence detection on semi-conductor basis, *Biomed. Technol.*, 31, 122, 1986.
21b. **Seiler, K., Wang, K., Kuratli, M. and Simon, W.,** Development of an ethanol-sensitive optode membrane based on a reversible chemical recognition process, *Anal. Chim. Acta*, 244, 151, 1991.
22. **Steffen, R. L. and Lytle, F. E.,** Multipoint measurement in optically dense media by using two-photon excited fluorescence and a fiber optic star coupler, *Anal. Chim. Acta*, 215, 203, 1988.
23. **Steffen, R. L. and Lytle, F. E.,** Remote sensing in a dense environment by using two-photon excited fluorescence and a multimode fiber optic, *Anal. Chim. Acta*, 200, 491, 1987.
24. **Zhujun, Z. and Seitz, W. R.,** A fluorescence sensor for quantifying pH in the range from 6.5 to 8.5, *Anal. Chim. Acta*, 160, 47, 1984.

25. **Parker, C. A. and Barnes, W. J.,** Some experiments with spectrofluorimeters and filter fluorimeters, *Analyst*, 82, 606, 1957.
26. **Wolfbeis, O. S.,** Fiber optic fluorosensors in analytical and clinical chemistry, in *Modern Luminescence Spectrometry. Methods and Applications* Vol. 2, Schulman, S. G., Ed., John Wiley & Sons, New York, 1988, chap. 3.
27. **Lutz, H. P. and Luisi, P. L.,** Correction for inner filter effects in fluorescence spectroscopy, *Helv. Chim. Acta*, 66, 1929, 1983.
28. **Street, K. W. and Tarver, M.,** Interdependence of primary and secondary inner filtering, *Analyst*, 113, 347, 1988.
29. **Eisinger, J. and Flores, J.,** Front-face fluorometry of liquid samples, *Anal. Biochem.*, 94, 15, 1979.
30. **Ratzlaff, E. H., Harfman, K. G., and Crouch, S. R.,** Absorption-corrected fiber optic fluorometer, *Anal. Chem.*, 56, 342, 1984.
31. **Lakowicz, J. R.,** *Principles of Fluorescence Spectroscopy*, Plenum Press, New York, 1983.
32. **Keizer, J.,** Nonlinear fluorescence quenching and the origin of positive curvature in Stern-Volmer plots, *J. Am. Chem. Soc.*, 105, 1494, 1983.
33. **Cukier, R. I.,** On the quencher concentration dependence of fluorescence quenching, *J. Am. Chem. Soc.*, 107, 4115, 1985.
34. **Trettnak, W., Leiner, M. J. P., and Wolfbeis, O. S.,** Fiber optic glucose biosensor with an oxygen optrode as transducer, *Analyst*, 113, 1519, 1988; see the appendix.
35. **Koller, E., Kriechbaum, M., and Wolfbeis, O. S.,** 1-Amino-pyrene- 3,6,8-trisulfonate: a fluorescent probe for thiamine, *Spectroscopy*, 3, 37, 1988.
36. **Wolfbeis, O. S. and Urbano, E.,** Fluorescence quenching method for determination of two or three components in solution, *Anal. Chem.*, 55, 1904, 1983.
37. **Wolfbeis, O. S., Posch, H. E., and Kroneis, H. K.,** Fiber optic fluorosensor for determination of halothane and/or oxygen, *Anal. Chem.*, 57, 2556, 1985.
38a. **Zhujun, Z. and Seitz, W. R.,** A carbon dioxide sensor based on fluorescence, *Anal. Chim. Acta*, 160, 305, 1984.
38b. **Perisset, P. M. J., Hauser, P. C., Tan, S. S. S, Seiler, K., Morf, W. E., and Simon, W.,** An ion-selective photodiode (ISP), *Chimia (Switzerland)*, 43, 10, 1989.
38c. **He, H. and Wolfbeis, O. S.,** Fluorescence based optodes for alkali ions based on the use of ion carries and lipophilic acid/base indicators, *Proc. SPIE*, 1368, 165, 1990.
38d. **Wolfbeis, O. S. and Li, H.,** Optical urea sensor with an ammonium optrode as the transducer, *Anal. Chim. Acta,* in press, 1991.
38e. **Luo, S. and Walt, D. R.,** Fiber-optic sensors based on reagent delivery with controlled release polymers, *Anal. Chem.,* 61, 174, 1989.
38f. **Berman, R. J. and Burgess, L. W.,** Renewable reagent fiber optic based ammonia sensor, *Proc. SPIE,* 1172, 206, 1989.
38g. **Zhou, Q. and Sigel, G. H.,** Porous polymer optical fiber for carbon monoxide detection, *Proc. SPIE,* 1172, 157, 1989.
38h. **Momin, S. A. and Narayanaswamy, N.,** Optosensing of chlorine gas using a dry reagent strip and diffuse reflectance spectrophotometry, *Anal. Chim. Acta,* 244, 71, 1991.
38i. **Inman, S. M., Stromvall, E. J., and Lieberman, S. H.,** Pressurized membrane indicator system for fluorogenic fiber optic chemical sensors, *Anal. Chem. Acta,* 217, 249, 1989.
39a. **Wolfbeis, O. S.,** The fluorescence of organic natural products, in *Molecular Luminescence Spectroscopy. Methods and Applications,* Vol. 1, Schulman, S. G., Ed., John Wiley & Sons, New York, 1985, chap. 3.
39b. **Scheper, T. and Bückmann, A. F.,** A fiber optic biosensor based on fluorometric detection using confined macromolecular NAD$^+$ derivatives, *Biosensors Bioelectronics,* 5, 125, 1990.
40. **Wolfbeis, O. S. and Posch, H. E.,** A fiber optic ethanol biosensor, *Z. Anal. Chem.*, 332, 255, 1988.
41. **Schaffar, B. P. S. and Wolfbeis, O. S.,** New optical chemical sensors based on the LB technique, *Proc. SPIE*, 990, 122, 1989.
42. **Cass, A. E. G., Davis, G., Francis, G. D., Hill, M. A. P., Higgins, I. J., Plotkin, E. V., Scott, L. D. L., and Turner, A. P. F.,** Ferrocene-mediated enzyme electrode for amperometric determination of glucose, *Anal. Chem.*, 56, 667, 1984.
43a. **Wolfbeis, O. S.,** Fiber-optic probe for kinetic determination of enzyme activities, *Anal. Chem.*, 58, 2874, 1986.
43b. **Zhang, Z., Seitz, W. R., and O'Connell, K.,** Amylase substrate based on fluorescence energy transfer, *Anal. Chim. Acta,* 236, 251, 1990.
44. **Wolfbeis, O. S.,** The development of fiber optic sensors by immobilization of fluorescent probes, in *Analytical Uses of Immobilized Biological Compounds for Detection, Medical, and Industrial Uses*, Proc. NATO Workshop, Guilbault, G. G. and Mascini, M., Eds., Reidel, Dordrecht, 1988, 219.
45. **Wolfbeis, O. S. and Koller, E.,** Remote fiberoptic detection of inhibitors of the enzyme acetylcholinesterase, work presented at the NATO Conf. A Forward Look into Detection and Characterization of Chemical and Biological Species, Salamanca, Spain, April 1 to 5, 1989.

46. **Burgess, L. W.,** A Fiber Optic Interferometer for Enthalpimetric Sensing in a Flowing Stream, Ph.D. dissertation, Virginia Polytechnic Institute and State University, Blacksburg, 1984.

47. **Knorre, W., Bergter, F., and Gira, G.,** East German Patent 243,351, 1987.

48a. **Badley, R. A., Drake, R. A. L., Shanks, I. A., Smith, A. M., and Stephenson, P. R.,** Optical biosensors for immunoassays: the fluorescence capillary-fill device, *Philos. Trans. R. Soc. London Ser. B,* 316, 143, 1987.

48b. **Parry, R. P., Love, C. A., and Robinson, G. A.,** Detection of *Rubella* antibody using an optical immunosensor, *J. Virol. Methods,* 27, 39, 1990.

49. **Hirschfeld, T. E.,** U.S. Patent 4,447,546, 1984.

50. **Wolfbeis, O. S. and Koller, E.,** Fluorescence immunoassay based on fluorescence quenching, European Patent Appl., 349, 520.

51. **Schultz, J. S., Mansouri, S., and Goldstein, I. J.,** Affinity sensor: a new technique for developing implantable sensors for glucose and other metabolites, *Diabetes Care,* 5, 245, 1982.

52a. **Meadows, D. and Schultz, J. S.,** Fiber optic biosensors based on fluorescence energy transfer, *Talanta,* 35, 145, 1988.

52b. **Rogers, K. R., Waldes, J. J., and Eldefrawi, M. E.,** Effects of receptor concentration, pH, and storage in a fiber optic biosensor, *Biosensors & Bioelectronics,* 6, 1, 1991.

52c. **Jähnig, F.,** Max-Planck-Institute for Biology, Tübingen (Germany), private communication, 1991.

53. **Cram, D. J.,** On molecular hosts and guests and their complexes, *Angew. Chem.,* 100, 1041, 1988.

53a. **He, H., Uray, G., and Wolfbeis, O. S.,** An enantioselective optode for the β-blocker propranolol *Proc. SPIE,* 1368, 175, 1990.

53b. **He, H. Uray, G., and Wolfbeis, O. S.,** Enantioselective optodes, *Anal. Chim. Acta,* 246 (2) 251, 1991.

53c. **Holy, P., Morf, W. E., Seiler, K., Simon, W., and Vigneron, J. P.,** Enantioselective optode membranes with enantiomer selectivity for (R)- and (S)- 1-phenethylammonium ions, *Helv. Chim. Acta,* 73, 1171, 1990.

54. **Fuh, M. R. S. and Burgess, L. W.,** Wavelength division multiplexer for fiber optic sensor readout, *Anal. Chem.,* 59, 1780, 1987.

55. **Smardewski, R. R.,** Multi-element optical waveguide sensor: General concept and design, *Talanta,* 35, 95, 1988.

56. **Rogers, A. J.,** Distributed optical fiber sensors, *J. Phys. D,* 19, 2237, 1986.

57. **Holland, W. R. and Hall, D. G.,** Waveguide mode enhancement of molecular fluorescence, *Opt. Lett.,* 10, 414, 1985.

58. **Showa Electric Wire & Cable Co.,** Japanese Patent Appl. 85/85, 356, 1985.

59. **Van Ryswyk, H. and Ellis, A. B.,** Optical coupling of surface chemistry: photoluminescent properties of a derivatized GaAs surface undergoing redox chemistry, *J. Am. Chem. Soc.,* 108, 2454, 1986, and references cited therein.

60. **Van Dyke, D. A. and Cheng, H. Y.,** Fabrication and characterization of a fiber optic based spectro-electrochemical probe, *Anal. Chem.,* 60, 1256, 1988.

61. **Fujiwara, K., Simeonsson, J. B., Smith, B. W., and Winefordner, J. D.,** Waveguide capillary flow cell for fluorimetry, *Anal. Chem.,* 60, 1065, 1988, and references cited therein on previous work.

62. **Griffin, J. W., Olsen, K. B., Matson, B. S., and Nelson, D. A.,** Fiber optic spectrochemical emission sensors, *Proc. SPIE,* 990, 55, 1989.

63. **Junker, B. H., Wand, D. I. C., and Hatton, T. A.,** Fluorescence sensing of fermentation parameters using fiber optics, *Biotechnol. Bioeng.,* 32, 55, 1988.

64. **Charlton, S. C.,** European Patent Appl. 125,554 and 125,555, 1984.

65. **Dix, J. P. and Vögtle, F.,** Ion-selective crown ether dyes, *Angew. Chem. Int. Ed. Engl.,* 17, 857, 1978.

66. **Wolfbeis, O. S.,** unpublished results, July 1986.

67. **Jones, J. E. and Spooncer, R. C.,** Two-wavelength referencing of an optical fiber intensity-modulated sensor, *J. Phys. E.,* 16, 1124, 1983.

68. **Lübbers, D. W. and Opitz, N.,** German Offen. 2,720,370, 1978 and U.S. Patent 4,306,877, 1981.

69. **King, R. R., Driver, I., Dawson, J. B., Ellis, D. J., and Feather, J. W.,** Fiber optic probe for in-situ measurements, *Proc. SPIE,* 906, 150, 1988.

70. **Harjunmaa, H.,** European Patent Appl. 174,722, 1986.

71. **Lübbers, D. W., Opitz, N., Speiser, P. P., and Bisson, H. J.,** Nano-encapsulated fluorescence indicator molecules measuring pH and $pO_2$, *Z. Naturforsch.,* 32C, 133, 1977.

72. **Wolfbeis, O. S.,** unpublished results, September 1988.

73. **Ramakrishnan, S. and Kersten, R. Th.,** A multi-purpose cross-talk sensor using multimode optical fibers, in Proc. 2nd Int. Conf. Opt. Fiber Sensors (OFS 1984), Stuttgart, September 5 to 7, 1984; VDE Publ., Berlin, 1984, 105.

74. **Lew, A., Depeursinge, C., Cochet, F., Berthou, H., and Parriaux, O.,** Single mode fiber evanescent wave spectroscopy, in Proc. 2nd Int. Conf. Opt. Fiber Sensors (OFS 1984), Stuttgart, September 5 to 7, 1984, 71.

75. **McGown, L. B. and Bright, F. V.,** Phase-resolved fluorescence in chemical analysis, *Crit. Rev. Anal. Chem.,* 18, 245, 1987.

76. **Demas, J. N. and Keller, R. A.,** Enhancement of luminescence and Raman spectroscopy by phase-resolved background suppression, *Anal. Chem.,* 57, 538, 1985.
77. **Wirth, M. J. and Chan, S. H.,** Comparison of time and frequency domain methods for rejecting fluorescence from Raman spectra, *Anal. Chem.,* 60, 1882, 1988.
78. **Skogerboe, K. J. and Yeung, E. S.,** Stray light rejection in fiber optic probes, *Anal. Chem.,* 59, 1812, 1987.
79. **Stueflotten, S., Christensen, Iversen, S., Hellvik, J. O., Almas, K., Wien, T., and Graar, A.,** An infrared fiber optic gas detection system, *Proc. SPIE,* 514, 87, 1984.
80. **Alarcon, M. C., Ito, H., and Inaba, H.,** All-optical remote sensing of city gas through methane absorption, *Appl. Phys.,* B 43, 79, 1987.
81. **Tai, H., Tanaka, H., and Yoshino, T.,** Fiber optic evanescent wave methane gas sensor, *Opt. Lett.,* 12, 437, 1987.
82. **Inaba, H., Chan, K., and Ito, H.,** All-optical remote gas sensor system over a 20 km range, *Proc. SPIE,* 514, 211, 1984.
83. **Simhony, S. and Katzir, A.,** Remote monitoring of ammonia using a carbon dioxide laser and infrared fibers, *Appl. Phys. Lett.,* 47, 1341, 1985.
84. Japanese Patent 85/100,743, 1985, through *Chem. Abstr.,* 103, 98095, 1985.
85. Japanese Patent 84/183,348, 1984, through *Chem. Abstr.,* 102, 124,798, 1985.
86. **Allison, S. W., Magnuson, D. W., and Cates, M. R.,** Use of fiber optics for remote uranium hexafluoride laser-induced fluorescence measurements, *Proc. SPIE,* 380, 369, 1983.
87. **Kobayasi, T., Hirama, M., and Inaba, H.,** Remote monitoring of $NO_2$ molecule by differential absorption using optical fiber links, *Appl. Opt.,* 20, 3279, 1981.
88. **Nguyen, Q. D. and Plaza, P.,** Possibilities of remote and multisite analysis by laser Raman spectroscopy with optical fibers, *Analusis,* 14, 119, 1986.
89. **Saturday, K. A.,** Absorption cell with fiber optics for concentration measurements in a flowing gas stream, *Anal. Chem.,* 55, 2459, 1983.
90. **Mazè, G., Carin, and Poulain, M.,** Fluoride glass IR fibers in medicine, *Proc. SPIE,* 576, 10, 1985.
91. **Groll, P. and Römer, J.,** An optical fiber laser photometer for on-line measurements, *Anal. Chim. Acta,* 190, 265, 1986.
92. **Gomy, C., Jouan, M., and Dao, N. Q.,** A laser Raman method with fiber optics for monitoring an alcoholic fermentation, *Anal. Chim. Acta,* 215, 211, 1988.
93. **Wolfbeis, O. S.,** Fiber optic sensors in bioprocess control, in *Sensors in Biopress Control,* Twork, J. V. and Yacynych, A. M., Eds., Marcel Dekker, New York, 1990.
94. **Reardon, K. F., Scheper, T. H., and Bailey, J. E.,** Use of a sensor for measurement of the NAD(P)H-dependent culture fluorescence of immobilized cell systems, *Chem. Ing. Tech.,* 59, 600, 1987.
95. **Levy, R. L.,** A new fiber optic sensor for monitoring the composite curing process, *Polym. Mater. Sci. Eng.,* 54, 321, 1986.
96. **Schirmer, R. E.,** Remote optical monitoring of polymer processing over long fiber optic cables, *Adv. Instrum.,* 43, 831, 1988.
97. **Freeman, J. E., Childers, A. G., Steele, A. W., and Hieftje, G. M.,** A fiber optic absorption cell for remote determination of copper in industrial electroplating baths, *Anal. Chim. Acta,* 177, 121, 1985.
98. **N. N.,** The measurement of low levels of water in solvent, Appl. Note A1-387, Guided Wave Inc., Helsingborg (Sweden), 1987.
99. **N. N.,** Determination of alcohols in mixtures using fiber optic spectroscopy and partial least squares regression, Appl. Note A4-188, Guided Wave Inc., Helsingborg (Sweden), 1988.
100a. **N. N.,** Rapid determination of hydroxyl number by NIR analysis, Appl. Note A2-687, Guided Wave Inc., Helsingborg (Sweden), 1987.
100b. **Mc Peters, H. L.,** On-line analysis of polymer melt processes, *Anal. Chim. Acta,* 238, 83, 1990.
100c. **Parisi, A. F., Nogueiras, L., and Prieto, H.,** On-line determination of fuel quality parameters, *Anal. Chim. Acta,* 238, 95, 1990.
100d. **Matson, B. S. and Griffin, J. W.,** IR fiber optic sensors for remote detection of hydrocarbons, *Proc. SPIE,* 1172, 13, 1989.
100e. **Saggese, S. J., Shariahri, M. R., and Sigel, G. H.,** Evaluation of an FTIR-fluoride fiber system for remote sensing of combustion products, *Proc. SPIE,* 1172, 2, 1989.
101a. **N. N.,** Determining the composition of solvent mixtures with NIR fiber optic spectrophotometry, Appl. Note A3-987, Guided Wave Inc., Helsingborg (Sweden), 1987.
101b. **Doyle, W. M. and Jennings, N. A.,** FTIR chemical reaction monitoring using an *in-situ* deep immersion probe, *Spectroscopy,* 5(1), 34, 1990.
102. **Lund, T.,** Simple and sensitive in-situ algae fluorescence sensor based on fiber optics, *IEE Proc.,* 131, 49, 1984.
103. **Snow, J. W., Paton, B. E., and Herman, A.,** A fiber optic remote sensing head for in-situ chlorophyll A fluorescence measurement in phytoplankton, *Proc. SPIE,* 838, 285, 1988.

104. **Kakui, Y., Nishimoto, A., Hirono, J., and Nanjo, M.,** Underway analysis of suspended biological particles with an optical fiber cable, *Adv. Chem. Ser.*, 209, 275, 1985.

105. **Zhong, X. and Li, J.,** Optical fiber sensor for dust concentration measurement, *Proc. SPIE*, 838, 285, 1988.

106. **Chudyk, W. A., Carrabba, M. M., and Kenny, J. E.,** Remote detection of groundwater contaminants using far-UV laser-induced fluorescence, *Anal. Chem.*, 57, 1237, 1985.

107a. **Costello, D. J.,** PCT Patent WO 87/0092, 1987.

107b. **Baldini, F., Falcai, R., Bechi, P., Cosi, F., Bini, A., and Milanesi, F.,** A portable optical fiber sensor for entero-gastric reflux detection, *Proc. SPIE*, 1510, xxx, 1991; in press.

107c. **Poscio, P., Depeusinge, Ch., Emery, Y., Parriaux, O., and Voirin, G.,** Biochemical measurement of bilirubin with an evanescent wave optical fibre sensor, *Proc. SPIE*, 1510, xxx, 1991; in press.

108. **Mayevsky, A. and Chance, B.,** Intracellular redox state measured in-situ by a multichannel fiber optic fluorometer, *Science*, 217, 537, 1982.

109. **Briggs, J., Fisher, M. L., Ghazarossian, V. E., and Becker, M. J.,** Fiber optic probe cytometer, *J. Immunol. Methods*, 81, 73, 1985.

110. **Swatland, H. J.,** Fiber optic reflectance and autofluorescence of bovine elastine, *J. Anim. Sci.*, 64, 1039, 1987.

111. **Manuccia, T. J. and Eden, J. G.,** U.S. Patent 4,509,522, 1985.

112. **Doiron, D. R., Keller, G. S., Profio, A. E., and Fountain, S. W.,** Fiber optic delivery and detection system for HpD photodynamic therapy, *Proc. SPIE*, 494, 56, 1984.

113. **Böck J., Gersing, E., Sundmacher, F., and Hellige, G.,** Intravascular fiber optic detection of $D_2O$ concentration in blood, *Proc. SPIE*, 906, 169, 1988.

114. **Demers, D. R.,** U.S. Patent 4,432,644, 1984.

115. **Jin, Z. J., Chan, Ch., and Whitaker, C.,** Plasma emission spectroscopy with an optical fiber probe, *Rev. Sci. Instrum.*, 59, 427, 1988.

116. **Mills, J. C. and Hodges, R. J.,** Fiber optics expand the range of a direct-reading spark optical emission spectrometer, *Appl. Spectrosc.*, 38, 413, 1984.

117. **Faires, L. M., Bieniewski, T. M., Apel, C. T., and Niemczyk, T. M.,** Optical fibers for remote spectrometry of alkali elements in the dc arc, *Appl. Spectrosc.*, 39, 9, 1985.

118. **Koehler, H. H., Redhead, D. L., and Nelson, M. A.,** Electron beam diagnostics using optical fibers, *Proc. SPIE*, 404, 119, 1984.

119. **Bergmann, K., Engelhardt, R., Hefter, U., and Witt, J.,** A detector for state-resolved molecular beam experiments using optical fibers, *J. Phys. E.*, 12, 507, 1979.

120. **Kychakoff, G., Kimball-Linnè, M. A., and Hanson, R. K.,** Fiber optic absorption-fluoresence probes for combustion measurements, *Appl. Opt.*, 22, 1426, 1983, and *Combust. Sci. Technol.*, 50, 307, 1986.

121. **Thomas, L. C. and Adams, A. K.,** Detection of fluorescent compounds by modified flame photometric GC detectors, *Anal. Chem.*, 54, 2597, 1982.

122. **Vurek, G. G. and Bowman, R. L.,** Fiber optic colorimeter for submicroliter samples, *Anal. Biochem.*, 29, 238, 1969.

123. **Kawabata, Y., Imasaka, T., and Ishibashi, N.,** Ultramicro flow cell for semiconductor laser fluorimetry, *Talanta*, 33, 281, 1986.

124. **Gluckman, J., Shelly, D., and Novotny, N.,** Laser fluorimetry for capillary column liquid chromatography, *J. Chromatogr.*, 317, 443, 1984.

125. **Fjeldsted, J. C., Richter, B. E., Jackson, W. P., and Lee, M. L.,** Scanning fluorescence detection in capillary supercritical fluid chromatography, *J. Chromatogr.*, 279, 423, 1983.

126. **Variano, B. F., Brenner, A. C., and Daniels, W. B.,** High-pressure cell for luminescence studies of condensed phases at low temperatures, *Rev. Sci. Instrum.*, 57, 497, 1986.

127. **Donati, S. and Tambosso, T.,** A fiber optic colorimeter for LC of amino acids, *Proc. SPIE*, 990, 70, 1989.

128. **Ward, E. H. and Hussey, C. L.,** Remote acquisition of spectro-electrochemical data in a room-temperature ionic liquid, *Anal. Chem.*, 59, 213, 1987.

129. **Imasaka, T., Nakanishi, K., and Ishibashi, N.,** A couple of optical fibers for thermal length spectrophotometry, *Anal. Chem.*, 59, 1554, 1987.

130. **Reichert, W. M., Ives, J. T., and Suci, P. A.,** Emission spectroscopy using an air-cooled argon laser and an optrode-based UV-VIS spectrophotometer, *Appl. Spectrosc.*, 41, 1347, 1987.

131. **Ackerman, S. B. and Kelley, E. A.,** Probe colorimeter for quantitating ELISA and other colorimetric assays performed with microplates, *J. Clin. Microbiol.*, 17, 410, 1983.

132. **Howard, W. E., Greenquist, A., Walter, B., and Wogoman, F.,** Automated instrumentation for fluorescence assays on reagent strips, *Anal. Chem.*, 55, 878, 1983.

133. **Fitch, P. and Gargus, A. G.,** Remote UV-VIS-NIR spectroscopy using fiber optic chemical sensing, *Int. Lab.*, pg. 100, Sep. 1986.

134. **Takashi, K. and Minoo, O.,** German Patent 3,511,758, 1985.

Chapter 4

# GUIDED WAVE ELECTROMAGNETISM AND OPTO-CHEMICAL SENSORS

## Oliver Parriaux

## TABLE OF CONTENTS

# I. INTRODUCTION

This chapter makes a survey of the basic electromagnetic concepts applying in the field of guided wave opto-chemical sensors. Guided wave optics is not only adding novel, compact, flexible, and cheap optical element. It allows new system architectures and offers for all parts of the system new ways of thinking, new routes to explore, and new problems also. These possibilities have not by far been exploited yet. Some of the reasons are: the field is new, there is a lack of high spatial coherence visible semiconductor sources, and the transparency spectrum of oxide glasses is quite restricted. Betting that this potential will be taken advantage of one day not too far off, it is the aim of what follows to provide the scientists and design engineers coming into this field with some useful mental tools, scaling rules, and practical examples for the development of systems which are not just reduced size flattened bulk optic systems, though useful these may be.

The basic electromagnetic concepts, equations, and properties of planar and circular optical waveguides will be reviewed in Sections I to IV, with emphasis on aspects which will prove especially meaningful for opto-chemical sensors. Section V is an introduction to the perturbation analysis which is an extremely useful and practical design tool. Sections VI, VII, and VIII describe the three main groups of functions that optical waveguiding can perform in an opto-chemical sensor system: light conveyance, optical processing, and acquisition of the chemical measurand. Section IX concludes with some comments on waveguide fluorescence sensing.

## A. THE RESPONSE OF A MATERIAL UNDER OPTICAL EXCITATION

An opto-chemical sensor basically makes an analysis of the response of a physical or chemical system under optical excitation. The electromagnetic response of a material is described by means of the polarization $\mathbf{P}$ induced by the electric field $\mathbf{E}$ of the incident optical wave on the various groups of dipoles that can be excited. It is contained in the constitutive equation for the electric displacement field $\mathbf{D}$:

$$\mathbf{D} = \varepsilon_0 \mathbf{E} + \mathbf{P}_l + \mathbf{P}_{nl}$$

where $\quad \mathbf{P}_l = \varepsilon_0 \chi_{(0)} \mathbf{E}$ (1)

and $\quad \mathbf{P}_{nl} = \varepsilon_0 \left( \chi_{(1)} \mathbf{E} + \chi_{(2)} \mathbf{EE} + ... \right) \mathbf{E}$

are the linear and nonlinear polarization terms, $\chi_{(o)}$ and $\chi_{(1)}, \chi_{(2)} ...$ are the linear and nonlinear susceptibility tensors, respectively.[1] $\varepsilon_o = 8.854.10^{-12}$ CN$^{-1}$ m$^{-2}$ is the absolute permittivity.

If the incident field E is not too strong, all polarization terms involving $\mathbf{E}$ to the power larger than 1 can be neglected. This will be assumed hereafter. In addition, only isotropic materials will be considered. $\chi_{(o)}$ is thereafter a scalar and (1) writes $\mathbf{D} = \varepsilon_o \varepsilon_r \mathbf{E}$ where $\varepsilon_r = 1 + \chi_{(o)}$ is the relative permittivity which is a complex function possessing a real part $\varepsilon_{rr}$ and an imaginary part $\varepsilon_{rj}$ ($j = \sqrt{-1}$ ):

$$\varepsilon_r = \varepsilon_{rr} - j\varepsilon_{rj}$$ (2)

The refractive index $n$ of the material is defined as the square root of the relative permittivity: $n = \sqrt{\varepsilon_r} = n_r - j\, n_j$.

In not too lossy materials ($\varepsilon_{rj} / \varepsilon_{rr} \ll 1$), including low loss metals like silver, the real and imaginary parts $n_r$ and $n_j$ of the refractive index are:

$$n_r \cong \sqrt{\varepsilon_{rr}}\left(1+\frac{1}{8}\left(\frac{\varepsilon_{rj}}{\varepsilon_{rr}}\right)^2\right) \cong \sqrt{\varepsilon_{rr'}}$$

which gives                                                                                              (3)

$$n_j \cong \frac{\varepsilon_{rj}}{2n_r}$$

Chemical species or reactions show themselves as an alteration of the dielectric constant $\varepsilon_r$ of the host medium which affects the characteristics of a wave propagating through it. The electric field $E(z,t)$ of a monochromatic wave of angular frequency $\omega$ propagating along z has a time and spatial dependence in the form:

$$E(z,t) = E_0 e^{-j\omega t}\, e^{-k_0\left(jn_r+n_j\right)z}$$                                             (4)

with $k_o = 2\pi/\lambda = \omega/c$; $k_o$, $\lambda$, and $c = (\varepsilon_o\,\mu_o)^{-1/2}$ are the vacuum propagation constant, wavelength, and light velocity, respectively. $\mu_o\ (= 4\pi\cdot 10^{-7}\,Ns^2C^{-2})$ is the absolute permeability. $E_o$ contains the field dependence on the transverse coordinates only.

From Equations 3 and 4, it follows that a change in the real part $\varepsilon_{rr}$ of $\varepsilon_r$ alters the phase $\phi = -k_o n_r z$ of the wave field, whereas a change of $\varepsilon_{rj}$ acts on its amplitude by the $-k_o n_j z$ term. $\varepsilon_{rj} > 0$ leads to an attenuation along z as in an absorptive medium at the wavelength considered, whereas $\varepsilon_{rj} < 0$ leads to an amplification as in the case of a fluorescent medium at the fluorescence wavelength. It is useful to relate the attenuation exponent $(k_o n_j z)$ to the penetration depth, i.e., the propagation length $L_e$ where the optical power, which is proportional to the squared modulus of the field, is decreased by a factor $1/e$: $L_e = 1/(2k_o n_j)$ or, from Equation 3:

$$L_e = \frac{n_r}{k_0 \varepsilon_{rj}}$$                                                                 (5)

## B. OPTICAL WAVEGUIDES AND OPTODE SYSTEMS

Optode systems do not, and will not, necessarily involve optical waveguides. The latter, however, can lead to a simplified and more rugged optical hardware, allow a spatial separation of the signal processing function from the optode transducer, and also offer new opto-chemical interaction topologies and mechanisms. The various functions an optical waveguide can perform in a sensor system can be distributed in three categories, referring to specific waveguide features:

1.  Light conveyance and routing to and from the transducer(s) where the important waveguide feature is the downlead sensitivity, i.e., the ability of the waveguide to transmit the spectral information supplied by the transducer without perturbation. Section VI will discuss the chromatic properties of multimode optical fibers and conclude that a fiber can be, to some extent, considered as an achromatic light conveyor regarding loss fluctuations.
2.  Optical signal preprocessing downstream or/and upstream from the optode transducer such as spatial multiplexing, spectral filtering, and analysis. These functions will be most adequately performed in an integrated optic form.[2] They will be discussed in Section VII.

3.  Opto-chemical sensing where the probe light interacts with the chemical system under investigation in its guided form. Section VIII will deal with the specific interaction configurations offered by guided wave optics for absorption and fluorescence spectroscopy, refractive index change monitoring, and will give the expectable sensitivities and establish some useful scaling rules.

## C. MATERIAL LIMITATIONS

All three types of functions mentioned above can be performed only within the spectral limits imposed by the material the waveguides are made of. Most existing fibers, related passive components, and planar circuits are oxide glasses. Their transmission spectrum covers a quite restricted wavelength range from practically 300 to 1700 nm. Other types of glasses such as zirconium fluoride[3] are being developed to extend the usable optical bandwidth beyond the IR silica band edge well within the region of molecule vibration modes. They can already be found on the market in fiber form.[4] However, silica based waveguide materials will not be ruled out for all that. A number of well developed fiber and waveguide technologies exist, possessing very good mechanical and chemical properties. A complete range of cheap connecting and branching devices is available, and III-V compound technologies provide semiconductor sources and detectors covering the entire transmission window except at below 630 nm, where semiconductor lasers do not exist yet; II-IV semiconductor technology is expected to cover the visible range one day. Meanwhile, hybrid devices using semiconductor laser-pumped second harmonic generation are becoming available.[113] There is therefore a complete technical and technological gear that can readily be used in optode systems.

As shown in Figure 1, the IR side of the transmission spectrum of silica based materials is limited by the combination of overtone bands of fused silica reinforced by the water stretching band at 2.73 μm.[5] In plastic waveguides or in plastic coated silica fibers (PCS), overtones of OH and CH vibrations bring the IR edge closer to the visible range.[6] Plastic fibers have a window of about 400 to 800 nm with an average 200 dB/km loss, whereas PCS fibers reach 10 dB/km as the light flows mainly in the pure silica core. At the UV side, the transmission is limited by two mechanisms,[5] namely the electronic absorption band edge which increases exponentially with frequency with an upper transmission limit slightly below 200 nm, and the Rayleigh scattering which increases as $1/\lambda^4$. The latter is the dominant limitation in the usable part of the spectrum. It results from frozen Brownian density and composition fluctuations. However, impurities play an important role and bring the UV edge closer to the visible range. This tendency can be somewhat thwarted in high OH content silica fibers.[7] Between the IR and UV absorption edges a number of artifacts add to the $1/\lambda^4$ scattering. Essentially, these are OH vibration overtones, atomic transitions between unfilled levels in transition metal impurities, and possible metal inclusions in reduced form.[5] They are no longer a limitation in fibers but remain a problem in other waveguide technologies where high purification is not performed.

In conclusion, direct spectroscopy of electronic transitions and molecular vibrations cannot be performed by means of oxide based waveguides, although a few interesting molecules exhibit absorption lines close to the IR band edge such as $NH_3$ (1.514 μm) and $CH_4$ (1.66 μm). Indirect spectroscopy of vibration states can be made using the Raman effect with silica based waveguides that can stand very high optical power density.[8] However, care must be taken for the separation of the expected Stokes lines from the high efficiency nonlinear effects generated in the waveguide itself.[9]

The transmission window of oxide based waveguides is therefore especially well adapted for direct transitions occurring in photochemistry and in the whole domain of dye absorption

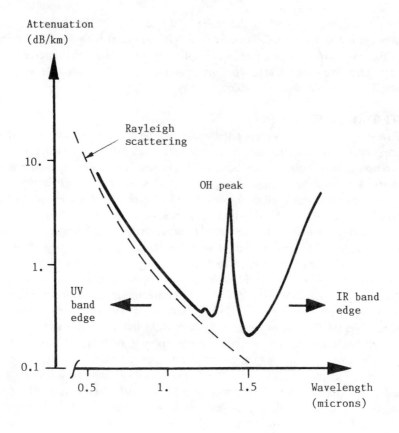

FIGURE 1. Transmission spectrum of silica based fibers.

and fluorescence.[10] Most examples given hereafter as an illustration of various waveguide characteristics and opto-chemical transducing mechanisms will refer to silica fibers[15] and oxide glass waveguides,[11] but the general scope of this chapter on guided wave concepts and principles is not restricted to structures pertaining to these technologies.

## II. OPTICAL WAVEGUIDE ELECTROMAGNETISM

### A. THE WAVE EQUATIONS

The electric and magnetic vector fields $\mathbf{E}$ and $\mathbf{H}$ of a wave at optical frequency $\omega$, propagating in an isotropic medium, of local relative permittivity $n^2$, containing no point charge and no current sheet or filament are governed by Maxwell's equations:

$$\nabla \times \mathbf{H} = j\omega\varepsilon_0 n^2 \mathbf{E} \qquad a) \qquad \nabla \cdot \mathbf{H} = 0 \qquad c)$$

$$\nabla \times \mathbf{E} = -j\omega\mu_0 \mathbf{H} \qquad b) \qquad \nabla \cdot \left(n^2 \mathbf{E}\right) = 0 \qquad d)$$

(6)

where the term of harmonic time dependence, $e^{j\omega t}$, is inferred in all field components. The propagating term $e^{-j\beta z}$ is also inferred because the dielectric structure is considered to be invariant along the direction of propagation z (Figure 2).

Taking the curl of (1 a) and using (1 c), one obtains:

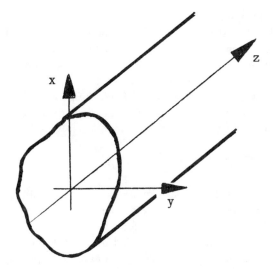

FIGURE 2. Cartesian coordinate system attached to an axially uniform waveguide structure. Propagation takes place along z.

$$\nabla^2 \mathbf{H} + k_0^2 n^2 \mathbf{H} = -j\omega\varepsilon_0 n^2 \frac{\nabla n^2}{n^2} x \mathbf{E} \tag{7}$$

The same operation on (1b) with the help of (1d) yields:

$$\nabla x \nabla x \mathbf{E} = \nabla(\nabla \cdot \mathbf{E}) - \nabla^2 \mathbf{E} = k_0^2 n^2 \mathbf{E}$$

which gives, since $\nabla \cdot \mathbf{E} = -\mathbf{E}\dfrac{\nabla n^2}{n^2}$

$$\nabla^2 \mathbf{E} + k_0^2 n^2 \mathbf{E} = -\nabla\left( \mathbf{E}\frac{\nabla n^2}{n^2} \right) \tag{8}$$

where $\nabla^2$ is the vectorial Laplace's operator. Equations 7 and 8 are the vectorial wave equations in an isotropic structure (n is scalar), that is longitudinally uniform along z ($e^{-j\beta z}$ field dependence), with an arbitrary transverse index distribution n(x,y). These equations are coupled by the term $\nabla n^2/n^2$ (= $\nabla \ln n^2$) which means that the field configurations satisfying them will normally be hybrid modes involving all six components. This makes the resolution of the wave equation difficult and physically not very meaningful as numerical methods have to be extensively used. Fortunately, most waveguides of practical interest met in the well developed field of optical communications and the more recent field of fiber sensors are dielectric structures where the cross-coupling term $\nabla \ln n^2$ can be neglected or be exactly zero piecewise. In these two cases, both E and H fields obey the same law separately:

$$\left( \nabla^2 + k_0^2 n^2 \right) \frac{\mathbf{E}}{\mathbf{H}} = 0 \tag{9}$$

The structures in which the fields obey the wave Equation 9 are thus of two types: the weakly guiding waveguides and the step index waveguides.

## 1. Weakly Guiding Waveguides

The transverse dielectric distribution in weakly guiding waveguides is usually composed of an infinite cladding of constant index $n_2$ embedding a transversally confined region of higher index $n_1(x,y)$ having a maximum, $n_o$. Under the assumption $\Delta = (n_o - n_2)/n_o \ll 1$, the $\nabla \ln n^2$ cross-coupling term can be neglected. If the field components are expressed in a Cartesian coordinate system x,y,z, the vectorial Laplace operator $\nabla^2$ does not couple field components and the vectorial wave equation becomes a scalar wave equation with the scalar Laplace operator acting on the transverse field components $\Phi_x$ or $\Phi_y$:

$$\nabla_t^2 \Phi(x,y) + \left(k_0^2 n^2(x,y) - \beta^2\right)\Phi(x,y) = 0 \tag{10}$$

where $\nabla = \nabla_t - j\beta$. $\nabla_t$ is the transverse scalar Laplace's operator. $\Delta$ is still high enough to ensure total internal reflection, but the longitudinal field components are very weak. As a result, the guided wave is quasi TEM and polarization effects will therefore be ignored. The requirements that $\Phi$ is continuous everywhere, and bounded, means that Equation 10 is an eigenvalue equation for the propagation constant $\beta$ of the modes allowed to propagate.

## 2. Step Index Waveguides

In a step index waveguide it is assumed here that the refractive index discontinuities between homogeneous regions where $\nabla \ln n^2 = 0$ have the symmetry of the coordinate system. This, in particular, is the case in step index slab waveguides with Cartesian coordinates and in circular fibers with polar coordinates where the vectorial wave Equations 7 and 8 can easily be handled and the field components expressed by means of known functions. Setting $\nabla \ln n^2 = 0$ in Equations 7 and 8 means that the index derivative only acts at the boundary between homogeneous regions. Instead of using a Dirac function in the right hand term of Equations 7 and 8, it is more appropriate to express the effect of the index discontinuity by means of the field continuity conditions derived from Maxwell's equations at a source-free interface.[12] These state that the electric and magnetic field components tangent to the interface are continuous. They also state that the D and B field components normal to the interface are continuous. The condition on **E** and **H** together with the requirement of transversely bounded fields yield the eigenvalue $\beta$ of the propagating vectorial eigenfields as seen below. The general vectorial wave Equations 7 and 8 with zero right hand term (as in Equation 9), will now be written in two important particular cases which are the circular waveguide and the slab waveguide.

### a. The Wave Equation in a Slab Waveguide

The basic step index planar structure is depicted in Figure 3. In a Cartesian coordinate system all vectorial field components obey the same scalar wave equation with the scalar Laplace operator and Equation 9 becomes:

$$\left[\frac{\partial}{\partial y^2} + \left(k_0^2 n^2 - \beta^2\right)\right]\Phi = 0 \tag{11}$$

where $\Phi$ stands for any x, y, or z component of **E** or **H**. The structure is invariant along x, therefore $\nabla^2$ reduces to $\partial^2/\partial^2_y - \beta^2$. Choosing the x component of Equation 9 as the wave equation gives for $\Phi_x$ solutions of the exponential type:

$$\Phi_x = Ae^{-jky} = Be^{jky} \text{ with } k^2 = k_0^2 n^2 - \beta^2$$

$$\text{in regions where } n(y) > n_e \left(= \beta/k_0\right) \tag{12}$$

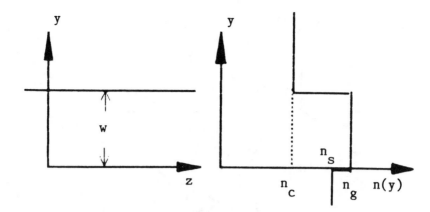

FIGURE 3. Cartesian coordinate system attached to a slab waveguide of refractive index profile n(y) and width w. $n_g$, $n_s$, and $n_c$: refractive index of waveguide, substrate, and cover.

$$\Phi_x = Ae^{-ky} + Be^{ky} \quad \text{with} \quad k^2 = \beta^2 - k_0^2 n^2 \tag{13}$$

in regions where $n(y) < n_e$

$$\Phi_x = Ae^{-ky} \qquad \text{in the upper half space,} \tag{14}$$

$$= Be^{ky} \qquad \text{in the lower half space}$$

where constants B, resp. A, were set to zero under the requirement of zero field at infinity.

$n_e = \beta/k_o$ is called the *effective index* of the propagating wave. These solutions correspond to a *bound field*. If $n_e$ is lower than the index n(y) in either lower or upper half space, or both, then the solution is a transverse oscillatory function like in Equation 12 and the wavefield is not bounded in the waveguide any longer. The wave is said to be a *radiation wave*.

Equation 6a and b give the other field components expressed in terms of $\Phi_x$. There are two sets of waves satisfying Equation 9 independently: transverse electric (TE) waves involving only ($E_x$, $H_y$, and $H_z$), and transverse magnetic (TM) waves involving ($H_x$, $E_y$, and $E_z$). Equation 6 leads to:

$$\Phi_y = \frac{\beta}{\omega\mu_0} \xi \Phi_x \tag{15a}$$

$$\Phi_z = \frac{1}{j\omega\mu_0} \xi \frac{\partial}{\partial y} \Phi_x \tag{15b}$$

where $\xi = 1$     for TE waves

and $\xi = \dfrac{-z_0^2}{n^2}$     for TM waves

$$z_0 = \sqrt{\frac{\mu_0}{\varepsilon_0}} = 377\Omega \text{ is the impedance of vaccum} \tag{16}$$

The matching of the tangential field components $\Phi_x$ and $\Phi_z$ yields the characteristic equation of the propagating modes (Section III.B).

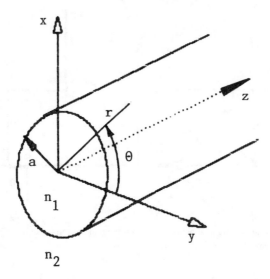

FIGURE 4.  Polar coordinate system attached to a circular waveguide. a, $n_1$, and $n_2$ are the radius, core, and cladding refractive index of a step index optical fiber.

The separation into sets of TE and TM waves of different polarization remains in cases of a graded index profile n(y). This type of profile will be discussed in Section III.E.

Planar waveguiding structures can be used as such as opto-chemical sensors in refractometry, interferometry, or spectrometry applications. However, transversely confined stripe structures will certainly be the type of waveguide used in practical opto-chemical chip transducers and associated optical preprocessing circuits. They will be discussed in Section III.F. The reason why step index planar structures are given a substantial attention hereafter is that they are simple enough to allow an easy understanding of basic waveguide properties, to give an introduction to the main design concepts, and, at the same time, realistic enough to offer a quantitative modelization of the expected effects.

### b. The Wave Equation in a Circular Waveguide

In a circularly symmetrical structure with polar coordinates r, θ, z as shown in Figure 4, the longitudinal component of the $\nabla \mathbf{E}$ or $\nabla \mathbf{H}$ vector in Equation 9 involves the longitudinal $E_z$ or $H_z$ field component only. Therefore, the longitudinal component of the vectorial wave equation is a second order decoupled equation in a single field component:

$$\nabla^2 E_z + k_0^2 n^2 E_z = 0 \qquad (17)$$

where $\nabla^2$ is the scalar Laplace's operator.

The same holds for $H_z$ but we keep on with $E_z$ for the sake of concision. In addition to the longitudinal invariance of the structure leading to the $e^{-j\beta z}$ propagation term, there is also an azimuthal invariance which imposes on the field an azimuthal dependence in the form $e^{jm\theta}$ where m is an integer. The field $E_z$ is now becoming an unknown function of the radial coordinate r only. Writing explicitly the θ and z derivatives and multiplying by $r^2$ yields the Bessel equation for $E_z(r)$ in either homogeneous region: in the core of radius a and index $n_1$ and in the cladding of index $n_2 < n_1$:

$$\left[ r^2 \frac{\partial^2}{\partial r^2} + r \frac{\partial}{\partial r} + \left( \frac{u^2 r^2}{a^2} - m^2 \right) \right] E_z = 0 \quad \text{in the core and} \qquad (18)$$

$$\left[\ldots\ldots\ldots\ldots - \left(\frac{w^2 r^2}{a^2} + m^2\right)\right] E_z = 0 \quad \text{in the cladding} \tag{19}$$

where $u = a\sqrt{k_0^2 - \beta^2}$ and $w = a\sqrt{\beta^2 - k_s^2}$, $k_g = n_1 k_0$, and $k_s = n_2 k_0$ are the propagation constants of plane waves that would propagate in an infinite medium of index $n_1$ and $n_2$ respectively.

$$V = \sqrt{(u^2 + w^2)} = a\,k_0\sqrt{(n_1^2 - n_2^2)} \text{ is the normalized frequency.}$$

The general solution for a bound wave is:[13]

$$E_z = A_1 J_m\left(\frac{ur}{a}\right) \quad \text{in the core and}$$
$$\tag{20}$$
$$E_z = A_2 K_m\left(\frac{wr}{a}\right) \quad \text{in the cladding,}$$

discarding $Y_m$ which is singular at $r = 0$ and $I_m$ which is a growing function of r. $J_m$ and $K_m$ are the Bessel and the modified Bessel functions,[13] $A_1$ and $A_2$ are two integration constants. All other field components can be written in terms of $E_z$ and $H_z$ from Equations 6 a and b expressed in polar coordinates:

let

$$(M) = -\frac{ja^2}{k^2}\begin{vmatrix} \dfrac{jm\beta}{r} & -\dfrac{\omega\mu_0 \partial}{\xi \partial r} \\[3mm] \beta\dfrac{\partial}{\partial r} & \dfrac{\omega\mu_0\, jm}{\xi r} \end{vmatrix} \tag{21}$$

where $k^2 = a^2(n^2 k_0^2 - \beta^2)$. Depending on the relationship between n and $n_e$ ($= \beta/k_0$), $k^2$ will be equal to $u^2$ (where $n > n_e$) or to $-w^2$ (where $n < n_e$). Then the other field components can be expressed in terms of the longitudinal components by means of:

$$\begin{pmatrix} E_\theta \\ E_r \end{pmatrix} = (M)\begin{pmatrix} E_z \\ H_z \end{pmatrix} \quad \text{with} \quad \xi = 1 \quad \text{in} \quad (M) \quad \text{and}$$
$$\tag{22}$$
$$\begin{pmatrix} H_\theta \\ H_z \end{pmatrix} = (M)\begin{pmatrix} H_z \\ E_z \end{pmatrix} \quad \text{with} \quad \xi = -\frac{z_0}{n^2} \quad \text{in} \quad (M) \quad \text{where} \quad z_0 = 377\Omega$$

The continuity conditions concern the longitudinal and azimuthal components $E_z$, $E_\theta$, and $H_z$, $H_\theta$ at $r = a$. They lead to an homogeneous system of four linear equations in the four integration constants involved in the solution for $E_z$ and $H_z$. Writing the condition for a nontrivial solution yields the characteristic equation of a step index fiber:

$$\left[ w\frac{K'_m(w)}{K_m(w)} + \frac{n_1^2}{n_2^2}\frac{w^2}{u}\frac{J'_m(u)}{J_m(u)} \right]\left[ w\frac{K'_m(w)}{J_m(w)} + \frac{w^2}{u}\frac{J'_m(u)}{J_m(u)} \right] = \left[ m\left(\frac{n_1^2}{n_2^2}-1\right)\frac{n_2^2 a\beta k_0}{u^2} \right]^2 \quad (23)$$

The value(s) of $\beta$ (or $n_e = \beta/k_o$) satisfying it are the propagation constants of the modal field satisfying the Maxwell's equation everywhere and the boundary conditions at the core cladding interface. For each azimuthal order m there is finite number n of roots $\beta_{mn}$ of Equation 23, i.e., of modal field configurations.[14] These are exact solutions to the vector wave equation which should in principle answer all questions on bending and transition loss, dispersion and coupling problems. However, it was found at the beginning of the 1970s that the propagation characteristics of multimode fibers can be better understood and easily calculated with a *weak guidance approach*[15] involving another set of propagation modes, the said *linearly polarized modes* $LP_{lm}$ which can be derived by a suitable combination of the true vectorial eigenmodes.[15]

## B. THE POWER OF A PROPAGATING MODE

The optical power P carried by a propagating mode is related to the flux of the Poynting vector $\mathbf{S} = \mathbf{E} \times \mathbf{H}^*$ through the infinite waveguide cross-section A orthogonal to its axis z:[12]

$$P = \frac{1}{2}\int Re\left(ExH^*\right)d\sigma \quad (24)$$

where $d\sigma$ is the cross-section surface differential oriented along z. Factor 1/2 accounts for the time averaging of $\mathbf{S}$. In a Cartesian coordinate system, Equation 24 writes as

$$P = \frac{1}{2}\int_A Re\left(E_x H_y^* - E_y H_x^*\right)xdy \quad (25)$$

which yields Expression 38 for TE and TM modes in a planar waveguide. In a polar coordinate system, Equation 24 leads to 70 which expresses the power propagated by a $LP_{lm}$ mode in a weakly guiding fiber. Expressions 25 and 70 concern the power carried by bound modes. They will be used to perform the normalization of the modal fields in various waveguide types. This physical quantity is of special interest in optode transducers using the evanescent modal field because it is the relative evanescent wave power which determines the sensitivity of the sensor or the efficiency of fluorescence excitation and capture; this will be dealt with in Section IX. The calculation of the power loss in the modelization of leakage-type transducers often makes use of the transverse flux of the Poynting vector $\mathbf{S}$ as shown in Section XIII.B.2 in the case of a tunneling power refractometric transducer.

# III. PROPAGATION IN PLANAR WAVEGUIDES

## A. REFLECTION AT A DIELECTRIC INTERFACE: PHASE SHIFT AND POWER DENSITY

A plane wave incident on a dielectric interface under the angle $\theta$ from medium g of index $n_g$, which will later be the waveguide, is shown in Figure 5. The x components of the field of the incident and reflected plane waves have the amplitude $A_g$ and $B_g$, respectively. Their propagation constant is $n_g k_o$ which has an horizontal projection $\beta = n_e k_o$ on z, where $n_e = n_g \cos\theta$ as illustrated in Figure 5. Boundary conditions at the interface between medium g and the cover medium c of index $n_c < n_g$ must be satisfied whatever the z abscissa. Therefore, the wave field in the cover must also have the same horizontal propagation constant $\beta$. The transverse projection of the propagation constant is therefore:

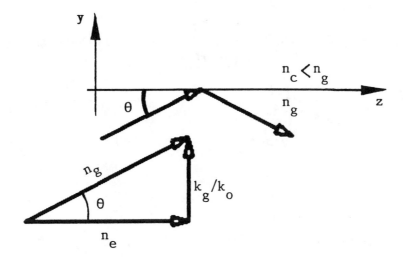

FIGURE 5. Reflection at a dielectric interface between an incident high index ($n_g$) and a lower index ($n_c$) material. $\theta$ is the angle of incidence; $n_e$ is the horizontal projection of $n_g$ directed along the incident beam direction.

$$k_g = k_0 \sqrt{n_g^2 - n_e^2} \ \text{ in medium } \ g \ \text{ and}$$

$$k_c = k_0 \sqrt{n_c^2 - n_e^2} \ \text{ or} \tag{26}$$

$$k_c' = k_0 \sqrt{n_c^2 - n_e^2} \ \text{ in medium } \ c$$

depending on the incident angle relative to the critical angle $\theta_{cr}$.

The critical angle $\theta_{cr}$, is that for which $n_e = n_c$, i.e., $\theta_{cr} = \arccos n_c/n_g$. At $\theta < \theta_{cr}$, $n_e > n_c$ and there is total reflection. At $\theta > \theta_{cr}$, $n_e < n_c$ and there is transmission.

Following Expressions 12 to 15 the TE and TM field components on either side of the interface are summarized in Table 1. For $\theta > \theta_{cr}$ the field in the cover propagates away whereas for $\theta < \theta_{cr}$ its amplitude decreases exponentially away from the interface.

Expressing the continuity of the tangential field components $\Phi_x$ and $\Phi_z$ yields the transmission and reflection coefficients t and r:

$$r = \frac{B_g}{A_g} = \frac{1 - a_c'}{1 + a_c'} \ ; \quad t = \frac{A_c}{A_g} = \frac{2}{1 + a_c'} \quad \text{for } \theta > \theta_c$$

$$r = \frac{1 + ja_c}{1 - ja_c} \ ; \quad t = \frac{2}{1 - ja_c} \quad \text{for } \theta < \theta_c \tag{27}$$

$$\text{with} \quad a_c' = \frac{\xi_c k_c'}{\xi_g k_g} \ ; \quad a_c = \frac{\xi_c k_c}{\xi_g k_g}$$

## TABLE 1
### Tangential Field Components of a Plane Wave in Either Side
### of a Dielectric Interface Located at y = 0

| Field component | Incident medium | Cover | |
|---|---|---|---|
| | | $\theta > \theta_{cr}$ | $\theta < \theta_{cr}$ |
| $\Phi_x$ | $A_g \, e^{-jk_g y} + B_g \, e^{-jk_g y}$ | $A_c \, e^{-jk_c' y}$ | $A_c \, e^{-k_c y}$ |
| $\Phi_z$ | $\dfrac{\xi_g k_g}{\omega \mu_0}\left( A_g e^{-jk_g y} + B_g e^{jk_g y} \right)$ | $-A_c \dfrac{k_c' \xi_c}{\omega \mu_0} e^{-jk_c' y}$ | $-A_c \dfrac{k_c \xi_c}{j\omega \mu_0} e^{-k_c y}$ |

with $\xi_{g,c} = 1$ for TE incidence and $-z_0^2/n_{g,c}^2$ for TM incidence.

*Note:* The transverse component $\Phi_x = E_x$ for a TE wave and $\Phi_x = H_x$ for a TM wave. The longitudinal component is $\Phi_z = H_z, E_z$, respectively. $z_0$ is the impedance of vacuum.

The power flux directed in the z direction is given by the z component of the Poynting vector S as in Equation 25. In the present symmetry $S_z$ is given, for both TE and TM waves, by the expression

$$S_z = \frac{\beta}{2\omega\mu_0}|\xi||\Phi_x|^2 \tag{28}$$

for either medium g and c.

The power density in the cover relative to that propagated by the incident wave, $I(n_e(\theta))$ is given by the ratio of the power flux:

$$I\big(n_e(\theta)\big) = \frac{S_c}{S_g} = \frac{|\xi_c|}{|\xi_g|}\frac{4}{\left(1+a'_c\right)^2} \qquad \text{for } \theta > \theta_c$$

$$\tag{29}$$

$$= \frac{|\xi_c|}{|\xi_g|}\frac{4}{1+a_c^2}e^{-2k_c y} \qquad \text{for } \theta < \theta_c$$

where $a_c^2$ and $a_c'^2$, given in Equation 27, can also be expressed further in terms of the incidence angle $\theta$ with the help of Equation 26:

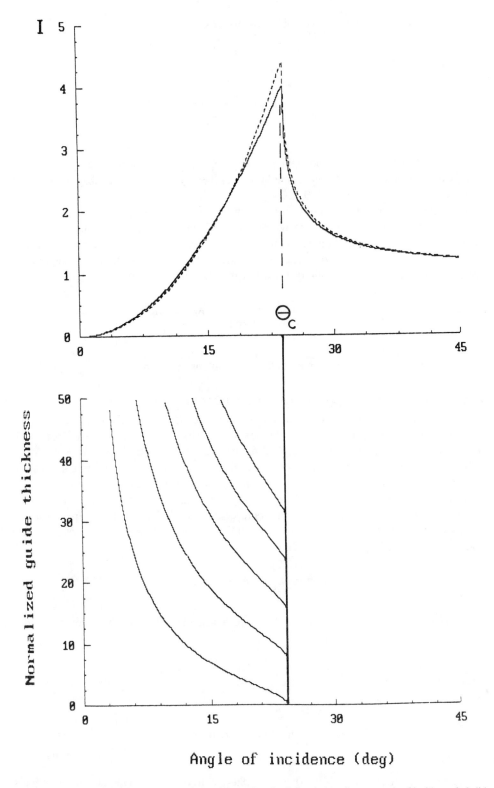

FIGURE 6.  (A) Light intensity I at the low index side y = 0+ of a dielectric interface vs. angle of incidence θ. Solid line: TE incidence; dashed line: TM incidence; $n_g = 1.46$, $n_c = 1.33$. (B) Angle of incidence θ of the TE trapped rays in a symmetrical slab of index $n_g$ and width w, bounded by two media of index $n_c$, vs. normalized width $k_o w n_g$.

$$a_c^2 = \frac{\xi_c^2}{\xi_g^2} \frac{\cos^2\theta - \cos^2\theta_{cr}}{\sin^2\theta}$$

(30)

$$a_c'^2 = \frac{\xi_c}{\xi_g} \frac{\sqrt{\cos^2\theta_{cr} - \cos^2\theta}}{\sin\theta}$$

Illustrated in Figure 6a is the light intensity $I(n_e[\theta])$ at the cover's side $y = 0+$ of the interface for both TE and TM incidence in the pertinent case of a silica substrate, $n_g = 1.46$, and water cover, $n_c = 1.33$, with $\theta_c \cong 24°$.

In Figure 6b is an anticipation: it gives, with the same abscissa scale for $\theta$, the TE mode lines in the dispersion diagram of the symmetrical waveguide obtained by bounding the silica slab of width w by a substrate identical to the cover. The ordinate is the normalized quantity $k_0 w\, n_g$. Figure 6b is obtained from the characteristic equation of a symmetrical waveguide (33) written here in terms of the angle $\theta$ instead of the effective index $n_e$:

$$k_0 w n_g = \frac{1}{\sin\theta}\left[ m\pi + 2\text{arc tg} \frac{\sqrt{\cos^2\theta - \cos^2\theta_{cr}}}{\sin\theta} \right]$$

(31)

where m is the mode index and $\cos^2\theta_{cr} = n_c^2 / n_g^2$

## B. THE TRANSVERSE RESONANCE IN A DIELECTRIC SLAB WAVEGUIDE

Let us return to Figure 5 and install a secondary boundary under the former. The slab of width w is now bounded by a substrate of index $n_s < n_g$. The incident wave now proceeds down the slab in the z direction with the longitudinal propagation constant $\beta$ at phase velocity $v_p = \omega/\beta$ by successive total reflections. For all reflected plane waves to interfere constructively within the slab, a condition on the transverse propagation constants involved must be satisfied. Similarly to the resonance condition in a Fabry-Perot interferometer, there is a transverse resonance condition which states that the transverse phase-shift $\phi$ of a round-trip must be an integer multiple of $2\pi$. $\phi$ is a result of two total reflections at the cover and substrate interfaces and of two transverse path lengths, i.e.,

$$\phi = 2k_g w + \phi_c + \phi_s = 2m\pi$$

(32)

$\phi_c$ and $\phi_s$ can be taken from Equation 27 where, at a single reflection on the cover boundary,

$$B_g = A_g r = A_g \frac{1 + ja_c}{1 - ja_c} = A_g e^{2j \text{ arc tg } a_c}$$

$$\text{with } \phi_c = \text{arc tg } a_c = \text{arc tg } \frac{\xi_c k_c}{\xi_g k_g}$$

The total reflection phase-shift $2\phi_c$ and the phase-shift of the incident plane wave progressing towards the boundary have opposite signs; this is the reason why the transverse resonance condition can be satisfied in a symmetrical waveguide ($\phi_s = \phi_c$) of zero width w as opposed to what occurs in a true Fabry-Perot resonator where the fundamental resonance order presents a non-zero cutoff width. The cutoff condition of waveguide modes is given in Equation 34. The total reflection phase-shift $\phi_{c,s}$ is therefore equivalent to a time lag at the reflection plane. It corresponds to an excursion of the wave into the lower index half-space. This is the part of the guided wave which will be used in evanescent wave sensing as seen in Sections VIII and IX. Introducing this so-called Goos-Hänchen shift[12] into Equation 32 yields:

$$wk_g - \text{arc tg } a_c - \text{arc tg } a_s = m\pi \tag{33}$$

which is the characteristic equation of a slab waveguide. Equation 33 assigns a finite number of discrete values to the effective index $n_{em}$ and defines the plane wave directions $\theta_m$ leading to a transverse resonance. As illustrated in Figure 6b, for each mode index m, the effective index goes from the higher of $n_s$, $n_c$ at its cutoff point and increases monotonically towards $n_g$ with increasing width or decreasing wavelength.

The cutoff condition $(wk_0)_c$, is that for which $\theta_m = \theta_c$ (i.e., $n_e = \max(n_s, n_c)$, usually $n_s$). Beyond it there is no longer total reflection and the wave leaks out as a radiated wave:

$$\left(wk_0\right)_c = \frac{1}{\sqrt{n_g^2 - n_s^2}} \left[m\pi - \text{arc tg } a_c\right] \tag{34}$$

The cutoff of the first mode, m = 0, is zero in a symmetrical guide, for which $a_s = a_c$.

## C. ELECTROMAGNETIC FIELD AND POWER IN A SLAB WAVEGUIDE

The ray optic treatment of the slab waveguide gives the notion of transverse resonance modes and their conditions of existence. An important question remains as one eventually measures optical power: how much power propagates, gets absorbed, couples, or simply leaks out? Although ray optics can give answers to these questions,[22] an electromagnetic approach allows a more homogeneous understanding of guided wave optics and gives more practical design tools. This is not to say that the exact electromagnetic treatment of actual guided wave structures is always necessary. It is necessary indeed, but often as a second priority, after a suitable mental representation of the problem on a simple, but not simplifying model, has come to a functional design. For most cases of interest in integrated optics, the step index slab waveguide represents this simple model, and its analysis will bring material that helps understand most guided wave optic principles and handle useful conceptual tools.

### 1. The Modal Field

The field components dependence in z and t in the form $\exp(j(\omega t - \beta z))$ is inferred in all three regions of the structure depicted in Figure 3. The refractive index of the substrate, guide of width w, and cover are $n_s$, $n_g$, and $n_c$, respectively. Following Equation 26 we define the transverse propagation constants as

$$k_g = k_0 \sqrt{n_g^2 - n_e^2}$$

$$k_s = k_0 \sqrt{n_e^2 - n_s^2} \tag{35}$$

$$k_c = k_0 \sqrt{n_e^2 - n_c^2}$$

## TABLE 2
### Modal Field Components $\Phi_i$ in a Step index Slab Waveguide

| Field components | Substrate | Waveguide | Cover |
|---|---|---|---|
| $\Phi_x$ | $B_s e^{k_s y}$ | $A_g e^{-jk_g y} + B_g e^{jk_g y}$ | $A_c e^{-k_c(y-w)}$ |
| $\Phi_z = \dfrac{1}{j\omega\mu_o}\xi\partial_y\phi_x$ | $\dfrac{k_s\xi_s}{j\omega\mu_o}B_s e^{k_s y}$ | $\dfrac{k_g\xi_g}{\omega\mu_o}(-A_g e^{-jk_g y} + B_g e^{jk_g y})$ | $\dfrac{k_c\xi_c}{j\omega\mu_o}A_c e^{k_c(y-w)}$ |
| $\Phi_y = \dfrac{\beta}{\omega\mu_o}\xi\phi_x$ | $\dfrac{\beta\xi_s}{\omega\mu_o}B_s e^{k_s y}$ | $\dfrac{\beta\xi_g}{\omega\mu_o}(A_g e^{-jk_g y} + B_g e^{jk_g y})$ | $\dfrac{\beta\xi_c}{\omega\mu_o}A_c e^{-k_c(y-w)}$ |

*Note:* $\Phi_y$ is normal to the interface planes. $\xi_i$ is defined as in Table 1.

and we are looking for bound waves which exhibit a transverse oscillatory behavior in the guide and decrease exponentially away from it.

Using Equations 12 to 14 for the x components either $E_x$ or $H_x$ of a TE or a TM mode, $\Phi_x(y)$, and 15a and b, we can write in Table 2 the field expression similarly to Table 1.

Equating the tangential components $\Phi_x$ and $\Phi_z$ at both boundaries $y = 0$ and $w$ yields a set of four homogeneous equations in the four integration constants $B_s$, $A_g$, $B_g$, and $A_c$. Suitable substitutions lead to

$$B_s = A_c e^{jk_g w}\frac{1-ja_c}{1+ja_s} = A_c e^{-jk_g w}\frac{1+ja_c}{1-ja_s}$$

$$\text{(36)}$$

$$\text{where} \quad a_s = \frac{k_s\xi_s}{k_g\xi_g} \quad \text{and} \quad a_c = \frac{k_c\xi_c}{k_g\xi_g}$$

Equation 36 has a nontrivial solution ($A_c \neq 0$) only if $e^{2jk_g w} = e^{2j\phi_s} e^{2j\phi_c}$ where $\phi_{s,c}$ are arc tg $a_{s,c}$ as in Equation 32, which leads to the characteristic equation for both TE and TM modes as given by Equation 33.

For any eigenvalue $\beta = k_o n_e$ satisfying Equation 33, three constants can then be expressed in terms of a single one, say, $A_c$, which is in turn determined by the total power propagating along z in a waveguide slice having a unit width along coordinate x:

$$B_s = \pm A_c\sqrt{\frac{1+a_c^2}{1+a_s^2}} \quad \text{with sign } + \text{ if m is even}$$

$$- \text{ if m is odd}$$

$$A_g = \frac{A_c}{2}\sqrt{1+a_c^2}\; e^{j(\phi_s + m\pi)} = B_g^*$$

$$\text{(37)}$$

## 2. Propagated Power and Field Normalization

Using Equation 25 and taking $\Phi_y$ from Table 2 in terms of $\Phi_x$ yields the power P carried by TE and TM modes:

$$P = \frac{\beta}{2\omega\mu_0} \sum_i |\xi_i| \int_i |\Phi_x|^2 dy, \quad i = s, c, g \tag{38}$$

$\Phi_x$ is given by Table 2. With the help of Equation 37 one obtains:

$$\Phi_x = \pm A_c \sqrt{1 + a_c^2} \cos\left(k_g y - \Phi_s\right) \quad \text{in the guide,}$$

$$= A_c \, e^{-k_c(y-w)} \quad \text{in the cover, and}$$

$$= \pm A_c \sqrt{\frac{1 + a_c^2}{1 + a_s^2}} \, e^{k_s y} \quad \text{in the substrate} \tag{39}$$

with sign $\pm$ for $m_{\text{odd}}^{\text{even}}$. Summing up all three contributions, regrouping terms and using the characteristic equations yields:

$$P = \frac{\beta |\xi_g| \left(1 + a_c^2\right) w_{\text{eff}}}{4\omega\mu_0} |A_c|^2$$

$$= \frac{\beta |\xi_g| w_{\text{eff}}}{4\omega\mu_0} |\Phi_M|^2 \tag{40}$$

where $\Phi_M$ is the maximum amplitude of $\Phi_x$ in the waveguide, taken from Equation 38: $\Phi_M = \pm A_c \sqrt{1 + a_c^2}$. For further use in evanescent wave sensing, we also give the power flow contribution of the cover, $P_c$:

$$P_c = \frac{\beta}{4\omega\mu_0} \frac{|\xi_c|}{k_c} |A_0|^2 \tag{41}$$

$w_{eff}$ is the effective waveguide width, i.e., the electromagnetic field size. Care should be taken here in distinguishing TE and TM cases were $w_{eff}$ is not given by the same expression:

$$w_{eff} = w + \frac{1}{k_c} + \frac{1}{k_s} \quad \text{in the TE case} \tag{42a}$$

$$= w + \frac{1}{k_c\left(n_e^2\left(\dfrac{1}{n_c^2} + \dfrac{1}{n_g^2}\right) - 1\right)} + \frac{1}{k_s\left(n_e^2\left(\dfrac{1}{n_s^2} + \dfrac{1}{n_g^2}\right) - 1\right)} \tag{42b}$$

in the TM case.

The effective width of TE and TM modes tend to be identical in the weak guidance case where $(n_g - n_s)/n_g$ and $(n_g - n_c)/n_g \ll 1$.

The constant $A_c$, or $\Phi_M$, normalized to total propagated power P, are therefore:

$$|A_c|^2 = \frac{4\omega\mu_0}{\beta|\xi_g|(1 + a_c^2)w_{eff}} P \quad \text{and}$$

$$\tag{43}$$

$$|\Phi_M|^2 = \frac{4\omega\mu_0}{\beta|\xi_g|w_{eff}} P$$

Evanescent wave sensing of very thin overlay films will essentially rely on the square modulus of the field, $|A_c|^2$, at the guide-cover interface. From Equation 39, one shows that the latter decreases approximately as $1/n_g{-}n_c$) in a structure where the index of the cover, $n_c$, is the varying quantity.

## D. THE PLASMON MODE AT A METAL-DIELECTRIC INTERFACE

A surface wave propagating along a single metal-dielectric interface can be of considerable interest in opto-chemical sensors for the spectroscopy of ultra-thin films[16a,b] and also, to some extent and more generally, for evanescent wave sensing. A comparison in this respect between dielectric and plasmon waveguiding can be found in Reference 16c.

Let us try and find a nontrivial solution to the characteristic equation resulting from field matching at the interface. As usual, the time and longitudinal dependence is inferred. Referring to Figure 7 and following Equation 15 and Table 2, the field components of a bound wave, which can only be of an exponential nature, write as in Table 3.

$$\text{with } k_m = k_0\sqrt{n_e^2 - \varepsilon_m}$$

$$k_c = k_0\sqrt{n_e^2 - \varepsilon_c}$$

$\varepsilon_m$ is the relative permittivity of the metal region and $\varepsilon_c = n_c^2$ that of the dielectric. For the time being $\varepsilon_m$ is real negative, i.e., the metal is lossless and the optical frequency is larger than the metal plasma frequency:[17]

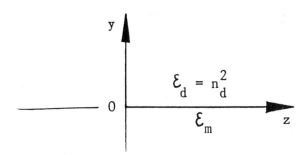

FIGURE 7. Cartesian coordinate system attached to a metal-dielectric interface. $\varepsilon_m$, $\varepsilon_c = n_c^2$ are the metal and dielectric relative permittivities.

$$\varepsilon_m < 0 \quad \text{and} \quad \xi_m > 0 \tag{44}$$

Equating the x and z components at y = 0 yields straightforwardly:

$$\frac{k_m \xi_m}{k_c \xi_c} = -1 \tag{45}$$

which can only be satisfied if $\xi_m$ and $\xi_c$ have opposite sign, i.e., in the TM case (44). The propagation constant $\beta$ of the TM wave can thus be explicitly expressed and it is unique: there is a single plasmon mode:

$$\beta^2 = k_0^2 \frac{\varepsilon_c \varepsilon_m}{\varepsilon_c + \varepsilon_m} \tag{46}$$

In Equations 45 and 46, $\varepsilon_m$ can actually be complex: $\varepsilon_m = \varepsilon_{mr} - j\varepsilon_{mj}$ with $\varepsilon_{mr} < 0$ and $\varepsilon_{mj} > 0$.

The integration constant is determined by the power P propagated by the plasmon obtained as in Equation 38 with $\xi$ expressed for a TM wave (43):

$$P = |A|^2 \frac{\beta}{4\omega\varepsilon_0} \left( \frac{1}{\varepsilon_c k_c} + \frac{1}{\varepsilon_m k_m} \right) \quad \text{which, by using (46) becomes :}$$

$$P = |A|^2 \frac{\sqrt{-\varepsilon_c \varepsilon_m}}{4\omega\varepsilon_0} \left( \frac{1}{\varepsilon_c^2} + \frac{1}{\varepsilon_m |\varepsilon_m|} \right) \tag{47}$$

**TABLE 3**

**Transverse, Longitudinal, and Normal Field Components
at a Metal-Dielectric Interface Located at y = o**

| Field components | Metal | Dielectric |
|---|---|---|
| $\phi_x$ | $A\,e^{k_m y}$ | $A\,e^{-k_c y}$ |
| $\phi_z$ | $\dfrac{k_m \xi_m}{j\omega\mu_o} A\,e^{k_m y}$ | $-\dfrac{k_c \xi_c}{j\omega\mu_o} A\,e^{k_c y}$ |
| $\phi_y$ | $\dfrac{\beta \xi_m}{\omega\mu_o} A\,e^{k_m y}$ | $-\dfrac{\beta \xi_c}{\omega\mu_o} A\,e^{k_c y}$ |
| $\xi$ TE | $\xi_m = 1$ | $\xi_c = 1$ |
| $\xi$ TM | $\xi_m = -\dfrac{z_o^2}{\varepsilon_m}$ | $\xi_c = -\dfrac{z_o^2}{n_c^2}$ |

*Note:* $\varepsilon_m$ is the real part of the metal relative permittivity.

when the first and second terms in brackets are the dielectric and metal contributions, respectively. Equation 47 can be further simplified with the often valid assumption $|\varepsilon_m| \gg \varepsilon_c$:

$$P \cong |A|^2 \; \frac{1}{4\omega\varepsilon_0} \; \frac{\sqrt{-\varepsilon_m}}{\varepsilon_c^{3/2}} \tag{48}$$

The integration constant of the transverse magnetic field, A, normalized to the total propagated power, is therefore:

$$A = \left[ 4\omega\varepsilon_0 \; \frac{\varepsilon_c^{3/2}}{\sqrt{-\varepsilon_m}} P \right]^{1/2} \tag{49}$$

From Equations 47 and 48 the relative power $P_m/P$ flowing in the metal is:

$$\frac{P_m}{P} \cong \frac{\varepsilon_c^2}{\varepsilon_m |\varepsilon_m|} \tag{50}$$

which is very low. Typically, with silver and silica at 633 nm wavelength, $P_m/P \cong 7.8\cdot10^{-2}$ using $\varepsilon_m \cong -15$. Therefore almost all power propagates in the dielectric, with the field maximum, A, at the interface. This feature is useful for surface wave sensing of species deposited on a metal surface.

A comparison can be made with the same ratio in step index dielectric waveguides, where the ratio being between the power propagating in the cover (41) and the total propagated power (40) is:

$$\frac{P_c}{P} = \frac{|\xi_c|}{k_c |\xi_g| \left(1 + a_c^2\right) w_{eff}} \tag{51}$$

which, in the simpler case of a TE mode, yields

$$\frac{P_c}{P} = \frac{n_g^2 - n_e^2}{k_0 \sqrt{n_e^2 - n_c^2} \left(n_g^2 - n_c^2\right) w_{eff}} \tag{52}$$

Equation 52 can be roughly estimated in a single TE mode waveguide not too far from the cutoff of the second TE mode, i.e., where $n_e \cong (n_g + n_s)/2$. Assuming weak guidance and $\tilde{n}_c$ not too close from $n_s$ gives

$$\frac{P_c}{P} < \frac{n_s \left(n_g - n_s\right)}{k_0 w} \tag{53}$$

Typically, $P_c/P < 10^{-3}$. This figure can be considerably increased in a symmetrical waveguide, $n_s = n_c$, in which case Equation 52 becomes

$$\frac{P_c}{P} = \frac{n_g^2 - n_e^2}{n_g^2 - n_s^2} \; \frac{1}{k_0 w \sqrt{n_e^2 - n_s^2} + 2} \tag{54}$$

If, in addition, the mode considered is close to cutoff, Equation 54 tends to $P_c/P \cong 1/2$.

The conclusion is that a plasmon mode allows a relative power concentration in a dielectric cover medium close to unity, whereas the relative power flowing into the same cover of a dielectric waveguide is practically 1 to 4 orders of magnitude smaller.

A plasmon mode exhibits however very high losses. The characteristic Equation 46 can also be solved exactly in presence of a complex metal dielectric constant. Introducing the complex propagation constant $\gamma = \alpha + j\beta$, and assuming $\varepsilon_{mj} \ll \varepsilon_{mr} (\varepsilon_c + \varepsilon_{mr})$, which is true for not too lossy metals, the attenuation coefficient $\alpha$ is given by

$$\alpha \cong \frac{k_0}{2} \; \frac{\varepsilon_c^{3/2}}{\sqrt{-\varepsilon_{mr}} |\varepsilon_c + \varepsilon_m|^{3/2}} \; \varepsilon_{mj} \tag{55}$$

which can be approximated by

$$\alpha \cong \frac{k_0}{2} \frac{\varepsilon_c^{3/2}}{\varepsilon_{mr}^2} \; \varepsilon_{mj} \tag{56}$$

and the propagation coefficient $\beta$ is given by

$$\beta^2 = k_0^2 \; \frac{-\varepsilon_c \varepsilon_{mr}}{|\varepsilon_c + \varepsilon_m|} \tag{57}$$

TABLE 4
Normalized Power Attenuation Coefficients $2\alpha/(k_o\,\varepsilon_c^{3/2})$
for a Few Metals at 633 nm Wavelength

| Metal | $\varepsilon_{mr}$ | $\varepsilon_{mj}$ | $2\alpha/(k_o\,\varepsilon_c^{3/2})$ |
|-------|--------|--------|--------|
| Ag | −15.037 | 1.017 | $4.5 \bullet 10^{-3}$ |
| Au | −8.257 | 1.117 | $16.4 \bullet 10^{-3}$ |
| Al | −56.59 | 21.267 | $6.65 \bullet 10^{-3}$ |
| Cu | −10.424 | 1.763 | $16.2 \bullet 10^{-3}$ |

Equation 56 shows that the power loss coefficient $2\alpha$ is less dependent on the choice of the dielectric than of the metal: $\alpha$ decreases approximately as $\varepsilon_{mr}^{-2}$. The normalized power attenuation coefficient $2\alpha/(k_o\,\varepsilon_c^{3/2})$ for a few metals at 633 nm wavelength[18] is given in Table 4.

As an example, the attenuation of a plasmon at wavelength 633 nm propagating at a silver-water interface would experience a power loss $2\alpha \cong 0.1\ \mu m^{-1} = -0.45\ dB/\mu m$.

There are two consequences of these very high losses as to the use of a plasmon as a probe wave. First, it should not be used as a propagating wave except in the 10 μm wavelength range where copper shows propagation distances of the order of 1 cm.[19] Second, the losses considerably broaden the linewidth of the mode; therefore, no sharp excitation conditions or synchronous resonant coupling can be expected. Therefore, the most interesting feature of a TM plasmon mode to be used in opto-chemical sensors is the very high field concentration at the metal-dielectric interface providing a very efficient excitation of, say, the fluorescence of marked species attached to the metal surface. The excitation wave could be a plane wave in the case of a very thin (of the order of 10 nm) metal film "sandwiched" between two dielectric halfspaces as shown in Figure 8a. The plane wave is incident from the high index substrate material. Its angle of incidence is adjusted so as to excite synchronously by tunneling the plasmon mode propagating at the lower index side where the species to be probed are located. If the medium containing or surrounding the species to be probed is transparent, the plane wave could then come from this side as shown in Figure 8b and excite the plasmon mode via a corrugation grating satisfying the synchronism condition as in Equation 141. The probe wave could also be a guided wave, instead of the plane wave of Figure 8a. The thin metal film can be deposited on a dielectric waveguide with an optional low index buffer between the latter and the metal film as shown in Figure 8c. The synchronism condition can be achieved by suitably setting the various refractive index and dielectric layer thickness.[20] This approach is interesting practically as it allows remote sensing and avoids the cumbersome arrangement of plane wave excitation.

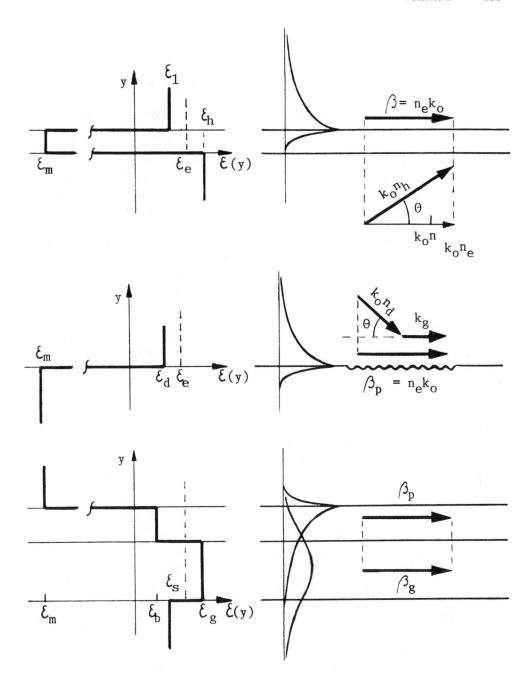

FIGURE 8. Three configurations for plasmon mode excitation: relative permittivity profile $\varepsilon(y)$, modal field(s), phase matching condition. (A) Tunneling through a very thin metal film from a high index $(n_h)$ dielectric to a lower index $(n_l)$ dielectric. $\theta$ is the angle of incidence. (B) Grating coupling. $k_g$ is the grating constant, $\varepsilon_d = n_d^2$ is the dielectric permittivity. (C) Resonant coupling between a dielectric waveguide of permittivity $\varepsilon_g = n_g^2$ and a plasmon mode through a buffer film of permittivity $\varepsilon_b = n_b^2$. $\varepsilon_s = n_s^2$ is the substrate permittivity. $\beta_g$ and $\beta_p$ are the propagation constants of the dielectric and plasmon modes.

It is worth mentioning that the restrictions in the use of plasmon waves due to their high losses can be notably released in case the metal film is very thin as in figure 8a): the degeneracy of the plasmon waves on either side of the metal film is then broken and the structure can propagate one even and one odd TM modes. The field of the former within the metal sheet, thus its attenuation, is smaller. The even mode is the so-called long range surface plasmon [19b] which may exhibit propagation lengths as high as a few hundreds of microns along silver films.

## E. PLANAR WAVEGUIDES OF VARIOUS INDEX PROFILES

So far only step index planar waveguides have been theoretically considered for sake of simplicity. They are not just idealizations of actual waveguide structures. A few technologies are available which lead to thin homogeneous waveguiding films. These include sputtering of oxides,[21] chemical vapor deposition on silica or silicon-based silica which leads to good quality films,[22] pyrolysis of solgel coatings (which offers a large variety of refractive index[23]) and spin-coating and dipping of all sorts of organic material in the form of solutions[24] or of Langmuir Blodgett molecular stacks.[25] The passive waveguide fabrication technology which has been the most successfully developed in practice is based on the ion exchange of monovalent ions in various oxide glasses.[26-28] The resulting graded index profile $n(y)$ is obtained through a diffusion process.

The exact vector wave Equations 7 and 8 written in the Cartesian coordinate system attached to the planar structure can still be separated into two independent TE and TM solutions. But the $\nabla (\ln n^2)$ term must be considered. It does not affect, however, the form of the wave equation for TE modes which remains exactly that of Equation 11 for the $E_x$ component:

$$\left[ \frac{\partial^2}{\partial y^2} + \left( k_0^2 \, n^2(y) - \beta^2 \right) \right] E_x = 0 \qquad (58)$$

The TM mode equation contains an additional term due to the index gradient:

$$\left[ \frac{\partial^2}{\partial y^2} - \frac{\partial}{\partial y} \left( \ln \, n^2 \right) \cdot \frac{\partial}{\partial y} + \left( k_0^2 \, n^2(y) - \beta^2 \right) \right] H_x = 0 \qquad (59)$$

In the weak guidance approximation, Equation 59 is again identical to Equation 58 and polarization properties are neglected.

There are few profile cases where an exact solution can be given to Equation 58 with known functions.[29] However, a more useful approach is to borrow asymptotic methods from quantum mechanics after realizing that, e.g., Equation 58 has the same form as that of a charged particle in a potential well $-k_o^2 \, n^2 (y)$ with eigenvalues $-\beta^2$ where $E_x$ stands for the wave function.[30] The WKB approximation[31] has been widely used in integrated[32] and fiber optics.[33] If the variation of $n(y)$ over one wavelength is not very small, numerical techniques must be used to solve Equations 58 and 59 considering, for instance, a staircase approximation of $n(y)$. This leads to very simple and usable algorithms.[34] It is, however, worth pointing out that most guided wave optic analysis and design problems, except those addressing pulse distortion questions (which are of little concern in opto-chemical sensors), can be very well understood and discussed with a step index waveguide simulation as an initial conceptual exploration.

## F. LATERALLY CONFINED STRIPE WAVEGUIDES

Future integrated optic optode transducers or systems will make use of transversally

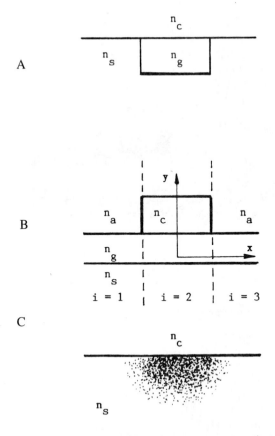

FIGURE 9. Types of laterally confined waveguides. (A) Channel guides (Reference 22 and related references). (B) Strip-loaded guides (Reference 89 and related references). (C) Diffused or ion exchanged guides (References 26 and 27).

confined probe light as stripe waveguides allow high integration density, nonambiguous sensing field, and spatial multiplexing. These are much more complex electromagnetic structures. Once again, however, much understanding can be gained by using a simple step index planar modelization. A very useful method allowing this simple representation, the effective index method,[35] is described in Section III.F.

Practical waveguides can be of the various types illustrated in Figure 9. We will only consider low mode number waveguides here. Deep channel optical plumbing can be discussed with the same approach as for highly multimode weakly guiding fibers[36] (Section VI). In a number of applications the opto-chemical transducer material will be placed as a cover on top of the waveguiding core within the evanescent field of the guided mode(s).

There are no analytical solutions of the vector and scalar wave Equations 7 to 10 for any of the structures of Figure 9. Piecewise homogeneity does not help much. Therefore, numerical techniques tackling the most general waveguide are advisable. Most practical structures are within the weak guidance approximation; there are two types of approaches for solving the scalar wave Equation 10. Those relying on an approximation of the operator $\nabla_t^2$, as the finite difference technique[37] and those relying on an approximation of the field as the finite element method.[38] Although less elegant, the former yields results which are easy to interpret and ready made programs exist.[39] An example is shown in Figure 10.

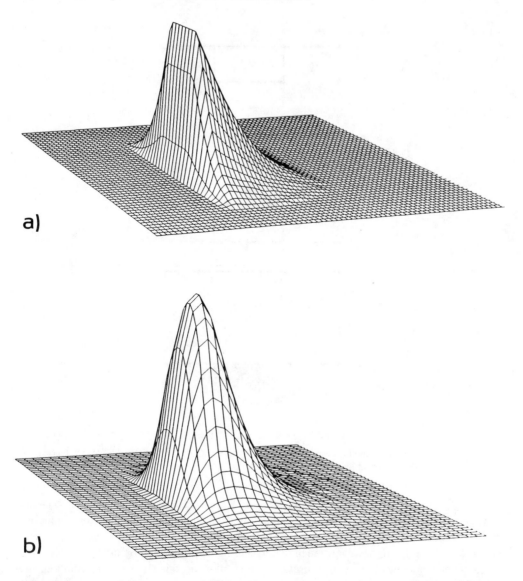

FIGURE 10.  Finite difference calculation (after Reference 39) of the first order modal field of an ion exchanged waveguide of graded-index profile. (a) Index profile $\Delta n(x,y)$. The sharp edge is at the substrate surface. (b) Transverse electric field distribution of the quasi-TE fundamental mode.

## 1. The Effective Index Method[40,41]

A quick outline of the method is given here with the example of lateral guidance in a strip loaded planar waveguide shown in Figure 9b. The cross-section is first chopped into three (or N in a more general case) vertical slices. Each vertical slice i is then considered as a uniform planar waveguide of infinite extension along x in which the propagation constant $\beta_{yi}$ is searched for by solving the characteristic Equation 33 (or Equations 58 and 59 in the more general graded case). Then the set of $\beta_{yi}$, or rather, the set of effective index $n_{eyi} = \beta_{yi}/k_o$ is looked at in the x direction and regarded as a planar waveguide with three (or N) regions of index $n_{eyi}$, $i = 1,2,3$ (or $i = 1,2...N$) having its interfaces normal to the x axis. The characteristic equation is then solved in this fictitious waveguide; the resulting propagation constant $\beta$ is expected to be that of the actual mode propagating in the transversally confined guide. It is now easy to understand why a nonguiding overlay strip of index $n_c$ placed on top of a planar guide having otherwise an index $n_a < n_c$ achieves lateral confinement: the effective index $n_{ey2}$ is higher than $n_{ey1} = n_{ey3}$ because of lesser asymmetry, ensuring a lateral guidance. More complex two-dimensional graded index profiles can similarly be considered.

This method fails to give reliable results close to cutoff, i.e., in conditions where, e.g., evanescent excitation and capture of fluorescence is the most efficient. Such problems should then be dealt with first by a planar simulation followed directly by a numerical method.[39]

# IV. WEAKLY GUIDING STEP INDEX FIBERS

Multimode fibers are now progressively substituted by single mode fibers in optical communications, but they remain of paramount importance in the field of opto-chemical sensors as they can capture and convey significant amounts of optical power coming from diffuse light emitters placed at either fiber end or distributed along a cladding stripped fiber core.[42] The weakly guiding fiber analysis is therefore well adapted to describe the properties which are specific to the optode application field. This is not to say that the vectorial analysis outlined above should be left aside. Although the weak guidance approach is also valid for usual single mode fibers, optode designers are likely to consider one day single mode or low mode number special fibers such as capillary fibers with high index jumps which would then require an exact vectorial mode analysis for taking into account polarization effects.

## A. THE CHARACTERISTIC EQUATION

The general scalar wave Equation 10 will now be written with its scalar Laplace's operator in polar coordinates acting on, say, $\Phi_y$ which represents the transverse electric wave field of $LP_{lm}$ mode directed along y:

$$\left\{ \frac{\partial^2}{\partial r^2} + \frac{1}{r}\frac{\partial}{\partial r} + \left[ k_0^2 n^2 - \left( \beta^2 + \frac{l^2}{r^2} \right) \right] \right\} \Phi_y = 0 \tag{60}$$

This is, as in Equations 18 and 19, the Bessel equation but the field component involved here is a transverse one.

The solution is:

$$E_y = A_1 \frac{J_l\left(\frac{ur}{a}\right)}{J_l(u)} \cos l\theta \quad \text{in the core with}$$

$$H_x = \frac{n_1}{z_0} E_y \quad \text{as in the plane wave case and} \tag{61}$$

$$E_y = A_2 \frac{K_l\left(\frac{wr}{a}\right)}{K_l(w)} \cos l\theta \quad \text{in the cladding with}$$

$$H_x = \frac{n_2}{z_0} E_y \tag{62}$$

$A_1 = A_2 = A$ is an integration constant to be determined. The way to the characteristic equation for $LP_{lm}$ modes will only be described hereafter.[43] The longitudinal components $H_z$ and $E_z$ are first obtained from Equation 15b in terms of $E_y$ and $H_x$ given by Equations 61 and 62 and neglecting the $E_x$ and $H_y$ components. Then the azimuthal components to be matched at the core boundary can be expressed from the just found longitudinal ones using Equation 22. Finally, the field matching under the weak guidance assumption $\Delta \ll 1$ leads to the $LP_{lm}$ characteristic equation:

$$u \frac{J_{l-1}(u)}{J_l(u)} = -w \frac{K_{l-1}(w)}{K_l(w)} \tag{63}$$

The field of $LP_{lm}$ modes is quasi-TEM. We could have chosen the polarization of the electric field in Equation 60 along x as well. In addition, we could also have chosen $\sin l\theta$ in Equation 61. The $LP_{lm}$ modes have therefore a twofold degeneracy in polarization and a twofold azimuthal degeneracy (except for $l = 0$).

## B. CUTOFF OF $LP_{lm}$ MODES

The cutoff condition of a propagation mode is $\beta = k_o n_2$, i.e., $w = 0$. The righthand term of Equation 63 is of the order of $w^2$ for small argument w.[44] Therefore the cutoff condition derived from Equation 63 is

$$J_{l-1}(V_c) = 0 \tag{64}$$

where $V_c$ is the cutoff value of the normalized frequency of all modes of azimuthal index l.

A representation of the condition of existence and field shape of a $LP_{lm}$ modal field can be obtained from the order index l and m: the $LP_{lm}$ cutoff occurs at the V-number values equal to the $m^{th}$ zero the Bessel function $J_{l-1}$ (V). $V_c = 0$ is not a solution of Equation 63 written at cutoff as all $J_l(0) = 0$ except when $l = 0$ as $J_{-1}(0) = -J_1(0)$. The $LP_{01}$ mode, with its zero cutoff, is thus the fundamental mode. The $LP_{lm}$ cutoffs occur therefore at the $m^{th}$ non-zero root of Equation 64 for $l > 0$ and at the $m^{th}$ of $J_1(V)$ (the first zero root being included in the count of m) for $l = 0$. This is suggested in Figure 11. All $LP_{lm}$ modal fields exhibit within the core cross-section (m–1) zeros along the radius r and 2l radial zero lines (i.e., 2l azimuthal optical power lobes). All $LP_{lm}$ modal fields with $l > 0$ have, in addition, zero field at the core center.

## C. THE NUMBER OF $LP_{lm}$ MODES

The characteristic equation at cutoff Equation 64 for large argument V writes[45] as:

$$J_{l-1}(V_c) = \sqrt{\frac{2}{\pi V_c}} \cos\left[ V_c - \frac{(l-1)\pi}{2} - \frac{\pi}{4} \right] = 0 \tag{65}$$

The cutoff condition for all modes of azimuthal index l and radial index m is therefore:

$$V_c \cong \frac{\pi}{2}(2m+1) \tag{66}$$

meaning that all modes having the same principal mode index $M = 2m + 1$ have the same cutoff. They can also be considered as nearly degenerate away from their cutoff.[45] The maximum value $m_l$ taken by m in Equation 66 in a fiber of normalized frequency V is the number of possible radial modes for each value of the azimuthal index l:

$$m_l = \frac{V}{\pi} - \frac{1}{2} \tag{67}$$

The maximum value allowed for l is $2V/\pi$. This can be represented in the (m,l) diagram of Figure 12 where all possible modes are dots located within the surface limited by the straight line Equation 67 the area of which is the total mode number N:

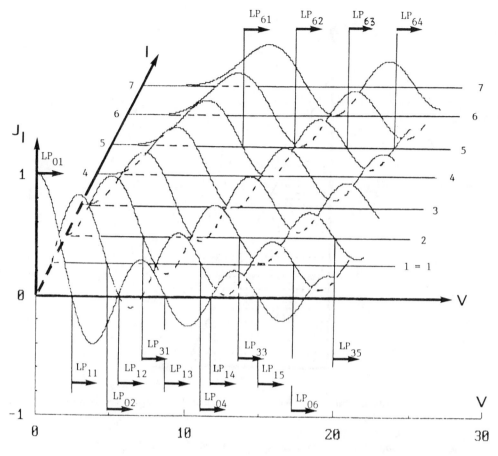

FIGURE 11.   Domain of existence of the $LP_{lm}$ modes in a step index fiber. The cutoff V-number $V = V_c$ is the zero of Bessel function $J_{l-1,m}(V)$ for $l \neq 0$ and of $J_{l,m-1}(V)$ for $l = 0$.

$$N \cong 4 \frac{V^2}{\pi^2} \tag{68}$$

where the factor 4 accounts for the fourfold degeneracy of all modes with $l \neq 0$. The number of modes close to cutoff, $N_c$, i.e., those for which the field extends deeply into the cladding, is proportional to V:

$$N_c \cong 4 \frac{2V}{\pi} \tag{69}$$

## D. THE POWER PROPAGATED BY $LP_{lm}$ MODES

The total power P propagated by a guided mode is given by Equation 24. This writes in a polar coordinate system for a y-polarized $LP_{lm}$ mode, using Equations 61 and 62:

$$P = \frac{1}{2} \int_0^\infty \int_0^{2\pi} rE_y H_x^* d\theta \, dr \tag{70}$$

$$= \frac{\alpha}{\kappa} \frac{V^2}{U^2} \tag{71}$$

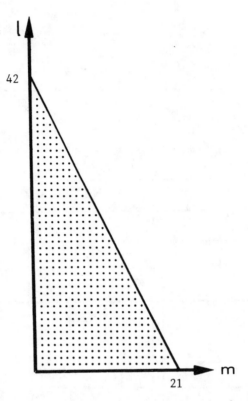

FIGURE 12.   Radial and azimuthal index diagram (l,m) of the bound $LP_{lm}$ modes of a step index silica fiber of 50-µm radius, core-cladding index difference 0.01 used at 800-nm wavelength. Each dot represents four degenerate modes.

The power propagating in the core, $P_{core}$, and in the cladding, $P_{clad}$, is:

$$P_{core} = \alpha\left[1+\left(\frac{w}{u}\right)^2 \frac{1}{\kappa}\right]; \quad P_{clad} = \alpha\left[\frac{1}{\kappa}-1\right] \tag{72}$$

with $\alpha = \dfrac{\pi a^2 nA^2}{4z_0}$

and $\kappa = \dfrac{K_1^2(w)}{K_{1-1}(w)K_{1+1}(w)} \cong 1 - \dfrac{1}{\sqrt{1^2+w^2+1}}$ (73)

Close to cutoff, with the approximation of small argument w,[44] $\kappa$ is approximated by

$$\kappa \cong 1 - \frac{1}{\sqrt{1^2+1}} \tag{74}$$

For evanescent wave sensing with PCS fibers using absorption or fluorescence,[42] it is particularly important to express the relative power of a $LP_{lm}$ mode propagating in the cladding:

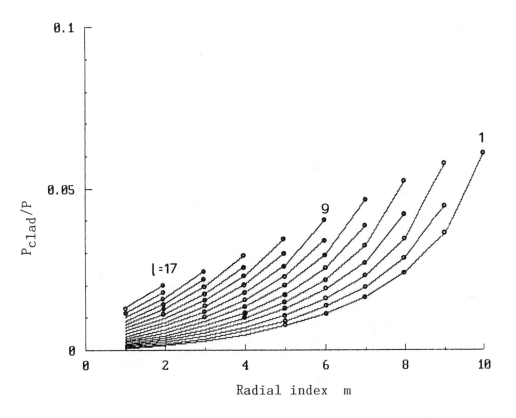

FIGURE 13.   Relative power $P_{clad}/P$ of a step index fiber propagating in the cladding vs. radial mode index m with the azimuthal mode index l as a parameter. Diameter $2a = 50$ μm, core-cladding index difference $\Delta n = 0.01$, wavelength $\lambda = 0.8$ um.

$$\frac{P_{clad}}{P} = \frac{u^2}{V^2}(1 - \kappa) \tag{75}$$

Close to cutoff, $u \cong V$ and $\kappa$ is given by Equation 74:

$$\frac{P_{clad}}{P} \cong \frac{1}{\sqrt{l^2 + 1}} \tag{76}$$

which means that the power confinement within the core is an increasing function of the azimuthal index l. The exact relative propagated power in the cladding for all the modes propagating in a step index fiber is given in Figure 13. It is useful to normalize the modal fields Equations 61 and 62 to the same propagated power P. From Equation 72 the integration constant A is:

$$A = 2\frac{u}{aV}\left[ P\frac{z_0}{\pi n}\kappa \right]^{1/2} \tag{77}$$

Fiber evanescent wave sensing is again concerned by this result: the interaction strength of a modal field in an ultra-thin film coating a PCS fiber is proportional to the average square modulus of the field at the core-cladding interface (see Section IX.C), $|E|^2$, using Equations 61 and 77:

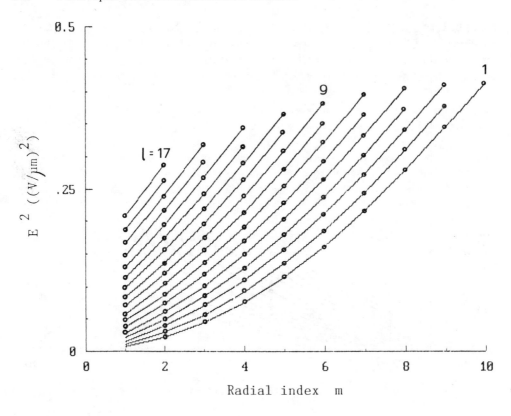

FIGURE 14.   Squared electric field in V/μm at the core-cladding interface of the fiber of Figure 13.

$$|E|^2 = \frac{A^2}{2} \tag{78}$$

The normalized field at the core-cladding interface of all the modes propagating is given in Figure 14.

## E. LEAKY MODES

This section will essentially be devoted to multimode step index fibers. A large amount of work has been done on the leaky mode properties in graded index fibers, their influence on propagation loss, dispersion in communication systems, and on the difficulties they bring in fiber characterization. Step index fibers will only be considered here as the majority of optode systems will make use of them either in the form of all silica fibers or in the form of hard clad, plastic clad, large silica core fibers. There are a few reasons for this choice: leaving aside exotic applications like high resolution time domain fiber spectroscopy, most systems will not be affected by the poor bandwidth of step index fibers (of the order of tens of MHz.km). Step index fibers can capture twice as much power from a diffuse source as a graded index fiber, and they can easily be made of high NA. Besides, in fibers like PCS fibers, the cladding can be stripped off and replaced by an opto-chemical transducer material.

## 1. The Electromagnetic Features of Leaky Modes

Looking back to the scalar wave Equation 10 for weakly guiding step index fibers, one would normally expect that a guided mode ($\beta > n_2 k_o$) immediately leaks out of the core at the cutoff wavelength ($\beta = n_2 k_o$ or $n_e = n_2$) as it does in the case of a planar waveguide when the condition of total refraction is suppressed. However, by looking at Expression 76 giving the ratio of power propagating in the core to the total guided power, one sees that there is still a significant amount of power in the core at cutoff for large azimuthal index l. This is not true for $LP_{om}$ modes which behave similarly to the modes of a planar waveguide and less true for $LP_{lm}$ with low azimuthal index l. This means that the eigenfield shape is not dramatically altered when $\beta$ crosses over its cutoff value $n_2 k_o$ and that it survives beyond it as a guided mode with, however, a leakage term which is the imaginary part of $\beta = \beta_r - j\beta_j$ ($\beta_r < n_2 k_o$). The guided power attenuation coefficient, $2\beta_j$, can be calculated by solving numerically the eigenvalue Equation 63 in the complex plane or by making suitable approximations[46] concerning the leaky modes of interest, i.e., those which contribute in power propagation over significantly long fiber lengths. It is worth pointing out here that a "significant fiber length" means ~1 km for a communication engineer, but ~10 m for an optode designer. This means that a larger part of possible leaky modes may be involved here. This calculation and results can be found in the literature,[46] and what follows is only an attempt to give the main features of these modes and thus help the optode designer take measures to cope with them. Modal analysis is too heavy to describe this leakage phenomenon in a cylindrical waveguide. However, brief recourse to the wave equation can bring a clear insight into this wave tunneling effect: making the change of variable $\Psi = R(r) \Phi$ in Equation 60 gives the wave equation a plane wave form:

$$\frac{d^2\psi}{dr^2} + K^2\psi = 0 \quad \text{where} \quad K^2 = k_0^2 \left( n^2 - n_e^2 - \frac{l^2 - \frac{1}{4}}{k_0^2 r^2} \right)$$

$$\text{and} \quad n = n_1 \text{ in the core}$$
$$n_2 \text{ in the cladding} \tag{79}$$

Let us select a guided $LP_{lm}$ mode of large l close to its cutoff, satisfying the characteristic Equation 63. The nature of its field is given by the sign of $K^2$. $K^2$ is positive (which corresponds to a transverse oscillatory field) in the core area limited by $r_o < r < a$ where $r_o \cong 1/(k_o \sqrt{n_1^2 - n_e^2}$ is the inner caustic point. $K^2$ is negative for $r < r_o$ and $r > a$ where the field decays exponentially. The field nature referring to the sign of $K^2$ is illustrated in Figure 15. If now the wavelength is increased beyond cutoff, the shape of the eigenfield does not change significantly, neither do the azimuthal bias term nor the position of the inner caustic $r_o$. However, with $n_e < n_2$, $K^2$ reaches another zero in the cladding at $r = r_c$:

$$r_c \cong \frac{1}{\left( k_0 \sqrt{n_e^2 - n_2^2} \right)} \tag{80}$$

meaning that, at $r > r_c$, the field becomes oscillatory again, propagating power away from the fiber core as suggested in Figure 15.

It is clearly apparent that the rate of power leakage is strongly dependent on the azimuthal index l as the position of the outside causing $r = r_c$ is proportional to l and that the field in the tunneling region $a < r < r_c$ decays like $(r/a)^{-1}$ close to cutoff.[47] As $n_e$ decreases, the outside caustic gets closer to a. At the value $\beta \cong \sqrt{n_2^2 k_o^2 - l^2/a^2}$ there is total leak and the mode becomes a refracted leaky mode losing its power very rapidly.

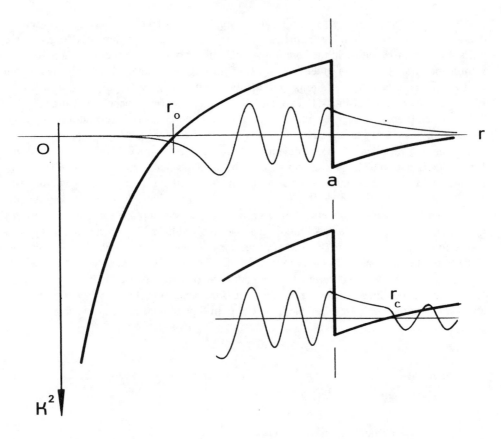

FIGURE 15. Radial dependence of the $K^2$ term in Equation 79 and representation of the modal field on either side of the cutoff condition. $r_o$, a, $r_c$: inner caustic, fiber radius, and outer caustic.

## 2. Propagation Capability of Leaky Modes

The refracted leaky modes can be discarded practically as only bound modes and tunneling leaky modes compete to convey light to and from an opto-chemical sensor. There is a finite number of tunneling leaky modes $N_{tlm}$ is a given structure. Considering large structures, or $\lambda \rightarrow 0$, the integration of all allowed transverse eigenvalues over the phase space pertaining to each mode family yields:[48]

$$N_{bm} \cong \frac{V^2}{2}$$

$$N_{tlm} \cong \frac{V^2}{2}\left(1+\frac{\sqrt{2\Delta}}{3\pi}\right) \quad \text{for } \theta_c \ll 1 \qquad (81)$$

$$\text{with } \theta_c = \text{arc cos } n_2/n_1$$

$$N_{tot} \cong \frac{V^2}{4\Delta}$$

where $N_{bm}$, $N_{tlm}$, $N_{tot}$ stand for bound, tunneling leaky, and total mode number and $\sqrt{2\Delta} =$

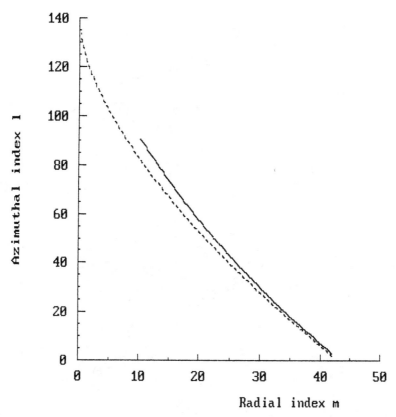

FIGURE 16. Radial and azimuthal index diagram of a PCS fiber with a = 50 μm, Δn = 0.04, V ≅ 140. Dashed line: exact bound mode limit. Solid line: limit of leaky modes exhibiting less than 1 dB/km attenuation.

NA/n. The total number of bound and tunneling modes is nearly equal. In low numerical aperture fibers the ratio of bound and tunneling modes to the total mode number is proportional to 4NA/n. The attenuation of tunneling modes is given by an approximation[49] of the therefore complex characteristic Equation 63. Following Marcuse,[50] the result of the latter can be vividly expressed in the (lm) mode diagram of Figure 12: Figure 16 shows two lines. The dashed line is the border line of the bound modes calculated exactly from Equation 64. The solid line is the border line of the leaky modes suffering a power loss lower than 1 dB/km; all modes below this line, with approximately the same mode density as below the bound mode line, contribute efficiently in conveying light over long lengths of fiber. As the fiber lengths considered in sensors often do not exceed a few meters, it is expected that a larger amount of leaky modes will be involved in the total optical power conveyance as compared to the optical communication case. The tunneling leaky mode attenuation will be considered hereafter using the more convenient ray approach.

### 3. Leaky Modes — Leaky Rays

In the large V limit, the field of a leaky mode can be represented by the family of rays impinging on the core-cladding surface making the same angle with the cylinder generating line and whose projection on the cross-section makes the same angle with the normal of the core-cladding boundary, therefore undergoing the same partial reflection loss. Referring to Figure 17, showing a fiber cross-section, the classification of rays is made in three groups:

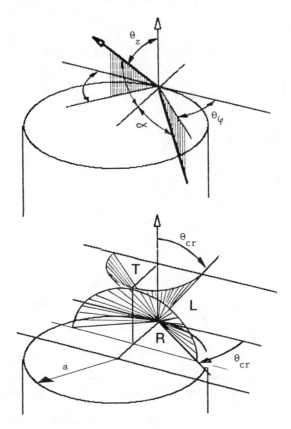

FIGURE 17.    Classification of rays at the core-cladding boundary of a step index fiber. Zones T, R, and L are for trapped, refracted, and leaky ray domains, respectively. $\theta_z$ is the angle between a ray and the core's generating line. $\theta_\phi$ is the angle in the cross-section plane between the ray projection and the tangent to the core.

trapped rays : $0 < \theta_z < \theta_c$  with

$$\beta = n_1 \, k_0 \cos\theta_z, \quad \cos\theta_c = n_2/n_1$$

$\theta_c$  being  defined with respect to the fiber axis

refracted rays : $\theta_N < \dfrac{\pi}{2} - \theta_c$

leaky rays :      $\theta_z > \theta_c$

$$\theta_N > \frac{\pi}{2} - \theta_c \tag{82}$$

The boundary between leaky and refracted rays is $\sin\theta_c = \sin\theta_z \cdot \sin\theta_\phi$. It can be seen that the existence of a leaky ray is necessarily associated with a ray skewness, i.e., to a modal azimuthal index $l \neq 0$.

We are now first examining how the attenuation of a ray depends on its input parameters $\theta_z$, $\theta_\phi$: The attenuation $\alpha$ of a leaky ray is given by the simple expression:[51]

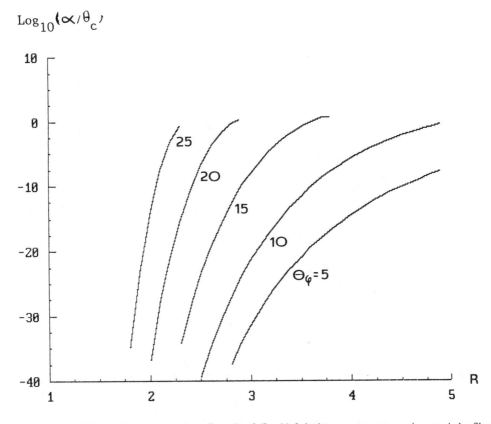

FIGURE 18. Relative leaky ray attenuation $\alpha/\theta_c$ vs. $R = \theta_z/\theta_c$ with $\theta_\phi$ in degrees as a parameter in a step index fiber with V = 1000.

$$\frac{\alpha}{\theta_c} \cong 2R^2 \sqrt{1 - R^2 \sin^2 \theta_\phi} \left[ \exp - \frac{2V}{3} \frac{\left(1 - R^2 \sin^2(\theta_\phi)\right)^{3/2}}{\left(R^2 - 1\right)} \right] \tag{83}$$

with $R \cong \theta_z/\theta_c$. It can be seen that $\alpha$ is very much dependent on the fiber radius a and on the wavelength involved in V, whereas the refracted ray attenuation is independent of them as a consequence of geometrical optics. The same leaky ray will experience a lower attenuation in a larger core fiber. The attenuation $\alpha/\theta_c$ vs. $\theta_z/\theta_c$ with $\theta_\phi$ as a parameter in the case of a fiber with V = 1000, typical of a PCS fiber of 250 μm radius used in the visible range is shown in Figure 18. The rays of lower attenuation are those of small $\theta_\phi$, i.e., the most skew rays. Figure 19 illustrates $\alpha/\theta_c$ vs. $\theta_\phi$ with $\theta_z/\theta_c$ as a parameter in the same fiber. What was said about an individual leaky ray attenuation can be used to express the fiber power transmission once the excitation distribution is defined. This is quite easy in the present field of interest as in most cases the excitation is performed with LEDs or fluorescent tip coatings into large core step index fibers.

As described in Section VI, expressions 120 to 124, such a diffuse source provides a Lambertian intensity distribution of brightness B. If the source is placed against the core cross-section, the total transmitted power is obtained by integrating Equation 83 over the core area and the solid angle intercepted by the various types of rays.[52] This yields the trapped and leaky ray power $P_{tr}$ and $P_{lr}$ respectively:

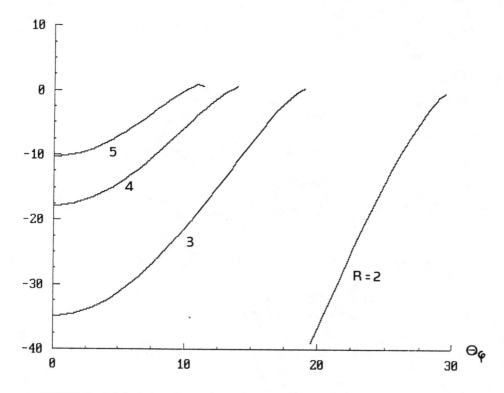

FIGURE 19.    Relative leaky ray attenuation $\alpha/\theta_c$ vs. $\theta_\phi$ with $R = \theta_z/\theta_c$ as a parameter V = 1000.

$$P_{tr} = \pi^2 a^2 B\, \theta_c^2 \tag{84a}$$

which remains constant along an ideal lossless fiber. $P_{lr}(z)$ is a function of z that must be evaluated numerically; however, one finds the launched power:

$$P_{lr}(0) = 8\pi a^2 B\, \theta_c^2 \left[ \frac{\pi}{8} - \frac{\sqrt{1-\theta_c^2}}{4\theta_c} + \theta_c\left(\frac{1}{4\theta_c^2} - \frac{1}{2}\right) \right] \tag{84b}$$

As stated previously and shown in Figure 20, the power ratio $P_{lr}(o)/(P_{tr} + P_{lr})$ of leaky rays excited at the fiber input is close to 50% for all practical fibers. If the source is transparent, i.e., consists of a fluorescent material placed at the fiber tip, the cos $\theta$ behavior of the Lambertian source should be somewhat modified: radiation coming from planes further away from the fiber tip is comparable to that of a partially diffuse source exciting more trapped rays than leaky rays.

The total power fraction remaining in the fiber vs. z is shown in Figure 21. The longitudinal dependence of the leaky ray total power exhibits an unconventional behavior which differs from an exponential decay: a first part (a), where the leaky ray power falls sharply, followed by a second part (b), where the remaining power leaks very slowly. The power loss decreases as V is increased.

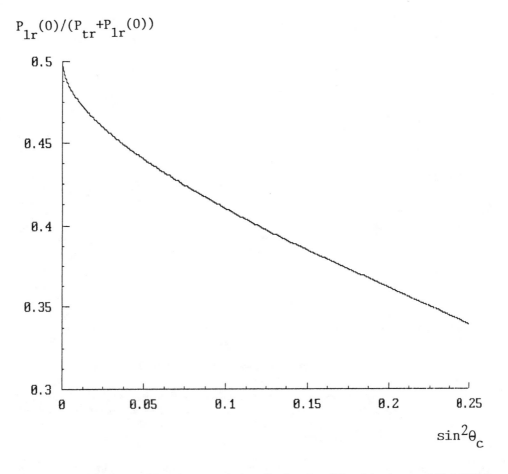

FIGURE 20. Relative power of the leaky rays excited at a fiber input by a diffuse light source vs. $\sin^2\theta_c$ $(=NA^2/n^2)$.

The overall transmission behavior of a step index fiber under diffuse illumination can be described by the fiber length $(z/a)_{1/2}$ after which the leaky ray total power has dropped by 50%.[52] $(z/a)_{1/2}$ can be very roughly represented for $\theta_c \geq 0.1$ in the condensed form of an empirical and simplistic formula:

$$\log_{10}\left(\frac{z}{a}\right)_{1/2} = 1 + 0.02(V - 25) \cdot \text{tg}\left[0.87 + 1.05\left(\theta_c - 0.1\right)\right] \qquad (85)$$

## F. CURVATURE LOSS

Giving a straight symmetrical step index planar waveguide a uniform curvature modifies the reflection conditions of the bouncing rays of Figure 3 on the core-cladding boundaries. The angle of incidence at the inner wall is decreased whereas that at the outer wall is increased provoking the leakage of all rays impinging on the latter at an angle larger than the critical angle. The loss mechanism of the modes propagating in an optical waveguide, e.g., an optical fiber, can be described by a conformal mapping[53] whereby a bent fiber of index profile $n_0(r)$ is equivalent to a straight fiber with a perturbed index profile:

$$n(r,\theta) = n_0(r)\left[1 + \frac{r}{R}\cos\theta\right] \qquad (86)$$

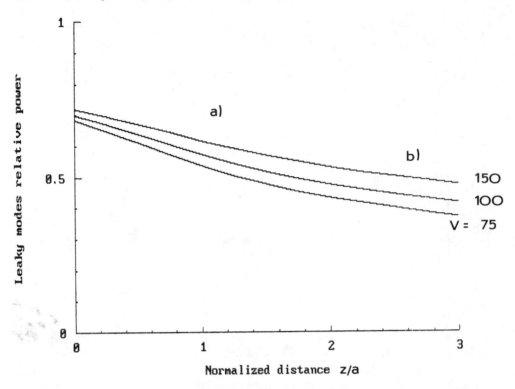

FIGURE 21.   Typical longitudinal dependence of the relative power $P_{lr}(z)/P_{lr}$ propagated by all leaky rays excited by a diffuse source vs. z/a in a $Log_{10}$ scale. The fiber is a PCS fiber with $\theta_c = 0.25$. The V-number is the parameter. In a 50-μm diameter fiber the three curves represent three spectral components at 800, 600, and 400 nm. (a) High leakage zone. (b) Beginning of the low leakage zone.

where R is the curvature radius of the bend and $\theta$ the azimuthal fiber coordinate having its origin in the plane of the bend. The resulting index profile is similar to that of Figure 15, i.e., of the case of a mode below its cutoff, leaking by tunneling into a high index power sink. The bending loss coefficient $2\alpha$ in an optical waveguide is proportional to:[54]

$$\exp\left[ -\frac{2}{3}n_2 k_0 R \left( \left(\frac{n_e^2}{n_2^2} - 1\right) - \frac{2a}{R} \right)^{3/2} \right] \tag{87}$$

For a given radius of curvature R, the loss becomes only significant when the effective index $n_e$ of a mode is below a "cutoff" value, set to be the value making the exponent of Equation 87 unity. If 2a/R is small enough regarding $(n_e^2/n_2^2 - 1)$, this sets a lower limit for R:[55]

$$R \gg \frac{3\lambda}{\pi\left(n_e^2 - n_2^2\right)^{3/2}} \tag{88}$$

Opto-chemical sensors mostly utilize large core step index fibers. The bending loss characteristics are thus best described by the number N of modes lost under curvature radius R relative to the total number N of modes propagating in a straight fiber:[56]

$$\frac{N_\infty - N_c}{N_\infty} = \frac{1}{2\Delta}\left[\frac{2a}{R} + \left(\frac{3\lambda}{4\pi n_2 R}\right)^{2/3}\right]$$

$$\text{with } \Delta = \frac{\Delta n}{n_2} = \frac{NA^2}{2n_2^2} \tag{89}$$

In large core fibers the wavelength dependent term exceeds the core radius term only for very large curvature radius, $R \gtrsim 300 \, \alpha^3/\lambda^2$, i.e., where the bending loss is negligible anyhow. R is practically limited by the maximum acceptable strain b/R in a curved fiber of outer radius b. Communication fibers are usually subject to a "screen test" corresponding to 0.2% strain, which is the limit that b/R, a fortiori a/R, should not exceed. Therefore $N/N_\infty < 0.2 \cdot 10^{-2}/\Delta$. It can then be said that the total number of bound modes remaining guided at the smallest possible curvature radius is higher than 80%. In addition, the wavelength dependence of bending loss is negligibly small. A multimode fiber can thus be considered as an achromatic bound mode conveyor with regard to bending loss.

In integrated optic waveguides the curvature is not associated with strain. The limitation is merely optical. From Equation 88 it can be seen that very small curvature radii can be considered for single mode waveguides. For instance, a single mode guide made on a glass substrate by ion exchange showing an index difference of 1% can in principle be subject to a bending radius as small as a few millimeters provided the guide path is a perfect circle. Heavily multimode light channels are, however, much more bending sensitive,[83] as shown in Figure 22. The bending losses increase with the waveguide radius and the numerical aperture at constant wavelength.

## G. MICROBENDING LOSS

Real fibers installed in sensor systems are not uniformly straight or curved. Core diameter or ellipticity fluctuations as well as nonuniformity of coating thickness are always present to some extent. Moreover, the fiber is subject to various winding or cabling strains and temperature bimorphous effects. All these effects lead to a distributed perturbation of the fiber straightness, the significant spectral components of which are between a fraction of millimeter and a few centimeters. The loss caused by these so-called microbending effects cannot be described as curvature losses but rather as a mode coupling effect. The most efficiently coupled modes, or groups of modes, i and j, of propagation constants $\beta_i$ and $\beta_j$, are those for which the phase matching condition is achieved by a strong spectral component of the microbending. Modes close to cutoff will be coupled in turn to leaky and radiation modes. This type of distributed mode coupling has been extensively studied in both multimode[57] and single mode[58] fibers. It is, however, difficult to quantify. The main characteristics to retain in the optode sensor field is the weak wavelength dependence of the microbending loss effect that was demonstrated in multimode fibers both theoretically[59] and experimentally on long fiber length.[60] Such a typical graded index multimode fiber characterization is shown in Figure 23. The transmission spectrum is measured with the fiber on its spool, subject to the microbends caused by the overlap of the windings (curve a). It is then represented versus $1/\lambda^4$ and compared with the straight line corresponding to the Rayleigh scattering loss alone (curve b). Figure 23 demonstrates that the microbending loss effect is basically wavelength independent in that it only amounts to a residual loss which translates curve b vertically. This result is typical of long fiber lengths. An extensive experimental study of the wavelength dependence

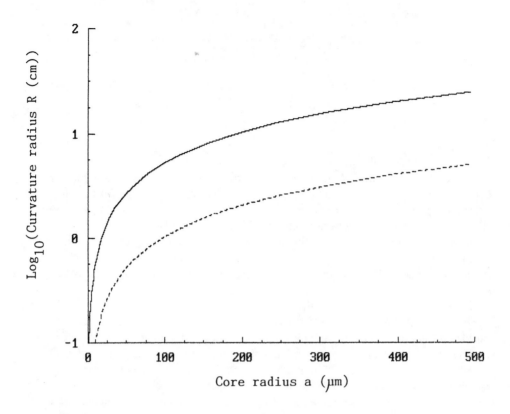

**FIGURE 22.** Bending radius R of a step index fiber provoking the leakage of 20% of the bound modes vs. core radius a in microns at 800 nm wavelength. Solid line: SBF (silica-based fiber with $\Delta n = 0.01$), dashed line: PCS (plastic-clad silica fiber with $\Delta n = 0.05$). R is expressed in centimeters on a $\log_{10}$ scale.

of the transmission spectrum of fibers for sensor application was recently performed.[61] It reveals that careful precautions must be taken in short fiber lengths if a simple two-wavelength sensor referencing is to be used. This is in accordance with the expectation of Section IV.E that leaky modes, and especially the wavelength dependence of their attenuation, play an important role and that a particular design effort is needed for a proper encoding of the transducer's information.

The microbending spatial spectrum of a fiber can be intentionally narrowed in order to selectively and efficiently couple modes[62] or groups of modes[63] as in the case of graded index multimode fibers. This scheme has been widely used in distributed physical sensors with Optical Time Domain Reflectometry (OTDR) readout. Chemical species could also be detected, located, and to some extent measured, if a transducing mechanism can be found that converts the chemical event into a mechanical effect.

## V. THE PERTURBATION ANALYSIS

Perturbation methods have long been used in the field of guided electromagnetic waves. The idea is to consider a complex structure as a perturbation of a simple one which has a known solution. The condition of validity for this is that the field of the perturbed structure does not differ significantly from that of the unperturbed structure. A large number of effects in fiber or integrated optics fulfill this condition for a number of reasons: transverse index distributions are smooth in general or exhibit small index jumps; as opposed to their micro-

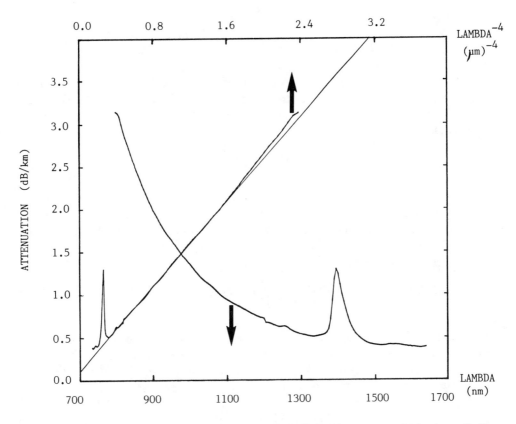

**FIGURE 23.** Transmission spectrum of a multimode graded index fiber subject to strong microbends near the fiber input, demonstrating the negligible wavelength dependence of the microbending loss. The microbends do not affect the general $1/\lambda^4$ Rayleigh scattering loss dependence and only contribute by a wavelength independent loss offset. (Courtesy of Cabloptic SA, CH-2016 Cortaillod.)

wave counterparts, guided wave optic coupling components, circuits, or transducers are often uniform or adiabatic with very weak interaction strength over lengths of hundreds or thousands of wavelengths.

## A. PERTURBATION FORMULA FOR THE PROPAGATION CONSTANT

We start by placing ourselves in the weak guidance condition ($\Delta \ll 1$). The unperturbed waveguide has an index profile $n(x,y)$ and a known eigenvalue $\beta$ and electric eigenfield $\Phi(x,y)$. The perturbed index distribution is $n_p(x,y)$ with unknown eigenvalue $\beta_p$ and electric eigenfield $\Phi_p(x,y)$. $\Phi$ and $\Phi_p$ are defined here as the transverse field function. The full spatial dependence must be completed by the propagating term $e^{-j\beta z}$. Both structures are described by the weak guidance wave Equation 10:

$$\left(\nabla_t^2 + k_0^2 n^2 - \beta^2\right)\Phi = 0 \tag{90}$$

$$\left(\nabla_t^2 + k_0^2 n_p^2 - \beta_p^2\right)\Phi_p = 0 \tag{91}$$

Multiplying Equation 90* by $\Phi_p$, Equation 91 by $\Phi^*$, then subtracting the two equations and integrating over the whole cross-section yields:

$$\beta_p^2 - \beta^2 = \frac{1}{\int \Phi^* \Phi_p d\sigma} \left\{ k_0^2 \int \left( n_p^2 - n^2 \right) \Phi^* \Phi_p d\sigma + \int \left( \Phi^* \nabla_t^2 \Phi_p - \Phi_p \nabla_t^2 \Phi^* \right) d\sigma \right\} \quad (92)$$

The last term of Equation 92 is zero for bound fields, as can be seen by applying the second Green's identity. Assuming a small relative perturbation, $\delta\beta/\beta \ll 1$ where $\delta\beta = \beta_p - \beta$, and that $\Phi_p \cong \Phi$, Equation 92 gives the general perturbation formula for weakly guiding waveguides:

$$\delta\beta = \frac{k_0^2}{2\beta} \frac{\int \left( n_p^2 - n^2 \right) \Phi^2 d\sigma}{\int \Phi^2 d\sigma} \quad (93)$$

where the dielectric perturbation can be transversely distributed or localized, real or complex. If $\Phi$ is the y electric field component, then, as in a plane wave, $n\sqrt{(\epsilon_o/\mu_o)}\,\Phi$ is the x magnetic field component of the weakly guiding structure. The propagated power P is therefore $P = 1/2\,n\sqrt{(\epsilon_o/\mu_o)} \int_\infty \Phi^2\,d\sigma$. Therefore, Equation 93 becomes:

$$\delta\beta = \frac{\omega\epsilon_0}{4P} \int \left( n_p^2 - n^2 \right) \Phi^2 d\sigma \quad (94)$$

This perturbation formula is very useful in practical sensor applications where the measurand acts as a dielectric perturbation of the waveguide over some length. This result is actually not limited to weakly guiding structures. Reciprocity considerations on the vector wave equation would show that, more generally:[64]

$$\delta\beta = \frac{\omega\epsilon_0}{4P} \int \left( n_p^2 - n^2 \right) |E|^2\ d\sigma \quad (95)$$

where E is the electric vector field. This formula should, however, be used with care when the various E-field components involved are not all continuous at the boundaries of the dielectric perturbation.[65]

## B. TWO-WAVEGUIDE COUPLING

Distributed coupling between two waveguides, or between two propagating modes, is an essential guided wave effect that has been widely used in passive as well as in active devices in optical communications. This mechanism will also find applications in guided wave optical preprocessing of the light signals returned by an optode transduced. It will probably even be used in the optode transducer itself as means of enhancing sensitivity and/or improving wavelength selectivity. A few suggestions will be described in sections VII.B.3 and VIII.A.3. What follows is not a sound development of the laws governing the coupling effects. This can be found very well done elsewhere.[66] We will just try and make plausible a few rules and properties.

Let us consider two weakly guiding identical waveguides $g_1$ and $g_2$ placed at some distance from each other. The case where coupling takes place between identical waveguides is termed "degenerate coupling". Although it is not a necessity, the waveguides are considered as single mode. They are not too far apart so that the eigenmode of each guide $g_i$ cannot ignore the boundary conditions imposed by the other guide $g_j$. They are not too close either so that the resulting eigenfield does not differ too much from the simple superposition of the eigenmodes of each waveguide $g_i$. We will establish the link between two complementary mental representations of the power exchange between the two waveguides:

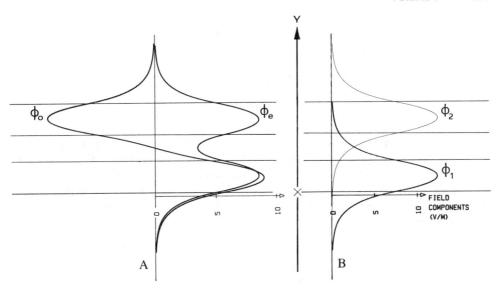

FIGURE 24. Modal fields involved in degenerate coupling between two neighboring waveguides (A) Even and odd eigenfields $\Phi_e$ and $\Phi_o$ of the complete two-waveguide structure. (B) Coupled single-waveguide eigenmode fields $\Phi_1$ and $\Phi_2$ overlapping, e.g., in guide 2.

1.  The coupling can be seen as the longitudinal beat of the two eigenmodes of the thus bimodal complete two-waveguide structure. These two modes are represented in Figure 24a. They have propagation constants $\beta_e$ and $\beta_0$ and eigenfield $\Phi_e$ and $\Phi_o$ for the even (dominant) and odd modes. Their beat length $L_b$ is the propagation length necessary for their phase difference to be $2\pi$:

$$L_b = \frac{2\pi}{\delta\beta} \quad \text{where} \quad \delta\beta = \beta_e - \beta_0 \tag{96}$$

2.  The coupling can also be represented as an interaction with a certain coupling strength, $\kappa$, between two transversely resonant fields $\Phi_1$ and $\Phi_2$ having the same propagation constant $\beta = \beta_1 = \beta_2$ and eigenfield shape as shown in Figure 24b.

The link between these two representations will be established by using a perturbation approach: the structure subject to the latter is the complete two-waveguide structure. The unperturbed structure is that with infinite distance between $g_1$ and $g_2$. The unperturbed eigenfields $\Phi_{e\infty}$ and $\Phi_{o\infty}$ of the full structure can be written exactly as a superposition of the eigenmode fields $\Phi_1$ and $\Phi_2$ of the individual guides $g_1$ and $g_2$ in isolation:

$$\Phi_{e\infty} = \Phi_1 + \Phi_2 \qquad \Phi_{o\infty} = \Phi_1 - \Phi_2 \tag{97}$$

The unperturbed eigenvalues $\beta_e$, of $\Phi_e$, and $\beta_o$, of $\Phi_o$, are obviously equal and equal to $\beta$. The perturbed structure is obtained from the unperturbed one by bringing $g_1$ and $g_2$ closer to each other. The unknown propagation constants of the perturbed structure, $\beta_{ep}$ and $\beta_{op}$, will be calculated using the perturbation formula 94 for the even and odd eigenfields. As the structure is symmetrical, the integral extends only over half of the structure:

$$\delta\beta_e = \beta_e - \beta = -\delta\beta_0 = \beta - \beta_0$$

$$= \frac{2\omega\varepsilon_0}{4(2P)} \int_{g_i} \left(n_p^2 - n^2\right)\Phi_1\Phi_2 \, d\sigma \tag{98}$$

where 2P is the total power propagated by the even or odd mode, i.e., twice the power P carried by the single waveguide $g_i$. $(n_p^2 - n^2)$ is the permittivity bump over the constant cladding value. It represents the presence of waveguide $g_i$. Equation 98 establishes that the propagation constants $\beta_e$ and $\beta_0$ of the even and odd modes can be expressed by means of the modal fields $\Phi_1$ and $\Phi_2$ of waveguides $g_1$ and $g_2$ in isolation in the form of a field overlap integral weighted by the permittivity load of one of the guides as illustrated in Figure 24b. The coupling coefficient $\kappa$ is defined as:

$$\kappa = \frac{\omega\varepsilon_0}{4P} \int_{g_i} \left(n_p^2 - n^2\right)\Phi_1\Phi_2 \, d\sigma \tag{99}$$

From Equation 96:

$$\kappa = \frac{\delta\beta}{2} = \frac{\pi}{L_b} \tag{100}$$

## C. THE COUPLED MODE EQUATION

The total field $\Phi(x,y)$ propagating in the two identical waveguide weakly guiding structure may now be expressed as a superposition of the field $\Phi_1, \Phi_2$ of the waveguides in isolation with coefficients $A_1(z)$ and $A_2(z)$ which can be slowly varying functions of z in order to formally allow power exchange between waveguides:

$$\Phi = A_1\Phi_1 + A_2\Phi_2 \tag{101}$$

$\Phi$ satisfies the scalar wave equation in the complete structure. $\Phi_i$ does not:

$$\nabla^2\Phi + k_0^2 n^2(x,y)\Phi = 0 \tag{102}$$

where n(x,y) is the index distribution of the complete structure. Introducing Equation 101 into Equation 102 yields

$$A_1\left[\nabla^2\Phi_1 + k_0^2 n^2\Phi_1\right] + A_2\left[\nabla^2\Phi_2 + k_0^2 n^2\Phi_2\right] - 2j\beta_1\left(\frac{\delta A_1}{\delta z}\Phi_1 + \frac{\delta A_2}{\delta z}\Phi_2\right) = 0$$

where  $\dfrac{\delta^2 A_i}{\delta z^2}$  were neglected $\tag{103}$

From last section the $\Phi_i$ can be expressed in terms of the even and odd eigenmode $\Phi_{e,o}$ as far as weak coupling is concerned:

$$\Phi_{1 \atop 2} = \frac{1}{2}\left(\Phi_e e^{\frac{-j\delta\beta_z}{2}} \pm \Phi_0 e^{\frac{+j\delta\beta z}{2}}\right) \quad (104)$$

Introducing Equation 104 into Equation 103, using the fact that $\Phi_e$ and $\Phi_o$ satisfy Equation 102, expressing explicitly the z derivatives and considering $\beta_e$, $\beta_o$, and $\beta_1$ as nearly equal, regrouping the terms multiplying the eigenmode $\Phi_o$ and $\Phi_e$, one is left with:

$$\frac{\partial A_1}{\partial z} = j\frac{\delta\beta}{2}A_2 = j\kappa A_2$$

$$\text{and} \quad \frac{\partial A_2}{\partial z} = j\kappa A_1 \quad (105)$$

Substitution yields $\partial^2 A_i/\partial z^2 = -\kappa^2 A_i$ which states that the field in either waveguide $g_i$ is a sinusoidal function of z. The power $P_i$ in either guide $g_i$ is proportional to $|A_i|^2$:

$$P_1 = P_0 \cos^2 \kappa z \quad (106)$$

Total guided power conservation implies that $P_2 = P_o (1-\cos^2 \kappa z)$.

## D. NONDEGENERATE COUPLING

The degenerate coupling configuration above is only a particular case of the more general coupling problem involving nonidentical waveguides or simply modes with nonequal eigenvalues $\beta_1 \neq \beta_2$. Substituting to $A_i$ in Equation 105 new variables $B_i = A_i e^{-j\beta_{1}z}$ gives Equation 105 the new form:

$$\frac{\partial}{\partial z}\begin{bmatrix} B_1 \\ B_2 \end{bmatrix} = M\begin{bmatrix} B_1 \\ B_2 \end{bmatrix} \quad \text{where} \quad M = j\begin{bmatrix} -\beta_1 & \kappa \\ \kappa & -\beta_1 \end{bmatrix} \quad (107)$$

This is the traditional formalism of coupled wave phenomena.[67] In its general expression for nonsynchronous waves, the diagonal contains two different $\beta$-values, $\beta_1 \neq \beta_2$; the coupling coefficients are designated by $\kappa_{12}$ and $\kappa_{21}$ and M becomes

$$M = j\begin{bmatrix} -\beta_1 & \kappa_{12} \\ \kappa_{21} & -\beta_2 \end{bmatrix} \quad (108)$$

where the expressions of $\kappa_{ij}$ may still remain the field overlap given by Equation 99 with $\kappa_{12} = \kappa_{21} = \kappa$. However, a recent discussion has shown that for $\kappa_{12} = \kappa_{21}$ to be truly satisfied, a more complex expression of $\kappa$ should be used.[68] In cases where $\beta_1$, $\beta_2$, and $\kappa$ are constant along z, the solution to Equation 107 is easily obtained after the new change of variables $C_i = B_i \exp [j(\beta_1 + \beta_2) z/2]$. M takes the form:

$$M = j\begin{bmatrix} j\dfrac{\Delta}{2} & \kappa_{12} \\ \kappa_{21} & -j\dfrac{\Delta}{2} \end{bmatrix} \quad (109)$$

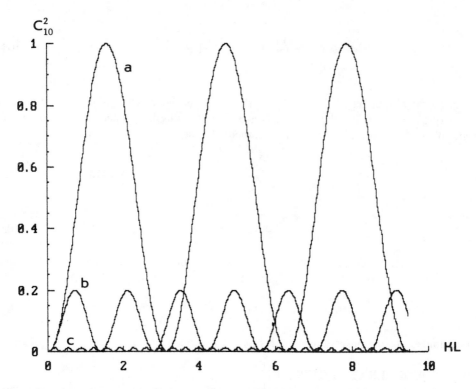

FIGURE 25. Normalized coupled power between two guided modes vs. normalized coupling length $\kappa z$ with the normalized phase mismatch $\delta = \Delta/2\kappa$ as a parameter. Curves a, b, and c are for $\delta = 0$, $\sqrt{(2)}$, and 3, respectively.

The solution is:

$$
\begin{bmatrix} C_1(z) \\ C_2(z) \end{bmatrix} = \begin{bmatrix} \cos z\mu + j\dfrac{\Delta/2}{\mu}\sin z\mu & j\,\dfrac{\kappa}{\mu}\sin z\mu \\[2ex] j\,\dfrac{\kappa}{\mu}\sin z\mu & \cos z\mu - j\dfrac{\Delta/2}{\mu}\sin z\mu \end{bmatrix} \begin{bmatrix} C_{10} \\ C_{20} \end{bmatrix} \tag{110}
$$

where $\mu = \sqrt{(F(\Delta,2)^2 + \kappa^2)}$, $\Delta = \beta_2 - \beta_1$; $C_{10}$, $C_{20}$ are the field amplitudes at the input of a waveguide coupling section. The meaning of Equation 110 is manyfold and can be appreciated easily after making $C_{20} = 0$, i.e., exciting waveguide $g_1$ with power $|C_{10}|^2$: the power exchange cannot be 100% any more as $\kappa/\mu < 1$ when the coupled waves are not in synchronism ($\beta_1 \neq \beta_2$). $\Delta = \beta_2 - \beta_1$ is the phase mismatch. The spatial period of the power exchange is

$$
L_b = \pi/\mu \tag{111}
$$

i.e., shorter than $L_b$ in the synchronous case where $L_b = \pi/\kappa$.

Various cases of power exchange with $\delta = \Delta/2\kappa$ as a parameter are shown in Figure 25. One important feature of such coupling regarding optode circuits where spectral sensing and encoding is used is its dependence on wavelength: $\lambda$ is contained in $\kappa$ as shown in an example in Sections VII.B.3.a and VII.B.3.b. $\lambda$ is also involved in $\Delta$ via $k_o = 2\pi/\lambda$. A directional coupler can therefore be a key element in future wavelength selective opto-chemical transducers and integrated wavelength demultiplexers.

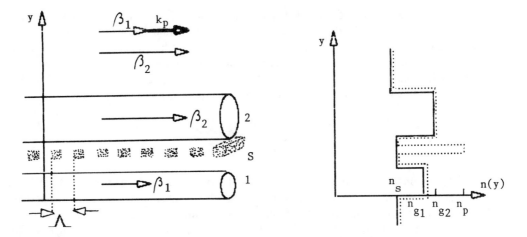

FIGURE 26. Nondegenerate coupling between modes with a periodical dielectric perturbation of transverse area S to restore phase matching.

## E. RESTORATION OF SYNCHRONISM WITH A PERIODICAL PERTURBATION

The last subject of this discussion on coupled waves is on the establishment of phase matching in an otherwise nondegenerate ($\beta_1 \neq \beta_2$) wave coupling. As above, its importance regarding opto-chemical sensors lies partially in its wavelength properties.

Reversing the change of variables C to A (107,108) gives the general form of Equation 110 in the nondegenerate case:

$$A'_1 = \kappa_{12} A_2 e^{-j\Delta z} \qquad (112)$$

The phase mismatch appears in the exponential term. It appears clearly that the synchronous coupling can be restored if $\kappa_{12}$ is given a periodical z dependence, say a cosine function:

$$\kappa_{12} = \kappa \cos k_p z = \kappa \frac{1}{2} \left( e^{jk_p z} + e^{-jk_p z} \right) \qquad (113)$$

where $k_p = 2\pi/\Lambda$ is the perturbation constant, $\Lambda$ is its spatial period, or pitch. Introducing Equation 113 into Equation 112 one sees that the resulting phase mismatch term can be cancelled if either $k_p$ or $-k_p$ satisfies the consequently called "phase matching condition":

$$-\Delta \pm k_p = 0 = \beta_2 - \beta_1 = \pm k_p \qquad (114)$$

The phase matching condition (114) restoring the degenerate coupling conditions is actually very wavelength dependent as $\Lambda$ is fixed and $\Delta$ contains $1/\lambda$. The phase matching condition can thus be described with an effective mismatch term $\Delta_e = \beta_2 - \beta_1 \pm k_p$.

The said perturbation in the optode field will most often be a corrugation grating etched into or deposited on the waveguide. The coupling coefficient can still be given by Formula 99 as described in section VII.B.3.c in the case of a corrugation grating.

As illustrated in Figure 26, let us consider the coupling of two modes that can be two modes of different waveguide or belonging to the same waveguide. n(x,y) is the unperturbed index distribution of the mismatched structure. Let us imagine a grating sheet, tube, or stripe along

z. The grating can be placed on a waveguide or at some distance from it or else between waveguides. This longitudinally periodical dielectric perturbation is considered as a dielectric load of index $n_p$ in place of the cladding index $n_s$. It gives a relative permittivity perturbation term $(n_p^2 - n_s^2) \cos k_p z$ occupying an area S on the cross-section. (In the case of corrugation grating etched into the waveguide $n_p$ is simply the waveguide index. The integral of Equation 94 extends only where the phase matching periodical perturbation acts, i.e., over S:

$$\kappa = \frac{\omega \varepsilon_0}{16P} \int_S \left( n_p^2 - n^2 \right) \Phi_1 \Phi_2 d\sigma \qquad (115)$$

where $\Phi_1$ and $\Phi_2$ are the two coupled electric fields normalized to propagated power P. Most often corrugation gratings are made quite efficient with high index $n_p$ ($Al_2O_3$, $TiO_2$). They can be very thin (less than 100 nm). Therefore, the surface integral of Equation 115 reduces to a line integral of the field product multiplied by $(n_p^2 - n^2)$ t, where t is the corrugation full thickness. In the case of high dielectric perturbation, some precaution must be taken when the modal electric field is not continuous at the perturbation boundary.[69]

## F. CONTRADIRECTIONAL COUPLING

A case where the wavelength dependence is very critical is the contra-directional coupling where a grating couples a forward propagating mode backwards to itself or to the backward propagating mode of a neighboring waveguide. Going back to the phase matching Equation 114 assuming $n_{e1}$ and $n_{e2}$ weakly wavelength dependent, the wavelength sensitivity of the phase mismatch term $\Delta_e$ is proportional to the effective index difference $(n_{e1} - n_{e2})$ which becomes $2n_{e1}$ in backward coupling. The condition of guided power conservation in the case of two coupled contra-propagating waves implies from Equation 105 that $\kappa$ is real.

The resolution of the coupled wave Equation 109 in this case yields hyperbolic sine and cosine functions of argument $\mu z$ where:[70]

$$\mu^2 = \kappa^2 - \left( \frac{\Delta_e}{2} \right)^2$$

with

$$\Delta_e = 2\beta - k_p \qquad (116)$$

If unit power $|C_1(0)|^2 = 1$ is incident on the grating, the reflected power at abcissa z under the grating is

$$|C_2(z)|^2 = |C_1(0)|^2 \operatorname{sh}^2 \frac{(\mu(z-L))}{(\operatorname{ch}^2 \mu L - \delta^2)} \qquad \text{with } \delta = \frac{\Delta_e}{2\kappa}$$

where the grating of length L starts at z = 0. The net reflected power at the grating input is:

$$|C_2(0)|^2 = |C_1(0)|^2 \left[ \delta^2 + \left( 1 - \delta^2 \right) \operatorname{cth}^2 (\mu L) \right]^{-1} \qquad (117)$$

As opposed to $\mu$ in the codirectional case, the radical can become negative if the phase mismatch term $\Delta_e$ becomes larger than $2\kappa$, or $\delta > 1$. This coupler works as a band rejection filter with two regimes which will be best described by introducing the optical frequency $\omega$

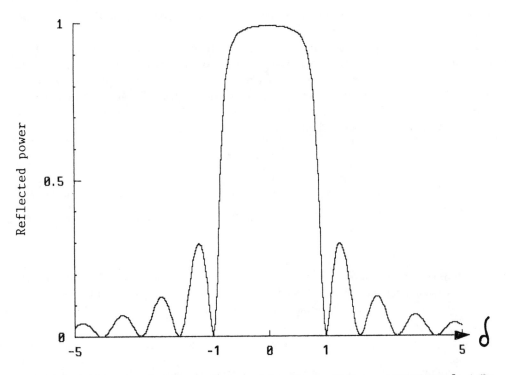

FIGURE 27. Guided power reflected by a reflection grating vs. the normalized phase mismatch term $\delta = \Delta_e/2\kappa$. $\delta = 0$ is for the central frequency. The rejection bond extends from $\delta = -1$ to $\delta = +1$. $\kappa \cdot L = 3$, corresponding to a power transmission of $-20$ dB only at the center frequency.

explicitly in Equation 116: $\mu = (\kappa^2 - n_e^2/c^2 \ (\omega - \omega_o)^2)^{1/2}$ with central frequency $\omega_o$ where the rejection is maximum, and half-width $\Delta\omega$:

$$\omega_0 = \frac{ck_p}{2n_e} = \frac{c\pi}{n_e\Lambda} \ ; \qquad \Delta\omega = \frac{c\kappa}{n_e} \tag{118}$$

For $\omega_o - \Delta\omega > \omega > \omega_o + \Delta\omega$, the argument of the hyperbolic function becomes imaginary and the modulus of the reflected wave becomes:

$$\left|C_2(0)\right|^2 = \left|C_1(0)\right|^2\left[\delta^2 - \left(1 - \delta^2\right)ctg^2(|\mu|)L\right]^{-1} \tag{119}$$

The rejection band of a lossless reflection grating of large length L is illustrated in Figure 27.

## VI. WAVEGUIDE LIGHT CONVEYANCE

Most existing optodes and reported experiments on opto-chemical sensors make use of multimode silica fibers. Their large core radius makes them tolerant to connection misalignment. For instance, a 0.5 dB connection loss is caused by a 3-μm lateral offset between two 50/125 graded index silica fibers[71] and 8.5 μm between two 100-μm core diameter PCS fibers, which is well within the tolerance of cheap connectors. This feature is especially valuable in the field of chemical sensors where the probe must often be disposable because of the short lifetime of the transducer or for preventing contamination.

Multimode fibers have also a quite large capture efficiency in end-firing from an LED or from a fiber tip fluorescent transducer. The expression of the capture efficiency $I_e$ is defined as the ratio between the total power $P_s$ emitted by the source's area $A_s$ with the power $P_c$ captured by a step index fiber of core area $A_c$ and relative index difference $\Delta$. Assuming the source to be Lambertian with brightness B, the power fraction $\Delta P$ emitted by the source element $dA_s$ into the solid angle $d\Omega$ making an angle $\theta$ with the fiber axis z is:

$$\Delta P = B \cos\theta \, dA_s \, d\Omega \qquad (120)$$

Integrating Equation 120 over the source area and the half space in front of the light radiator gives the full emitted power, $P_s$:

$$P_s = \pi A_s B \qquad (121)$$

Performing the integration over the core area and the solid angle intercepted by the core limited by the critical angle $\theta_{cr} = \arccos n_2/n_1$ and assuming the source and fiber surfaces to be in contact gives the power $P_c$ of the bound rays:

$$P_c = 2\pi A_c B\Delta \qquad (122)$$

which is identical to the Expression 84a for the trapped ray power as $\theta_{cr}^2 \cong 2\Delta$. The capture efficiency is therefore:

$$I_e = \frac{P_c}{P_s} = 2\frac{A_c}{A_s}\Delta \qquad (123)$$

Introducing the numerical aperture NA as the sine of the fiber acceptance cone angle in air, $I_e$ writes as:

$$I_e = \frac{A_c}{A_s}\frac{NA^2}{n_2^2}$$
$$\text{with} \quad NA = \sqrt{n_1^2 - n_2^2} \cong n_2\sqrt{2\Delta} \qquad (124)$$

which, for equal source and core area, in the case of a typical PCS fiber, gives an efficiency of a few percents.

Considering now the source power captured by leaky rays and referring to Expressions 81 or 84b, it can be said that the excitation efficiency at the fiber input is given by a value very close to Expression 123. The fraction of the total captured power propagated by leaky modes falls very rapidly along the first meters, then maintains over a long distance with small leakage attenuation as illustrated in Figure 21; it is therefore very dependent on the fiber lead length, especially in the first meters. As an example we are considering a PCS fiber of radius a = 50 $\mu$m, index difference $\Delta n = 0.023$, i.e., a critical angle $\theta_c \cong 10°$. Applying the empirical formula of Equation 85 shows that the fiber length $z_{1/2}$ where the leaky ray total power has dropped by 50% is about 60 mm at 80 nm wavelength (V = 100) and about 0.8 m at 400 nm wavelength (V = 200).

The most important feature of an optical fiber for optical signal conveyance is the down-lead sensitivity regarding the type of encoding supplied by the opto-chemical transducer. Two types of transducing mechanisms are considered hereafter: those causing an alteration of the optical spectrum and those relying on a single wavelength intensity modulation.

## A. TRANSMISSION OF AN ABSORPTION OR FLUORESCENCE SPECTRUM ALTERATION

Let $P_i(\lambda)$ the power spectrum injected into a multimode fiber at the optode's location. During the propagation down-leads each spectral component is subject to the wavelength dependent fiber attenuation. This attenuation along the fiber axis z is usually described by an exponential term $\exp(-2\alpha(\lambda)z)$ where $\alpha(\lambda)$ is the field attenuation coefficient. The power $P_m(\lambda)$ reaching the analyzer and detection system is:

$$P_m(\lambda) = P_i(\lambda)e^{-2\alpha(\lambda)L} \qquad (125)$$

where L is the length of the fiber lead. Setting $\lambda_o$ as a reference, ratioing the power of all spectral components to $P_m(\lambda_o)$ yields:

$$\frac{P_m(\lambda)}{P_m(\lambda_0)} = \frac{P_i(\lambda)}{P_i(\lambda_0)} e^{-2L\left[\alpha(\lambda)-\alpha(\lambda_0)\right]} \qquad (126)$$

A variation of the chemical measurand induces a modification of the injected and measured spectra which now become $P_i'(\lambda)$ and $P_m'(\lambda)$. Writing Equation 126 again with $P_i'$ and $P_m'$ and ratioing the two expressions eliminates the fiber dependent term which yields:

$$\frac{P_i'(\lambda)}{P_i'(\lambda_0)}\left[\frac{P_i(\lambda)}{P_i(\lambda_0)}\right]^{-1} = \frac{P_m'(\lambda)}{P_m'(\lambda_0)}\left[\frac{P_m(\lambda)}{P_m(\lambda_0)}\right]^{-1} \qquad (127)$$

It is noteworthy that Expression 127 remains true if the fiber attenuation is not exponential as in the case of short sections of multimode fibers where the attenuation is dependent on the modal power distribution, i.e., on the excitation conditions and on the fiber environment.[72] The steady state modal power distribution in multimode fibers is only reached after hundred meters of propagation,[73] which will rarely be the case in optode systems, or by using a mode scrambler device inducing microbends somewhere on the fiber.[74] Writing Equation 127 assumes the time stability of the fiber dependent term, i.e., of the relative attenuation coefficient $\alpha(\lambda) - \alpha(\lambda_o)$. This characteristic has been studied on long multimode communication fibers as mentioned in Section IV.G. The conclusion can be expressed by saying that microbending losses and their fluctuations only amount to shifting the fiber attenuation spectrum of, say, Figure 1 or 23 parallel to itself vertically, keeping the difference term $\alpha(\lambda) - \alpha(\lambda_o)$ unchanged. Time variations of bending effects would be wavelength dependent but, as already stated in Section IV.F, these are negligible. Multimode fiber components such as connectors and couplers are also basically achromatic.

The same cannot be said about short multimode fiber sections where effective bound mode coupling takes place and leaky modes propagate a substantial amount of power. The tunneling power leakage is wavelength dependent and affected by bends and microbends[75] and their variations. The spectrum perturbation by leaky modes is likely to be very dependent on the length of the fiber leads because of their peculiar attenuation longitudinal dependence which is far from the bound mode exponential decay as described in Section IV.E.3. Except in single mode fibers, leaky mode excitation in fiber tip optodes cannot be avoided, especially in fluorescence type sensors as skew rays will always be generated. The filtering out of leaky modes can be attempted on the lead by using usual communication mode filters.[76a] However, this is a difficult operation without secure criteria for its achievement on short fiber lengths. The contribution of leaky modes to the propagated signal would be significantly reduced if the fiber cross-section could be made noncircularly symmetric.[76b] Such fibers can now be found on the market but no experimental evidence has been reported yet on their possible better light spectrum conveyance properties.

It can therefore be concluded that high performance opto-chemical sensors should not rely on too elementary a spectrum inspection such as two wavelength measurement systems. Microelectronics allows compact and low cost integrated detector and preamplifier arrays which will help make a finer spectrum analysis. Besides, efforts should be made towards encoding systems, such as time resolved measurement,[77] that would be intrinsically independent from perturbations on the fiber lead.

## B. TRANSMISSION OF AN INTENSITY MODULATED SIGNAL

The effects of single wavelength power absorption, leakage, or outcoupling mechanisms have to be distinguished from source fluctuation, detector aging, and from the attenuation variations on the lead to and from the transducer. This problem is also that of physical sensors where various referencing schemes have been proposed: two mode sensing using a differential attenuation effect, both modes being similarly affected by the fluctuations on the fiber;[78] conversion of optical intensity modulation into the phase modulation of a baseband signal by means of an RF interferometric head containing the transducer;[79-81] spectral encoding can be used with two or more wavelengths if the transducing mechanism can be made sharply wavelength dependent[82] as compared to the width of the fiber optical transfer function.

# VII. WAVEGUIDE OPTICAL PREPROCESSING FUNCTIONS

Guided wave optics can lead to optode system integration at two levels. On one hand it allows serial or parallel spatial multiplexing of a number of transducers. On the other hand, it lends itself to the integration of a number of optical preprocessing functions such as wavelength multiplexing within a rather wide optical spectrum in an all-fiber form or on a single planar chip. So far little use has been made of this potential, and we are reviewing hereafter a few possibilities among passive functions.

## A. SPATIAL MULTIPLEXING

The already well-established market of optical fiber communication components offers a wide range of connectors, branching devices which could readily be used in multisensing optode systems.

However, planar integrated optics allows much better spatial confinement and lends itself to batch techniques and technologies. A few examples of possible configurations are described hereafter:

1. Multiple waveguide tip transducer with a single 1 to N port star power divider as suggested in Figure 28. The most suitable waveguide fabrication technology is ion exchange in glass.[11]

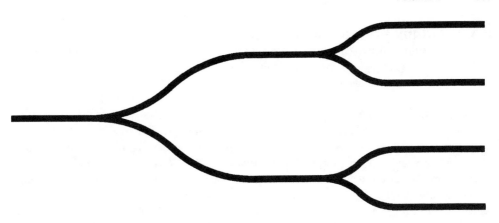

FIGURE 28.   Integrated optic 1 to N port power splitter.

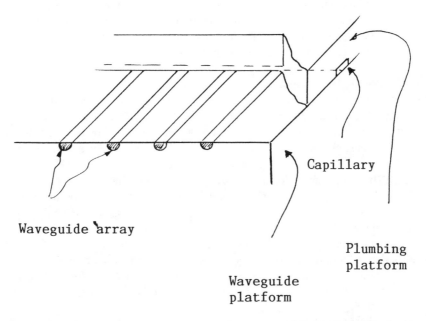

FIGURE 29.   Association of a planar light distribution network with fine fluid distribution plumbing.

2.   The same pattern can be used to etch deep and smooth circular grooves into a glass substrate. They will later be filled by an organic transducing material which is used as the waveguide core. Very high efficiency in absorption and fluorescence sensors is expected. The guide geometry is compatible with plugged PCS input and output filters.[83]

3.   The same type of pattern can also be defined with single mode waveguides. The sensing material is placed as an overlay which is probed by the modal evanescent wave as described in more details in Section VIII.

A planar waveguide configuration  can very well be compatible with fine plumbing of gas or liquids in surface etched channels in glass, as in 2, or in silicon as suggested in Figure 29. Inorganic bonding, such as anodic bonding, would attach the integrated optic chip to the fluid circulation chip with good spacing precision.

## B. WAVEGUIDE INTEGRATED FUNCTIONS

A few functions that are usually performed by micro-optic components can nowadays be implemented in a waveguide form. This brings compactness, monolithicity, self-alignment, and can possibly be achieved by batch processes.

### 1. Sources in All-Fiber Form

Light generation and excitation is one of the most cumbersome parts of a sensor system. Semiconductor LED and lasers have long been available in the main communication windows (780 to 900 nm and 1250 to 1550 nm) in a fiber pigtailed form. The fiber excitation head for visible semiconductor laser[84] (670 nm) and visible LEDs must still be assembled especially. Semiconductor pumped rare earth doped fiber lasers offer a large variety of luminescence lines which tend to fill the gaps left by semiconductor lasers.[85a] In the blue and green part of the spectrum, white light excited fluorescent plastic fibers[86] give an excellent all-fiber source that can be connected with heavily multimode fibers. The lack of high spatial coherence semiconductor sources in the blue side of the spectrum is still a considerable obstacle which certainly hampers the development of the domain of single mode optodes using the full potential of guided wave optics. However, other markets also push in the same direction, and the availability of such sources is only a question of time.[85b,c]

### 2. Opto-Electronic Detection

The wavelength range of interest for optode systems is well provided with a large variety of semiconductor detectors.[87]

Optoelectronic detection in fiber systems is often best achieved by plugging the fiber onto a commercial detector. Planar optical geometries can make a wider use of the possibilities offered by microelectronics. The geometry of a single chip integrated optic network is characterized by a high integration density transversally to the optical axis and by very smooth and long bends, junctions, and coupling sections along it. The lateral confinement can be as low as a few wavelengths in single mode circuits, whereas waveguide routing or coupling elements are thousands of wavelengths long. Waveguide tip or grating coupling[88] can therefore be used to direct the guided optical signals to a dedicated single chip detector and preamplifier array. Silicon-based integrated optics[89] can involve both optical and electronic circuits on the same substrate with a quite wide compatibility of technologies;[90] however, high integration density is possible in one dimension only for both circuits.

### 3. Optical Spectral Analysis

Optical spectral analysis is an essential function in optode systems. Simple multimode fiber[91] or even integrated[92] systems make use of interference filters that are introduced across the light path. Applications requesting the analysis of a large number of spectral components can use dedicated monochromators similar to the micro-optic grating demultiplexers that have been developed in the field of optical communications.[93]

Planar optical geometries can bring further possibilities for future systems propagating spatially coherent optical fields. Various guided wave coupling mechanism are wavelength selective. The simplest one is a planarization of the classical monochromator grating as illustrated in Figure 39: a wide band guided wave will be coupled out to radiation waves by the corrugation grating and dispersed according to the grating formula of Equation 141. Fully integrated wavelength selective coupling mechanisms have been thoroughly investigated in single mode optical communications.[94] A planar step index modelization is made in Section V. Three cases will be described hereafter.

### a. Codirectional Degenerate Coupling

Forward power coupling between the modes of two identical waveguides with spacing d is described by Expression 106. The wavelength dependence is contained in the coupling coefficient, $\kappa$, given by Equation 99. Performing the overlap integral over one of the guides using the field expressions given by Table 2 and the integration constants given in Equations 39 and 40 in terms of the total power P yields for TE modes:

$$\kappa = 2 \frac{k_g^2 k_s}{\beta \left( k_s^2 + k_g^2 \right) w_{eff}} e^{-k_s d} \tag{128}$$

and for TM modes:

$$\kappa = 2 \frac{k_g^2 k_s}{\beta \left[ w \left( \dfrac{n_s^4 k_g^2 + n_g^4 k_s^2}{n_g^2 n_s^2} \right) + \dfrac{2}{k_s} \left( k_g^2 + k_s^2 \right) \right]} e^{-k_s d} \tag{129}$$

The wavelength dependence of $\kappa$ is largely dominated by the exponential term. In the single mode regime away from cutoff the preexponential term is a slowly decreasing function of $\lambda$ due to the $1/k_s$ term contained in $w_{eff}$. The coupling coefficient $\kappa$ is therefore an increasing function of $\lambda$. The wavelength dependence increases with increasing spacing d. From Equation 106 the spectral response of a degenerate coupler is an oscillatory function in $\lambda$ with a wavelength period decreasing for increasing wavelength. The comb filter period can be adjusted by choosing the waveguide spacing d. It is worth pointing out that such coupler can also be used as a wide band device: bringing the waveguides closer to each other (d $\rightarrow$ 0) drastically decreases the wavelength dependence.

### b. Codirectional Nondegenerate Coupling

A band-pass filter character can be obtained in a codirectional coupling between two waveguides of differing dispersion. The principle is illustrated in Figure 30. The dispersion curves $n_e(k_o)$ of waveguides $g_1$ and $g_2$ of width $w_2 < w_1$, guide index $n_{g1} < n_{g2}$ and same cladding index $n_s$ intersect at $k_o = k_{oc}$, or $\lambda = \lambda_c$, where synchronism occurs with $n_e = n_{ec}$.

The coupling coefficient $\kappa_c$ at the synchronism wavelength between the two modes of same order m propagating in $g_1$ and $g_2$ is obtained here in the TE case by integrating Equation 99 over one of the waveguides $g_i$ as in the degenerate case. One finds:

$$\kappa_c = 2 B_1 B_2 \frac{\sqrt{n_{ec}^2 - n_s^2}}{n_{ec}} \exp \left( -k_0 d \sqrt{n_{ec}^2 - n_s^2} \right) \tag{130}$$

$$\text{where} \quad B_i^2 = \frac{n_{gi}^2 - n_{ec}^2}{\left( n_{gi}^2 - n_s^2 \right) w_{effi}} \quad , \quad i = 1, 2$$

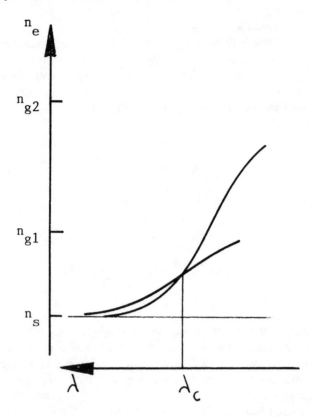

FIGURE 30.  Principle of a wavelength band-pass filter in a co-directional coupling between two waveguides of differing dispersion characteristics. The phase matching condition is only satisfied at, and in the neighborhood of the crossover wavelength $\lambda_c$.

We will later assume all terms in Equation 130, except $k_o$, to be wavelength independent within the optical bandwidth where the coupling is significant. The power coupled from one guide to the other in a coupling section of length L is then given by the off-diagonal terms of Equation 110. The phase mismatch term $\Delta = \beta_1 - \beta_2$ in $\mu$ is the difference between the propagation constants of the two coupled modes. It increases as $\lambda$ goes away from $\lambda_c$. We are assuming here that significant power coupling only takes place within a narrow wavelength neighborhood around $\lambda_c$, i.e., that the phase mismatch term $\Delta$ takes over rapidly on $\kappa_c$. $\Delta(\lambda)$ can therefore be well approximated by the difference between the tangents of the two dispersion curves at the coupling point. These are given by taking the derivatives $\partial \beta_i / \partial k_o$ in the characteristic Equation 33:

$$\Delta\left(k_o\right) \cong \frac{1}{n_{ec}}\left[\left(n_{g2}^2 - n_{ec}^2\right)\frac{w_2}{w_{eff2}} - \left(n_{g1}^2 - n_{ec}^2\right)\frac{w_1}{w_{eff1}}\right]\left(k_o - k_{oc}\right) \qquad (131)$$

The transmission spectrum of a nondegenerate coupler is shown in Figure 31 vs. $\lambda$ in a specific example. A basis waveguide of width $w_1$ and index $n_{g1}$ is coupled to a second waveguide of width $w_2$ and index $n_{g2}$. At $\lambda = \lambda_c$ these two waveguides are still in the single mode regime and are set to couple over one coupling length, i.e., $\kappa L_c = \pi$. Curves a and b are for different waveguide spacing d and show how the wavelength selectivity improves for increasing d.

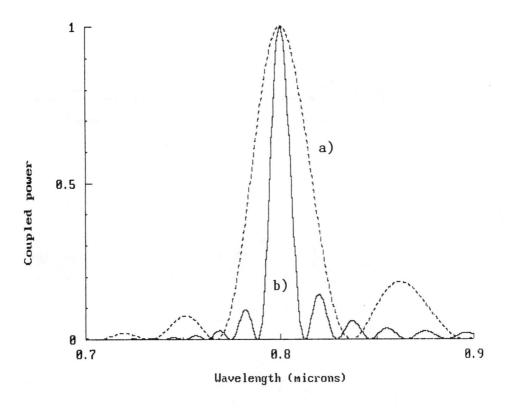

FIGURE 31. Coupled power spectrum in the configuration of Figure 30 in differing conditions: curve a: waveguide spacing d = 4λ, curve b: d = 5λ. Substrate index $n_s$ = 1.46. Waveguide 1: $w_1$ = 1.906 μm $n_{g1}$ = 1.475. Waveguide 2: $w_2$ = 0.625 μm $n_{g2}$ = 1.49. The synchronism wavelength $\lambda_c$ is 800 nm.

The transmission bandwidth of a four-port nondegenerate waveguide coupler can therefore be centered by adjusting the relative width or the relative index of the waveguides involved. Once these are fixed the bandwidth essentially depends on the coupling coefficient, i.e., on the waveguide spacing. The scheme of what could be a serial demultiplexer based on the principle just described is sketched in Figure 32a where the shorter wavelength components are the first coupled out.

### c. Wavelength Dependence in Contradirectional Grating Coupling

Much higher wavelength selectivity can be obtained by using a waveguide reflection grating. A waveguide grating used as a reflector is equivalent to a band rejection filter with center optical frequency $\omega_0$ and bandwidth $2\Delta\omega$ as expressed by Equation 119 and illustrated in Figure 27. The rejection can be made partial or total depending on the grating length L. The center frequency $\omega_0$ in a given grating structure is only affected by the waveguide effective index $n_e$ which may vary under cover and waveguide change as seen in Section VIII.C.1 according to expressions 114 and 147. The rejection band half-width $\Delta\omega = c\kappa/n_e$, given by Equation 118, is affected by both effective index $n_e$ and coupling coefficient $\kappa$, but mostly by $\kappa$. The coupling coefficient in the case of thin grating, t << w, is given by Equation 115 where the integration on the grating thickness is replaced by $(n_p^2 - n_c^2)$ t and $\Phi_1 = \Phi_2$ is the value of the field normalized to power P at the guide-cover boundary given by Equation 43 in the case of a TE mode:

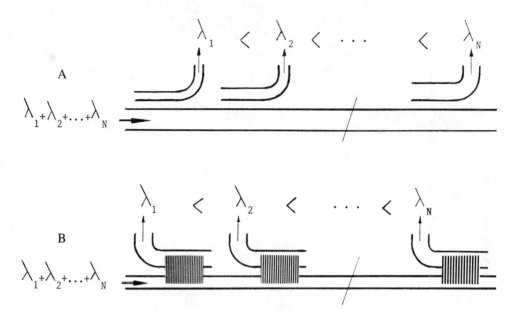

FIGURE 32.    Serial wavelength demultiplexers with coupling wavelength $\lambda_c$ increasing downstream: (A) using the band-pass coupling principle of Figure 31; (B) using the grating contra-directional principle of Figure 33.

$$\kappa = \frac{\pi\left(n_g^2 - n_e^2\right)\left(n_p^2 - n_c^2\right)}{2\lambda n_e\left(n_g^2 - n_c^2\right)w_{eff}}\, t \tag{132}$$

where $n_p$ is the index of the grating material. If the grating is a simple corrugation of the waveguide surface, then $n_p = n_g$ and Equation 132 becomes:

$$\kappa = \frac{k_0^2}{4}\,Qt \quad \text{and} \quad \frac{\Delta\omega}{\omega_0} = \frac{k_0 Q}{4n_e}\, t$$

$$\text{where} \quad Q = \frac{n_g^2 - n_e^2}{\beta w_{eff}} \tag{133}$$

Expression 132 reveals that the coupling efficiency does not depend on the side on which the grating is made. We will now illustrate how the efficiency, and therefore the rejection bandwidth, of a reflection grating on a waveguide is affected by the cover index in the case of a simple corrugation waveguide (Expression 133). It can be foreseen that a decrease of the cover index below that of the substrate increases $\kappa$ as it increases $(n_g^2 - n_e^2)$ and decreases $w_{eff}$. Figure 33 represents the normalized coupling coefficient Q vs. the cover index $n_c$ in the case of a single mode waveguide on a glass substrate, $n_s = 1.51$, $n_g = 1.52$; $k_0 w = \pi/\sqrt{(n_g^2 - n_s)^2}$ is the cutoff condition of the $TE_1$ mode in the symmetrical structure $(n_s = n_c)$.

In order to also allow a spatial separation of the coupled spectral components, the grating can be achieved on top of a two-waveguide coupler which is intentionally put out of synchronism so as to impeach forward degenerate coupling. The grating of selected pitch couples an incoming wave to the reflected wave of the other waveguide. A scheme of what a serial-

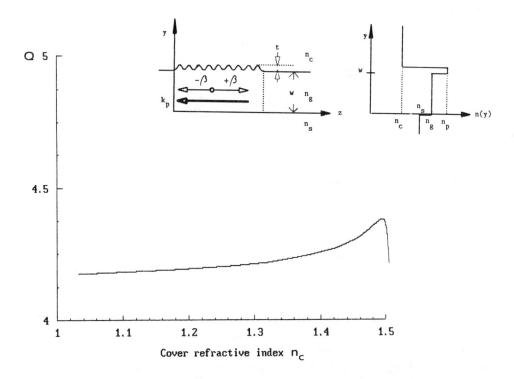

FIGURE 33. TE contra-directional coupling efficiency of a corrugation grating vs. the cover's index $n_c$ made on a step index waveguide with $n_g = 1.52$, $n_s = 1.51$. $k_o$ and w are such that the $TE_1$ mode is at its cutoff. Q is in units of $10^{-4}$.

waveguide demuliplexer could be is sketched in Figure 32b where the shorter wavelengths are first coupled out.

## 4. Guided Wave Interference

The most accurate technique for the measurement of refractive index changes is interferometry. As shown in Figure 34, one of the branches of a waveguide interferometer is subject to an effective index change induced by the variation of a chemical measurand. The signal and reference waves recombine in a waveguide coupler whose output ports deliver optical power signals that depend on the input optical phase shift. The recombination function can very well be performed in a guided wave form and various structures can be used.

If the recombining coupler is a four port element like a two branch degenerate 3-dB coupler of length $L_c/2$ (Section V.B), the two output signals are out of phase and no information is given on the sign of the input phaseshift. The same happens in a hybrid Y coupler where the second output port is the power leaking into the substrate. In order to perform fringe counting and interpolation without sign ambiguity, a coupler with more than two output power signals must be used. As explained in Section V.C in the case of a two branch coupler, the beat of the three or four eigenmodes of a three or four branch coupling section can be made to deliver three or four output power signals respectively which have a prescribed phase relationship between each other ($2\pi/3$ and $\pi/2$ respectively).[95] The same type of function can be performed by using the beat of the whole mode collection of a multimode box-waveguide as illustrated in Figure 35.[96] The principle used is the self-imaging property of multimode step index waveguides. The field distribution at the box-waveguide input repeats itself downstream

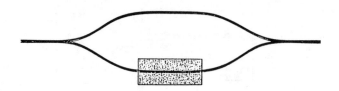

FIGURE 34.   Integrated optic Mach-Zehnder interferometer for refractive index measurement.

periodically with intermediate multiple images. The object field is given here as the output field of the two incoming waveguides. After a propagation length equal to a fourth of the mode collective beat length leading to the first inverted image, a fourfold image is obtained. A suitable choice of the two object field lateral position makes the sub-image fields to overlap with a quadrature relationship between sub-image power.[97] These are only a few examples of the specific possibilities offered by spatially coherent single mode guided wave optics. The next section deals with opto-chemical transducing mechanisms which can be associated in a planar form with the described optical preprocessing functions.

## VIII. THE OPTICAL WAVEGUIDE AS THE SENSOR HEAD

We will focus here on the various opto-chemical transducing mechanisms involving the optical wave probe in its guided form. We are leaving aside micro-optic fiber tip configurations where the waveguiding properties essentially concern the lead to and from the transducer, as seen in Section VI.

### A. ABSORPTION MECHANISMS

An absorptive material is characterized by a non-zero imaginary part of its relative permittivity, $\varepsilon_{rj}$ (2). In the transducing mechanism considered here, $\varepsilon_{rj}$ is a function of a chemical measurand. The absorptive material can be the waveguide core itself or the waveguide cover or cladding. In all cases the power absorption coefficient $2\alpha$ can be calculated using the perturbation Formula 94 and integrating the modal field $\Phi$ over the part of the cross-section where the permittivity perturbation, here $\varepsilon_{rj}$, is non-zero. In most cases the $\varepsilon_{rj}$ term in the absorptive region is homogeneous. Therefore, from Formula 94, the attenuation coefficient is proportional to the relative power flowing in this region.

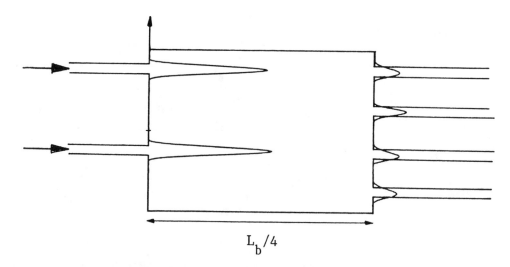

$$L_b/4$$

FIGURE 35. Multiple imaging in a multimode step index waveguide for nondegenerate interference of two phase-shifted optical signals. The two sets of four-fold sub-images superpose so as to deliver four power signals.

### 1. Fibers with Absorptive Cladding

The main configuration of practical interest presently is that of a multimode step index fiber embedded into a low index absorptive cladding such as a PCS fiber with its cladding stripped off and replaced by an absorptive liquid or organic coating.

The optical attenuation is very mode-dependent. Only those modes close to cutoff and, among them, mainly those of low azimuthal mode index l are subject to significant attenuation, as discussed in Section IV. The relative power $P_{cl}/P_{tot}$ of $LP_{lm}$ modes flowing in the cladding in the case of a fiber of 50 $\mu$m diameter is shown in Figure 13. Figure 13 was obtained by solving the characteristic equation in its approximate form (Equation 63) and calculating the relative power using Equation 75.

Assuming all bound modes excited with equal power and considering only the modes close to cutoff as bringing a significant contribution to the absorption, the average relative power flowing in the absorbing cladding is approximately proportional to $1/a$. Therefore, the modulation depth of the transmitted optical power under the action of the measurand on a given interaction length is a decreasing function of the fiber diameter. Modes providing the highest sensitivity are also those which are the most subject to bending loss. Therefore, the overall efficiency of such a multimode transducing mechanism is inherently dependent on the modal power distribution and presents a high downlead sensitivity. Evanescent wave absorption sensing has a much lower efficiency than bulk sensing, especially in practical cases where the cover index has to be significantly lower than that of the substrate as with water and most solvents. A possible solution is to force the field out by designing electromagnetic waveguide cross-sections of the capillary type.[114]

A much better efficiency can be gained if the core is composed of the absorptive material. This is an interesting approach for multimode heavy plumbing where the light conveyance and routing function can be made compatible with the topology and technology of the waveguide probe.

### 2. Single-Mode Waveguide Absorptive Overlay

The problem of modal power distribution is completely avoided if the optical field probe is that of a single-mode waveguide. Considering a TE mode of a step index planar waveguide,

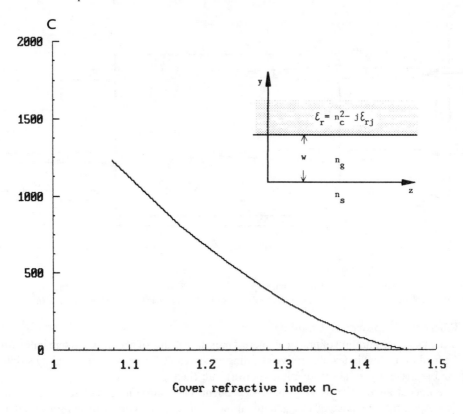

FIGURE 36.   Ratio C of bulk ($L_e$) and guided wave ($L_e'$) propagation lengths in a step index waveguide with low index lossy overlay vs. cover index $n_c$. Waveguide opto-geometrical parameters are: $n_s$ = 1.46, $\Delta n$ = 0.015, $\lambda$ = 633 nm, w = 1.5 μm.

the perturbation formula yields the power attenuation coefficient $2\alpha$ in Neper per unit length after integrating Equation 95 in the cover:

$$2\alpha = \frac{1}{C \cdot L_e} \quad \text{or} \quad \frac{L_e'}{L_e} = C \tag{134}$$

$$\text{where} \quad C = \frac{k_0 w_{eff} n_e \sqrt{n_e^2 - n_c^2} \left(n_g^2 - n_c^2\right)}{n_c \left(n_g^2 - n_e^2\right)}$$

$L_e$ is the propagation length of a plane wave in the cover's material given by Equation 5 in term of $\varepsilon_{rj}$ of the latter; $L_e'$ = $1/2\alpha$ is the guided wave propagation length. The efficiency is strongly dependent on the field confinement on the cover side. The more symmetric the waveguide, the higher the field penetration in the cover. The dependence of term C in the case of the $TE_0$ mode of a silica based slab waveguide vs. the cover index is shown in Figure 36. Its width w is taken as the cutoff width of the $TE_1$ mode in the symmetrical case ($n_c = n_s$). As expected from Equation 134, the absorption sensitivity is the highest at the mode cutoff in the symmetrical case or very close to it otherwise.

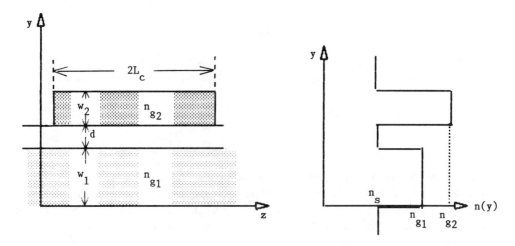

FIGURE 37.   Waveguide transducer head using resonant coupling between a basis waveguide $g_1$ and a sensor waveguide $g_2$. $L_c$ is the coupling length at synchronism.

Single-mode evanescent wave spectroscopy has not received much attention so far, mainly because of the lack of high spatial coherence sources. However, it allows sensitivities one to three orders of magnitude larger than the conventional total internal reflection spectroscopy. A review of first achievements and possible applications can be found in Reference 98a,b.

### 3. Resonant Coupling to an Absorptive Film

A potentially interesting combination of an absorptive transducing material and a solid-state single-mode basis circuit with all its integrated functions can be thought of if the waveguide substrate index is equal to or lower than that of the transducing material. As illustrated in Figure 37, the sensing material of index $n_{g2}$ is deposited on the basis guide $g_1$ of index $n_{g1}$ with an optional low index buffer in between. If the film thickness $w_2$ is chosen and adjusted so as to bring the two guiding structures in synchronism for a certain wavelength $\lambda_c$, then the structure has become a section of a two-branch directional coupler. Now if the length of the deposited film is made equal to twice the coupling length $L_c$ at the working wavelength, we have achieved a high sensitivity probe: the whole power has made an excursion into the sensing medium over a length which can possibly be made large by using a low index buffer which reduces the coupling coefficient.

This approach would be particularly suitable in association with silica based waveguides[99] as a number of solvents or polymers have a higher index. Such waveguide could, for instance, be a polished fiber probe where the sensing material is deposited onto the polished section.[100] If the waveguide refractive indices are very dissimilar, an accurate sensing film thickness should then be achieved by using technologies of the Langmuir-Blodgett type, for instance.[101]

### B. POWER LEAKAGE MECHANISMS

Opto-chemical measurements relying on refractive index change, such as mixture concentration, oil degradation, temperature, etc., can be performed by optical power leakage measurement if the index of the probe medium is not lower than that of the cladding or substrate material.

### 1. Truncation of a Multimode Fiber Profile

The number of modes propagating in a large core step index fiber of index $n_g$ is given by Equation 68. If the cladding index is linearly dependent on a chemical quantity $\chi$, the substrate or cladding index $n_s$, expresses:

$$n_s(\chi) = n_s(\chi_0) + n_\chi(\chi - \chi_0) \tag{135}$$

where $n_\chi$ is the slope of $n_s$ $(\chi)$ vs. $\chi$. Using Equation 68, and assuming all bound modes to propagate the same power, the power remaining in the fiber $P(\chi)$ relative to $P(\chi_0)$ is:

$$\frac{P(\chi)}{P(\chi_0)} = 1 - \frac{n_\chi}{n_g - n_s(\chi_0)}(\chi - \chi_0) \tag{136}$$

Considering the temperature of a liquid as an example, $n_c = -10^{-3}$ typically, and a silica fiber with a 0.2 numerical aperture NA at temperature 44°C, the transmitted power will be zero at 30°C. The accuracy of this fairly simple transducing mechanism is, however, limited by the dependence of the modal power distribution on the excitation conditions and on various perturbations on the fiber lead.

## 2. Tunneling Leakage from a Single-Mode Waveguide

The ambiguity on the optical power transverse distribution is suppressed if a single mode is involved in the leakage into a higher index medium. The rate of power leakage will be calculated using a perturbation technique in a planar step index waveguide model. The type of structure considered is depicted in Figure 38: a step index waveguide with substrate, guide, and cover indices $n_s$, $n_g$, and $n_c$, respectively. The chemical system to be investigated is placed on top of the cover of finite width d. It is assumed that the real part of the analyte index $n_h$ is higher than the effective index $n_e$ of the probe mode and that its imaginary part can be neglected.

Keeping in mind the ray picture of a guided mode (Figure 5) it can be said that the ray now undertakes a frustrated total reflection at the guide upper boundary suffering, therefore, a power loss by leakage into the high index overlay. Although there is no more full guidance, the leaky mode still propagates with an attenuation rate which depends on the buffer optical width and on the external index $n_h$. The measurement of the power remaining in the guide after going through the transducer section of finite length gives an analog determination of the overlay index, therefore of the chemical measurand after calibration. This effect can be used with laterally confined single-mode waveguides. Care should be taken of the monomodicity: if more than one mode propagate, the overall power attenuation will give again ambiguous measurand determination as it will depend on the power propagated by each mode, i.e., by the excitation conditions and the differential mode attenuation down the guide. Such a transducer structure can be even achieved in a fiber form: by tangentially polishing a single-mode fiber down to the core neighborhood, the modal field tail can be made to expand into the high index medium where it leaks, as demonstrated in Reference 102 in the case of a fiber evanescent field refractometer. The expression of the leakage loss is established hereafter for the TE and TM modes of a slab waveguide following the procedure of Marcuse.[103]

### Slab Waveguide Analysis

We will consider the presence of an external semi-infinite medium of higher index $n_h$ as a perturbation of the single waveguide structure of Figure 5. The outer load-cover interface is located at distance d from the waveguide-cover interface. The modal effective index $n_e$ as well as the substrate and waveguide part of the modal field $\Phi_{x,z}$ in the perturbed structure are considered to be the same as in the unperturbed substrate-waveguide-semi-infinite cover structure. Field matching conditions at the waveguide cover and cover-high index load interfaces will be used to express the field components at the load side of the outer interface

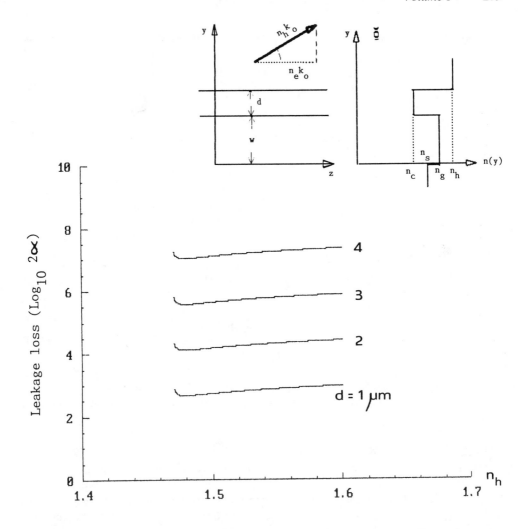

FIGURE 38. Slab waveguide with high index ($n_h$) power sink (window). The leakage occurs through the buffer region of low index $n_c$. The graph shows the Logarithm of the leakage coefficient $2\alpha$ vs. the overlay's high index ($n_h$) through a buffer cover of thickness d. $n_s = n_c = 1.46$, $n_g = 1.475$, $\lambda = 633$ nm, $w = 1.5$ $\mu$m.

at $y = (w + d)_+$. These field components $\Phi_x$ and $\Phi_z$ will be finally used to derive the y-component of the Poynting vector which describes the power flux radiating away from the waveguide, i.e., the attenuation $2\alpha$ of the propagated power.

The tangential field components in the waveguide, cover and semi-infinite load can be written similarly to Table 2. The field components in the substrate and waveguide are taken as if the high index load was not present and can be expressed by means of the normalized constants given by Equation 39 with the plus and minus sign for even and odd modes.

Matching the x and z field components at $y = w$ and $y = w + d$ gives the expression of the transverse field amplitude $A_h$ at the cover-load interface in terms of $\Phi_M$:

$$A_h = \frac{2\Phi_M}{\sqrt{1 - a_c^2 \left(1 - ja_h\right)}} e^{-k_c d}$$

$$(137)$$

where $a_c = \dfrac{k_c \xi_c}{k_g \xi_g}$ and $a_h = \dfrac{k_h \xi_h}{k_c \xi_c}$

The y-component of the Poynting vector, $S_y$, is non-zero unlike in the case of a true guided mode:

$$S_y = \frac{k_h |\xi_h|}{2\omega\mu_0} |A_h|^2 \tag{138}$$

The rate of power leakage $2\alpha$ is given by the ratio $S_y/P$ where P is the power flowing in the guide without high index load:

$$2\alpha = \frac{8k_h |\xi_h|}{\left(1 + a_c^2\right)\left(1 + a_h^2\right)\beta |\xi_g| w_{eff}} \cdot e^{-2k_c d} \tag{139}$$

where $w_{eff}$ for TE and TM modes is given by Equation 42a and b. In the case of a TE mode, Equation 139 can be further developed:

$$2\alpha = 8 \frac{\sqrt{n_h^2 - n_e^2}\left(n_g^2 - n_e^2\right)\left(n_e^2 - n_c^2\right)}{\left(n_h^2 - n_c^2\right)n_e w_{eff}\left(n_g^2 - n_c^2\right)} e^{-2k_c d} \tag{140}$$

where $n_e$ is the TE mode effective index in the waveguide without its high index load. Expression 140 in the case of a silica based symmetrical waveguide versus $n_h$ with d as a parameter is illustrated in Figure 38.

## C. PHASE CHANGE MEASUREMENT

Chemical species or processes that translate into a refractive index change can be monitored with much higher accuracy by means of an optical phase measurement technique. Integrated optics offers a variety of possible configurations that can be distributed into two broad categories which are synchronism tuning techniques and interferometric techniques.

### 1. Synchronism Tuning of Waveguide Grating Excitation

The change of refractive index in any region composing a waveguide alters the effective index of its propagating modes. An index change can be caused, for instance, by a variation of the cover concentration, by the penetration of selected molecules into the guide material, by the adsorption of chemical species, or the deposition of ultra-thin monomolecular films at the guide surface.

If distributed phase matched coupling is used, e.g., with a grating, and if the coupling location is within the area where the index perturbation occurs, the latter will cause an alteration of the excitation conditions. Maximum coupling efficiency can be restored by changing one of the opto-geometrical parameters governing the phase matching condition. This is often made by changing the angle of the incident plane wave.[104]

We will show how the perturbation analysis can lead to simple expressions for the expectable sensitivity in the whole category of plane wave coupled guiding film sensors with wavelength monitoring. The two cases where the index perturbation occurs in the waveguide and in its cover will be treated hereafter. The case of an ultra-thin waveguide load can be studied similarly using the perturbation Expression 95 leading to Expression 153.

### a. Plane Wave to Guided Wave Coupling

The phase matching condition on the excitation from the substrate of a TE mode of

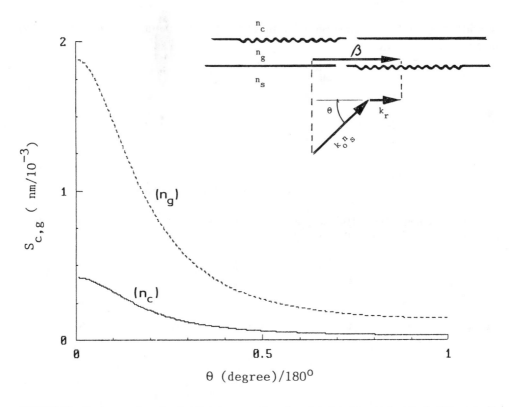

FIGURE 39.   Grating coupling of an incident plane wave from the substrate to the guided mode of a slab waveguide. Sensitivity $S_{c,g}$ of the phase matching condition to a change of refractive index of the cover ($n_c$) and waveguide ($n_g$), respectively. $S_{c,g}$ is expressed in nanometers per $10^{-3}$ change of index.

propagation constant $\beta$ in a waveguide by means of a corrugation grating at its surface or at the substrate-guide interface (Figure 39) is:

$$n_s k_0 \cos \theta + k_r = \beta \qquad (141)$$

where $k_r = 2\pi/\Lambda$ is the grating constant, $\Lambda$ is the grating pitch, $\theta$ the incident plane wave angle in the substrate. Propagation is along positive z. Equation 141 can also be written:

$$k_0 = \frac{k_r}{n_e - n_s \cos \theta} \qquad (142)$$

$n_e$ satisfies the characteristic Equation 33. A change of the cover or guide index brings a change of the effective index, therefore, of the excitation conditions: in a laboratory experiment, the phase matching condition would be restored by changing the angle of incidence $\theta$. However, this is technically impracticable in a working sensor because of the mechanics involved and the resulting vulnerability to the outside world. It is more advisable to restore the phase matching condition in a solid structure by tuning the optical frequency. This is now achievable since the advent of single-mode semiconductor lasers by acting on the injection current or on the temperature.[105] The estimate of the achievable resolution is made hereafter.

**Sensitivity to Cover Index**

The effective index change $\delta n_e$ due to a small variation $\delta n_c$ of the cover index can be found by using the perturbation formalism (Equation 95) where the perturbation is $\Delta\varepsilon = 2n_c \cdot \delta n_c$ integrated all over the cover:

$$\delta n_e = \frac{\omega}{4k_0 P}\varepsilon_0 \int_w^\infty \Delta\varepsilon \left|E_x\right|^2 dy \tag{143}$$

The integration yields:

$$\delta n_e = r_c \delta n_c \quad \text{where} \quad r_c = \frac{n_c\left(n_g^2 - n_e^2\right)}{n_e\sqrt{n_e^2 - n_c^2}\left(n_g^2 - n_c^2\right)k_0 w_{eff}} \tag{144}$$

Besides, the tuning offset $\delta\omega$ of the optical frequency also brings a change of effective index, which can be obtained from the characteristic equation:

$$\delta n_e = r_k \delta k_0 \quad \text{where} \quad r_k = \frac{\partial n_e}{\partial k_0} = \frac{n_g^2 - n_e^2}{\beta}\frac{w}{w_{eff}} \tag{145}$$

Introducing Equations 144 and 145 into Equation 142 allows the net change $\delta k_0$ to be expressed vs. $\delta n_c$:

$$\frac{\delta k_0}{\delta n_c} = -\frac{k_0^2 r_c}{k_r - k_0^2 r_k} \tag{146}$$

**Sensitivity to Guide Index**

The dielectric perturbation $\Delta\varepsilon = 2b_g \delta n_g$ is now introduced into the perturbation Formula 95 and the integration is performed from 0 to w. The integration followed by some algebraic manipulations yields:

$$\delta n_e = r_g \delta n_g \quad \text{where} \quad r_g = \frac{n_g}{n_e}\frac{w_g}{w_{eff}}$$

$$w_g = w + \frac{b_s}{k_s} + \frac{b_c}{k_c} < w_{eff}$$

with

$$b_s = \frac{n_e^2 - n_s^2}{n_g^2 - n_s^2} \quad ; \quad b_c = \frac{n_e^2 - n_c^2}{n_g^2 - n_c^2} \tag{147}$$

Introducing Equations 147 and 145 into Equation 142 yields the corresponding change $\delta k_0$ vs. $\delta n_g$:

$$\frac{\delta k_0}{\delta n_g} = -\frac{k_0^2 r_g}{k_r + k_0^2 r_k} \tag{148}$$

Let us define the sensitivity $S_c$ and $S_g$ in terms of the actual optical frequency $\nu$ in Hertz:

$$S_{c,g} = \frac{\delta(v)}{\delta n_{c,g}} = \frac{c}{2\pi} \frac{\delta k_0}{\delta n_{c,g}}$$

(149)

$\dfrac{\delta k_0}{\delta n_c}$ and $\dfrac{\delta k_0}{\delta n_g}$ being given by Equations 146 and 148

The sensitivities $S_{c,g}$ vs. the angle of incidence $\theta$ in the case of a single-mode step index waveguide is illustrated in Figure 39. The example chosen is a silica based pyrolyzed $TiO_2$-$SiO_2$ solgel film of index 1.8 and thickness 0.167 μm. The cover index is 1.33. The nominal excitation wavelength is 780 nm. The effective index is $n_e = 1.55$. The nominal grating constant $k_r$ is always taken so as to satisfy the phase matching condition. In the example of a 780 nm GaAs laser beam at vertical incidence ($\theta = 90°$), a change of 1 mA (resp. 4.5 mA) of the injection current, or a change of 0.1°C (resp. 0.45°C) of the laser temperature, are requested to tune the phase matching condition under a $1 \cdot 10^{-4}$ change of the cover index (resp. guide index). The frequency characteristics of the laser considered in this example are –3 GHz/mA and –30 GHz/°C.

## 2. Synchronism Tuning of Guided Mode Coupling

The incident plane wave of former section can be replaced by a guided mode to be coupled to the counter-propagating wave in the same waveguide, or in a neighboring waveguide if spatial separation is wanted, in a configuration similar to that of Figure 32b. This sensor configuration leads to a much more compact measurement set and its sensitivity is higher than in the case of plane wave.

The phase matching condition in a contra-directional coupling is:

$$k_r = 2n_e k_0$$

(150)

Taking the derivative of Equation 148 and using Equations 141, 142, and 146 yields:

$$\frac{\partial k_0}{\partial n_{c,g}} = -\frac{k_0^2 r_{c,g}}{\dfrac{k_r}{2} + k_0^2 r_k}$$

(151)

## 3. Interferometric Phase Change Measurement

Refractometric applications where minute index changes must be measured will be best served by interferometric techniques. Passive integrated optics offers various types of interferometric circuits with a high flexibility in the design of geometric configurations. We will limit ourselves here to the modelization of the most practical scheme where the medium or the thin film to be monitored is placed as an overlay on top of a waveguide within the evanescent wave of a guided mode. The expected phase change $\Delta\phi = k_0 n_e L$, where L is the transducer length, will be estimated by calculating the effective index change using the perturbation approach of Section V. Two cases of particular interest will be treated: the case of a semi-infinite overlay of varying index $n_c$ and that of a very thin overlay film of index $n_f$ and varying thickness t immersed in a cover of index $n_c$.

**Semi-Infinite Medium**

Integrating Expression 95 for a TE mode in the external medium yields the change of effective index $\delta n_e$ under the effect of a change of $n_c$, $\delta n_c$:

$$\frac{\delta n_e}{\delta n_e} = \frac{1}{C} \tag{152}$$

with C given by Equation 134 and Figure 36. In the example of the waveguide of Figure 36, taken symmetrical, a change $\delta n_c = 10^{-4}$ of the cover index causes a phaseshift of 1 radian in a 13-mm long waveguide section at 633 nm wavelength.

**Ultra-Thin Film Overlay**

Again using Expression 95 for a TE mode and replacing the integration by the product of the film thickness t by the square of the field amplitude at the waveguide boundary y = w yields similarly:

$$\delta n_e = \frac{\left(n_g^2 - n_e^2\right)\left(n_f^2 - n_c^2\right)}{n_e\left(n_g^2 - n_c^2\right)w_{eff}} t \tag{153}$$

Figure 40 represents the effective index variation $\delta n_e$ normalized to $(n_f^2 - n_c^2) \cdot t$ in the case of a silica based waveguide with a water-like cover ($n_c = 1.33$). In contrast to the case of semi-infinite medium, the sensitivity reaches here a maximum corresponding approximately to the maximum modal field confinement, i.e., where $w_{eff}$ is minimum. This actually corresponds to the condition of maximum field at the interface which occurs at $V \cong 1.75$. This condition depends on the degree of guide asymmetry.[106] The example is now given of a silica based waveguide of width w = 1.5 µm close to the cutoff of the $TE_1$ mode immersed in a water-based solvent. The solvent contains organic molecules which are adsorbed at the guide surface in the form of a thin monomolecular layer of index $n_f = 1.4$.[104] Applying Expression 153 shows that the presence of a 10-nm thick organic film increases the effective index $n_e$ so as to cause a phaseshift of 1 radian in a 1.7-mm long waveguide section at 633 nm wavelength. The same phaseshift would be obtained with an air cover ($n_c = 1$) and the same film ($n_f = 1.4$ and t = 10 nm) in a 0.9-mm long waveguide section. The sensitivity would even be higher in the case of a higher index waveguide.

# IX. WAVEGUIDE FLUORESCENCE SENSORS

The nature of fluorescence radiation does not speak necessarily in favor of its use in association with optical waveguides. As long as spontaneous emission is concerned the expectable fluorescence capture efficiency in a waveguide tip configuration is very low, and it is even lower in evanescent wave configurations. In addition, silica-based waveguides are close to their UV transmission edge for a number of interesting excitation lines. However, optical waveguides offer a number of practical advantages such as remote sensing, point sensing in well-confined areas, or, in the case of evanescent wave sensing, very large interaction lengths and short response times. In a planar form the substrate can play the double role of being the optical cladding of the surface wave probe and the material support of various species to be analyzed with an inherent capability for the spatial multiplexing of a number of sensors.

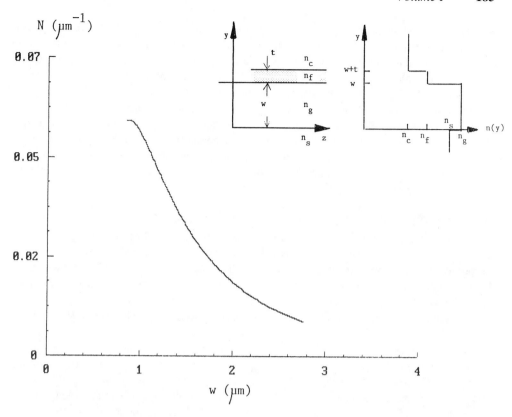

FIGURE 40. Relative variation of the effective index, $N = \delta n_g/((n_f^2 - n_c^2) \cdot t)$, vs. the geometrical width w of a slab step index waveguide with $n_s = 1.46$, $n_g = 1.475$, $n_c = 1.33$, $\lambda = 633$ nm.

## A. FLUORESCENCE EXCITATION

Fluorescent radiation does not necessarily have to be trapped into the waveguide conveying the excitation signal. Detection and signal processing can be performed locally using a microelectronic chip and micro-optic elements to filter out the excitation signal. The opto-chemical sensing and opto-electronic preprocessing head located at a fiber tip can involve some degree of waveguiding transversally to the fiber axis in applications which do not request a too small cross-section area. The excitation light of the fiber can be directed perpendicularly using a waveguide diffraction grating. As in bulk optics, the overall excitation efficiency depends on the fraction of guided power absorbed by the fluorescent material. In a fiber tip or multimode tubular waveguide configuration, all rays including leaky rays get absorbed. The type of configuration is discussed at the beginning of Section VI.

The main subject of the present section will be evanescent wave fluorescence excitation. A practical example of fluorescence transducer configuration is that of a PCS fiber with a low index fluorescent cladding propagating the excitation signal downleads, then capturing and conveying the fluorescence signal upstream or downstream to a detection unit performing spectrum analysis. In the spontaneous emission regime the transmission and reflection configurations are equivalent.

At first sight a ray emitted in the low index cladding medium surrounding a higher index core cannot be trapped. Those rays entering the core would get out again after a few partial reflections. From a wave optics standpoint, however, there is some degree of overlap between the fluorescence radiation field in the core neighborhood and the eigenfield pertaining to the set of modes that the core may propagate at the fluorescence wavelength.[107] The resulting non-zero coupling efficiency is very mode dependent. As illustrated in Figure 6a of Section III, in the case of a total reflection at a dielectric interface, the highest field penetration depth occurs at the critical angle, i.e., for modes close to cutoff. This holds for the excitation wave as well as for the resulting emitted fluorescence wave by reciprocity considerations.[108a,b] Going back to fibers and following Marcuse,[107] two types of fluorescent source distribution will be distinguished. In both cases, the guided optical excitation is assumed to be provided by a diffuse source of area larger than the fiber core. Consequently, all possible trapped modes are excited and propagate the same power down to the transducer. The latter is a long fiber section with low index fluorescent cladding. The fluorescent material is assumed to be lossless at the fluorescence wavelength.

## B. FLUORESCENT CLADDING OF INFINITE THICKNESS

A typical configuration is a PCS fiber with a thick plastic cladding containing a homogeneous concentration of dye.[109,110] In this case, the generated fluorescence power density is proportional to the sum of the squared values of the modal evanescent fields everywhere the latter is significant. There are two differing excitation regimes along the transducer abscissa: along the first section most absorption of the excitation power is due to modes close to cutoff whose field extends deeply into the fluorescent cladding, as discussed in Section VIII.A.1 and illustrated by Figure 13. Most fluorescence power is wasted. A large part of the trapped fluorescence radiation is captured by the modes close to cutoff at the fluorescence wavelength. The excitation power attenuation and guided fluorescence amplification rates are maximum. In this first regime, an increase of the fiber V-number increases the number of cutoff modes approximately linearly, as given by Equation 69 and the absolute captured fluorescence power accordingly. As most captured power is propagated by nearly cutoff modes, the sensor's signal will be very sensitive to bending and micro-bending effects. The leaky mode contribution in this regime should be rather weak. High loss leaky modes mainly will be excited as their field extends deeply into the cladding. Low loss leaky modes (i.e., modes with large l) are weakly excited as their power is well confined in the core according to Expression 76 and the considerations of Section V.B.

The beginning of the second transducer regime can be located where the excitation power of nearly cutoff modes is exhausted. Further downleads, the absorption as well as the fluorescence capture are due to all remaining modes which are better confined. The absorption and amplification rates are lower. However, the absolute fluorescence power captured by the core can still be significant over long interaction lengths. The absolute trapped power increases with the fiber V-number proportionally to the number of modes involved, i.e., approximately to $V^2$. The fluorescence signal will be less affected by bending and microbending effects as propagated by well confined modes.

## C. ULTRA-THIN FLUORESCENT COATING

In fluorescent opto-chemical sensors involving ultra-thin films, the capture efficiency does not depend on the total guided power flowing in the cladding, but rather on the power density at the core-cladding interface, i.e., on the square of the modal field at the boundary. Typical sensor configurations belonging to this case are marked antigen-antibody binding immunoassays,[111] Langmuir-Blodgett monolayers[112] immobilized or deposited on a PCS fiber core immersed in a water or solvent cladding.

The expression of the field of a $LP_{lm}$ mode at the core boundary, normalized to total propagated power P, is given by Expressions 78 and 77. Using the approximate Expression 73 for $\kappa$ yields:

$$|E(a)|^2 = 2 \, \frac{u^2}{a^2 V^2} \, P \, \frac{z_o}{\pi n} \left( 1 - \frac{1}{\sqrt{w^2 + l^2} + 1} \right) \tag{154}$$

The distribution of the square of the normalized field at the core-cladding interface in a fiber of 50-μm diameter and index difference $\Delta n = 0.01$, propagating about 450 modes at 800 nm wavelength, is shown in Figure 14. The most significant difference between the excitation and capture efficiencies of Equation 154 with Equation 75, giving $P_{clad}/P$, represented in their respective Figures 14 and 13, is that all modes are now more equally involved, even those with large l, inclusive of low loss leaky modes. There will also be a larger contribution of well-confined modes in the conveyance of the captured fluorescence.

The fluorescence capture efficiency is lower than in the former case as a smaller part of the evanescent power is used. However, Marcuse's calculation show that an efficiency of a few percents is expectable with a fiber of reasonable V number.[107] As in the second regime in former case the efficiency is a growing function of $V^2$.

## D. SINGLE-MODE EVANESCENT FIELD FLUORESCENCE EXCITATION

It can be concluded from the last sections that evanescent wave fluorescence excitation and trapping by a large core multimode fiber can be quite efficient if long interaction length transducers are used. Very high sensitivities are therefore expectable, though the accuracy is likely to be quite poor. The transducer configuration whose sensitivity is the least dependent on the incident modal power distribution and on the fluctuations on the fiber lead is that described in Section IX.C concerning ultra-thin fluorescent coatings.

If such sensors are to be used not only for the identification of the signature of species or for the detection of traces of a chemical element, but also for quantitative measurement, referencing schemes must be used and perhaps found, whereby the transducer information is encoded in such a way as not to be perturbed by the nature and circumstances of fluorescence excitation and capture as well as of download propagation. Such would be the case, for instance, with fluorescence time decay measurement which is a widely used technique in multimode physical sensors in temperature sensing.[77]

Intensity measurements would be more reliable with single mode excitation and capture as the power distribution only depends on the waveguide structure. Depending on the cover refractive index $n_c$ a significant amount of power can flow in the cladding. The excitation efficiency can therefore be rather high, but the captured power is very low as one mode only takes part in the coupling between the radiation modes and the propagating field. Most fluorescent power is wasted, but a long interaction length will still give rise to a significant guided fluorescence power provided the guided excitation signal has enough power. This, however, imposes that high spatial coherence sources are used. Present systems can only be operated with large size lasers or frequency doubled lasers[113] and are limited to laboratory use. This domain will only develop when the visible wavelength range is covered by semiconductor lasers down to the blue-green range.[85b] Quite promising is the high efficiency generation of fluorescence in the blue-green part of the spectrum by means of up-conversion in rare earth doped fluoride glasses using IR semiconductor laser excitation.[85c]

The ratio between a single guided mode to the number of emitted fluorescence radiation modes sets quite a low upper limit to the core capture efficiency. Various types of electromagnetic structures can be thought of to bring the actual efficiency closer to this limit. Four effects will be briefly mentioned hereafter as their study is beyond the scope of the present survey:

**1. Forcing the evanescent field out of the guide for achieving a better overlap with the fluorescence radiation field** — Single-mode waveguide cores in a planar or fiber form having a capillary central region of depressed index filled by a circulating fluorescent fluid can have a very high relative power in the evanescent wave. In case of a planar capillary filled with air, the evanescent wave power can be as high as 15%.[114]

**2. Fiber bending or microbending** — Long-range perturbations of the fiber axis increase the core capture efficiency of radiation waves propagating in the cladding co-linearly with the axis.

It was shown in Section VI, Expression 86, that a bent waveguide section is equivalent to a straight waveguide with an increased outer cladding index leaking its power approximately co-linearly along the tangent of the bend. The reverse is true whereby a radiation wave can be coupled to the guided mode. This effect is currently used in the field of communications when no access to the fiber end is possible such as in fiber splicing machines.[115] The radiation to guided wave coupling is also due to transitions between sections of different bending radius where transverse field matching necessarily involves radiation waves.[116] Bending and microbending effects can be used in both single- and multimode fiber transducers.

**3. Grating coupling** — Short-range periodical perturbation of the waveguide surface helps radiation wave coupling to a guided mode as described in Section VIII.C.1. A grating would also increase the outcoupling of the excitation wave. This would lead to a waste in the ultra-thin film case of Section IX.C. In the case of Section IX.B, the outcoupled wave would excite fluorescent sites located deeper in the cladding and shorten the interaction length. An interesting configuration could be that of a corrugation grating in which the period is adjusted so as to phase match fluorescence waves directed along the waveguide axis to the guided mode in the same direction. Under these conditions the grating cannot couple the shorter excitation wavelength to the fluorescent cladding, as the grating phase matching vector is too short and because the mode effective index is larger at the excitation wavelength. Grating can be used in both single- and multimode fiber transducers.

**4. Resonant coupling** — The fluorescent film can be considered as a waveguide if it is thick enough and has a refractive index higher than the effective index of the single-mode propagating in the conveyance guide. As in Section VIII.A.3, the synchronism condition can be achieved to transfer most of the excitation power into the probe guide. If the synchronism condition is also achieved at the fluorescence wavelength, a large part of the trapped fluorescence power will be coupled to the conveyance guide if the transducer waveguide length corresponds to the coupling length.[117]

As a conclusion, single-mode fluorescence sensors systems appear to be fundamentally of low efficiency. Their main interests lie in the nonambiguous shape of the guided field and especially in the multiple possibilities offered by single mode integrated optics for signal preprocessing functions described in Section VII. Quantitative measurements would therefore be much more reliable than with their multimode counterparts.

These features will not, however, find numerous practical applications before the advent of reliable and cheap semiconductor lasers, or semiconductor laser-pumped microlasers, for the whole visible wavelength range.

## ACKNOWLEDGMENTS

The author would like to thank the Swiss Center for Electronics and Microtechnology (CSEM) for allowing him to do this work. He is grateful to Mrs. Moser for her patience in typing the manuscript.

Dr. George Kotrotsios is especially acknowledged for providing most computing and drawing material appearing in this chapter.

# REFERENCES

1. **Shen, Y. R.,** *The Principle of Nonlinear Optics*, John Wiley & Sons, New York, 1984.
2. **Suhara, T. and Nishihara, H.,** Integrated optics components and devices using periodic structures, *IEEE J. Quantum Electron.*, QE22, 845, 1986.
3. **Ohishi, Y. and Sakaguchi, S.,** Spectral attenuation measurements for fluoride glass single-mode fibers by Fourier transform techniques, *Electron. Lett.*, 23, 272, 1987.
4. **Sakoguchi, S. and Takahashi, S.,** Low-loss fluoride optical fibers for midinfrared optical communication, *J. Lightwave Technol.*, LT-5, 1219, 1978.
5. **Beals, K. J., Day, C. R., Duncan, W. J., Midwinter, J. E., and Newns, G. R.,** Preparation of sodium borosilicate glass fiber for optical communication, *Proc. IEE*, 123, 591, 1976.
6. **Kaino, T.,** Influence of water absorption on plastic optical fiber, *Appl. Opt.*, 24, 4192, 1985.
7. **Boucher, D., Guerder, P., and Aldebert, P.,** Ultraviolet (UV) transmission of plastic clad silica (PCS) fibers: characteristics, measurements, methods and production, *Proc. SPIE*, 279, 153, 1981.
8. **Rabolt, J. F., Schlotter, N. E., Swaler, J. D., and Santo, R.,** Comparative raman studies of molecular interactions at a dye/polymer and a dye/glass interface, *J. Polymer Sci.*, 21, 1, 1983.
9. **Shibata, N., Horigudhi, M., and Edahiro, T.,** Raman spectra of binary high-silica glasses and fibers containing $GeO_2$, $P_2O_5$ and $B_2O_3$, *J. Non-Cryst. Solids*, 45, 115, 1981.
10. **Wolfbeis, O. S.,** Fiber Optical Fluorosensors in Analytical and Clinical Chemistry, in *Molecular Luminescence Spectroscopy: Methods and Applications — Part II*, Schulman, S. J., Ed., John Wiley & Sons, 1988, chap. 3.
11. **Ramaswamy, R. V. and Srivastava, R.,** Ion-exchanged glass waveguides: a review, *J. Lightwave Technol.*, LT-6, 984, 1988.
12. **Born, M. and Wolf, E.,** *Principles of Optics*, Section 1.1, Pergamon Press, Oxford, 1980.
13. **Gradshteyn, I. S. and Ryzhik, I. M.,** *Table of integrals, series, and products*, Section 8.4, Academic Press, New York, 1980.
14. **Midwinter, J. E.,** The Prism-Taper coupler for the excitation of single modes in optical transmission fibers, *Opt. Quantum Electron.*, 7, 297, 1975.
15. **Midwinter, J. E.,** *Optical Fibers for Transmission*, John Wiley & Sons, New York, 1979, chap. 5.
16a. **van Gent, J., Kreuwel, H. J. M., Lambeck, P. V., Gerritsma, G. J., Sudhölter, E. J. R., Reinhoudt, D. N., and Popma, T. J. A.,** Optochemical sensors based on chromoionophores, in Proc. SPIE, Enschede, The Netherlands, November 2 to 4, 1988.
16b. **Cullen, D. C. and Lowe, C. R.,** A direct Surface Plasmon-Polariton Immunosensor: Preliminary Investigation of the Non-specific Adsorption of Serum Components to the Sensor Interface, *Sensors and Actuators,* to be published, 1989.
16c. **Parriaux, O. and Voirin, G.,** Plasmon Wave Versus Dielectric Waveguiding for Surface Wave Sensing, Proc. Transducers'89, Montreux, Switzerland, June 25-30, 1989, Paper D16.1.
17. **Lopez-Rios, T.,** Modification of the dispersion relation for surface plasmons by very thin surface films in the vicinity of their plasma frequency, *Opt. Commun.*, 17, 342, 1976.
18. **Palik, E. D.,** *Handbook of Optical Constants of Solids*, Academic Press, New York, 1985.
19a. **Schoenwald, J., Burstein, E., and Elson, J. M.,** Propagation of surface polarizations over macroscopic distances at optical frequencies, *Solid State Commun.*, 12, 185, 1973.
19b. **Quail, J. C., Rako, J. G., and Simon, H. J.,** Long-Range Surface-Plasmon Modes in Silver and Aluminium Films, *Optics Lett.*, 8, 377, 1983.
20. **Parriaux, O., Gidon, S., and Cochet, F.,** Fiber-optic polarizer using plasmon — guided wave resonance, in 7th ECOC, Copenhagen, P6-1, 1981.
21. **Suzuki, T., Yamakazi, T., Yoshioka, H., and Hikichi, K.,** Influence of thickness on $H_2$ gas sensor properties in polycrystalline $SnO_x$ films prepared by ion-beam sputtering, *J. Mater. Sci.*, 23, 1106, 1988.
22. **Takato, N., Jinguji, K., Yasu, M., Toba, H., and Kawachi, M.,** Silica-based single-mode waveguides on silicon and their application to guided-wave optical interferometers, *J. Lightwave Technol.*, LT-6, 1003, 1988.
23. **Thomas, I. M.,** Optical Coatings by the Sol-Gel Process, *Optics News*, 18, August 1986.
24. **Selvarej, R., Lin, H. T., and McDonald, J. F.,** Integrated optical waveguides in polyimide for wafer scale integration, *J. Lightwave Technol.*, LT-6, 1034, 1988.
25. **Haga, H. and Yamamoto, S.,** Precise control of phase constant of optical guided-wave devices by loading Langmuir-Blodgett films, *J. Lightwave Technol.*, LT-6, 1024, 1988.
26. **Ross, L., Fabricius, N., and Oeste, H.,** Single mode integrated optical waveguides by ion exchange in glass, in Proc. EFOC/LAN, Basel, June 1987, 99.
27. **Béguin, A., Dumas, T., Hackert, M. J., Jansen, R., and Nissim, C.,** Fabrication and performance of low loss optical components made by ion exchange in glass, *J. Lightwave Technol.*, LT-6, 1483, 1988.

28. **Ikeda, Y., Okuda, E., and Oikawa, M.,** Graded-index optical waveguides and planar microlens arrays and their applications, in Proc. EFOC/LAN, Basel, June 1987, 103.

29. **Snyder, A. W. and Love, D.,** *Optical Waveguide Theory,* Chapman and Hall, New York, 1983, chap. 12.

30. **Menzel, D. H.,** *Fundamental Formulas of Physics,* Dover Publications, 1960, chap. 21.

31. **Kim, C. M. and Ramaswamy, R. V.,** WKB analysis of asymmetric directional couplers and its application to optical switches, *J. Lightwave Technol.,* LT-6, 1109, 1988.

32. **White, J. M. and Heidrich, P. F.,** Optical waveguide refractive index profiles determined from measurement of mode indices: a simple analysis, *Appl. Opt.,* 15, 151, 1976.

33. **Marcuse, D.,** *Light Transmission Optics,* Van Nostrand Reinhold, 1982, chap. 11.

34. **Decotignie, J. D., Parriaux, O., and Gardiol, F. E.,** Wave propagation in lossy and leaky planar optical waveguides, *AEÜ,* 35, 201, 1981.

35. **Knox, R. M. and Toulios, P. P.,** Integrated circuits for the millimeter through the optical frequency range, in Proc. Symp. Submillimeter Waves, Brooklyn, March 31 to April 2, 1970, 497.

36. **Snyder, A. W. and Love, J. D.,** *Optical Waveguide Theory,* Chapman and Hall, New York, 1983, chap. 13.

37. **Decotignie, J. D., Parriaux, O., and Gardiol, F. E.,** Birefringence properties of twin-core fibers by finite differences, *J. Opt. Commun.,* 3, 8, 1982.

38. **Rahman, B. M. A. and Davies, J. B.,** Finite-element solution of integrated optical waveguides, *J. Lightwave Technol.,* LT-2, 682, 1984.

39. **Elewaut, L.,** Program LEAPOW, Ph.D. thesis, Labor voor Electromagnetisme en Acustica, Ghent University, St. Pietersnieuwestraat 41, B-9000 Ghent, Belgium.

40. **Van de Velde, K., Thienpont, H., and Van Geen, R.,** Extending the effective index method for arbitrarily shaped inhomogeneous optical waveguides, *J. Lightwave Technol.,* LT-6, 1153, 1988.

41. **Marcatili, E. A. J. and Hardy, A. A.,** The azimuthal effective-index method, *IEEE J. Quantum Electron.,* QE-24, 766, 1988.

42. **Newby, K., Reichert, W. M., Andrade, J. D., and Benner, R. E.,** Remote spectroscopic sensing of chemical adsorption using a single multimode optical fiber, *Appl. Opt.,* 23, 1812, 1984.

43. **Gloge, D.,** Weakly guiding fibers, *Appl. Opt.,* 10, 2252, 1971.

44. **Angot, A.,** Compléments de Mathématiques, Masson, Paris, Section 7.5.26, 1982.

45. **Gloge, D.,** Optical power flow in multimode fibers, *Bell Syst. Tech. J.,* 51, 1767, 1972.

46. **Snyder, A. W. and Love, J. D.,** *Optical Waveguide Theory,* Chapman and Hall, New York, 1983, chap. 24.

47. **Midwinter, J. E.,** *Optical Fibers for Transmission,* Section 5.3, John Wiley & Sons, New York, 1979.

48. **Snyder, A. W. and Mitchell, D. J.,** Leaky rays on circular optical fibers, *J. Opt. Soc. Am.,* 64, 599, 1974.

49. **Snyder, A. W. and Mitchell, D. J.,** Leaky mode analysis of circular optical waveguides, *Opto-electronics,* 6, 287, 1974.

50. **Miller, S. E.,** *Optical Fiber Telecommunications,* Section 3.4, Academic Press, New York, 1979.

51. **Snyder, A. W., Mitchell, D. J., and Park, C.,** Failure of geometric optics for analysis of circular optical fibers, *J. Opt. Soc. Am.,* 64, 608, 1974.

52. **Park, C. and Snyder, A. W.,** Illumination of multimode optical fibers-leaky ray analysis, *Opto-electronics,* 6, 297, 1974.

53. **Heiblum, M. and Harris, J. H.,** Analysis of curved optical waveguides by conformal transformation, *IEEE J. Quantum Electron.,* QE-11, 75, 1975.

54. **Miller, S. E.,** *Optical Fiber Telecommunications,* Section 3.5, Academic Press, New York, 1979.

55. **Marcatili, E. A.,** Bends in optical waveguides, *Bell Syst. Tech. J.,* 48, 2103, 1969.

56. **Gloge, D.,** Bending loss in multimode fibers with graded and ungraded core index, *Appl. Opt.,* 11, 2506, 1972.

57. **Das, S., Englefield, C. G., and Goud, P. A.,** Power loss, modal noise, and distortion due to microbending of optical fibers, *Appl. Opt.,* 24, 2323, 1985.

58. **Petermann, K.,** Theory of microbending loss in monomode fibers with arbitrary refractive index profile, *AEÜ,* 30, 337, 1976.

59. **Olshansky, R. and Nolan, D. A.,** Mode-dependent attenuation of optical fibers: excess loss, *Appl. Opt.,* 15, 1045, 1976.

60. **Gardner, W. B.,** Microbending loss in optical fibers, *Bell Syst. Tech. J.,* 54, 457, 1975.

61. **Jones, R. and Jones, K. W.,** Investigation of the wavelength-dependent transmission characteristics of optical fiber sensor systems, *Opt. Eng.,* 27, 23, 1988.

62. **Kotrotsios, G., Dénervaud, P., Falco, L., and Parriaux, O.,** High dynamic dual mode fiber transitometry, in Proc. SPIE, Hamburg, September, 1989.

63. **Wood, L. F. and Romero-Borja, F.,** Optical attenuation by periodic micro-distortions of a sensor fiber, *Opt. Lett.,* 10, 632, 1985.

64. **Snyder, A. W. and Love, J. D.,** *Optical Waveguide Theory,* Sections 18 to 21, Chapman & Hall, New York, 1983.

65. **Kogelnik, H.,** Theory of dielectric waveguides, in *Integrated Optics*, Tamir, T., Ed., Springer, 1979, 69.
66. **Marcuse, D.,** *Light Transmission Optics*, Van Nostrand Reinhold, 1982, chap. 10.
67. **Miller, S. E.,** Coupled wave theory and waveguide applications, *Bell Syst. Tech. J.*, 33, 661, 1954.
68. **Vassallo, C.,** About coupled-mode theories for dielectric waveguides, *J. Lightwave Technol.*, LT-6, 294, 1988.
69. **Weller-Brophy, L. A. and Hall, D. G.,** Local normal mode analysis of guided mode interactions with waveguide gratings, *J. Lightwave Technol.*, LT-6, 1069, 1988.
70. **Yariv, A. and Nakamura, M.,** Periodic structures for integrated optics, *J. Quantum Electron.*, QE-13, 233, 1977.
71. **Adams, M. J., Payne, D. N., and Sladen, F. M. E.,** Splicing tolerances in graded-index fibers, *Appl. Phys. Lett.*, 28, 524, 1976.
72. **Olshansky, R., Blankenship, M. G., and Keck, D. B.,** Length dependent attenuation measurements in graded index fibers, in 2nd ECOC, Paris, 1976, 111.
73. **Zemon, S. and Fellows, D.,** Characterization of the approach to steady state properties of multimode optical fibers using LED excitation, *Opt. Commun.*, 13, 198, 1975.
74. **Seikai, S., Tokuda, M., Yashida, K., and Uchida, N.,** Measurement of baseband frequency response of multimode fibers by using a new type of mode scrambler, *Electron. Lett.*, 13, 146, 1977.
75. **Coppa, G., Di Vita, P., and Potenza, M.,** Theory of scattering in multimode optical fibers, *Opt. Quantum Electron.*, 14, 177, 1982.
76a. **Ankiewicz, A.,** Ray theory of graded non-circular optical fibers, *Opt. Quantum Electron.*, 11, 197, 1979.
76b. **Petermann, K.,** Leaky mode behaviour of optical fibers with noncircularly symmetric refractive index profile, *AEU*, 31, 201, 1977.
77. **Grattan, K. T. V., Palmer, A. W., and Willson, C. A.,** A miniaturized micro-computer-based neodymium "decay-time" temperature sensor, *J. Phys. Instrum.*, 20, 1201, 1987.
78. **Kotrotsios, G.,** Ph.D. thesis, Swiss Center for Electronics and Microtechnology, CH-2007 Neuchatel, Switzerland, 1989.
79. **Davies, D. E. N., Chaimowicz, J., Economou, G., and Foley, J.,** Displacement sensor using a compensated fiber link, in Proc. OFS, Stuttgart, 1985, 387.
80. **Sixt, P., Kotrotsios, G., Falco, L., and Parriaux, O.,** Passive fiber Fabry-Perot filter for intensity-modulated sensors referencing, *J. Lightwave Technol.*, LT-4, 926, 1986.
81. **Sakai, I. and Parry, G.,** Multiplexing interferometric fiber sensors by frequency modulation techniques, in OFC/OFS, San Diego, 1985, 128.
82. **Dakin, J. P. and Croydon, W. F.,** Applications of fiber optics in gas sensing, EFOC/LAN, Amsterdam, 1988, 239.
83. **Voirin, G., Scheja, B., and Parriaux, O.,** Applications of glass etching to guided wave optics, in Int. Cong. Optical Science and Engineering, Paris, April 24 to 28, 1989.
84. **Asahi, H., Kawamura, Y., and Nagai, H.,** Molecular beam epitaxial growth of In Ga AlP visible laser diodes operating at 0.66-0.68 μm at room temperature, *J. Appl. Phys.*, 54, 6958, 1983.
85a. **Ainslie, B., Craig, S. P., and Davey, S. T.,** The Absorption and Fluorescence Spectra of Rare Earth Ions in Silica-Based Monomode Fiber, *J. Lightwave Tech.*, 6, 287, 1988.
85b. **Gunshor, R. L., Kolodziejski, L. A., Otsuka, N., and Nurmikko, A. V.,** Growth and characterization of wide gap V-VI heterostructures, *Proc. SPIE*, 797, 158, 1987.
85c. **Tong, F., Risk, W. P., MacFarlane, R. M., and Lenth, W.,** 551 nm Diode-laser-pumped Upconversion Laser, *Electron. Lett.*, 25, 1389, 1989.
86. **Chiron, B., Gauthier, F., and Bourdinaud, M.,** Capteurs à Fibres Optiques Plastiques, in Proc. SEE Conf. Optical Fibers, Paris, 1985, 63.
87a. **Melchior, H.,** Demodulation and photodetection techniques, in *Laser Handbook*, Arecchi, F. T., Ed., Elsevier, Amsterdam, 1989.
87b. Optical Detector Wall Chart, Laser & Optronics, 6, 49, July 1987.
88. **Seshadri, S. R.,** Coupling of guided modes in thin films with surface corrugation, *J. Appl. Phys.*, 63, R115, 1988.
89. **Lizet, J., Gidon, P., and Valette, S.,** Integrated optics displacement sensor achieved on silicon substrate, in Proc. ECIO, Glasgow, 1987, 210.
90. **Ura, S., Suhara, T., and Nishihara, H.,** An integrated-optic disk pick-up device, *J. Lightwave Technol.*, LT-4, 913, 1986.
91. **Minowa, J. and Fijii, Y.,** Dielectric multilayer thin-film filters for WDM transmission systems, *J. Lightwave Technol.*, LT-1, 116, 1983.
92. **Seki, M., Sugawara, R., Okuda, E., Wada, H., Yamasaki, T., and Hamada, Y.,** Low-loss, guided-wave multi/demultiplexer using embedded gradient-index ion exchange waveguides, in 12th ECOC, Barcelona, 1986, 439.

93. **Laude, J. P.,** Wavelength division multiplexers: review of some devices proposed recently, in EFOC/LAN, Montreux, Switzerland, 1985, 231.

94. **Wahlen, M. S., Divino, M. D., and Alferness, R. C.,** Demonstration of a narrowband Bragg-reflection filter in a single-mode fiber directional coupler, *Electron. Lett.*, 22, 681, 1986.

95. **Burns, W. K. and Milton, A. F.,** $3 \times 2$ channel waveguide gyroscope couplers: theory, *IEEE J. Quantum Electron.*, QE-18, 1790, 1982.

96. **Niemeier, Th. and Ulrich, R.,** Quadrature outputs from fiber interferometer with 4 ¥ 4 coupler, *Opt. Lett.*, 11, 677, 1986.

97. **Roth, P. and Voirin, G.,** Integrated optic coupler for interferometric mixer, in Proc. SPIE, Hamburg, September 1988.

98a. **Olivier, M.,** Guided-wave optical spectroscopy of thin films, in *New Directions in Guided Wave and Coherent Optics,* Vol. 2, Series E, No. 79, Ostrowsky, D. B. and Spitz, E., Eds., Martinus Nijhoff, 1984, 639.

98b. **Stewart, P., Norris, J., Clark, D., Tribble, M., Andonovic, I., and Culshaw, B.,** Chemical Sensing by Evanescent Field Absorption: the Sensitivity of Optical Waveguides, SPIE, 990, Chemical, Biochemical, and Environmental Applications of Fibers, 188, 1988.

99. **Verbeek, B. H., Henry, C. H., Olsson, N. A., Orlowsky, K. J., Kazarinov, R. F., and Johnson, B. H.,** Integrated four-channel Mach-Zehnder multi/demultiplexer fabricated with phosphorous doped $SiO_2$ waveguides on Si, J. Lightwave Technol., LT-6, 1011, 1988.

100. **Jaccard, P., Scheja, B., Berthou, H., Cochet, F., Parriaux, O., and Brugger, A.,** A new technique for low cost all-fiber device fabrication, *Proc. SPIE*, 476, 16, 1984.

101. **Blodgett, K. B. and Langmuir, I.,** Built-up films of barium stearate and their optical properties, *Phys. Rev.*, 51, 964, 1937.

102. **Falco, L., Spescha, G., Roth, P., and Parriaux, O.,** Non-ambiguous evanescent-wave fiber refractive index and temperature sensor, *Opt. Acta*, 33, 1563, 1986.

103. **Marcuse, D.,** *Light Transmission Optics*, Van Nostrand Reinhold, 1972, 420.

104. **Lukosz, W. and Tiefenthaler, K.,** Sensitivity of integrated optical grating and prism couplers as (bio) chemical sensors, *Sensors Actuators*, 15, 273, 1988.

105. **Strzelecki, E. M., Cohen, S. A., and Coldren, L. A.,** Investigation of tunable single frequency diode lasers for sensor applications, *J. Lightwave Technol.*, LT-6, 1610, 1988.

106. **Kogelnik, H. and Ramaswamy, V.,** Scaling rules for thin-film optical waveguides, *Appl. Opt.*, 13, 1857, 1974.

107. **Marcuse, D.,** Launching light into fiber cores from sources located in the cladding, *J. Lightwave Technol.*, LT-6, 1273, 1988.

108a. **Lee, E. H., Benner, R. E., Fenn, J. B., and Chany, R. K.,** Angular distribution of fluorscence from liquids and monodispersed spheres by evanescent wave excitation, *Appl. Opt.*, 18, 862, 1979.

108b. **Carniglia, C. K., Mandel, L., and Drexhage, K. H.,** Absorption and emission of evanescent photons, *J. Opt. Soc. Am.*, 62, 479, 1972.

109. **Lieberman, R. A., Blyler, L. L., and Cohen, L. G.,** Distributed fluorescence oxygen sensor, in Proc. OFS, January 27 to 29, New Orleans, Tech. Digest Ser. 2, 1988, 346.

110. **Blyler, L. L., Ferrara, J. A., and MacChesney, J. B.,** A plastic-clad silica fibre chemical sensor for ammonia, in Proc. OFS, January 27 to 29, New Orleans, Tech. Digest Ser. 2, 1988, 369.

111. **Andrade, J. D., VanWagenen, R. A., Gregonis, D. E., Newby, K., and Lin, J. N.,** Remote fiber-optic biosensor based on evanescent-excited fluoro-immunoassay: concepts and progress, *IEEE Trans. Electron Devices*, ED-32, 1175, 1985.

112. **Rabolt, J. F., Santo, R., Schlotter, N. E., and Swalen, J. D.,** Integrated Optics and Raman Scattering: molecular orientation in thin polymer films and Langmuir-Blodgett Monolayers, *IBM J. Res. Dev.*, 26, 209, 1982.

113. **Werth, D.,** Microlaser: small lasers with a big future, *Photonics Spectra*, 143, April 1988.

114. **Parriaux, O., Kotrotsios, G., and Newman, V.,** Sensitivity enhancement of evanescent wave integrated optic transducers, in Proc. Transducers '89, Montreaux, Switzerland, June 25 to 30, 1989, Paper D14.4.

115. **Miller, C. M.,** *Optical Fibre Splices and Connectors*, Sections 5.5.3 and 6.4.3, Marcel Dekker, New York, 1986.

116. **Gambling, W. A., Payne, D. N., and Matsumara, H.,** Radiation from curved single-mode fibers, *Electron. Lett.*, 12, 567, 1976.

117. **Poscio, P., Depeursinge, C., Voirin, G., and Parriaux, O.,** Realization of a miniaturized optical sensor for biomedical application, in Proc. Transducers '89, Paper D14.2, Montreux, Switzerland, June 25 to 30, 1989.

Chapter 5

# INTRINSIC FIBER OPTIC CHEMICAL SENSORS

**Robert A. Lieberman**

## TABLE OF CONTENTS

## ABSTRACT

Fiber optic techniques are rapidly gaining popularity in chemical sensing. This chapter describes the intrinsic approach to chemical sensing.

Most fiber optic chemical sensors are extrinsic devices, i.e., the transduction of information on chemical concentration or state takes place in a region outside the fiber, which acts as a simple "light pipe" to guide the resulting optical signal to a detection system. Intrinsic sensors utilize the fiber itself as the transduction element by relying on chemically-induced changes in the optical properties in the fiber core, cladding, or jacket materials.

The chapter begins by giving an overview of the main types of intrinsic fiber optic chemical sensors (evanescent-field spectroscopic probes, core-based sensors, cladding-based sensors, and jacket-based sensors) and giving examples of some applications. Following this introduction, specific sensor types are discussed in detail, using illustrative examples. Among the topics covered are transduction techniques, detection techniques, sensitivity, specificity, sensor length, and sensor manufacture. A review of the relevant literature is also given for each sensor type.

## I. INTRODUCTION

The most common types of fiber optic chemical sensors are those which may be called "extrinsic," i.e., sensors that rely on fibers as simple "light pipes" to carry optical energy to and from test regions having chemically dependent optical properties.[1] In contrast to extrinsic sensors, an intrinsic sensor uses the fiber itself as a sensitive component. The fiber, rather than being a passive element in the sensing system, is an integral part of the detection process; this means that it must have optical characteristics that depend in some way upon the chemical environment in which it finds itself. Because of this requirement, intrinsic fiber optic chemical sensors must rely on specially developed or highly-modified fibers: most commercially available fibers have been specifically engineered to *reduce* the effect of the environment on the transmission of light by the fiber. Despite this barrier, there are several distinct advantages that make it worthwhile to pursue the intrinsic approach. Among these advantages are the capability of intrinsic sensors to perform distributed measurements, the improved specificity obtainable using intrinsic sensors for some measurements, and the ability to produce sensors that operate without physically interrupting the path of light in the fiber. The advantages of the technique have motivated workers in several laboratories to initiate research and development efforts in intrinsic fiber optic sensing over the past few years; many different approaches have been used to make optical fibers into chemical sensors, and a wealth of new possibilities for sensing is just beginning to be explored.

Although many of the optical detection techniques and chemical methodologies developed for extrinsic sensors may be adapted for use in intrinsic fiber optic sensing, there are several important differences between the two different sensor types, particularly in practical applications. The relationship between intrinsic and extrinsic fiber sensors and their bulk-optic progenitors is shown in a schematic way in Figure 1*.

Optical chemical sensors fall into two broad classes: "direct" (spectroscopic), and chemically assisted or "indicator-mediated."[2] Direct methods such as bulk optic spectrophotometry and colorimetry have been used in chemical sensing for more than a century; the earliest fiber optic chemical sensors simply used bundles of fibers to carry light to and from a spectroscopic test cell located at some distance from a conventional spectral analysis instrument. The most common extrinsic "direct" fiber sensors are still composed of two fibers or fiber bundles arranged at either end of a well-defined optical path in the test medium and connected to the output and input ports of a spectrophotometer, fluorimeter, Raman spectrophotometer, or other spectral analysis instrument.

Just as in the case of direct spectroscopic extrinsic fiber sensors, direct intrinsic sensors can take advantage of all of the spectroscopic analysis techniques which have been developed for bulk-optic measurements. A direct intrinsic fiber optic sensor, however, must rely on the interaction of the light carried by the fiber with the medium to be measured without using a spectroscopic cell: the fiber completely defines the optical path of the light interacting with the environment being studied. Many fiber configurations exist for which the spectral properties of the fiber's surroundings will have a relatively large effect on the spectral absorbance ("loss") of the fiber itself. For example, fibers which are completely permeable can let chemical species in the fiber's environment diffuse directly into the core region, where these species interact directly with guided light, causing the optical properties of the fiber to depend directly on the optical properties of the chemicals in the surrounding medium.

Another means by which the optical properties of the surrounding medium can affect the optical properties of the fiber is through the evanescent field of light guided by the fiber core. Even for "ideal fibers" (with no loss and perfect index of refraction profiles) the electromagnetic field of light carried by guiding modes in the fiber core has a non-zero component in the cladding.[3] By interacting with this field, particularly in the region nearest the core, materials

---

* All figures will appear at the end of this chapter.

in the cladding can affect the optical properties of the fiber. This phenomenon, which in telecommunications causes the well-known (though onerous) phenomenon of "cladding loss",[4] gives rise to the most common type of intrinsic fiber optic chemical sensor. The spectral properties of a measurand in the evanescent field of the fiber are be superimposed on the "native" loss spectrum of the fiber.

The use of indicator dyes in chemical sensing is almost as old as the use of direct methods. pH test paper, for example, may be thought of as a "bulk-optic" (or "eyeball-optic") indicator-mediated chemical sensor. The relatively recent invention of "opt(r)odes"[5] and the adaptation of this concept for use with optical fibers[6] has brought with it a renewed interest in indicator-mediated sensing.

In "chemically assisted," or indicator-mediated sensors, an opto-chemical intermediary system is chosen so that, in contact with the substance to be detected, it undergoes a large change in some optical property (such as optical absorbance, fluorescence output intensity, or optical path length). In extrinsic sensors this change is detected, either via bulk optics or through optical fibers, by a system tuned to measure not the optical properties of the chemical species being sensed, but the optical properties of the intermediary medium. Intrinsic sensors using chemical transduction also operate in this way: the chemical sensitivity of the fiber is not caused by a change in the optical properties of the measurement environment, but by the interaction of the environment with an intermediary substance present in or on the fiber itself. In many cases the same sensor reagents used in optodes can be adapted for use in active intrinsic sensors: both absorbance-based indicator dyes and chemically sensitive fluorescent compounds have been used in intrinsic fiber sensors.

The same optical principles used for direct intrinsic sensing can be applied to chemically assisted intrinsic fiber optic chemical sensors. Thus, the optical absorbance of a permeable fiber may be caused to change if a measurand diffuses into the core of the fiber and interacts with a chemically sensitive dye in the core. Alternatively, a fluorescence- or absorbance-based indicator dye in the cladding of a fiber will interact with the evanescent field of light guided by the core; when a measurand molecule changes the optical properties of the dye, the optical properties of the fiber itself also change. Light can be coupled into or out of guided modes in this way. Unlike direct intrinsic fiber optic chemical sensors, chemically assisted sensors can also make use of "jacket" materials that do not come in contact with the electromagnetic field of light propagating in the fiber. For example, a sensor jacket deposited directly outside the cladding, can cause changes in the optical propagation characteristics of the fiber (e.g., optical path length) by changing the physical properties of the fiber in the presence of a measurand.

The main justification for using indirect ("chemically assisted") methods in intrinsic sensing is the same as for extrinsic sensing: chemically transduced optical signals can be several orders of magnitude greater than associated changes in the optical properties of the native system under study. Chemically assisted sensor fibers may thus be used to measure chemical systems for which it would be impractical to use other optical techniques. A proper choice of chemical system can also impart much greater specificity to individual measurands, a benefit which is particularly useful in measurement environments where there are many different chemical species present.

Perhaps the most important characteristic of intrinsic fiber sensors is that they are capable of being used in distributed sensing systems — systems in which the measurement region is not "pointlike" (defined by a relatively small optrode), but is linear, extending over a path defined by the fiber. Linear sensors can be installed in straight or "area-covering" serpentine paths, for example, along the entire circumference of an O-ring seal, back and forth across the ceiling of a room, or entirely around a toxic waste dump in a closed path. An intrinsic fiber optic sensor can, therefore, accomplish tasks that would require several (in some case hundreds of) point-like sensors, using a single set of monitoring electronics coupled to a single optical detector. Distributed sensors can be used as "alarm" style sensors, alerting personnel that a measurand has been detected somewhere along the fiber path. Intrinsic sensors can also be used to measure the spatial average of chemical concentration along the path defined by the fiber. Techniques such as optical time domain reflectometry (OTDR) even offer the possibility of obtaining information about the chemical state at each point along the sensor fiber, with spatial resolution on the order of a few centimeters or less.

## II. REFRACTOMETRIC SENSORS

The earliest reports of intrinsic optical waveguide sensors describe simple refractometers, where the propagation constants of fibers immersed in test media (most often liquids) depend on the refractive index of the liquids themselves.

Intrinsic fiber optic refractometry and fiber optic evanescent wave spectroscopy (discussed in the following section) are closely related, in that both methods of chemical detection involve the direct modification light carried in the fiber by a chemical measurand. In both techniques the interaction of the light with the environment is achieved by means of the electromagnetic field that exists outside a fiber core guiding light; the difference between them is that, in one technique the primary quantity of interest is the refractive index of the medium around the fiber, whereas in the other the optical absorbance is measured. Thus, if the properties of the medium are represented by the complex refractive index,

$$n' = n + i\kappa$$

then changes in the real part ($n$) of the complex refractive index are measured by refractometry, while it is mainly changes in $\kappa$ that are detected by spectroscopic absorbance measurements.

Changes in the $n$ or $\kappa$ of a medium in which a fiber is immersed can affect light propagating in the fiber when the medium is physically close enough to the fiber core to become, in effect, part of the fiber cladding. This interaction between light in the fiber core and chemical species in the cladding takes place because light guided by the fiber core is actually not fully contained in the core: the evanescent field of the core-bound light extends into the fiber cladding.

For materials where $\kappa$ is small ("clear" materials) refractometric sensors are used, and a relatively straightforward application of ray optics can illustrate the effect of cladding index on the light guiding properties of a fiber. The critical angle for internal reflection, $\theta_{crit}$, is completely determined by the relative indices of refraction of the medium in which the ray is traveling ($n_1$) and the medium on the other side of the interface from which the ray is reflected ($n_2$):

$$\theta_{crit} = \arccos\left[\frac{n_2}{n_1}\right]$$

where $\theta_{crit}$ is measured from the interface. Rays incident at angles less than $\theta_{crit}$ are totally reflected. Thus if the refractive index of the reflecting medium decreases relative to the index of the medium of incidence, there is a wider range of angles that result in total internal reflection and, for uniform illumination, more rays will be reflected at the interface. This means that the fiber is capable of guiding more light when the index difference is larger; in fact, the half-angle of the "entrance cone" of light guided by a fiber is directly related to the critical angle defined above:

$$\sin(\theta_{entrance}) = 2 \cdot NA$$

where

$$NA = \sqrt{n_{core}^2 - n_{clad}^2}$$

or

$$NA = n_{core} \cdot \sin\theta_c$$

NA is called the numerical aperture of the fiber. The numerical aperture of a fiber is dependent only on the relative indices of the core and the cladding materials; for example, the numerical aperture of a quartz fiber in pure water, independent of fiber size, is approximately 0.532, corresponding to an acceptance cone half-angle of roughly 35°.

In the mode-theoretic description of optical propagation in fibers, the total number of modes in which light can propagate is completely determined by the "waveguide parameter" ("V number") of the fiber:

$$V = \left[2\frac{\pi}{\lambda}\right] r_{core} \cdot NA$$

where $r_{core}$ is the fiber core radius, and $\lambda$ is the wavelength of the light. For a step-index multimode fiber, the number of modes is equal to the smallest integer just exceeding $V^2/2$. Although in general not all modes get equally illuminated by a source, it is true that, all other things being equal, fibers with more modes available (e.g., fibers with a greater refractive index difference between core and cladding) will carry more light than fibers with less modes. When the cladding refractive index becomes equal to (or larger) than the core refractive index, the fiber does not guide light at all.

Intrinsic fiber optic refractometric sensors can be used in direct measurements of refractive index, in liquid level sensors, and in the measurement of chemical concentration when materials of known refractive index are present in simple mixtures in the test medium. Although fiber refractometers tend to be "optrode-like" rather than extending over large distances, it is still appropriate to begin our survey of intrinsic sensors with these devices, the first to perform chemical measurements using the interaction of the measurand directly with light being carried in a waveguide.

The earliest reports of optical waveguides used as intrinsic chemical sensors do not actually describe fibers, but rods of glass or quartz, employed as the optical elements of refractometric measurement systems. As the field evolved the rods chosen for use became thinner, and reports of the use of true optical fibers quickly appeared. This evolutionary pattern has been repeated in the development of intrinsic fiber optic chemical sensors based on other principles (direct spectroscopic sensors as well as coated fiber and core-based sensors of all types). Karrer and Orr[7] described a simple refractometer based on the use of a 12 mm diameter rod with a U-shaped bend immersed in the liquid to be tested (Figure 2a). The bend served to increase the angle of incidence of rays with the core/cladding (rod/test medium) interface, thereby increasing sensitivity to refractive index changes. Karrer and Orr also described a technique to measure the difference in refractive index between two liquids by using two such rods coupled to a single light source and having their outputs impinging on two different photovoltaic cells connected to a simple nulling circuit. One rod was immersed in each of the solutions to be compared, and a potentiometer was used to balance the voltages produced by the detectors. A variation of this technique, wherein the "reference" photocell was replaced by a potentiometer and a battery, allowed Karrer and Orr to measure a voltage change of −19.52 mV when the index of refraction of the test solution was changed from 1.3336 (water) to 1.4607 (carbon tetrachloride). Although the intrinsic U-rod sensor was only able to measure refractive index changes on the order of $10^{-3}$, compared with $10^{-7}$ for bulk-optic devices, the simplicity of the intrinsic device and its usefulness as a "dipping probe" were stressed by Karrer and Orr, who also forecast the use of such bent-waveguide structures in liquid level monitors.

Though the bent rod technique was demonstrated to have sensitivity to changes in index of refraction acceptable for many purposes, the mathematical treatment of light propagation in straight rods (Figure 2b) is considerably easier. Kapany and coworkers, in a series of papers[8-10] presented a relatively detailed analysis of the performance of linear-rod refractometers, based on straightforward application of the ray approximation to propagation in optical waveguides. The calculations were compared with experimental results obtained using rods having different indices of refraction. Waveguides diameters ranged from 1 to 2 mm, almost thin enough to be considered true "fibers".

To simplify the analysis, Kapany et al. assumed a uniform distribution of light across the input face of the rod (Lambertian illumination); this assumption unfortunately led to a considerable lack of quantitative agreement between the theoretical predictions and experimental results. Despite this, three important qualitative features of straight-guide refractometers were brought out by the analysis: (1) the sensitivity of such refractometers (and indeed, of waveguide refractometers in general) is greatest when the index of refraction of the guide is very close to the index of refraction of the medium being measured; (2) absorbance in the waveguide has very little effect on the sensitivity of the refractometer, as long as sufficient light reaches the photodetector; and (3) relatively high aspect ratios (length-to-diameter ratios greater than 1000) are required for the dependence of power on refractive index difference to be linear for differences on the order of 0.001 or greater.

Kapany and Pontarelli[10] also obtained some experimental results on sensors with non-straight geometries (U-shaped bends, "kinks", helices), and appear to be the first to attempt to measure the properties of evanescent-wave refractometers in the presence of absorbing species, as discussed in the next section of this chapter.

The early work of Kapany et al. was actually incorporated in a practical refractometer used in liquid chromatography detection.[11] The "fiber" in this system was a 1 mm diameter glass rod, on the order of 10 mm long, and having a refractive index of 1.675. A lens system was used to focus a HeNe laser onto the end of the rod, and a movable prism allowed the angle of incidence to be adjusted by means of a digitally controlled stepper motor. Optically, this system approximated the ray-based model analyzed by Kapany.[9] Far from being a multimode waveguide with its optical energy having reached the equilibrium distribution among modes of propagation, the fiber in this system operated as a medium for producing multiple (approximately ten) reflections of a light beam at the glass-liquid interface. Refractive indices in the range 1.32 to 1.50 were measured with an accuracy on the order of $\Delta n = 10^{-5}$ by scanning the incident beam to determine the critical angle for total internal reflection. A similar system has been described by Ross and Mbanu for use in measuring glucose concentrations[12] although absolute transmission rather than measurement of critical angle was used to determine refractive index. Ross and Mbanu's work is interesting in that it compares a square cross-section fiber to a round fiber and finds the square fiber to be more sensitive to glucose concentration.

A sensor based on a 1 mm diameter quartz rod with multiple bends has been described by Sharma and Brooks,[13] whose main goal was to monitor the level of cryogenic propellants in liquid-fueled rockets. Light was conducted from an LED into one end of the "W" shaped sensor element by a bundle of fibers; a similar bundle carried light from the output end of the "W" to a p-i-n photodector. The authors stated that the purpose of the sharp bends in the sensor was to convert "the large-angle rays ... that are instrumental in measurement of the presence of low-index fluids ... from and into low-angle rays that can be guided by the connecting optical fibers."[13] The shape of the sensor (Figure 2d) was empirically determined to give the best sensitivity to the liquids measured (liquid oxygen, n = 1.221, and liquid hydrogen, n = 1.1097). One suspects that the induction of higher order modes (much more sensitive to cladding refractive index changes) at the bends, and the suppression of the "whispering gallery" modes associated with single-bend sensors, were actually the most important effects of the "kinked" shape of the sensor. This sensor was tested successfully at cryogenic temperatures, along with sensors for temperature and pressure based on other optical principles.

Straight sections of thin-walled quartz capillary waveguide (essentially an "annular core" fiber with air cladding) have been used by Giuliani and Jarvis at the U.S. Naval Research Laboratory in a refractometric hydrocarbon gas sensor.[14] The waveguide, a modification of a coating-based device originally developed for ammonia monitoring (see the section on coating-based sensors in this chapter), had to be suspended in a humidity-controlled chamber for use. In a two-step measurement cycle, the sensor was first exposed to humid (80% relative humidity) air to form a thin (probably less than 0.01 μm) coating of adsorbed water on the waveguide, causing its transmittance to rise as water filled light-scattering irregularities in the glass surface. For measurement of gas concentration, the waveguide was next exposed to humid air containing a controlled amount of an alkane gas such as methane or propane, which displaced the adsorbed water layer. The transmission of 560 or 660 nm light (provided by LEDs) by the fiber was monitored using a phototransistor, and increases in transmittance that resulted from the presence of the gases were observed. These increases were presumably due to decreases in the effective refractive index of the adsorbed layer, leading to larger waveguide numerical apertures and therefore greater light-carrying capacity. When the gases were admitted to the sensor region under dry conditions, no changes in transmittance were observed; likewise, the gas-induced changes in transmittance were completely reversible when the measurement chamber was flushed with dry air. The change in transmittance at 560 nm upon exposure to hydrated methane was observed to be a linear function of gas concentration over the range 1000 to 35,000 ppm (partial pressures of 0.76 to 26.6 torr at atmospheric pressure), an order of magnitude lower than the lower explosive limit (50,000 ppm, or 38 torr) for methane. It was noticed in the initial study of this sensor that, for identical gas concentrations, the amount of transmittance increase experienced by the waveguide upon exposure to hydrated gas was a monotonically decreasing function of the molecular weights of the gasses tested. This presumably comes about because the molecular complexes formed by the interaction of gas molecules with water molecules cause some scattering of the guided light near the surface of the waveguide, with larger molecules inducing larger scattering centers. The capillary structure has consequently been used to study the detailed properties (thickness, clathrate structure, etc.) of thin layers of hydrated alkanes and noble gasses on free silica surfaces.[15,16] Giuliani and Bey[17] have also devised a computer model that successfully predicts the behavior of the capillary waveguide structure when it is immersed in fluids of varying refractive index.

A sensor reported by Takeo and Hattori[18] can lay claim to being the first truly "fiber optic" refractometer. The sensor element was created by removing the cladding from a short section at the beginning of a bend in a waveguide made of poly(methyl methacrylate) (PMMA — refractive index = 1.495) and clad with a fluoropolymer (refractive index = 1.402) (Figure 2c). Waveguides with diameters as small as 0.5 mm were used. The detection system was schematically similar to that used decades earlier by Karrer and Orr, with a HeNe laser in place of the filament light source, and a phototransistor in place of the photovoltaic cells. Takeo and Hattori were able to achieve accuracies on the order of $\Delta n = 10^{-3}$ for refractive indices near 1.4, and obtained qualitative agreement with a ray-based theoretical model that predicted greater attenuation, and therefore greater sensitivity to changes in refractive index, for smaller radii of curvature in the bent section of fiber. Reproducibility of the measurements was poor, due to contamination of the fibers by the measurands (two different types of oil), but the authors showed that this sensor could be useful in monitoring for oil spills in an otherwise dry environment, or as a liquid level sensor. They also proposed industrial process control applications for improved versions.

Similar results (resolution of $\Delta n \sim \cong \sim 10^{-3}$ near $n = 1.4$) were achieved by Spenner et al.,[19] using step-index silica core fibers ranging in diameter from 0.4 to 1.0 mm, with the (unspecified) cladding material removed over a 3 cm U-shaped measuring region. A red LED and a p-i-n photodetector were used for illumination and detection, and bend radii from 1 to 3 mm were applied to the fibers in the stripped region. Spenner et al., used a variety of clear organic liquids (turpentine, glycerine, propanol, benzene) to vary refractive index and, using a white light source and a monochromator, found that signal intensities were remarkably independent of wavelength in the visible and near-IR range studied, depending in a linear and reproducible way on on refractive index. Spenner et al. also observed dramatic changes in the behavior of the sensor when ink was used to color the test solutions, illustrating the fact that media must be "clear" (i.e., optical absorbance less than 0.01 absorbance units) for refractive indices to be accurately measured by evanescent-wave technique.

An interesting description of a refractometer designed for automotive battery measurements is presented in a paper by Harmer.[20] This sensor used plastic fibers (with either polystyrene or PMMA cores) with the cladding still intact, in a configuration with three alternating bends (Figure 2e) akin to the "W" configuration of Sharma and Brooks. The fiber was relatively short, and the LED and p-i-n detector used for illumination and detection were included in a single unit along with the fiber. The bends induced light to enter the highly attenuated "cladding modes" of the fibers, and Harmer achieved sensitivities as high as $\Delta n = 10^{-6}$ in some versions of the probe. The probe was designed to operate in the range n = 1.35 to n = 1.38 (the range commonly found in lead-sulfur batteries). As one method of compensating for the effect of temperature on refractive index, which could cause spurious readings of the battery's charge state, Harmer devised a two-probe system, where a second probe, made of glass, was used to give a signal having a different temperature dependence than that of the plastic probe. With the appropriate signal processing, a temperature-independent refractive index figure was obtained. A similar probe, based on a simple U-shaped bend in a clad 0.4 mm diameter polystyrene/PMMA fiber, was used by Bergman et al.[21] to investigate thermal diffusion in stratified salt-water solutions.

Belkerdid et al.[22] also have reported a bend-based refractometric sensor making use of a 1 mm diameter fluoropolymer-clad PMMA fiber. The clad fiber was wound around 14 semicircular bends of diameters ranging from approximately 11 to approximately 33 mm, arranged tangential to one another in a plane with their centers on a common line. Light was coupled into cladding modes by each bend and was subsequently coupled out of the fiber placed in contact with a high-index liquid. The operating principle is thus the same as that used in the sensor reported by Harmer et al., and the sensor depends on the ability of the fiber to carry light for short distances (approximately 1 m) in cladding modes. This design was conceived as a liquid level sensor, with the line of centers of the semicircles mounted vertically, and was tested in ethylene glycol which has a refractive index (1.4318) considerably higher than that of the cladding (approximately 1.40). As the liquid level reached each of the tangential junctions between the first ten bends, transmitted intensity dropped by several percent, remaining essentially constant for depth changes in the regions between the first seven bends, giving the sensor a "digital" readout capability.

Another geometrical method of increasing sensitivity to changes in the index of the medium surrounding a multimode fiber core is to change the cross section of the fiber in the interaction region. Inducing a taper in the fiber from its initial diameter in the clad region to a smaller diameter in an unclad interaction region (Figure 2f) essentially lowers the number of modes available for propagation and "overpopulates" the higher-order modes (which are the most sensitive to changes in cladding index). Kumar et al.[23] showed that by carefully controlling the taper geometry, peak sensitivity could be shifted to the vicinity of any desired refractive index within a range defined by the index of the core itself, and that the ratio of the output power to the input power was proportional to the square of the refractive index of the test medium. A later study of this technique by Bobb et al.[24] compared a tapered clad fiber with a tapered fiber that had some of the cladding removed by etching. By carefully controlling the launch angle of the input light, Bobb et al. were able to selectively excite various propagation modes of the tapered sensor, effectively allowing sensor response to be "tuned" to different ranges of refractive index. With this system, refractive index differences on the order of $5 \times 10^{-5}$ were resolved. In a practical application of tapered fiber intrinsic refractometric sensors, workers at the U.S. National Aeronautics and Space Administration (NASA) have created a series of several tapered sections along the same fiber. This fiber, coupled to an OTDR apparatus, was used as a "single-ended" intrinsic fiber optic leak-detection system with multipoint spatial resolution.[25]

Temperature dependence can cause problems in the accurate measurement of refractive index, but in the field of sensing, "one person's problem is another person's sensor," and there are several reports of fibers that have been modified to become intrinsically sensitive to temperature by making use of the variation of refractive index of materials in the cladding region.[26-34] While many of these are not strictly intended as intrinsic *chemical* sensors, they do measure refractive index, and may be (in fact, in some cases are) used as refractometric chemical sensors.

One intrinsic temperature sensor described by Scheggi et al.[27,28] was an optrode-style sensor created from a silicone-clad multimode fiber with a 200 μm diameter silica core. The cladding was stripped for 8 mm on one end of the fiber, and replaced with a liquid having a refractive index close to that of the cladding, but with a very large temperature coefficient. The tip of the fiber was metallized (Figure 2g) so that the sensor could be employed in a reflective mode. Using this design, temperatures in the medical range (30 to 50°) could be measured with a resolution better than 0.1°C.

A transmission-type refractometric sensor was studied in detail by Arie et al.[35,36] who constructed a very successful theoretical model of a plastic clad 300 μm silica core fiber with the cladding stripped and replaced by "index-matching fluid" (silicone oil) in sections ranging between 5 and 20 mm in length. Light was launched into the fiber by means of a microscope objective coaxially centered on the fiber, the effective numerical aperture of the launch was varied by displacing the end of the fiber from the focal point of the lens, and the index of refraction of the stripped section was varied by varying the local temperature. The theoretical model, based on ray optics, was used to describe the behavior of the sensor when the refractive index of the modified cladding section varied, and good agreement with experimental results was achieved. Energy loss in the modified section was at a maximum (between 4 and 12 dB, for launch NAs ranging from 0.085 to 0.25) when the index of refraction matched that of the core, since there are essentially no reflections at the core/cladding "boundary" in this case. For indices lower than the core's, transmission losses due to presence of the modified section dropped rapidly, becoming essentially zero when the difference in refractive index was more than 0.01. For indices above the core index some decrease (about 1 dB) in loss, relative to the matched-cladding case, was observed. This is because there was still partial reflectivity at the discontinuity of refractive index between core and cladding, even though there were no longer any rays experiencing total internal reflection. (In the mode-theoretic description: leaky modes still existed in the modified section, and could transmit light for short distances, coupling it into guiding modes in the waveguide on the distal side of the section.) Arie et al. were able to theoretically predict the experimentally determined dependence of transmitted power on the effective numerical index of the launch and on the length of the stripped section (see Figure 3). Situations where the fiber was illuminated by focussed white light, or by monochromatic light incident through a butt-coupled fiber with a smaller core were also successfully modeled.[36] The sensor showed the lowest sensitivity to cladding index variations when light was launched into it from the small core fiber, since only low-order propagation (i.e., tightly corebound) modes are excited by such a launch.

Afromowitz[37] has described a sensor that uses the loss of transmissivity that occurs when the fiber cladding index matches the core index to monitor the curing of clear polymeric materials. A sensor fiber manufactured from the polymer to be monitored is attached at both ends to convenient lengths of commercial multimode fiber and allowed to fully cure. The sensor section is then placed in the uncured material, and a simple source/detector pair is connected to the ends of the fiber "leads". Monomers, and partially cured polymers, generally have refractive indices far lower than fully cured polymers of the same composition, so the sensor section starts out as a very good waveguide, with relatively low transmission loss. As the polymer cures, transmission loss increases dramatically, and the sensor section becomes nonguiding when the sample is fully cured. Almost no light will be coupled from the input fiber to the output fiber if the sensor section is bent so that the input and output fiber ends are not aligned. Afromowitz constructed a 500 to 1000 μm diameter S-shaped sensor fiber from Devcon epoxy that successfully demonstrated this concept, and proposed the use of the technique in monitoring the cure of composite materials or other systems where the hardening of clear resins is of interest.

Single mode fibers have also been employed in intrinsic refractive index sensors, using a technique proposed by Pavlath et al.[29] and studied by the group of Parriaux.[30-32] In single-mode fibers the evanescent field extends much farther into the cladding than in multimode fibers, but the geometry of commercially available single mode fibers (core/cladding ratios less than 1:15, concentric design), and the necessity for the cores to be very small (less than 10 μm in diameter) makes it difficult or impossible to completely strip the cladding from the core so this field can interact the surrounding medium. Likewise, "cladding light" induced by bending cannot be readily coupled back into the single propagation mode, so schemes such as that used by Harmer[20] are also impractical for single mode fibers. For this reason, Pavlath et al.[29] proposed using a method for "thinning" the cladding of single mode fibers that was first introduced by workers at Stanford University.[38] This method involves stripping the protective coating from the fiber, bonding it into a groove in a small "polishing block", and grinding or polishing away the cladding (see Figure 2h). The interaction of the guided light with the medium surrounding the fiber takes place when light "tunnels" through the thinned cladding. To create a sensor based on this principle Falco et al.[31,32] employed a more refined polishing technique that used semiconductor device processing technology to create "polishing blocks" of silicon with grooves having a precisely controlled cross section.[39] Using a sensor made in this way, a refractive index sensitivity of $\Delta n \sim \cong \sim 10^{-4}$ was achieved; coupling the sensor to "index matching oil" (Cargille, Inc.), which has a known temperature coefficient of refractive index resulted in temperature sensitivities of ±0.1°C over the range 20 to 50°C. The same group has proposed the use of a single-fiber Fabry-Perot geometry incorporating such a sensor[33] to increase the sensitivity and stability of the system. The use of polished fiber sections in index oil as temperature sensors was also investigated by Tomita and Walker at Bell Laboratories[34] who demonstrated that, depending on geometry, such systems could be used either as thermometric devices or as "set-point" devices which drastically change fiber attenuation over a very limited temperature range.

A refractive index sensor that could be being used in a practical distributed measurement system is the "twisted-pair" sensor of Smela and Santiago-Aviles[40] based on earlier work by El-Sheriff and Zemel.[41] In this sensor (Figure 2i) two silicone-clad silica fibers with clear nylon jackets still intact are twisted around one another and fixed in place, either with clamps or by insertion into a tube. One fiber is connected to a (HeNe) light source, and the other to a (p-i-n) photodetector. Light passes from the "transmitter" fiber to the "receiver" fiber (the definition has been stretched a bit to include this device in a chapter on "intrinsic" sensors!), with the amount of transmitted light depending almost linearly on refractive index over the range $n = 1.32$ to $n = 1.44$. It is interesting to note that, unlike most evanescent field based intrinsic fiber optic sensor configurations (for which most of the light in a spatially extended sensor would be lost), this sensor can readily be used in situations where the medium to be measured has a refractive index higher than that of the fiber core. Another interesting point is the fact that the sensor is actually more sensitive when it is simply coated with a thin film of the liquid being measured than when it is immersed in the liquid, because of the "cladding" effect of air surrounding the optical waveguide structure formed by the liquid-coated fiber pair.

A radical departure from all of the above techniques, which require manipulation or modification of standard fibers, involves a fiber actually manufactured by Yoshikawa et al.[42] in a form suitable for distributed refractive index sensing. This fiber was a single mode structure produced with an off-center core (7.5 μm diameter) located very near (within 4 μm) the surface of the cladding (Figure 2j). This design allowed the evanescent field of the light carried by the core to interact strongly with the fiber's surroundings over its entire length. The fiber is thus uniquely suitable for such applications as wide-area spill monitoring. Initial tests showed that the attenuation in this fiber went from a very low value (below 1 dB/km) to nearly 5 dB/cm when the index of refraction of the medium just outside the cladding was changed from 1.460 to 1.463. The effect of bending on sensor performance has not been reported, but it is possible that subjecting this waveguide to periodic corrugations along its length could significantly increase its sensitivity. This improvement would occur because of resonant reflections caused by the optical Bragg effect in the single mode waveguide[42] and would be particularly useful if the fiber were to be used in a "single-ended" detection system.

It is clear that there is a wide variety of refractive index measurement techniques based on the direct interaction of light guided in an optical fiber with the environment around the fiber. Although the vast majority of these techniques make use of fibers that have been modified in some way to encourage this interaction, they still may be classified as intrinsic, since there is no break in the guided light path, and since the interaction with the environment takes place through the "side" of the fiber. As an illustration, consider the sensor shown in Figure 2g: although originally conceived as an "optrode", it could be used as a distributed sensor (and would be more clearly seen as an intrinsic sensor) if the stripped section were extended for several meters. Many of the sensors discussed here have the same potential for distributed measurement, but the work of Yoshikawa et al.[42] points the way to fibers which are actually refractometer-ready at the time of their manufacture. Multipoint techniques based on OTDR[25] can also be used with virtually any concatenation of the sensors described above to achieve quasi-distributed measurements.

# III. EVANESCENT WAVE SPECTROSCOPY

The absorption of light by molecules in the evanescent field of what would otherwise be a totally reflected ray is a well known phenomenon, studied for many years.[44,45] The phenomenon, under the name "total internal reflection" (TIR) or "attenuated total reflection" (ATR) spectroscopy has more recently been used with great success to investigate the optical spectra of thin films and highly absorbing materials.[46] In general terms in the ray-optics picture, the effect of $\kappa$, the imaginary part of the complex refractive index in the cladding region is to decrease the magnitude of the reflection coefficient, particularly at angles above the critical angle, "smearing out" the otherwise very sharp dependence of reflectivity upon angle of incidence.[3]

It should be noted that the optical absorbance spectrum obtained by evanescent-wave techniques does not, in general, coincide with spectra obtained by transmission methods. One cause of the difference between transmission spectra and evanescent-wave spectra obtained with fibers is related to the fact that for fibers the fraction of optical power in the evanescent field depends on the waveguide parameter (V number) of the fiber. When light traveling in fibers has reached its equilibrium intensity distribution among the modes of propagation, fibers with lower V numbers have a larger percentage of their light carried by the evanescent field; higher V number fibers carry their light "closer to the core."[3,4,46] The fraction of the power carried by the evanescent field, and therefore the cladding absorbance, always increases with decreasing V-number. For multimode fibers with even moderately high V numbers (above 20), the relationship between cladding power fraction and V approximates an inverse proportionality,[47,48]

$$\alpha_{cladding}/\alpha_{core} \approx (const) \cdot 1/V$$

For single mode fibers that are not operating near their cut-off wavelength, the cladding power fraction is proportional to $1/V^2$.[47] The intensity of the evanescent field falls off exponentially from the core-cladding boundary, with a "decay constant" proportional to the inverse of the optical wavelength.[3] This wavelength dependence means that the measurand spectrum may show exaggerated absorbance in the longer wavelength regions, since a greater percentage of the light is carried in fields outside the core (i.e., by the evanescent field in the measurand) for these wavelengths.

For evanescent-wave spectroscopic techniques that rely on "DC" illumination of multi-mode fibers, with the cladding being the medium in which the measurand resides, another factor that can affect a sensor fiber's V number is spectral dispersion, the wavelength dependence of the index of refraction. Dispersion is a well known and extensively studied effect in optical fibers, primarily because of its distorting influence on the high-speed pulses used in digital telecommunications. If dispersion in the core causes the index of refraction to lessen in a particular wavelength band, the fiber will be more sensitive to measurand optical absorbance in that band, since more power penetrates into the cladding (measurand) region. A measurand spectrum recorded using such a fiber would show exaggerated optical absorbance in the wavelength band where the core refractive index was lower.

Even if the core is dispersion-free, in spectra from measurands with strong absorption bands, peak absorbance will appear to be shifted to longer wavelengths. This is because of the "anomalous dispersion," inherently associated with any optical absorbance,[49] which lowers the index in the medium around the fiber for wavelengths shorter than the peak absorbance wavelength (raising the effective V number of the fiber), and raises the index for longer wavelengths (lowering the effective V number). Since higher V numbers mean that a smaller percentage of the light is carried by the evanescent field, and vice versa, the anomalous dispersion will enhance the amount of power absorbed per unit of optical density for wavelengths above the peak, and reduce the amount of power absorbed at the shorter wavelengths, causing the apparent position of the peak to shift upward.

The dependence of cladding absorbance on the waveguide parameter, V, holds in general: the effect of changes in the concentration of absorbing measurands in spectroscopic evanescent-wave sensors will always be greatest for fibers with the lowest V-numbers.

The dependence of absorbance on V also holds when fibers with thin claddings are used as intrinsic spectroscopic sensors, as shown by a study of the behavior of fibers brought into contact with absorbing media carried out at McGill University.[50] This study modeled a step-index single mode fiber with a cladding-to-core ratio of 6:5 (a fiber that would have an outside diameter of less than 20 μm at visible and near-IR frequencies!) and studied three cases: cladding index lower than core, but higher than measurand medium; cladding index higher than both core and medium index; and cladding index lower than both other indices. In all cases the core index was assumed to be higher than that of the surrounding medium. The results showed that for this structure, just as for the simple core/cladding structure, the waveguide parameter, V, should be minimized to achieve maximum sensitivity to changes in the absorbance of the measurand. The structure with cladding index lower than both core and measurand indices is particularly sensitive to this requirement. The only lower limit on V is provided by the fact that it must not drop so low as to bring the waveguide near "cutoff" (the point where the fiber will no longer guide light in the wavelength range of interest). In practice, this means that the cladding should be made as thin as possible, the core diameter as small as possible, and the core index as close as possible to the refractive index of the surrounding medium to achieve maximum sensitivity.

The evanescent field can also provide a means of coupling light *into* the fiber.[51,52] This process, which is in a sense the inverse to the "cladding-loss" (evanescent-wave absorbance) phenomenon, is very important in intrinsic fiber optic sensing, since it provides a means by which fluorescent energy generated in the cladding region can be captured and guided by the fiber. An important difference between cladding loss and cladding fluorescence collection is that mode-theoretic calculations predict that the percentage of light incident on the core from the cladding that is coupled to guided modes in the fiber core actually *increases* with fiber V number.[52]

The earliest study in evanescent-wave absorbance spectroscopy using thin cylindrical waveguides reported results obtained with a 3 mm diameter quartz rod bent into an approximately circular shape.[53] The sensitivity of the configuration was investigated using solutions of potassium permanganate in water, and an approximately linear dependence of optical absorbance on concentration was observed for permanganate concentrations ranging from 0.1 to 1.0 $M$. This system successfully detected a 0.083 $M$ solution of potassium permanganate with a signal to noise ratio greater than 1000, implying an ultimate sensitivity to permanganate in the micromolar or submicromolar concentration range. Hansen[53] also mentions using a "bundle of fine glass fibers", presumably made of unclad silica, in place of the single rod. No data is presented for the fiber system, but Hansen may, in fact, be the first author to have reported the use of optical fibers as intrinsic chemical sensors.

Another early report of intrinsic fiber optic evanescent-wave spectroscopy comes from Kapany and Pontarelli[10] who, although primarily interested in refractometric measurements, also recognized the potential of optical waveguides for intrinsic chemical absorbance measurements, and investigated the performance of several rod-type evanescent wave refractometers in the presence of absorbing species. Using a simple multi-reflection ray-propagation approach, these authors modeled the propagation of light rays incident at the interface between the rod and the measurand solution at precisely the critical angle. By immersing the rods in test solutions of methylene blue dye in buffered water, Kapany and Pontarelli found that results obtained using a multi-turn helical rod geometry showed the highest sensitivities to changes in dye concentration, and most closely approximated the theoretically predicted sensitivity curve. Using this probe, methylene blue concentrations were successfully measured over two orders of magnitude; possible extensions of the technique from 1 mm diameter rods to true optical fibers were mentioned but not explicitly reported.

Another refractometric sensor for which absorbance spectroscopy applications have been reported is the twisted-pair sensor illustrated in Figure 2i.[41,54] Pairs of unclad plastic fibers were twisted together and immersed in aqueous solutions containing bromothymol blue, and the transfer of 632.8 nm radiation from the input fiber to the output fiber was investigated for dye concentrations in the range from 0.1 to 1.0 m$M$. pH-induced changes in the absorbance of the bromothymol blue were measured; photocurrent in the silicon p-i-n diode was a nearly linear function of pH in the weak base range (pH 7 to pH 9).

The behavior of a stripped-cladding multimode humidity sensor has been investigated as a model for intrinsic chemical sensors aimed at possible use in automotive applications.[55] This sensor used a silicone clad 200 μm core quartz fiber, with stripped sections 3 and 5 cm long. One end of the sensor was connected to a broadband LED emitting at 940 nm, and the other led to the input of a 1 × 2 fused fiber coupler. The detection system was based on a dual-wavelength technique, using interference filters with peak transmittances at 920 and 960 nm (wavelengths near a spectral absorbance line of water) placed in front of a matched pair of p-i-n photodectors on the output arms of the coupler. A roughly linear dependence of differential attenuation on relative humidity was observed, with a calibration coefficient of 0.001 dB per percent change in relative humidity. When the system was compared with a simple single-wavelength system using a monochromator in conjunction with the same test fiber, similar response was seen. Data from the monochromator system indicated that much better sensitivity to humidity could have been obtained if longer wavelengths were used in the dual wavelength system.

To confirm the theoretical prediction of the linear dependence of cladding absorbance on the reciprocal of the waveguide parameter, Paul and Kychakoff[48] investigated the behavior of a 3-m long piece of 50 μm core fused silica fiber, with the cladding removed from the central 1 m, mounted in a liquid flow cell (see Figure 4). The V number of the stripped section was controlled by using liquids of differing indices of refraction in the flow cell, and cladding absorbance was controlled by precisely adding various amounts of Rhodamine 6G dye to the flowing fluid. Sensor illumination was provided by an argon-ion laser, and the resulting output was detected with a silicon p-i-n. Dependence of absorbed power on dye concentration (as measured by using a conventional spectrophotometer) was linear over five orders of magnitude in concentration, for four different V numbers ranging from 24 to 190. As predicted by the theory, the proportionality constants were themselves inversely proportional to V number (with a small deviation, proportional to $1/V^2$), having a proportionality constant near 2; the deviation from linearity tended to reduce the proportionality constant for higher V numbers. For the lowest V (24), the fiber absorbance change due to the presence of the dye was 0.095 times the absorbance measured in the spectophotometer. Similar results were obtained for coiled fibers, as long as the radius of curvature of the coils exceeded 2.5 cm. This result is somewhat surprising, given the fact that the approximations used by Paul and Kychakoff were only meant to be valid for large V numbers. They concluded that the ultimate sensitivity of intrinsic fiber chemical sensors such as theirs can be comparable to that obtainable by conventional methods based on transmission spectroscopy.

Another study of a rather long intrinsic sensor based on evanescent field attenuation was performed at the University of Washington's Center for Process Analytical Chemistry.[56,57] This sensor was unique in that it did not rely on a stripped fiber, but employed a permeable cladding that allowed measurand molecules to diffuse into the region near the fiber core. The nylon protective jacket was removed from a series of 200/300 µm (core/clad) silicone-clad silica fibers, for lengths ranging from 0.25 to 1.2 m. The optical absorbance of these fibers was studied for both straight and coiled geometries in solutions of oxazine perchlorate dye dissolved in chloroform/toluene mixtures that easily permeated the cladding, carrying the dye with them. One rather surprising result was the absence of any noticeable distortion of the dye spectrum when comparing the optical absorbance obtained by using a 1.2 m permeable-cladding fiber (together with a monochromator and a tungsten source) with a conventionally measured transmission spectrum. Using a fixed wavelength (564 nm) provided by an LED, the effects of fiber length, bending radius, and cladding index on sensitivity were studied. As might be expected, longer fibers showed a greater dependence of absorbance on dye concentration than shorter fibers, and curved fibers performed better than straight fibers (sensitivity doubled when the radius of curvature was 6 mm). The dependence of absorbance on dye concentration was remarkably linear for all concentrations studied, from 0.002 to 0.1 m$M$. Although it was not possible to accurately measure the refractive index of the silicone/solvent medium which composed the cladding of the sensor fibers, increasing the solvent refractive index from 1.440 to 1.4646 (and thus presumably decreasing the V number of the fiber) did increase sensitivity, by roughly a factor of 5. As a final test, DeGrandpre and Burgess[57] stripped the cladding from a 25 cm piece of the fiber, and compared its performance with a similar length of clad "swelled" fiber. Responses were approximately three times larger for the stripped fiber, but a significant nonlinearity in the dependence of absorbance on concentration was seen (see Figure 5).

A thinned-fiber style (see Figure 2f) evanescent wave spectroscopic sensor design has been investigated for use in methane detection by Tanaka, et al.[58,59] Sensors were made from step-index 50/125 µm (core/clad) multimode fiber, stripped of buffer for short (5 to 10 mm) sections and drawn down to radically small diameters, ranging from 1.8 to 7 µm. The cladding was etched from the stripped sections before or after heating and pulling the fibers. At the mid-IR wavelength (3.392 µm) used, the tapered sections operated as "few-mode" waveguides, with the smallest diameter sections actually behaving as single mode fibers. This achieves a great increase in sensitivity over conventional multimode intrinsic sensor designs, since the fraction of optical power carried in the evanescent field is much higher for singlemode fibers. This power fraction was actually estimated from the experimental results for three taper diameters, namely 7.0, 2.8, and 1.8 µm, and found to be 5, 12, and 40%, respectively. The tapered sections were calculated to have seven, two, and one independent modes, respectively. The dependence of optical loss (dB) on the concentration of methane in nitrogen was linear over the range from 0 to 100%. For the most sensitive (singlemode) tapered section, methane concentrations near the lower explosive limit for methane (5%) could be detected. A severe limitation on the use of this fiber for remote methane detection is the fact that there are no practical optical fibers capable of transmitting 3.392 µm radiation over appreciable distances. Attenuation at this wavelength for the silica-based fiber used by Tanaka et al. was on the order of 1 dB/cm: even a 500 W laser (which would be cumbersome, would present difficulties in coupling to the fiber, and might cause nonlinear behavior in the taper) would deliver less than a nanowatt of power at the end of a single meter of this fiber, neglecting the loss added by the taper and methane absorbance. Creating such extreme tapers in the field and protecting the resulting microscopically thin fibers of unjacketed glass would also present quite a problem. Nevertheless, the work of Tanaka et al. provides an exceptional demonstration of the ability of single-mode structures to detect gaseous species directly using intrinsic fiber optic sensing techniques.

Other sensors designed for operation in the mid-infrared region of the spectrum are discussed by Simhoney et al.,[60-62] who have reported work on ATR cells based on fibers of silver halides (AgBr/AgCl), which have a much lower attenuation (on the order of 1 dB/meter) in the 2 to 20 μm range than the quartz-glass fibers used by Tanaka et al. Straight, unclad fibers were used (see Figure 2b), either suspended in a flow cell, cupped in a split capillary tube, or placed at the bottom of a V-shaped silica trough. Infrared energy was coupled into and out of the fiber with ZnSe lenses, and spectra were recorded using a commercial Fourier Transform Infrared (FTIR) spectrometer. Tests of system response to organic molecules dissolved in water showed response to both acetone (20% in water) and glycine (2.22 *M*); an experiment with a bundle of several fibers implied an ultimate glycine sensitivity of roughly 0.03 *M*. The fibers may also be used to record spectra from thin films of solid organic compounds formed by allowing solvents to evaporate from the test solutions, an attractive alternative to solution-based methods, considering the high signal-to-noise ratios obtained,[62] or to study solutions and suspensions of protein molecules that are too viscous for use in standard ATR cells.[61] The relatively low cost of the fibers indicates possible uses in disposable sensors for such applications as *in situ* monitoring of the curing of organic adhesives. Although these fibers are far superior to quartz-glass fibers for IR transmittance, they are physically much more difficult to work with. For example, it would be extremely difficult to produce fibers with the extremely thin tapered regions used by Tanaka et al. This precludes the use of silver halide fibers as single-mode sensors and limits their utility to measurands such as liquid solutions or condensed solids that have relatively high optical absorbances. Another limitation arises from the fact that, since the fibers are actually polycrystalline, it is impractical to produce fibers much longer than a few meters. This means that if silver halide fibers were to be used as part of a remote monitoring system, they would have to rely on some other long-wavelength transmitting fiber (possibly chalcogenide-based) to link the relatively short sensor sections.

The comparative sensitivities of straight, unclad, tapered multimode fibers and straight, clad, tapered single mode fibers have been investigated by a group at the U.S. Naval Research Laboratory.[63] For multimode experiments, 100/140 μm (core/clad) silica glass fiber was used, etched over a 37 cm length so as to have a diameter continuously tapering down to 30 μm in the middle of the etched section, and back to 140 μm at the other end. The tapered single mode sensor was fabricated from a 2.3 mm section of 6/80 μm fiber, pulled to an extension length of roughly 20 mm.[64] The response of these fibers to changes in the concentration of methylene blue dye in water was tested at 632.8 nm; changes in the absorbance of the multimode sensor were approximately 0.35 times the absorbance changes measured by a conventional spectrophotometer, the corresponding ratio for the single mode sensor was approximately 0.1 (see Figure 6). Villaruel et al.[63] concluded that fiber sensors based on these designs could ultimately have sensitivities to methylene blue (or other species with similar absorbance coefficients) on the order of one part in $10^9$.

Although it is not the purpose of this chapter to discuss instrumentation techniques, mention must be made of the innovative work of Sixt et al.,[65,66] who have proposed a technique based on the use of single-fiber Fabry-Perot sensors to convert the intensity-based signals characteristic of absorbance type sensors to phase-based signals. This conversion would be accomplished by illuminating the fibers with amplitude-modulated light and monitoring the phase of the output signal at the modulation frequency. The Fabry-Perot structure, a section of fiber with partially reflective surfaces normal to the axis of the fiber on both ends, would cause some of the guided light to be reflected, making multiple "passes" through the sensor region before being transmitted to the detector. The final phase of the amplitude-modulated signal detected at the output would be the weighted vector sum of the phase of the directly transmitted light with the phases of the portions of the light that have made three, five, seven, or more passes through the sensor. Since the weighting would depend on the attenuation experienced by the beams, changes in attenuation would change the weighting factor for each multiply reflected beam, thereby causing a change in the total phase of the output signal. For example, raising the attenuation would cause the phase to become closer to the phase of the "direct" beam. Similar phase changes would also be apparent in the light reflected from the sensor. A preliminary test of the idea, using a free-space Fabry-Perot cavity formed by the polished end of a fiber and a mirror, demonstrated modulation frequency phase shifts of almost 180° when the reflectance of the cavity was changed by changing the distance between the fiber and the mirror. Although not yet demonstrated with Fabry-Perot structures made of evanescent-field based sensor fibers, this technique could eliminate many of the difficulties (incidental losses due to fiber bending or vibration, problems associated with source fluctuations, etc.) now faced by the designers of intensity-based detection systems for intrinsic fiber optic chemical sensors.

In contrast with the loss-based intrinsic sensor mechanisms discussed above, it is possible for fibers to actually *gain* energy through the evanescent field, if the cladding medium is generating light.[67] The coupling of fluorescence excited by the evanescent fields of totally reflected rays into the medium from which the rays are incident has been used for analytical purposes for many years in Total (Internal) Reflection Fluorescence, or T(I)RF,[68] spectroscopy to study the fluorescence of organic and biochemical compounds, as well as to detect and investigate the properties of fluorescent-labeled compounds. Because the excitation and emission take place very near the interface of two media of different refractive indices, TIRF has found its main applications in the study of thin layers, often adsorbed onto the surface of transparent materials.[69] With the growth of interest in fiber optic sensing techniques, groups interested in TIRF began to study total reflection fluorescence systems that make use of fibers, rather than the traditional prisms or plates, as the primary transmission medium.[70-73] Fiber evanescent-wave fluorimetry is a two-step process, relying on evanescent-wave absorption of excitation energy from the core by the fluorescent species being studied and subsequent evanescent-field capture by the core of the resulting fluorescent emission. Because of its reliance on two evanescent-field coupling steps, the efficiency of evanescent-wave fiber fluorimetry is rather low, when compared with conventional fluorimetry. A further complication arises from the fact that, according to mode-theoretic calculations for long fibers[52] the efficiency of fluorescence collection increases with increasing fiber V number, while the efficiency of fluorescence excitation (an evanescent field absorption process) decreases with increasing fiber V number. Thus, the fibers which are most suitable for excitation are least suitable for energy collection, and vice versa. Despite these problems, fluorescence excitation and collection can be appreciable at the same time for fluorescent species in the region very near the fiber surface. For this reason, and because of the history of TIRF as a tool for probing surface and interfacial effects, much of the work on intrinsic fiber fluorimetry has involved the use of fibers with specially prepared surfaces which are designed to react with the measurand and enhance the fluorescence collection and emission (see the next section of this chapter).

Work on direct fiber fluorimetry, without the aid of special coatings, still has produced some interesting results. A stripped-core fiber very similar to the refractometer/temperature sensor design shown in Figure 2g was one of the first to explicitly use the evanescent field of light confined in an optical fiber (as opposed to a rod) to perform spectroscopic measurements.[70] Newby et al.[70,74] prepared a short (approximately 3 cm) section of cladding-stripped multimode PCS fiber, capped the end with an absorbing material, and investigated its performance in measuring fluorescence returned from various solutions when 488 nm argon laser light was coupled into the fiber. Fluorescence intensity collected from a 1 m$M$ solution of rhodamine dye depended in a very linear way on the depth to which the sensor was immersed in the solution, indicating that very little of the excitation light is actually coupled out of the fiber into the surrounding medium. Newby et al. also showed that the presence of 0.8 mg/ml of rhodamine-labeled gamma-globulin (IgG) in water could be detected with the probe after allowing a 20-min incubation time with the sensor immersed in the solution, presumable to allow the labeled IgG to adsorb onto the surface of the fiber. When the sensor was immersed in benzene, a solvent with an index of refraction greater than fused silica, some of the Raman-scattered light from the benzene was detectable. This effect could be explained without recourse to evanescent-field considerations, since rays from the benzene could be refracted into the short section of stripped silica core at all angles, including those suitable for propagation in the plastic-clad fiber connected with the stripped section.[75] A later version of this sensor was used in a study of the binding of IgG to the surface of the fiber which showed that roughly half of the fluorescence signal from the probe arose from protein that was irreversibly bound to the surface of the fiber.[76] A study of the relative advantages of "backscattering" and transmissive modes of operation for intrinsic fiber sensors based on direct fluorescence was also carried out,[77] with the conclusion being that, except for the necessity of using more sophisticated band-rejection filters in the transmissive mode (to eliminate excitation light), the two methods were capable of achieving similar performance. The limit of detection for IgG using the sensor was shown to be 14 nmol/l.

In the simplest ray models, light crossing a boundary from lower to higher index of refraction will always be refracted into rays that lie within a cone defined by the critical angle for total internal reflection at the boundary. For the ray description of propagation in fibers, this would mean that no light from the cladding could ever be coupled into true bound rays. Despite this, the double process of coupling light out of and into the cores of intrinsic fiber optic fluorosensors has been modeled using ray optics by some groups.[75,78,79]

Wang et al.[75] performed a strict ray analysis of the sensitivity of fluoroprobes composed of short sections of stripped PCS fibers with very thin (10 nm) layers of fluorescent immounochemicals adsorbed on their exposed silica surfaces. In this model, which entirely ignored the evanescent field as a means of "back-coupling" rays into the fiber core, fluorescence could be carried by the fiber only if the effective index of the adsorbed layers was greater than the index of the unstripped cladding. This requirement is easily met by gamma globulin (IgG, refractive index > 1.5)[80] the immunochemical most commonly used in tests of fiber optic immunofluorescence probes. By placing 3 cm stripped sections of 600 μm core fibers into a flow cell carrying buffered fluorescein-labeled IgG solutions Wang et al. showed that, for IgG concentrations lower than $10^{-9}$, the amount of fluorescence coupled into the fiber was very small (below background levels). For concentrations between $10^{-8}$ and $10^{-6}$, the fluorescence intensity depended linearly on the logarithm of IgG concentration; the signal reached an apparent "plateau" (presumably due to saturation of the fiber surface by labeled protein molecules) at $10^{-5}$ $M$ IgG. Proof that at least some of the increase in fluorescence is due to the change in local refractive index caused by a layer of protein molecules near the fiber surface was provided by an experiment in which a fiber that was washed in running buffer after exposure to labeled IgG was placed in a solution of *unlabeled* IgG: after washing, some labeled protein remained irreversibly bound to the probe surface, producing a remanent fluorescence signal; upon immersion in the solution of unlabeled protein, the fluorescence signal increased by 15%. Since the unlabeled IgG was incapable of producing more fluorescent photons, this increase in signal could only have been caused by a change in the coupling efficiency of the probe tip caused by the binding of a layer of unlabeled protein.

In an effort to explain experimental observations of the transmission of fluorescence generated in aqueous solutions (refractive index $\approx$ 1.33) by silica fibers (refractive index $\approx$ 1.45), augmented ray-based theories have been constructed that predict significant coupling into bound rays from sources located in low-index media around the core. These treatments owe their validity to the work of Carniglia and Mandel,[51] who devised a classical model of evanescent field "in-coupling" that predicts the existence of rays outside the critical cone (an effect actually observed more than 75 years ago by Selenyi[67]) and conducted experiments that verified the validity of the model. Glass et al.[78] used a ray-approximation approach to investigate the effect of varying the numerical aperture of light launched into short reflective evanescent field intrinsic sensors with large diameters, with the main conclusion being that the numerical aperture of the launch system should match the numerical aperture of the sensor rod for optimum performance. Love and Button[79] also used the ray approximation in a somewhat more comprehensive treatment of optical coupling in intrinsic fiber fluorosensors based on fibers with V > 1000; in this study launch NA and fiber NA were varied, as were illumination spot size and concentration of fluorophore. An empirical model, developed using data collected with a stripped fused silica fiber in fluorescein/sucrose solutions, was compared with the theoretical model. The results confirmed that numerical apertures of fiber and launch optics should be matched, and showed that collection intensity varies as the eighth power of launch NA for values near the fiber NA. Fluorescence yield was observed to depend on the square of the spot size, as predicted by numerical integrations of the theoretical model. A dependence of yield on the inverse fourth power of fiber NA was observed, predicted by the model for fluorophores closely bound to the surface of the fiber, but not for bulk samples, where the theoretical model predicted a dependence on the inverse seventh power of NA. Love and Button deduced from this that the fluorescein/glucose combination adsorbed preferentially onto the surface of the fiber.

A comparison of the sensitivity of stripped fibers with planar waveguides has been carried out by Stewart et al.,[81] who found that singlemode planar structures had sensitivities roughly three times that of stripped multimode fibers with moderate V numbers (V = 145).

As a final note, it should be mentioned that most of the theoretical analyses of intrinsic optical fiber sensors based on evanescent-field effects fall into two categories:

1. Ray-approximation calculations, which basically ignore the mode structure associated with light traveling in multimode fibers, assuming that the fiber behaves like a large diameter glass rod. This is equivalent to assuming very large V numbers (V >> 100).
2. Mode-theoretical calculations, which in general assume that an equilibrium distribution of optical energy among the fiber's modes of propagation has been achieved by the light traveling in the fiber.

Both of these types of calculation have shortcomings: the ray-based calculations lose validity for fibers with core diameters smaller than approximately 100 µm for most cladding materials; the mode-theoretical calculations, particularly those analyzing multimode fibers, ignore the fact that many intrinsic fiber sensors use fiber lengths much shorter than the length (approximately 500 m, for a 50 μm core silica fiber with V = 30 operating at visible wavelengths) required to allow light to reach its equilibrium mode distribution. This may explain the discrepancy between the prediction of the ray-based calculations[78,79] that the efficiency of coupling of light into the core should decrease with increasing V number, and the mode-theoretic result that the coupling increases with increasing V.[52]* It should be noted that, in virtually all the experiments reported to date on stripped-section fibers immersed in fluorescent media, partial guidance of energy by "leaky" (unbound) rays could not be ruled out, since the leads connecting the stripped sections to detectors have all been very short.

## IV. COATED FIBER SENSORS

With a few notable exceptions, intrinsic fiber optic chemical sensors whose sensitivity is based on a chemically active coating are directly related to the "passive" spectroscopic and refractometric intrinsic fiber optic chemical sensors discussed above. In fact, *any* of the evanescent-field absorbance techniques described in the previous section could be modified by the application of chemically sensitive coatings to become "chemically assisted" sensors. In coated spectroscopic sensors, the sensitive coating usually has an optical property that is affected by the presence or concentration of the chemical measurand of interest, and the measurand is sensed because of the interaction between the coating and the evanescent field of light guided by the fiber core. The same considerations that govern sensitivity in direct (passive) spectroscopic sensors apply to coated-fiber spectroscopic sensors, with the stipulation that in some cases the sensory coating is actually so small that the spatial extent of the evanescent field encompasses all of the active material and extends beyond it, rather than being completely contained by the optical medium being probed. Because of their enhanced specificity and sensitivity, coating-based intrinsic fiber optic chemical sensors have been studied by a number of groups, and currently constitute one of the most actively investigated classes of fiber optic chemical sensors.

As with passive sensors, the development of coating-based "chemically assisted" intrinsic fiber optic chemical sensors began with studies of structures that are almost fibers: thin rods. Hardy, et al.[82] treated the surface of a 1 × 20 mm quartz rod with an aqueous solution of polyvinyl alcohol and sodium picrate (1 and 0.1% by weight, respectively) and allowed the solution to dry, forming a cyanide-sensitive coating on the surface of the rod. Subsequent optical measurements at a variety of wavelengths showed a linear drop in transmittance when the rod was treated with varying amounts of $CN^-$, with the effect saturating for amounts greater than 0.4 μg.

The rod technique was also used in later experiments to measure low concentrations (0.01 to 1.0 ppm) of ammonia gas in air.[83] The sensor dye chosen was ninhydrin (triketohydrindene hydrate), which undergoes an irreversible color change in the presence of small quantities of ammonia. The dye was applied in a permeable polymer layer made of either polyvinly alcohol (PVA) or polyvinylpyrrolidine (PVP). In the presence of ammonia, the transmittance of the sensor decayed rapidly, with the time constant of this decay determined by the concentration of ammonia. The moisture content of the ambient air also had a large effect on the decay constant, and the PVA films were essentially insensitive to ammonia concentration when the relative humidity was below 50%. In tests where humidity was controlled, sensor response

---

* Note added in proof: A recent study by Egalon (C. Egalon, "Injection Efficiency of Bound Modes," NASA Cotractrot Report CR 4333, NASA Langley Research Center, Hampton, VA, 1990), employing a more complete mode theoretic calculation, claims that the difference between core and cladding refractive indices, rather than V, is the best predictor of coupling efficiency, and that coupled power can in fact either increase or decrease with increasing V."

was reproducible, with PVP films having an almost linear dependence of decay time on ammonia concentration for concentrations ranging from 10 to 200 parts in $10^{-9}$. A later variation of this technique that used actual (100 μm diameter) fibers coated for 2 cm with ninhydrin-doped PVA was used to perform measurements of the concentration of ammonia dissolved in blood.[84] Although not strictly "sensors", since they incorporated irreversible reactions, these devices did show that coated guided-wave optical devices can be used to measure very low concentrations of environmental pollutants in the atmosphere, and presaged some commercial chemical analysis products. Hardy et al. also demonstrated the feasibility of using such probes for blood chemistry determinations, such as measurements of blood urea nitrogen (BUN), serum CPK activity, creatinine, and L-amino acid levels.

Early work at the U.S. Naval Research Laboratory[85,86] took a different tack, utilizing thin-walled soda-glass capillary tubes (ID × OD = 0.8 × 1.1 mm) coated on the outside with reversible sensor reagents (progenitors of the capillary-based refractometric sensor mentioned in Section II of this chapter). Capillary cylindrical waveguides were used "to maximize the number of reflections," experienced by light rays transmitted from the source to the detector. In reality, since an unfocussed 560 nm LED was used for a light source, all possible angles of incidence propagate in the capillary, and it may be more useful to use the mode description of waveguide propagation to explain the signal enhancement: using a capillary geometry is akin to removing the most tightly bound (lowest order) modes of a multimode rod- or fiber-type light guide, leaving only the higher order modes, which have the largest evanescent-field components. The practical effect is to increase the percentage of guided light that is carried by the evanescent field. In one set of experiments, a thin (less than 1 μm) coating of dried oxazine perchlorate dye crystals was applied to the outside of the tube to make it sensitive to ammonia vapor.[85] Oxazine perchlorate undergoes a drop in absorbance at 560 nm in the presence of ammonia. Sensor response was observed to be completely reversible, but some cross-sensitivity to humidity was also seen (possibly due to hydration of the dye crystals). A lower limit of detectability of 60 ppm was implied from sensitivity experiments using a 9 cm capillary tube. Despite the extremely thin coating, full response to the presence of ammonia took approximately 1 min, although signal strength dropped much more rapidly when ammonia was removed. A sensor for organophosphonates (nerve gas components) was prepared in a similar fashion, by coating a tube with crystals of another dye, 4,4-bis diethyl-amino benzophenone (known as "EMKO"),[86] which has an absorbance band in the red that increases with toxic vapor concentration. Using a red (660 nm) LED, this sensor was tested in various levels of an organophosphonate simulant, methane sulfonyl chloride (MSC). The minimum measurable concentration of MSC in dry air at room temperature was on the order of 10 ppm; for humid air, sensitivity, response time, and sensor lifetime were severely degraded. Another version of the capillary sensor used a PVP coating containing cobalt chloride, rather than a thin solid film of sensor dye, to make a moisture sensor.[87] The PVP coating, which was either air-brushed onto the surface or applied by dip-coating, had a index of 1.53, higher than that of the capillary (1.49). Thus, the air-clad annular waveguide actually had a two-layer core, with more light traveling in the sensor coating than in the glass "substrate". The polymer section of the core was permeable to both air and water vapor. The sensor, monitored with light from a red LED, was quite responsive to relative humidity changes in the range 60 to 80%, but exhibited little sensitivity outside that range. Related work[88] used the capillary lightguide arrangement to study the solubility of organic vapors in polyethylene maleate and polyfluoropolyol by observing changes in optical transmission that were induced when the solvents permeated and softened the polymer films, affecting the interface between the cladding and the core.

An early mention of intrinsic sensors based on the use of short coated fibers is the important patent held by by Buckles.[89] This patent, which primarily covers irreversible sensors ("probes") for biochemical measurands, and which is most specifically concerned with coatings meant to enhance the response of guiding and nonguiding permeable-core fibers, mentions a number of possible single- and multi-layer coating strategies that might be used in chemical monitoring. Potential measurands mentioned included oxygen (measured by fluorescence quenching) and pH, as well as a wealth of antibodies and antigens. A few specific chemical systems are mentioned, but no evidence of reduction to practice is cited.

Russell and Fletcher[90] reported a humidity sensor based on a coating of cobalt chloride-doped gelatine applied to a 12 cm stripped section of 600 μm core PCS fiber. Gelatine was chosen as the coating "carrier" because other permeable materials (PVA, PMMA, and nitro-cellulose) developed scattering centers and turned cloudy in high humidities. The gelatine/dye mixture has a higher index of refraction than the silica fiber, so the sensor section was essentially an "enlarged core" fiber, with the outer layer containing the sensitive reagent. The sensor was interrogated with light at 680 nm, and showed sensitivity over the range 30 to 80% relative humidity. Sensor response was reversible, but response time was very long (on the order of one hour), even though a very thin (0.1 μm) coating of gelatine was used.

Following the work of Giuliani et al.[85] researchers at AT&T Bell Laboratories investigated the first optical fiber specifically drawn for the purpose of intrinsic chemical sensing.[91-94] Rather than being a rod or tube, or a short section of fiber where an inert cladding had been removed and replaced with sensor material, this fiber had a 125 μm diameter fused silica core continuously clad during draw clad to 230 μm total diameter with a silicone acrylate elastomer containing 0.2% (wt.) oxazine perchlorate dye. The rubbery nature of the silicone acrylate cladding made it permeable to ammonia (and other gases) so that the entire fiber was sensitive to the contents of the atmosphere around it. When a 280 m piece of the sensor was tested by exposure to ammonia, a dramatic drop in optical transmission loss (from 450 to 90 dB/km) was observed at the dye absorbance peak, 665 nm. The sensor responded fairly rapidly (within 20 sec) to the change in atmosphere from air to ammonia, but exhibited a longer-term "drift" extending for several minutes. An irreversible loss change was also observed each time the ammonia concentration in the fiber environment was changed. This effect, which was not seen in the dye-coated capillary tubes of Giuliani et al.[85] could be explained by a chemical reaction between the ammonia and the silicone acrylate: if such a reaction yielded a product to which the silicone acrylate was impermeable, it would remain in contact with the dye and could affect the fiber loss. This fiber, which probably remains the longest intrinsic optical fiber chemical sensor ever tested, demonstrated that intrinsic sensors may be mass produced. Other cladding/dye systems suitable for the process used to manufacture this fiber could be employed to produce extended absorbance-based intrinsic sensors.

Sawada et al.[94] also reported on an intrinsic fiber optic ammonia sensor based on cladding absorbance. The fiber consisted of a 500 μm PMMA core, clad with a 5 μm thick layer of PVA into which thymol blue had been dissolved. A short section (10 cm) of this fiber was mounted in a controlled-atmosphere chamber, and illuminated by shining a tungsten lamp on a fluorescent-core fiber spliced to one end of the sensor fiber; the other end was connected to a photodetector by means of a clear plastic fiber. The fluorescent fiber had an emission peak near the wavelength (610 nm) where thymol blue undergoes a dramatic increase in absorbance in the presence of basic species such as ammonia gas. When the test chamber was filled with ammonia gas, the sensor responded in approximately 1 min, a relatively long time considering the thinness of the sensor coating; the response was completely reversible, but complete recovery took more than 5 min. At 45% relative humidity, an ammonia gas concentration of 500 ppm caused a drop in signal intensity of approximately 15%. Sawada et al. also observed that the change in fiber loss was directly related to the logarithm of the ammonia concentration over the range 10 to 1000 ppm and concluded that the ultimate sensitivity of the 10 cm long sensor was better than 5 ppm.

A large body of research has been devoted to the use of optical fibers as TIRF "cells". Much of the impetus for this work has come as a result of the interest of biomedical researchers and commercial enterprises in creating fiber-optic probes based on fluoroimmunoassay (FIA) techniques. Although they are not true "sensors", since the effect of the measurand is not generally reversible without a washing step between exposures,* immunoassay-based optical probes can be very sensitive, and deserve mention here.[96] In its fiber optic version, FIA uses a fiber coated with an antibody, antigen, or immunochemical complex which then reacts with and binds another immunochemical present in the solution to be tested.[71,72] A number of fluorescent detection schemes are possible: a "sandwich" immunoassay technique has been used by Sutherland et al.[97,98] to detect the presence of human immunoglobulin G (IgG). In this technique, anti-IgG was first bound to the fiber (a short section of stripped 600 μm fused silica), and then the probe was exposed to the sample to be measured for a specific time, during which IgG bound to the antibody on the fiber. The probe was then removed from the sample and placed in a solution containing fluorescent-labeled anti-IgG, which bound to the IgG on the fiber. TIRF was used to measure the amount of fluorescent label present, thus giving a quantitative reading of the IgG bound to the probe. The estimated limit of detection of this probe was on the order of 1.5 mg/l of IgG, roughly twice as sensitive as a quartz-slide planar waveguide TIRF probe using the same "sandwich" system. Other work by the same group demonstrated the feasibility of basing TIRF (and, by analogy, intrinsic fiber optic) probes on the native optical absorbance of a measurand (the drug, methotrexate) bound to an antibody coating on the surface of the probe.[99]

Other immunological approaches may be taken in the design of intrinsic fiber optic biochemical probes.[100] One possible approach is to base a probe on competitive binding between the measurand and a labeled antibody or antigen in a solution into which the probe is immersed in a "readout" step after exposure to the solution being tested. It should also be possible in some cases to measure the native fluorescence of antibodies that bind directly onto antigen-treated fiber surfaces. Probes have been reported for insulin,[101] human serum albumin,[102] and digoxin.[103]

The use of nonimmunological reagents in chemical detection schemes may offer a means by which biochemicals can be incorporated into true reversible intrinsic optical sensors. Krull et al.[104-108] have investigated the use of lipid membranes deposited directly onto the fiber surface as a means of immobilizing biochemical receptors. As with immunochemical coatings, these membranes are extremely thin (on the order of 10 nm), making them ideal candidates for evanescent-wave based optical interactions. Binding of measurand molecules to receptor molecules will cause changes in the state of the membrane, and there is a wide choice of fluorescent probes of bilayer state that may be used to transduce the state change to an optical signal. Early experiments[104] showed that phosphatidylcholine/cholesterol monolayers doped with fluorophore and deposited on borosilicate glass surfaces exhibited dramatic changes in evanescent-coupled fluorescence upon exposure to such membrane-disturbing reagents as valinomycin and phloretin. For good adhesion, it was necessary to alkylate ("silanize") the surface of the glass by treating it with octadecyltrichlorosilane before applying the sensor layers. In the first actual demonstration of energy coupling from lipid membranes to the evanescent field of fiber propagation modes, fluorescence spectra were obtained from a 400 μm diameter quartz fiber that had been surface-coated with a monolayer of fluorophore-doped stearic acid.[106,107] Krull et al. have also deposited up to 200 layers of phosphati-

---

* Andrade et al.[95] have proposed a method (based on photoisomerization of polymers co-immobilized with sensor material) that might be able to remove measurands between measurements without a washing step, but there are as yet no experimental reports of its use.

dylcholine/cholesterol and stearic acid on quartz and borosilicate glass surfaces, to improve the adhesion of fluorescent-doped layers and to investigate the possibility of using such relatively thick structures as true fiber "claddings". Membrane fluorophores used in the experiments of Krull et al. included 1-anilinonaphthalene-8-sulphonate (ANS), 12-(9-an-throyl-oxy)stearic acid (12-ASA), and trans-4-dimethylamino-4'-(1-oxybutyl)-stilbene (DOS). The effect of these probes on membrane structure has been studied by performing pressure-area phase measurements using a hydrostatic loading cell;[107] from the results of these experiments it appears that the local environment of the probes may be sufficiently compli-cated that only total fluorescence, rather than lifetime- or polarization-based measurements, will be useful for monitoring sensors based on lipid-layer coated fibers.

Evanescent fluorescence based intrinsic fiber sensors that do not make use of biochemicals are rather rare in the literature. Two different groups have reported intrinsic fiber optic oxygen sensors based on the immobilization of a fluorescent molecule, 9,10-diphenylanthracene (9,10-D), in polymer coatings around a fused silica core. 9,10-D is a fluorophore well known to optrode researchers[109] for its excellent fluorescence-quenching properties in the presence of oxygen partial pressures near the atmospheric range. The (completely reversible) drop in fluorescence emission intensity follows the well-known Stern-Volmer equation,[110]

$$\frac{I_0}{I} = 1 + (\text{const}) \times P_{O_2}$$

where $I_0$ is the fluorescence intensity measured in the absence of oxygen and $I$ is the intensity when the oxygen partial pressure is $P_{O_2}$.

In one effort,[111] coatings of an oxygen-permeable hydrophilic polymer, the "hydrogel" poly(2-hydroxyethylmethacrylate) (PHEMA), doped with 9,10-D, were cast onto short sec-tions of stripped 1 mm core PCS fiber that had been prepared by treatment with a silanizing reagent. When this sensor was tested in oxygenated aqueous solutions in a "transmissive" mode; oxygen diffused from the water into the permeable PHEMA causing the fluorescence output from the fiber to drop in agreement with the Stern-Volmer relation. Response times on the order of a few minutes were obtained for fibers with thin (50 μm) coatings, and the signal dropped by roughly 20% when going from oxygen-free to oxygen-saturated solutions. Inves-tigations were also performed on bulk samples of PHEMA that had been co-doped with 9,10-D and glucose oxidase (an enzyme that catalyzes the reaction of glucose with oxygen) in preliminary experiments aimed at using this system to create an intrinsic fiber optic glucose sensor.

Almost all intrinsic fiber optic chemical sensors based on evanescent-wave fluorescence make use of very short lengths of fiber that are hand-prepared from stripped conventional fibers. Lieberman et al.[93,112,113] have reported drawing an optical fiber with an intrinsically fluorescent cladding composed of 0.1% (wt.) 9,10-D dissolved in polydimethylsiloxane (PDMS), a rubbery silicone. The entire fiber, which had a 125 μm core and a 50 μm thick cladding, was sensitive to oxygen concentration and, due to its length, provided the first experimental verification that optical energy generated in an optical fiber cladding can be coupled to true core-guided modes of the fiber. Sections of the fiber were tested in gaseous mixtures of nitrogen and oxygen in different concentrations by side-illuminating various lengths and collecting transmitted fluorescence through an unilluminated segment. Fluores-cence emission intensity was found to follow the Stern-Volmer relation, with response times on the order of a few seconds or less (see Figure 7), and the sensor was inferred to have an ultimate sensitivity of approximately $P_{O_2} = 1$ torr for 18 m of fiber. The efficiency of coupling the fluorescence incident on the fiber core to bound modes of propagation was found to be approximately 0.02%, in agreement with the value predicted by Marcuse[52] for a core in a fluorescent cladding of "infinite" extent. For lengths up to a few meters, the total collected fluorescence intensity was roughly proportional to total illuminated length. For longer illuminated sections, intensity did not increase as rapidly, due to the effect of transmission loss in the fiber itself. The "saturation length" (length at which 90% of the total possible coupled fluorescence is achieved) for a fiber such as this, with "gain" due to fluorescence balanced against intrinsic loss, is ten times the reciprocal of the loss (in decibels per unit length) at the fluorescence emission wavelength, or 18 m for this fiber.

The main problem in evanescent-coupled fluorescence based intrinsic fiber optic chemical sensors is their reliance on a process (evanescent-field coupling) that is essentially very inefficient for the transfer of optical energy from the sensory coating to the core. Preliminary work on a solution to this problem has been done by Lieberman and Brown,[114] who studied a "two-stage" coupling process involving fluorophores in both the core and a permeable sensory coating on the outside of the fiber. 9,10-D doped PDMS was cast to an approximate thickness of 200 μm on the outside of short (10 cm) sections of 500 μm plastic (PMMA) core fibers clad with a very thin (3 to 4 μm) layer of poly (4-methyl-pentene). The fibers had been manufactured with cores highly doped with a fluorophore[94] whose excitation spectrum overlapped the emission spectrum of the 9,10-D. In this system, fluorescence from the sensor coating is transmitted through the fiber cladding, and would normally pass through the core without being trapped in guided modes, was strongly absorbed by the secondary fluorophore which then acted as a source in the fiber core, emitting energy that was guided by the fiber. The quantum efficiency with which photons incident on the core of such a fiber are absorbed and re-emitted as photons guided by the core can approach 100%,[115] so this two-stage fluorescence sensor exhibited much stronger signals than coated-fiber sensors based on evanescent coupling. Sensor response times were still very rapid (on the order of seconds) since the coating was a thin layer on the outside of the fiber. The transmission loss of the commercially manufactured fluorescent-core fiber at the emission wavelength was 4.0 dB/m, meaning that the "saturation length" of this particular sensor was only 2.5 m. It is possible that other fluorophores could be used in the fiber core to alleviate this limitation.

A coating-based intrinsic fiber optic chemical probe that worked on refractometric principles has been reported by Silvus et al.[116-118] The initial design of the probe resembled the refractometer shown in Figure 1c, with the entire U-shaped section of 130 μm diameter fiber stripped and treated with an organophilic material; later versions used a longer stripped section held in a helical shape. Coating materials included octadecyltriethoxysilane, octadecyltrichlorosilane, and trimethylchlorosilane; some coatings also contained hydrogenchloride scavenging reagents. Fibers were tested by suspending them in capillary tubes and flowing various water/hydrocarbon mixtures though the tubes; as the organic materials encountered the treated fibers, they adhered to the surface, causing a change in the refractive index of the cladding (initially water), thereby causing a progressive increase in fiber transmission loss as the amount of bound hydrocarbon increased. The rate of change of loss was directly proportional to the concentration of the hydrocarbon substances tested. One drawback of this approach arose from the fact that many organic materials have refractive indices lower than that of most optical fiber materials (e.g., 1.458 for fused silica) and so would act as a cladding when bound to the fiber, causing little or no effect on its guiding properties. The sensitivity of this probe to a wide range of higher-index organic contaminants was tested (see Figure 8) and it was shown that the sensor could detect crude oil concentrations on the order of 3 mg/l, and diesel fuel levels near 15 mg/l. To demonstrate the repeated use of these treated fiber "probes", a prototype sensor system was built that provided for cleaning the fiber surface between measurements by flushing the capillary with a solvent mixture.

Zemel et al.[119] have proposed a refractometric intrinsic sensor based on an optical fiber containing a Bragg grating built into it by inducing periodic changes in the refractive index of the core or cladding, or by bending or twisting the fiber. A sensory coating that underwent a change in refractive index in the presence of the measurand, applied in the grating area, would cause a variation in the back-reflected light induced by such a grating when the measurand was present. Theoretical calculations based on a short section (2 mm) of waveguide with a very small core (2 μm) showed that appreciable changes in reflectance (from 0 to 80%) could be induced for cladding index variations on the order of $\Delta n = 0.05$; Zemel proposed using the anomalous dispersion of absorbance-based indicator dyes to create the desired index change. Practical limitations (for example, the extremely small core size) seem to have precluded the actual construction of fiber sensors based on this effect.

Another phenomenon that can be used to transduce chemical concentration information to optical signals in intrinsic sensor fibers is scattering of light by the fiber cladding. Ogawa et al.[120] described an intrinsic humidity sensor based in part on the increase in scattering loss that occurs when a microporous silica cladding is permeated by water vapor. Sensor fibers were created by coating a silica core with microporous silica; because the density of the microporous silica coating is lower than that of fused silica, its refractive index is also lower, and it acts as a cladding material, making the sensor a true step-index optical waveguide. When a 4-cm section of sensor fiber was tested in air with relative humidities between 20 and 95%, transmission through the sensor decreased monotonically with increasing humidity. The total transmission loss reached a value of 4dB at the highest humidity. Although part of the loss increase was probably due to water displacing air and causing the cladding refractive index to rise (and the fiber numerical aperture to fall), measurements of backscattered light showed a monotonic increase with increasing humidity. Backscattering increased by nearly 14 dB when relative humidity rose from 0 to 90%, indicating that at least some of the power loss is caused by scattering. When OTDR was used to investigate a multipoint humidity sensor made of three sensor sections spliced into a 130 m length of 200/300 μm (core/cladding) PCS fiber, it was observed that the scattered light from each of the sensor sections caused peaks to appear in the OTDR traces, with each peak followed by a drop associated with the loss induced by the presence of the sensor section (see Figure 9). As the relative humidity in the vicinity of one of the sensor segments was increased, the system clearly showed increases in the height of the peak and the depth of the loss drop associated with that segment. Ogawa et al. concluded that a quasi-distributed sensor system could be constructed using this multipoint technique, but the relatively large loss (roughly 25 dB/meter at 50% relative humidity) precludes the use of porous-clad fibers as long-length intrinsic sensors.

A novel coating based sensor that does not make use of evanescent-field coupling has been proposed by workers at Mitsubishi International, Inc.[121] This sensor would use a conventional clad and buffered fiber enclosed in a stainless steel capillary tube filled with a material that undergoes a change in elasticity in the presence of hydrocarbons. The stainless steel outer jacket could be punctured at the points where oil concentration was to be monitored, or the jacketed fiber could be mass produced with holes at regular intervals. The physical change experienced by the filler material in the presence of oil would cause changes in the local microbending loss experienced by the fiber which could be detected by simple transmission loss measurement procedures, or by OTDR. Lest this idea seem far-fetched, it should be pointed out that stainless-steel jacketed fiber is commercially available, and there are in fact several polymeric materials that undergo changes in stiffness or density in the presence of hydrocarbons.

Another coating-based intrinsic sensor that does not depend on evanescent-field coupling is the interferometric hydrogen sensor first investigated by Butler at Sandia Laboratories.[122] Although Leslie et al.[123] had earlier used a fiber-optic Mach-Zender transducer in a "spectro-phone" to measure pressure changes induced by the absorbance of mid-IR radiation by methane in a closed container, Butler's is the first report of an interferometer incorporating an optical fiber intrinsically sensitive to changes in chemical concentration. The chemically sensitive coating in this case was palladium, applied by sputtering and subsequent electrode-position to a thickness of approximately 10 μm on the surface of a 3 cm stripped section of single mode graded index fiber. When hydrogen diffused into the palladium the coating expanded, causing the fiber to lengthen slightly, and also causing a stress-induced change in the index of refraction of the fiber. This change was completely reversible, though the time required for the hydrogen to diffuse into the coating was much longer (more than 10 s) than the time required for it to diffuse out (under a tenth of a second). When the coated fiber was placed in one arm of a Mach-Zender interferometer (see Figure 10a) and optically compared with a similar piece of fiber in the other arm, phase shifts due to the hydrocarbon induced change in optical path length were detected as changes in the output of the interferometer. Butler performed the optical comparison by allowing light from the reference fiber and the sensor fiber to fall on a photodector and observing changes in the resulting interference fringes. Using a single p-i-n diode, Butler was able to detect partial pressures of hydrogen in nitrogen as low as 4 torr. In later work[124] Butler used a 128-element diode array in the output plane to follow fringe shifts, and was able to detect hydrogen concentrations over a range of six orders of magnitude, from 0.01 ppm (roughly $8 \times 10^{-5}$ torr at atmospheric pressures) to 10% (76 torr). Investigation of the details of the fiber's response to hydrogen concentration indicated that electrodeposited palladium had a far greater (by a factor of 7) response to hydrogen than sputter-deposited palladium, whose properties resemble that of "normal" (bulk) palladium metal. Butler and Ginley[125,126] used the same sensor geometry to monitor the strain state of metallic coatings during electrodeposition of titanium, nickel, and palladium.

Other work on palladium based fiber hydrogen sensors has been carried out by Farahi et al.[127,128] who used an all-fiber Michelson interferometer (see Figure 10b) with a short (6 cm) piece of 0.5 mm palladium wire glued to a stripped section of one of its legs. Taking advantage of abundant prior expertise in processing signals from interferometric-based intrinsic fiber optic sensors, a pseudo-heterodyne detection scheme was used to monitor optical phase shifts in the sensor by converting them to radio-frequency phase shifts in a signal used to modulate the light input to the interferometer. Despite the rather crude nature of the sensor itself, the sophisticated processing scheme allowed the sensor to measure hydrogen concentrations as low as a 2 ppm (0.0015 torr). Using the same detection system and an interferometer having one leg containing a 10 cm section of fiber that had been stripped and coated with a 3 μm layer of platinum, Farahi et al.[129] created a sensor for flammable organic gases in air. This sensor differed from the hydrogen sensors discussed above in that in this case heat, rather than pressure, was used to transduce the chemical concentration change into a change in the optical path length in the sensing arm of the interferometer. The platinum coating acted as a catalyst for the reaction of the gases with ambient oxygen; the resultant heat of this reaction caused a physical lengthening of the fiber through thermal expansion, as well as inducing changes in the refractive index of the fiber's core and cladding. Since the optical fiber thermometer upon which this sensor was based was sensitive to changes on the order of 0.1 millidegrees for a 10 mm segment,[130] fairly low concentrations (on the order of 7 torr for butane and 14 torr for methane, well below the lower explosive limit for these gasses) could be measured with this device.

# V. CORE-BASED INTRINSIC SENSORS

Despite the prevalence of evanescent-field based techniques in intrinsic fiber optic chemical sensing based on spectroscopic or refractometric principles, there are a few reports of sensors that use the fiber core itself to transduce chemical concentration changes into changes in optical signal levels. The main advantage of this technique is that, compared to evanescent field techniques, the optical changes take place in the region of highest optical energy density. This means that signal levels for core-based sensors are usually much higher than for other types of sensors based on the same chemical-to-optical transduction schemes (whether direct or chemically assisted). A disadvantage is that the measurand must actually get into the core, a process usually accomplished in one of two ways: either the measurand diffuses through the cladding and into a permeable core region, or the measurand (in this case usually a liquid) is pumped into a hollow cladding and *is* the fiber core. No matter what method is used to place the measurand in the core region, response times for core-based intrinsic sensors tend to be longer than for coating-based sensors, since permeation of the entire fiber (or replacement of the core) takes longer than permeation of a thin cladding or coating region.

Presaging the use of hollow core fibers as sensor elements per se, workers at Bell Laboratories reported spectroscopic measurements performed using optical fiber cores. When these early experiments were performed, hollow silica fibers filled with transparent liquids of higher refractive index had the lowest optical transmission loss of any dielectric waveguide structure and were being seriously considered for use in long-haul optical communication links. Because of this, the first chemical measurements performed were detailed spectroscopic absorbance studies of candidate core materials such as carbon tetrachloride and tetrachlorethylene.[131,132] It was very rapidly realized that the extraordinarily long optical path lengths achievable with fluid-filled fibers provided an opportunity for investigating low-level optical effects in liquids that were previously very difficult to observe. Fluid-core waveguides proved to be particularly successful in the detailed study of the Raman effect in such liquids as carbon disulfide, benzene, toluene, and tetrachloroethylene.[133-135] The filled-fiber technique has been used to investigate the spectral properties of many types of samples, using path lengths up to 25 m.[136,137]

The use of fluid-core optical fibers as intrinsic chemical sensors grew naturally out of the use of capillary optical absorbance cells for on-line colorimetric monitoring of contaminants in fluids.[138,139] Fuwa et al.[140] reported a spectrophotometric sensor system based on the use of Pyrex lightguides fibers with inside diameters of 250 or 450 μm. Commercially available liquid chromatography column components were used for fluid handling (e.g., pumping the test solutions through the fiber core region). With a 4 m fiber, system sensitivity to iodine, phosphorous (in phosphomolybdenum heteropoly blue), and mercury and chloride ions (in dithizone complexes) in carbon disulfide was greater by approximately a factor of 3000 than the sensitivity of a system based on a 1 cm cuvette. With a 50 m long section of sensor fiber, sensitivity to molybdenum blue was increased by a factor of 30,000. A difficulty with this technique was the need to fill the entire tube with samples; although volumes are very small (total volume for a 50 m long section of fiber is only a few milliliters), colorimetric results were not always reproducible, possibly due to contamination of the inner surface of the fiber. One means of minimizing this problem is the use of flow-injection analysis techniques for sample delivery, as reported by Fujiwara and Fuwa.[141] In the flow injection analysis system, carbon disulfide flowed through the intrinsic sensor fiber in a slow, continuous stream. Small amounts of sample solutions were introduced into the stream before it reached the sensor fiber, and when the sample-containing "pulses" of fluid entered the optical waveguide, a simple colorimetric detection system responded to the changes in fluid composition. Fujiwara et al.

demonstrated that, with a 5 m long intrinsic sensor fiber and a flow rate of about 4 ml/min, this system was capable of detecting iodide ion concentrations as low as 0.12 mg/l. The minimum practical time between sample injection pulses was approximately 5 min. It should be noted that flow injection analysis is not limited to simple injection of measurand-bearing samples into the spectroscopic detection cell; it is also possible to inject reagents that react with the measurand to produce chemically enhanced spectral changes. Fluorescence measurement can also be used as an optical detection method in flow injection analysis systems; using a simple flow system with quartz fibers having inside diameters on the order of a few hundred microns, Fujiwara et al.[142] investigated the fluorescence of perylene in carbon disulfide. The fiber was illuminated coaxially using the output of a tunable dye laser, and perylene concentrations as low as $4 \times 10^{-13}$ (wt.) could be detected. This represents a factor of 5 improvement over the best previously reported detection system.[143]

Optical fiber chemical detection using an indicator-mediated approach based on fibers with reactive species in a permeable core was first proposed by Buckles,[122] who also envisioned the use of multiple permeable coatings to provide a series of reaction and separation steps as the measurand diffused its way into the core. A practical realization of the simplest embodiment of this technique has been discussed by Shariari et al.,[144-146] who reported a humidity sensor based on porous optical fibers containing cobalt-chloride. The porous fibers were created specifically for use as intrinsic sensor elements: borosilicate glass was drawn to diameters ranging from 150 to 300 μm, and then heat treated to separate borate-rich and silicate rich phases of the glass. After phase separation, a short (5 mm long) section of the fiber had its alkalai-rich phase leached away by immersion in acid, leaving a highly porous region. The porous section was then immersed in concentrated cobalt chloride solution and allowed to dry, before being placed in a humidity-controlled chamber for testing. As humidity was increased, the optical transmittance at 690 nm was observed to increase in a reversible and reproducible way. The response time of the sensor was rather long, on the order of a few minutes, due to the need for water vapor to fully permeate the fiber. By varying the cobalt chloride concentrations used in the doping solution, sensor response can be tailored to different humidity ranges (see Figure 11). For example, a sensor treated with 10 mg/ml of cobalt chloride exhibited a fairly linear response to relative humidities in the range 1 to 5%. Other concentrations resulted in sensors linear over the ranges 10 to 15% and 25 to 40%. It should be noted that relative humidities this low have not be measured by other intrinsic sensors.[86,89,120] The cobalt chloride dye response is extremely sensitive to ambient temperature,[146] but this presents no fundamental barrier to the use of the sensor in practical applications.

Enlarging on the same general theme, Shahriari[147] created an ammonia-vapor sensor by doping a 5 mm section of a 200 μm diameter porous borosilicate fiber with bromocresol purple, a pH indicator dye. The response time of this sensor was on the order of 10 min or more, possibly because of a somewhat different pore size than that of the moisture sensor. By monitoring optical absorbance at 580 nm, and comparing with intensity measured at a reference wavelength of 700 nm (where the dye undergoes no absorbance change in the presence of ammonia), ammonia concentrations in nitrogen as low as 1 ppm could be measured. Presumably, porous fibers could be soaked in solutions containing almost any indicator dye, meaning that this system could find a wide range of applications in gas monitoring situations that do not require short response times. In order to use core-based chemically assisted sensors to measure chemical concentrations in solution, it is necessary to immobilize the sensor reagents so that they do not diffuse out of the core during the measurement process (as would be the case with the sensors described by Zhou et al.).

Macedo et al.[149] described a porous-core pH sensor that addressed this problem. After forming a porous region on the end of a borosilicate fiber using phase separation and hot-acid leaching, sensor dye (either cresol red or phenol red) was covalently bound to the glass matrix in a silanization process that used amine-bearing alkylating reagents such as 3-aminopropyltri-ethoxysilane (APT), N-2-aminoethyl-3-diaminopropyltrimethoxysilane (AAPT), N-methylaminopropyltrimethoxysilane (MAPT), and bis[3-(triethoxysibyl)-propyl]-amine (BAPT). After dye immobilization, a thin layer of gold was evaporated onto the sensor to increase the optical signal measured when the device was tested in "backreflection" mode with a system incorporating a fused fiber coupler. Round-trip "loss" for the probe was a linear and reproducible function of pH over the range pH 5 to pH 8, with the calibration constant being on the order of 1.4 pH/dB at 675 nm. The same process was also used to produce a pinacyanol chloride-bearing tip that acted as a temperature sensor, albeit one with a very nonlinear response. Macedo et al. also pointed out that porous-core fibers can be used as passive core-based intrinsic fiber optic chemical sensors for measurands with strong absorption, fluorescence, or Raman spectral features.

## VI. CONCLUSION

Research into the use in optical fibers as chemically sensitive components of measurement and monitoring systems has been carried out for more than 30 years. The earliest sensitive fibers, used as components in simple evanescent-wave refractometers, are the direct "ancestors" of sensors that have found wide acceptance in commercially available liquid level monitoring systems. As scientific tools, evanescent-wave spectroscopic absorbance cells have already proven very useful in the study of the adsorption of chemical substances on silica surfaces, and in the general study of spectra from absorbing media that are difficult to measure with conventional transmitted-light techniques. The main advantage of using fibers in these applications is the fact that very large interaction lengths may be achieved with relative ease. Other intrinsic sensing techniques based on the use of the evanescent field to couple optical energy through the side of the fiber have begun to reach maturity, with coated-fiber im-munosensitive probes leading the way. The development of this technique into a true sensor technology that will allow continuous monitoring, rather than temporal "snapshots" of the concentration of a biologically interesting measurand, and is the subject of active research and development efforts at several laboratories around the world.

Coated-fiber techniques now being investigated point the way to a future where intrinsi-cally sensitive fibers will be capable of measuring more measurands, with greater sensitivity, and in larger sample areas, than can be achieved with other techniques. Particularly notewor-thy are efforts to create intrinsic chemical sensors by combining the highly sensitive interfer-ometric techniques developed originally for physical measurands with coated fibers that transduce chemical changes into optical phase shifts. These sensors have the potential of reaching sensitivities hitherto impossible to achieve with optical techniques. The rapid prog-ress being made in the application of chemically sensitive indicator dyes to optical sensing, primarily stimulated by efforts to develop better chemical optrodes, also offers new possibili-ties for coating-based intrinsic sensors.

The range of applications outside the laboratory for the "new generation" of intrinsic fiber optic chemical sensors now being investigated is very broad. Intrinsic fiber optic chemical sensors with interaction lengths on the order of several meters, or tens of meters, are particularly suitable for application in systems designed to monitor chemical concentration over broad areas. Environmental air-quality monitoring, toxic waste site containment assur-ance, and effluent monitoring are but a few of the situations where such systems will find use. The distributed nature of the measurements made by long-path intrinsic chemical sensors may also make them suitable for use in chemical process control applications, for example in monitoring average reactant concentrations in large chemical baths or fermentation vessels. Immunosensors and other biochemical-based intrinsic fiber optic probes are likely to join the related optrode-based chemical biosensors in biotechnological process control situations, and short (less than 1 m) intrinsic sensors may even be used as optrodes for some medical applications. The monitoring of indoor chemical concentrations, particularly those related to air quality in the "microenvironments" of rooms or buildings is another application where the properties of intrinsic fiber optic chemical sensors can offer unique advantages.

Though the use of the intrinsic properties of optical fibers for chemical sensing has been studied for many years, new applications and new sensing methodologies are still being found. As this progress proceeds, intrinsic fiber optic chemical sensors will continue to find applications in science, industry, medicine, and environmental studies; the unique advantages of the intrinsic approach to sensing may lead to entirely new fields of application.

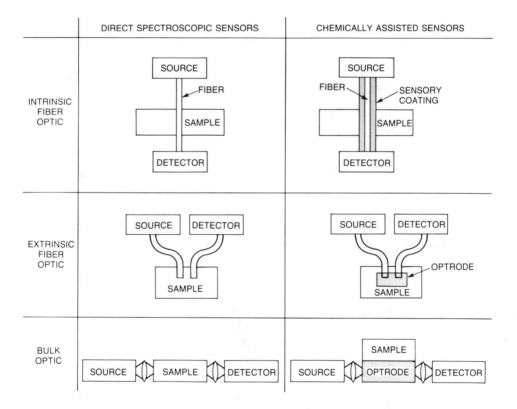

FIGURE 1. Optical sensor types. Schematic examples illustrating the similarities between bulk optical, extrinsic fiber optic, and intrinsic fiber optic chemical sensors. Both direct spectroscopic and "chemically assisted" signal transduction methods are shown.

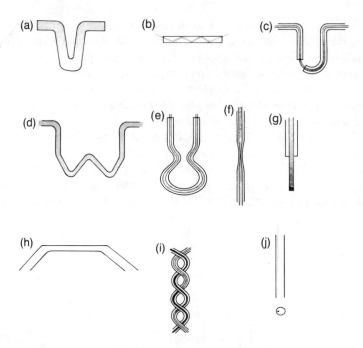

FIGURE 2. Intrinsic fiber optic refractive index sensors. (Shading indicates light path) (a) U-shaped quartz rod waveguide,[7] (b) straight quartz rod waveguide,[9] (c) U-shaped bent multimode fiber with cladding section removed,[18] (d) W-shaped quartz rod waveguide,[13] (e) "compound bend" plastic multimode fiber with cladding intact,[20] (f) stripped-tip multimode fiber with reflector on end,[27] (g) bent single mode fiber with cladding region thinned,[30] (h) twisted multimode fiber pair,[41] (i) single-mode fiber with off-center core.[42]

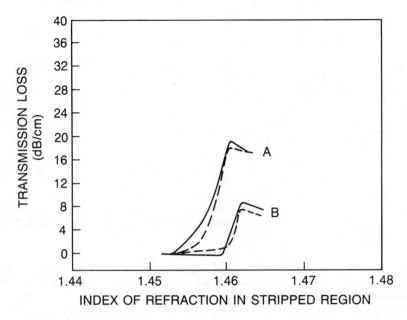

FIGURE 3. Dependence of stripped-core refractive index sensor sensitivity on numerical aperture of launch.[36] Fiber: 300 mm diameter silica core with cladding stripped from 5 mm section. Wavelength: 632.8 nm. A — Launch NA = 0.15; B — Launch NA = 0.085. Solid lines — Theory. Broken lines — experiment.

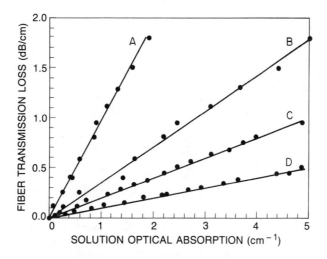

FIGURE 4. Dependence of stripped-core absorbance-based sensor response on waveguide parameter (V).[48] Fiber: 50 mm diameter silica with cladding stripped, immersed in solution having different index refraction an different Rhodamine 6 dye concentration. A — V = 24; B — V = 51; C — V = 99; D — V = 190. Wavelength: 514.5 nm. Solution absorption measured using test solutions in conventional spectrophotometer.

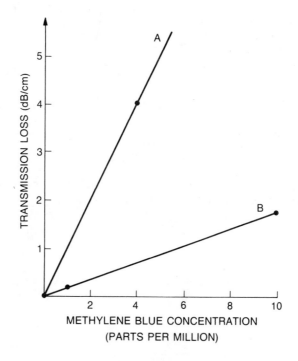

FIGURE 5. Difference in response between unclad and porous-clad absorbance-based sensors.[57] Fibers: A—25 cm length of stripped 200 mm (core) silica. B — 25 cm straight length of 200/300 mm (core/cladding) silicone clad silica. Test solutions: oxazine 4 perchlorate in chloroform, various concentrations. Solution absorption measured using test solutions in conventional spectrophotometer. Wavelength: 564 nm.

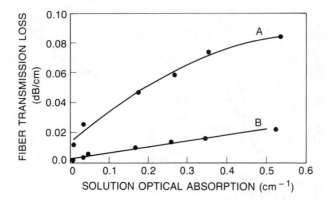

FIGURE 6. Sensitivity of tapered single-mode fiber absorbance-based sensor.[63] A — transmission loss (dB/cm) for 1 cm path length, as measured by conventional spectrophotometer. B — transmission loss (dB/cm) for tapered-fiber sensor. Fiber: 6/80 mm (core/cladding); taper length: 20 mm. Wavelength: 632.8 nm. Note: for this solution, abcsisa can be convered to units of optical absorption (in cm$^{-1}$) simply by dividing by 10.

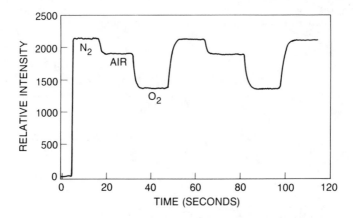

FIGURE 7. Oxygen response of fluorescent-clad fiber.[113] Fiber: 130 cm length of 125 mm silica core, clad to 230 mm with polydimethylsiloxane doped with 0.1% (wt) 9,10-diphenylanthracene.

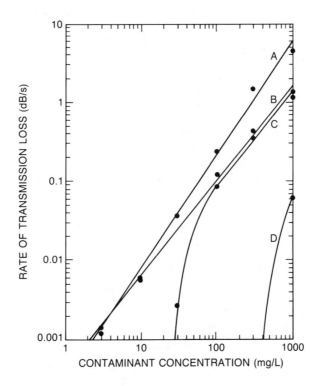

FIGURE 8. Response of coiled silanized fiber to hydrocarbons in water.[117] Fiber: 130 mm diameter core, coated with octadecyltrichlorosilane to facilitate adsorption of hydrocarbons. Hydrocarbon: A — crude oil; B — p-xylene; C — tetralin; D — n-hexylbenzene.

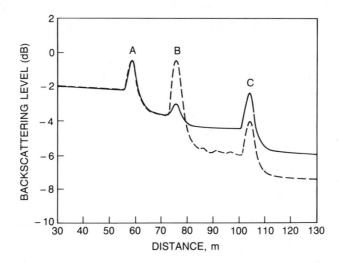

FIGURE 9. OTDR response of multipoint humidity sensor.[120] Fiber: 200/300 mm (core/clad) plasic clad silica. Core stripped and re-clad with porous silica at three locations: A — 60 m from input end; B — 78 m; C — 105 m. Stripped sections maintained at different humidity levels. (Solid curve: A — 40% R.H.; B — 20% R.H.; C — 40% R.H.) (Dashed curve: A — 40% R.H.; B — 80% R.H.; C — 40% R.H.)

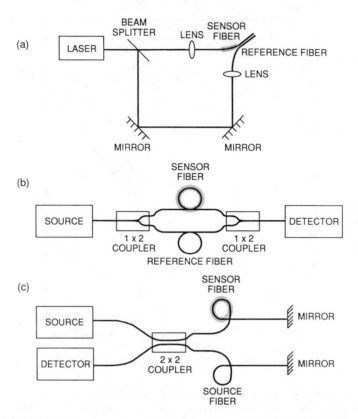

FIGURE 10.   Interferometric detection schemes for intrinsic sensors. (a) Mach-Zender interferometer used with palladium-coated fiber in sensor arm for hydrogen detection.[122] (Note: sensor and reference fibers fixed in place on a quartz plate.) (b) All-fiber Mach-Zenter interferometer. (c) All-fiber Michelson interferometer used for hydrogen or hydrocarbon gas detection with palladium or platinum coated fibers, respectively, in sensor arm.[128]

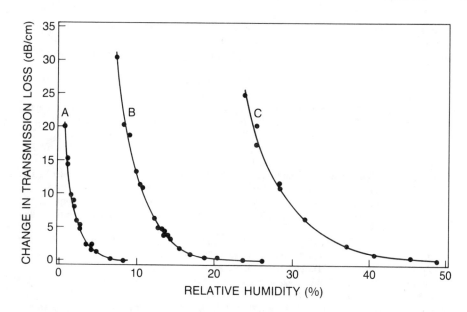

FIGURE 11.   Humidity response of porous silica fiber.[145] Fiber: air-clad 150 mm borosilicate:alkalai glass. Sensor: 5 mm section of porous core at the fiber tip, immersed in aqueous solutions of $CoCl_2$ (A — 10 mg/ml; B — 139 mg/ml; C — 213 mg/ml) and then allowed to dry. Change in transmission loss at 690 nm is measured relative to the loss at the highest-humidity point for each curve.

# REFERENCES

1.  **Wolfbeis, O. S.,** Chapter 1, this book
2.  **Wolfbeis, O. S.,** Chapter 3, this book
3.  **Parriaux, O.,** Chapter 4, this book.
4.  **Snyder, A. W. and Love, J. D.,** *Optical Waveguide Theory*, Chapman and Hall, New York, 1983.
5.  **Lübbers, D. W. and Opitz, N.,** Die $P_{co2}$-$P_{o2}$ optode: eine new $PCO_2$ bzw.$PO_2$ messonde zur messung des $P_{co2}$ order $P_{o2}$ von gasen und flussigkeiten, *Z. Naturforsch.*, 30C, 532, 1975.
6.  **Peterson, J. I. and Vurek, G. G.,** Fiber-optic sensors for biomedical applications, *Science*, 224, 123, 1984.
7.  **Karrer, E. and Orr, R. S.,** A photoelectric refractometer, *JOSA*, 336(1), 42, 1946.
8.  **Kapany, N. S. and Pike, J. N.,** Fiber optics IV. A Photorefractometer, *JOSA*, 47(12), 1109.
9.  **Kapany, N. S. and Pontarelli, D. A.,** Photorefractometer I., extension of sensitivity and range, *Appl. Opt.*, 32(4), 425, 1963.
10. **Kapany, N. S. and Pontarelli, D. A.,** Photorefractometer II. Measurement of n and k, *Appl. Opt.*, 2(10), 1043, 1963.
11. **David, D. J., Shaw, D., Tucker, H., and Unterleitner, F. C.,** Design, development and performance of a fiber optics refractometer: application to HPLC, *Rev. Sci. Instrum.*, 47(9), 989, 1976.
12. **Ross, I. N. and Mbanu, A.,** Optical monitoring of glucose concentration, *Opt. Laser Technol.*, 31, 1985.
13. **Sharma, M. and Brooks, M. E.,** Fiber optic sensing in cryogenic environments, in *Proc. S.P.I.E., 224, Fiber Optics for Communications and Control*, 1980, 46.
14. **Giuliani, J. F. and Jarvis, N. L.,** Detection of simple alkanes at a liquid-glass interface by total internal optical scattering, *Sens. Acta*, 6, 107, 1984.
15. **Giuliani, J. F. and Jarvis, N. L.,** A systematic investigation and characterization of interfacial hydrated n-alkane films by total internal multiple reflection, *J. Chem. Phys.*, 82(2), 1021, 1985.

16. **Giuliani, J. F.,** Optical detection and comparison of rare gas clathrate hydrate interfacial films with those of the simple n-alkane series, *J. Chem. Phys.*, 83(11), 5998, 1985.

17. **Giuliani, J. F. and Bey, P. P.,** Computer modeling for optical waveguide sensors, *Appl. Opt.*, 27(7), 1353, 1988.

18. **Takeo, T. and Hattori, H.,** Optical fiber sensor for measuring refractive index, *Jpn. J. Appl. Phys.*, 21(10, 1509, 1982.

19. **Spenner, K., Singh, M. D., Schulte, H. and Boehnel, H. J.,** Experimental investigation on fiber optic liquid level sensors and refractometers, in *Proc. 1st Int. Conf. Optical Fiber Sensors,* Institute of Elect. Eng., London, 1983, 96.

20. **Harmer, A. L.,** Optical fiber refractometer using attenuation of cladding modes, in *Proc. 1st Int. Conf. Optical Fiber Sensors,* Institute Elect. Eng., London, 1983, 104.

21. **Bergman, T. L., Incropera, F. P., and Stevenson, W. H.,** Miniature fiber-optic refractometer for measurement of salinity in double-diffusive thermohaline systems, *Rev. Sci. Instrum.*, 56(2), 291, 1985.

22. **Belkerdid, M. A., Ghanderioun, N., and Brennan, B.,** Fiber optic fluid level meter, in *Proc. S.P.I.E. 566 Fiber Optic and Laser Sensors III*, Moore, E. L. and Ramer, O. G., Eds., Soc. Photoinst. Eng., Bellingham, WA, 1985, 153.

23. **Kumar, A., Subrahmanyam, T. V. B., Sharma, A. D., Thygarajan, K., Pal, B. P. and Goyal, I. C.,** Novel refractometer using a tapered optical fibre, *Elect. Lett.*, 20(13), 534, 1984.

24. **Bobb, L. C., Krumbolt, H. D., and Davis, J. P.,** An optical fiber refractometer, in *Proc. S.P.I.E. 990, Chemical, Biochemical, and Environmental Applications of Fibers*, Lieberman, R. A. and Wlodarczyk, M. T., Eds., S.P.I.E., Bellingham, WA, 1988, 164.

25. **Anon.,** NASA Technical Support Package KSC-11311, Laser-Pulse/Fiber Optical Liquid-Leak Detector, J. F. Kennedy Space Center, Cape Canaveral, FL, 1984.

26. **Kopera, P. M. and Tekippe, V. J.,** Transmission of fibers with a short section of modified cladding, *Opt. News*, 7(3), 44, 1981.

27. **Scheggi, A. M., Brenci, M., Confortoi, G., and Falciai, R.,** Optical-fibre thermometer for medical use, *I.E.E. Proc.*, 131-H(4), 270, 1984.

28. **Brenci, M., Conforti, G., Falciai, R., Mignani, A. G., and Scheggi, A. M.,** All-fibre temperature sensor, *Int. J. Opt. Sens.*, 1(2), 163, 1986.

29. **Pavlath, G. A., Moore, E. L., and Suman, M. C.,** Applications of all-fiber technology to sensors, in *Proc. S.P.I.E. 412, Fiber Optic and Laser Sensors*, Moore, E. L. and Ramer, O. G., Eds., S.P.I.E., Bellingham, WA, 1983, 70.

30. **Lew, C., Depeursinge, C., Cochet, F., Berthou, H., and Parriaux, O.,** Single-mode fiber evanescent wave spectroscopy, in *Proc. S.P.I.E. 514, Optical Fibre Sensors '84*, Kersten, R. Th. and Kist, R., Eds., S.P.I.E., Bellingham, WA, 1984, post-deadline paper.

31. **Falco, L., Berthou, H., Cochet, F., Scheja, B., and Parriaux, O.,** Temperature sensor using single mode fiber evanescent field absorption, in *Proc. S.P.I.E. 586, Fiber Optic Sensors*, Arditty, J. J. and Jeunhomme, L. B., Eds., S.P.I.E., Bellingham, WA, 1985, 114.

32. **Falco, L., Spescha, G., Roth, P., and Parriaux, O.,** Non-ambiguous evanescent-wave refractive index and temperature sensor, *Opt. Acta*, 33(12), 1563, 1986.

33. **Sixt, P., Kotrotsios, G., Falco, L., and Parriaux, O.,** Passive fiber Fabry-Perot filter for intensity-modulated sensors, *J. Lightwave Technol.*, 4(7), 926, 1986.

34. **Tomita, A. and Walker, K. L.,** Personal communication, 1986.

35. **Arie, A., Karoubi, R., Gur, Y. S., and Tur, M.,** Measurement and analysis of light transmission through a modified cladding optical fiber with applications to sensors, *Appl. Opt.*, 25(11), 1754, 1986.

36. **Arie, A., Tur, M., and Goldsmith, S.,** Measurement and analysis of a modified cladding optical fiber with various input illuminations, in *Proc. S.P.I.E. 718, Fiber Optic and Laser Sensors IV*, DePaula, R. P. and Udd, E., Eds., S.P.I.E., Bellingham, WA, 1986, 160.

37. **Afromowitz, M. A.,** Fiber optic polymer cure sensor, in OFS '88, post deadline paper.

38. **Bergh, R. A., Kotler, G., and Shaw, H. J.,** Single mode fiber optic directional coupler, *Electron. Lett.*, 16, 260, 1980.

39. **Jaccard, P., Scheja, B., Berthou, H., Cochet, F., and Parriaux, O.,** A new technique for low cost all-fiber device fabrication, in *Proc. S.P.I.E. 479, Fiber Optic Couplers, Connectors, and Splice Technology*, 1984, 16.

40. **Smela, E. and Santiago-Aviles, J. J.,** A versatile twisted optical fiber sensor, *Sens. Acta*, 13, 117, 1988.
41. **El-Sherif, M. and Zemel, J. N.,** Twisted-pair optical fiber pH sensors, in *Proc. 3rd Int. Conf. Solid State Sensors and Actuators,* I.E.E.E., Philadelphia, 1985, 434.
42. **Yoshikawa, H., Watanabe, M., and Ohno, Y.,** Distributed oil sensor using eccentrically cladded fiber, in Proc. 4th Int. Conf. Opt. Fiber Sensors, Tokyo, 1986, 105.
43. **Wlodarczyk, M. T.,** Bragg effect refractometer, in *Proc. S.P.I.E. 990, Chemical, Biochemical, and Environmental Applications of Fibers*, Lieberman, R. A. and Wlodarczyk, M. T., Eds., S.P.I.E., Bellingham, WA, 1988, 181.
44. **Quincke, G.,** Optische Experimental-untersuchungen: Über das Eindringen des total reflectirten Lichtes in das dunnere Medium, *Ann. Phys. Chem.*, 127, 1, 1866.
45. **Fröhlich, P.,** Die Gültigkeitzgrenze des geometrischen Gesetzes der Lichtbrechung, *Ann. Physik*, 65(4), 577, 1921.
46. **Harrick, N. J.,** *Internal Reflection Spectroscopy*, John Wiley & Sons, New York, 1967.
47. **Gloge, D.,** Weakly guiding fibers, *Appl. Opt.*, 10(10), 2252, 1971.
48. **Paul, P. H. and Kychakoff, G.,** Fiber-optic evanescent field absorption sensor, *Appl. Phys. Lett.*, 51(6), 12, 1987.
49. **Kramers, H. A. and Heisenberg, W.,** *Z. Phys.*, 31, 681, 1925.
50. **Safaai-Jazi, A., Jen, C. K., and Farnell, G. W.,** Optical fiber sensor based on differential spectroscopic absorption, *Appl. Opt.*, 24(15), 1341, 1985.
51. **Carniglia, C. K., Mandel, L., and Drexhage, K. H.,** Absorption and emission of evanescent photons, *JOSA*, 62(4), 479, 1972.
52. **Marcuse, D.,** Launching light into fiber cores from sources located in the cladding, *J. Lightwave Technol.*, 6(8), 119, 1988.
53. **Hansen, W. P.,** A new spectrophotometric technique using multiple attenuated total reflection, *Anal. Chem.*, 35(6), 1963.
54. **Zemel, J. N., VanderSpiegel, J., Fare, T., and Young, J. C.,** Recent advances in chemically sensitive electronic devices, in *Proc. ACS Symposium Ser. No. 309, Fundamentals and Applications of Chemical Sensors,* Am. Chem. Soc., 1986, 2.
55. **Wlodarczyk, M. T.,. Vickers, D. J., and Kozaitis, S. P.,** Evanescent field spectroscopy with optical fibers for chemical sensing, in *Proc. S.P.I.E. Vol. 718, Fiber Optic and Laser Sensors IV*, DePaula, R., and Udd, E., Eds., Soc. Phot. Inst. Eng., Bellingham, WA, 1986, 192.
56. **DeGrandpre, M. D. and Burgess, L. W.,** All-fiber spectroscopic probe based on an evanescent wave sensing mechanism, in *Proc. S.P.I.E. Vol. 990, Chemical, Biochemical, and Environmental Applications of Fibers*, Lieberman, R. A. and Wlodarczyk, M. T., Eds., S.P.I.E., Bellingham, WA, 1988, 170.
57. **DeGrandpre, M. D. and Burgess, L. W.,** Long path fiber-optic sensor for evanescent field absorbance measurements, *Anal. Chem.*, 60(23), 2582, 1988.
58. **Tanaka, H., Ueki, T., and Tai, H.,** Fiber-optic evanescent wave gas spectroscopy, post-deadline paper PDS2-1 in *Proc. Conf. on Optical Fiber Communications and 3rd Int. Conf. on Optical Fiber Sensors*, Opt. Soc. America, Washington, D.C., 1985.
59. **Tai, H., Tanaka, H., and Yoshino, T.,** Fiber-optic evanescent-wave methane-gas sensor using optical absorption for the 3.392-mm line of a He-Ne laser, *Opt. Lett.*, 12(6), 437, 1987.
60. **Simhoney, S., Kosower E. M., and Katzir, A.,** Novel attenuated total internal reflection spectroscopic cell using infrared fibers for aqueous solutions, *Appl. Phys. Lett.*, 49(5), 253, 1986.
61. **Simhoney, S., Kosower E. M., and Katzir, A.,** Fourier transform infrared spectra of aqueous protein mixtures using a novel attenuated total internal reflectance cell with infrared fibers, *Biochem. Biophys. Res. Commun.*, 142(3), 1059, 1987.
62. **Simhoney, S., Kosower E. M., and Katzir, A.,** Fourier transform infrared spectra of organic compounds in solution and as thin layers obtained by using an attenuated total internal reflectance fiber-optic cell, *Anal. Chem.*, 60(18), 1910, 1988.
63. **Villaruel, C. A., Dominguez, D. D., and Dandridge, A.,** Evanescent wave fiber optic chemical sensor, in *Proc. S.P.I.E. vol 798, Fiber Optic Sensors II*, Scheggi, A. M., Ed., 1987, 225.
64. **Burns, W. K., Adebghe, M., and Villaruel, C. A.,** Parabolic model for shape of fiber tapers, *Appl. Opt.*, 24(17), 2753, 1985.

65. **Sixt, P., Falco, L., Jeanneret, J. P., and Parriaux, O.,** General self-referencing single-fiber head for intensity modulated sensors, in *Proc. I.O.O.C./E.C.O.C. 1985,* 1985, 797.

66. **Sixt, P., Kotrotsios, G., Falco, L., and Parriaux, O.,** Passive fiber Fabry-Perot filter for intensity-modulated sensors referencing, *J. Lightwave Technol.,* 4(7), 926, 1986.

67. **Selenyi, P.,** Sur l'existence et l'observation des ondes lumineuses spheriques inhomogenes, *C. R.* 157, 1408, 1913.

68. **Hirschfeld, T.,** Total reflection fluorescence (TRF), *Can. Spectrosc.,* 10, 1'28, 1965.

69. **Rockhold, S. A., Quinn, R. D., VanWagenen, R. A., Andrade, J. D., and Reichert, M.,** Total internal reflection fluorescence (TIRF) as a quantitative probe of protein adsorption, *J. Electroanal. Chem.,* 150, 261, 1983.

70. **Newby, K., Reichert, W. M., Andrade, J. D., and Benner, R. E.,** Remote spectroscopic sensing of chemical adsorption using a single multimode optical fiber, *Appl. Opt.,* 23(11), 1812, 1984.

71. **Block, M. J. and Hirschfeld, T. B.,** Assay Apparatus and Method, U. S. Patent 4,558,014, 1984.

72. **Hirschfeld, T. B. and Block, M. J.,** Fluorescent Immunoassay Employing Optical Fiber in Capillary Tube, U. S. Patent 4,447,546, 1984.

73. **Block, M. J. and Hirschfeld, T. B.,** Apparatus Including Optical Fiber for Fluorescence Immonoassay, U. S. Patent 4,582,809, 1984.

74. **Newby, K. E.,** A Remote Interfacial Chemical Sensor Using a Single Multimode Optical Fiber, M.S. thesis, University of Utah, Salt Lake City, 1984.

75. **Wang, J., Christensen, D., Brynda, E., Andrade, J., Ives, J., and Lin, J.-N.,** Sensitivity analysis of evanescent fiber optic sensors, in *Proc. S.P.I.E. 1067, Optical Fibers in Medicine IV,* S.P.I.E. Bellingham, WA, 1989.

76. **Newby, K., Andrade, J. D., Benner, R. E., and Reichert, W. M.,** Remote sensing of protein adsorption using a single optical fiber, *J. Colloid Interface Sci.,* 111(1), 280, 1986.

77. **Yoshida, D. E., Ives, J. T., Reichert, W. M., Christensen, D. A., and Andrade, J. D.,** Development of fiber optic fluoroimmunoassay: proximal vs. distal end collection geometries of a fiber sensor, in *Proc. S.P.I.E. 904, Microsensors and Catheter-Based Imaging Technology,* West, A. L., Ed., 1988, 57.

78. **Glass, T. R., Lackie, S., and Hirschfeld, T. B.,** Effect of numerical aperture on signal level in cylindrical waveguide evanescent fluorosensors, *Appl. Opt.,* 26(11), 2181, 1987.

79. **Love, W. F. and Button, L. J.,** Optical characteristics of fiber optic evanescent wave sensors, in *Proc. S.P.I.E. 990, Chemical, Biochemical, and Environmental Applications of Fibers,* 1988, 175.

80. **Arwin, H.,** Optical properties of thin layers of bovine serum albumin, g-globulin, and hemoglobin, *Appl. Spectrosc.,* 40(3), 313, 1986.

81. **Stewart, G., Norris, J., Clark, D., Tribble, M., Andonovic, I., and Culshaw, B.,** Chemical sensing by evanescent field absorption: the sensitivity of optical waveguides, in *Proc. S.P.I.E. 990, Chemical, Biochemical, and Environmental Sensors,* Lieberman, R. A. and Wlodardzyk, M. T., Eds., S.P.I.E., Bellingham, WA, 1988, 188.

82. **Hardy, E. E., David, D. J., Kapany, N. S., and Unterleitner, F. C.,** Coated optical guides for spectrophotometry of chemical reactions, *Nature (London),* 257, 666, 1975.

83. **David, D. J., Willson, M. C., and Ruffin, D. S.,** Direct measurement of ammonia in ambient air, *Anal. Lett.,* 9(4), 389, 1976.

84. **Smock, P. L., Orofino, T. A., Wooten, G. W., and Spencer, W. S.,** Vapor phase determination of blood ammonia by optical waveguide technique, *Anal. Chem.,* 51(4), 505, 1979.

85. **Giuliani, J. F., Wohltjen, H., and Jarvis, N. L.,** Reversible optical waveguide sensor for ammonia vapors, *Opt. Lett.,* 8(1), 54, 1983.

86. **Giuliani, J. F., Wohltjen, H., and Jarvis, N. L.,** General theory and design considerations for optical waveguide chemical vapor sensors, in U.S. Nav. Res. Lab. Memorandum Rep. 5457, Washington, D.C., 1984.

87. **Ballantine, D. S. and Wohltjen, H.,** Optical waveguide humidity detector, *Anal. Chem.,* 58(13), 2883, 1986.

88. **Giuliani, J. F.,** An investigation of the solubility of organic vapors in polymer films using an optical waveguide interfacial probe, *J. Polymer Sci. B: Polymer Phys.,* 26, 2197, 1988.

89. **Buckles, R. G.,** Method for quantitative analysis using optical fibers, U.S. Patent 4,321,057, 1982.

90. **Russell, A. P. and Fletcher, K. S.,** Optical sensor for the determination of moisture, *Anal. Chim. Acta*, 170, 209, 1985.

91. **Blyler, L. L.,** Personal communication.

92. **Blyler, L. L., Ferrara, J. A., and MacChesney, J. B.,** A plastic-clad silica fiber chemical sensor, in *Proc. Int. Conf. Opt. Fib. Commun./Opt. Fib. Sensors ("OFC/OFS 88")*, Opt. Soc. America, 1988, 369.

93. **Blyler, L. L., Lieberman, R. A., Cohen, L. G., Ferrara, J. A., and MacChesney, J. B.,** Optical fiber chemical sensors utilizing dye-doped silicone polymer claddings, *Polymer Eng. Sci.*, 29(17), 1215, 1989.

94. **Swada, H., Takahashi, E., Tanaka, A., and Wakatsuki, N.,** Plastic optical fiber doped with organic fluorscent material, in *Proc. S.P.I.E. 989, Fiber Optic Systems for Mobile Platforms II*, Lewis, N. E. and Moore, E. L., Eds., S.P.I.E., Bellingham, WA, 1988.

95. **Andrade, J. D., Lin, J.-N., Herron, J., Reicjhert, M., and Kopecek, J.,** Fiber optic immunodetectors: sensors or dosimeters?, in *Proc. S.P.I.E. 718, Fiber Optic and Laser Sensors III*, Udd, E., and DePaula, R. P., Eds., S.P.I.E., Bellingham, WA, 1986, 280.

96. **Place, J. F., Sutherland, R. M., and Dähne, C.,** Opto-electronic immunosensors: a review of optical immunoassay at continuous surfaces, *Biosensors*, 1, 321, 1985.

97. **Dähne, C., Sutherland, R. M., and Place, J. F.,** Detection of antibody-antigen reactions at a glass-liquid interface: a novel fibre-optic sensor concept, in *Proc. S.P.I.E. v. 514, International Conf. on Opt. Fiber Sensors*, S.P.I.E., Bellingham, WA, 1984.

98. **Sutherland, R. M., Dähne, C., Place, J. F., and Ringrose, A. R.,** Immunoassays at a quartz-liquid interface: theory, instrumentation and preliminary application to the fluorescent immunoassay of human immunoglobulin G, *J. Immunol. Methods*, 74, 253, 1984.

99. **Sutherland, R. M., Dähne, C., Place, J., and Ringrose, A. S.,** Optical detection of antibody-antigen reactions at a glass-liquid interface, *Clin. Chem.*, 30(9), 1533, 1984.

100. **Andrade, J. D., VanWagenen, R. A., Gregonis, D. E., Newby, K., and Lin, J. N.,** Remote fiber optic biosensors based on evanescent-excited fluoro-immunoassay: concept and progress, *I.E.E.E. Trans. Electron Devices*, 32(7), 1175, 1985.

101. **Hirschfeld, T. F.,** Photometric sensors, in *Proc. NSF/IEEE Symp. on Biosensors*, Potvin, A. R. and Neuman, M. R., Eds., I.E.E.E., New York, 1984.

102. **Kooyman, R. P. H., deBruijn, H. E., and Greve, J.,** A fiber-optic fluorescence immunosensor, in *Proc. S.P.I.E. 798, Fiber Optic Sensors II*, Scheggi, A. M., Ed., S.P.I.E., Bellingham, WA, 1987, 270.

103. **Love, W. T. and Slovacek, R. E.,** Fiber optic evanescent sensor for fluoroimmunoassay, in *Proc. 1986 Int. Conf. on Optical Fiber Sensors (OFS '86)*, Optoelect. Ind. and Tech. Devel. Assoc., Tokyo, 1986, 143.

104. **Krull, U. J., Bloore, C., and Gumbs, G.,** Supported chemoreceptive lipid membrane transduction by fluorescence modulation: the basis of an intrinsic fibre-optic biosensor, *Analyst*, 111, 259, 1986.

105. **Krull, U. J. and Brown, R. S.,** Fiber-optic remote chemical sensing, *Chem. Anal. (Warsaw)*, 505, 1988.

106. **Krull, U. J., Brown, R. S., and Safarzadeh-Amiri, A.,** Optical transduction of chemoreceptive events: towards a fiber-optic biosensor, in *Proc. S.P.I.E. 906, Optical Fibers in Medicine III*, Katzir, A., Ed., S.P.I.E., Bellingham, WA, 1988, 49.

107. **Krull, U. J., Brown, R. S., DeBono, R. F., and Hougham, B. D.,** Towards a fluorescent chemoreceptive lipid membrane-based optode, *Talanta*, 35(2), 129, 1988.

108. **Krull, U. J.,** Chapter 21, this book.

109. **Cox, M. E. and Dunn, B.,** Detection of oxygen by fluorescence quenching, *Appl. Opt.*, 24(14), 2114, 1985.

110. **Stern, O. and Volmer, M.,** Über die Abklingungszeit der Fluoreszenz, *Phys. Z.*, 20, 183, 1919.

111. **Shah, R., Margerum, S. C., and Gold, M.,** Grafted hydrophilic polymers as optical sensor substrates, in *Proc. S.P.I.E. 906, Optical Fibers in Medicine III*, S.P.I.E., Bellingham, WA, 1988, 65.

112. **Lieberman, R. A., Blyler, L. L., and Cohen, L. G.,** Distributed fluorescence oxygen sensor, in *Proc. Int. Conf. Opt. Fib. Commun./Opt. Fib. Sensors ("OFC/OFS 88")*, Opt. Soc. America, 1988, 346.

113. **Lieberman, R. A., Blyler, L. L., and Cohen, L. G.,** A distributed fiber optic sensor based on cladding fluorescence, *J. Lightwave Technol.*, 8(2), 212, 1990.

114. **Lieberman, R. A. and Brown, K. E.,** Intrinsic fiber optic chemical sensor based on two-stage fluorescence coupling, in *Proc. S.P.I.E. 990, Chemical, Biochemical, and Environmental Fiber Sensors*, Lieberman, R. A. and Wlodarczyk, M. T., Eds., S.P.I.E., Bellinigham, WA, 1988, 104.

115. Plastifo Fluorescent Plastic Optical Fibers, product brochure, Optectron, Inc., Les Ulis, France, 1988.

116. **Silvus, H. S., Jr., Newman, F. M., Fodor, G. E., and Kawahara, F. K.,** Development of a novel hydrocarbon-in-water monitor, in *Energy and Resource Development of Continental Margins*, Pergammon, New York, 1980, 147.

117. **Silvus, H. S., Jr., Newman,a F. M., and Frazar, J. H.,** Development of oil-in-water monitor, phase II, U.S. Env. Prot. Agency Rep. 600/4-80-040, National Tech. Info. Ser. Doc. No. PB80226251, Washington, D.C., 1980.

118. **Kawahara, F. K., Fuitem, R. A., Silvus, H. S., Newman, F. M., and Frazar, J. H.,** Development ofa novel method for monitoring oils in water, *Anal. Chim. Acta*, 151, 315, 1983.

119. **Zemel, J. N., Keramati, B., and Spivak, W.,** Non-FET chemical sensors, *Sens. Acta*, 1, 427 1981.

120. **Ogawa, K., Tsuchiya, S., Kawakami, H., and Tsutsui, T.,** *Electron. Lett.*, 24(1), 42, 1988.

121. **Kala, M.,** Personal communication, 1988.

122. **Butler, M. A.,** Optical fiber hydrogen sensor, *Appl. Phys. Lett.*, 45(10), 1007, 1984.

123. **Leslie, D. H., Trusty, G. L., Dandridge, A., and Giallorenzi, T. G.,** Fibre optic spectrophone, *Electron. Lett.*, 17(7), 581, 1981.

124. **Butler, M. A. and Ginley, D. S.,** Hydrogen sensing with palladium-coated optical fibers, *J. Appl. Phys.*, 64(7), 3706, 1988.

125. **Butler, M. A. and Ginley, D. S.,** In situ measurement of strain during electrodeposition, *J. Electrochem. Soc.*, 134(2), 510, 1987.

126. **Butler, M. A. and Ginley, D. S.,** New technique for measurement of electrode strain during electrochemical reactions, *J. Electrochem. Soc.*, 145(1), 45, 1988.

127. **Farahi, F., Akhavan Leilabady, P., Jones, J. D. C., and Jackson, D. A.,** Fibre optic interferometric hydrogen sensor, in *Proc. 1986 Int. Conf. on Optical Fiber Sensors (OFS '86)*, Optoelect. Ind. and Tech. Devel. Assoc., Tokyo, 1986, 127.

128. **Farahi, F., Akhavan Leilabady, P., Jones, J. D. C., and Jackson, D. A.,** Interferometric fibre-optic hydrogen sensor, *J. Phys. E*, 20, 432, 1987.

129. **Farahi, F., Akhavan Leilabady, P., Jones, J. D. C., and Jackson, D. A.,** Optical-fibre flammable gas sensor, *J. Phys. E*, 20, 435, 1987.

130. **Corke, M., Kersey, A. D., Jackson, D. A., and Jones, J. D. C.,** All fiber "Michelson" thermometer, *Electron. Lett.*, 19, 0.471, 1983.

131. **Stone, J.,** Optical transmission loss in liquid-core hollow fibers, *IEEE J. Quantum Electron.*, 8, 386, 1972.

132. **Stone, J.,** Optical transmission in liquid-core quartz fibers, *Appl. Phys. Lett.*, 20, 239, 1972.

133. **Ippen, E. P.,** Low-power quasi-cw Raman oscillator, *Appl. Phys. Lett.*, 16, 303, 1070.

134. **Wahlrafen, G. E. and Stone, J.,** Intensification of spontaneous Raman spectra by use of liquid core optical fibers, *Appl. Spectrosc.*, 26(6), 585, 1972.

135. **Stone, J.,** Inverse Raman scattering: continuous generation in optical fibers, *J. Chem. Phys.*, 69(10), 4349, 1978.

136. **Ross, H. B. and McClain, W. M.,** Liquid core optical fibers in Raman spectroscopy, *Appl. Spectrosc.*, 35(4), 439, 1981.

137. **Schaefer, J. C. and Chabay, I.,** Generation of enhanced coherent anti-Stokes Raman spectroscopy signals in liquid-filled waveguides, *Opt. Lett.*, 4(8), 1979.

138. **Wei, L., Fujiwara, K., and Fuwa, K.,** Determination of phosphorous in natural waters by long-capillary-cell absorption spectrometry, *Anal. Chem.*, 55(6), 951, 1983.

139. **Dasgupta, P. K.,** Multipath cells for extending dynamic range of optical absorbance measurements, *Anal. Chem.*, 56(8), 1401, 1984.

140. **Fuwa, K., Wei, L., and Fujiwara, K.,** Colorimetry with a total-reflection long capillary cell, *Anal. Chem.*, 56(9), 1640, 1084.

141. **Fujiwara, K. and Fuwa, K.,** Liquid core optical fiber total reflection cell as a colorimetric detector for flow injection analysis, *Anal. Chem.*, 57(6), 1012, 1985.

142. **Fujiwara, K., Simeonsson, J. B., Smith, B. W., and Winefordner, J. D.,** Waveguide capillary flow cell for fluorimetry, *Anal. Chem.*, 60(1), 1065.

143. **Jurgensen, A., Inman, E. L., Jr., and Winefordner, J. D.,** Comprehensive analysis of figures of merit for fluorimetry of polynuclear aromatic hydrocarbons, *Anal. Chim. Acta*, 131, 187, 1981.

144. **Shahriari, M. R., Sigel, G. H., and Zhou, Q.,** Porous fiber optic for a high sensitivity humidity sensor, in *Proc. Int. Conf. Opt. Fib. Commun./Opt. Fib. Sensors ("OFC/OFS 88")*, Opt. Soc. America, 1988, 373.

145. **Zhou, Q., Shahriari, M. R., Dritz, D., and Sigel, G. H.,** Porous fiber-optic sensor for high-sensitivity humidity measurements, *Anal. Chem.*, 60(20), 2317, 1988.
146. **Zhou, Q., Shahriari, M. R., and Sigel, G. H.,** The effects of temperature on the response of a porous fiber optic humidity sensor, in *S.P.I.E. Proc. 990, Chemical, Biochemical, and Environmental Fiber Sensors,* Lieberman, R. A. and Wlodarczyk, M. T., Eds., S.P.I.E., Bellingham, WA, 1988, 153.
147. **Shahriari, M. R.,** Personal communication, 1987.
148. **Shahriari, M. R., Zhou, Q., and Sigel, G. H.,** Porous optical fibers for high-sensitivity ammonia-vapor sensors, *Opt. Lett.*, 13((5), 407, 1988.
149. **Macedo, P. B., Barkatt, Aa., Feng, X., Finger S. M., Hojaji, H., Laberge, N., Mohr, R., Penafiel, M., and Saad, E.,** Development of porous glass fiber optic sensors in *Proc. S.P.I.E. 986, Fiber Optic Structures and Smart Skins*, Udd, E., Ed., S.P.I.E., Bellingham, WA, 1989.

Chapter 6

# INSTRUMENTATION FOR FIBER OPTIC CHEMICAL SENSORS

**Douglas N. Modlin and Fred P. Milanovich**

## TABLE OF CONTENTS

# I. INTRODUCTION

Recent developments in the sensing chemistries utilized in fiber optic chemical sensors (FOCS) have fueled the need for advances in the accuracy and reliability of the instrumentation to which they are attached. In addition, many applications require the instrumentation to be increasingly portable and impose other constraints as well. For example, sensors and instrumentation designed for use in monitoring water quality in underground wells must be ruggedized and easy to transport to remote locations, whereas systems for use in biomedical applications must be sterile and highly compact.

This chapter presents an introduction to the principles of instrument design for FOCS and introduces the reader to the basic set of tools necessary for successfully dealing with the requirements and constraints of practical applications. Once acquainted with the fundamentals, the reader should be able to approach the design of optimized instrumentation for FOCS in a systematic fashion.

Section II begins with a discussion of basic concepts in remote fiber sensing. These include intensity modulation, time decay, and interferometric sensing techniques as well as an introduction to the operating principles of optical fibers and descriptions of commonly used sensor geometries. In addition, a generic instrumentation system is discussed at the block diagram level.

In Section III, analytical models are presented for the basic building blocks of FOCS instrumentation. These include light sources, optical systems, photodetectors, and the sensor transfer function. Section III concludes with a worked-out example for a prototype fiber optic pH sensor in which the models for each sub-system are worked out and combined to produce a complete analytical expression for its performance. The analysis draws upon a mathematical derivation of the transfer function for the prototype pH sensor which is presented in the Appendix.

Since instruments must be constructed from real components, an overview of the applicable light sources, lenses, optical filters, optical splitters, and detectors is presented in Section IV. Armed with the introductory background and modeling techniques presented in Sections II and III and an awareness of the available components, the reader should be equipped with enough basic tools and information to embark upon the analysis and successful development of FOCS instrumentation.

# II. BASIC CONCEPTS OF FIBER OPTIC INSTRUMENTATION AND SENSORS

In the interest of simplifying the discussion that follows, it is useful to define a terminology to distinguish between what is being sensed and what is doing the sensing. The substance which the sensor or probe is designed to sense will be called the "analyte". The term "transducer" will be used to refer to the geometrical arrangement of materials and/or chemical compounds which make up the sensing element.

## A. SENSOR TYPE CLASSIFICATION

Fiber optic sensors can be used in either monitoring or analytic applications, each of which can have entirely different instrumentation requirements. If the purpose of the measurement is to determine the chemical identity or some other property of the analyte, then the application is considered to be analytic. Examples of analytic applications include remote wavelength-scanned fluorescence, absorbance spectroscopy for species identification, and excited-state lifetime measurements.[1-7]

The sensor is considered to be a monitor if the chemical identity of the analyte (or other property of interest) is known. In a monitoring application, it is desired to know the amount of a given substance present above or below a predetermined threshold value. Examples of monitoring applications include physiological measurements of blood oxygen, carbon dioxide, and pH and the monitoring of toxic substance storage areas for leakage of toxic substances.[8-14]

Because of the many possible physical implementations and accompanying instrumentational requirements, it is difficult to simply classify the various types of sensors. There are, however, at least four basic ways in which the transduction system can modulate the signal carrying light in the optical fiber. These four schemes will be referred to as intensity, phase, time decay, and modal modulation.

In intensity modulated sensors, the amplitude (or intensity) of the signal detected by the instrument is designed to vary in a known way with the measured property of the analyte, e.g., concentration, refractive index, pH, etc. Intensity modulated sensors are quite attractive because of the large base of existing indicators. For example, there are literally hundreds of colorimetric dyes and fluorescent indicators that have been used for many years in scientific, industrial, and medical applications.[15,16] Although it is quite simple in principle to use these indicators in fiber optic sensors, intensity modulated sensors suffer from an intrinsic sensitivity to interference from any factor that can affect the intensity of the signal carrying light. For example, variations in the intensity of the light source, connector losses, photobleaching of the indicator dye, and fiber bending losses can produce spurious responses in intensity modulated sensors. It is often necessary to implement self-referencing schemes to circumvent these problems. Self-referencing schemes are discussed further in Chapter 3 of this volume.

Phase modulated (e.g., interferometric) and time decay sensors are intrinsically insensitive to factors which affect the intensity of the signal carrying light. In this class of sensors, the instrumentation is designed to sense only changes in the phase or decay time of the signal returning from the sensor and, in most cases, these sensors can be designed to be self-referencing. Numerous examples of fiber optic sensors based on phase modulation have been described in the literature. However, most of these are for sensing nonchemical quantities such as force, pressure, and temperature, or to function as a gyroscope.[17-19]

The final (and least developed sensor modality) is modal modulation.[20] In this technique, the transducer is designed to modulate the mode volumes in the fiber or, in other words, the numbers of light rays traveling at different angles. For example, such a transduction system would modulate the amplitude of light traveling at steep angles within the fiber by a different amount than it would for light traveling at shallow angles. The instrument must include means to separate and detect light traveling in the different modes. Modes are discussed further in Section II.B.1.b and in Chapter 4 of this volume.

It is not difficult to envision a vast array of potential applications and implementation schemes for FOCS and the accompanying instrumentation. Given the need for self-referencing schemes and multiple analyte sensing capability, it may even be desirable to develop hybrid sensors which utilize multiple transduction systems. Since further discussion of the various types of fiber optic sensors is beyond the scope of this chapter, readers interested in more infromation are referred to References 21 to 28 and to the other chapters in this volume.

## B. BASIC OPERATING PRINCIPLES

As indicated above, fiber optic sensors, much like their electrical counterparts, can be operated in a great variety of physical configurations. In general, one or more fibers are involved in the transmission of light to and from the instrument and sensor, and it is the involvement of light in the measurement that distinguishes fiber optic from electrical sensors.

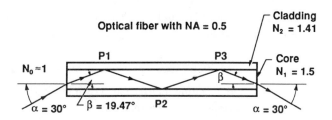

FIGURE 1. Illustration of the ray path through an optical fiber with a numerical aperture of 0.5.

In most cases, the light emanating from the instrument (the optical excitation) stimulates some physical system whose response is somehow modulated by the presence of the analyte. It is possible, however, that the optical excitation is not at all involved in the actual measurement. This would be the case if the purpose of the optical fiber was to provide power to a remote electrical circuit and a pathway for the information-carrying return signal.[29] In any case, it is not possible to begin a discussion of FOCS instrumentation without providing an introduction to the basic operating principles of the optical fiber itself.

The following section is designed to provide the reader with an intuitive feel for the operating principles of optical fiber which pertain to FOCS instrumentation. The use of mathematics is kept to a minimum so as not to obscure the basic operating principles with complex formalism. The mathematics presented here is designed to assist the reader in building a practical knowledge base which is compatible with the kind of information found in the handbooks and data sheets of manufacturers of fiber optic products.[30,31]

## 1. Optical Fiber Basics

Many opportunities exist for exploiting the multi-faceted functionality of optical fibers (or alternatively optical waveguides) in FOCS instrumentation as well as in the sensors themselves. In order to appreciate the variety of possibilities, it is necessary to understand the propagation of light in optical fibers from both the "ray" and "mode" perspectives. While the notion of light rays traveling inside an optical fiber provides the most simple and intuitive perspective, it provides little insight into phenomena such as the evanescent field, modal noise, and light propagation in single mode fibers. For this reason, a discussion of the optical fiber from both the "ray" and "mode" perspectives will be presented in this section with the aim of transferring to the reader an intuitive feel for the way the optical fiber actually works. With this understanding, the reader will gain an appreciation for the great utility of the optical fiber in sensor and instrument applications. The reader is directed to References 32 to 36 for more detailed treatments of light propagation in optical fibers.

### a. Ray Theory of Light Propagation in Optical Fibers

The trajectory of a "ray" of light within an optical fiber indicates the direction of travel of a planar electromagnetic wave. The ray approximation provides an accurate description of light propagation in the multimode fiber used in FOCS since the fiber diameter (usually 100 μm or greater) is hundreds of times larger than the wavelength of the light propagating within it.

Optical fibers are generally constructed with a thin cylindrical cladding material surrounding a central core of light transmitting material such as glass or plastic. The operation of an optical fiber is illustrated in Figure 1. Light traveling outside the fiber in a material with an index of refraction $N_0$ is incident on the core of the optical fiber (with index $N_1$) at an angle $\alpha$ from the normal to the surface. For an optical fiber in air, $N_0 < N_1$ and the light is bent (or refracted) toward the normal inside the fiber and travels at an angle $\beta$. This relationship is normally expressed as Snell's law where

$$N_0 \sin \alpha = N_1 \sin \beta \qquad (1)$$

Snell's law also applies at the core/cladding interface as shown in Figure 1. If the index of refraction of the cladding material is made to be less than that of the core ($N_1 > N_2$), then light rays traveling in the core at angles less than the critical angle, ($B \leq \theta_c$), will be totally internally reflected and will continue to propagate down the optical fiber. Given that $\beta$ and $\theta_c$ are measured relative to the fiber axis, the critical angle for rays traveling in the fiber is given by

$$\theta_c = \cos^{-1} \left[ \frac{N_2}{N_1} \right] \qquad (2)$$

Light rays which exceed the critical angle are partially transmitted into the cladding each time the light refracts at the core/cladding interface. If the cladding is surrounded by a material of higher index, then the energy coupled into the cladding will be radiated. If the cladding is surrounded by material of a lower index, then the energy will be bound and propagate in the cladding when the ray angle is less than the critical angle determined by the materials at the outer cladding interface. It is important to be aware of this phenomenon when choosing materials such as fiber buffers and other coatings for FOCS since light propagation in the cladding can appreciably affect sensor performance in certain cases.

A commonly used measure of the throughput of an optical fiber is the numerical aperture (NA). The numerical aperture can be defined in terms of the half-angle of the cone of acceptance, $\alpha_c$, or alternatively in terms of the index of refraction of the materials used in the construction of the optical fiber as

$$NA = N_0 \sin \alpha_c = \sqrt{N_1^2 - N_2^2} \qquad (3)$$

When operated in air, where $N_0 \approx 1$, light rays launched at angles greater than $\alpha_c$ will be partially refracted into the cladding and therefore will propagate with a gradual decrease in amplitude over several meters depending on the incident angle and the fiber NA. For this reason, light is efficiently coupled into optical fibers only at angles equal to or less than $\alpha_c$.

### b. Mode Theory of Light Propagation in Optical Fibers

As has been previously pointed out, the ray perspective alone does not provide a complete picture of light propagation in optical fibers. For example, it is not possible to completely explain the operation of sensors based on modal modulation or evanescent field coupling with ray theory alone. To analyze these cases, the optical fiber is traditionally treated as an electromagnetic waveguide and Maxwell's equations are solved given the appropriate boundary conditions for the electric and magnetic fields at the core/cladding interface. The solutions to Maxwell's equations show that there are specific "modes" of propagation (and associated ray angles) which are determined by the physical dimensions and material properties of the optical fiber core, cladding, and surroundings. Since the purpose of this discussion is to provide the reader with an intuitive feel for those aspects of modal theory which are relevant to FOCS instrumentation, the mathematical details of solutions to Maxwell's equations are not discussed further, and readers interested in more information are referred to Chapter 4 of this volume and References 32 to 36.

Waveguide modes represent the radial (or transverse) electric and magnetic field intensity distributions of standing electromagnetic waves within the fiber cross section. For example, Figure 1 illustrates the ray path associated with a planar wavefront traveling down an optical fiber from point P1 to P3. Although only one ray path is shown, it is assumed that a family of rays with the same angle are traveling simultaneously within the optical fiber. Constructive and destructive interference of the electromagnetic fields associated with each family of rays results in a standing wave (or modal) pattern in the transverse direction to the axis of the fiber.

The standing wave patterns associated with the different modes result in highly distinct and sometimes nonuniform distributions of light within the cross section of an optical fiber. Individual modes are generally identified by a reference to the dominant transverse component(s) of the electric and/or magnetic fields and a subscripted index referring to the order of a particular mode. For example, the lowest order mode in a step index multimode fiber (which is made up of transverse electric and magnetic field components) is called the $HE_{11}$ mode. This mode has a relatively uniform cross sectional intensity profile, whereas the higher order modes, i.e., those associated with rays traveling at steeper angles, generally result in more complex modal patterns.

The solutions to Maxwell's equations result in unique values for the wave propagation constant and hence the ray angle associated with each mode. This is because, for a wave to be guided, the transverse fields must drop off to zero in the vicinity of the core/cladding interface, and this can only happen at certain ray angles. More specifically, when the transverse field components contain an integral number of wavelengths, then the field intensity must approach zero at the core/cladding interfaces on either side of the fiber. For this reason, light rays traveling at shallow angles with respect fiber axis are associated with low order modes, whereas light rays traveling at steep angles are associated with higher order modes.

It is important to point out that the connection implied above between rays and modes is not be taken literally since, by the principle of superposition, a ray or wave traveling in any specified direction can be represented as the sum of waves with the appropriate amplitudes and phases traveling in all directions. This is somewhat analogous to the Fourier series. Therefore, although the preferred ray direction is that associated with the wave propagation constant for each mode, a given mode can be thought of as containing rays traveling in other directions within the NA of the fiber as well. While this apparent ambiguity is aesthetically unappealing, problems of this sort often arise when attempting to express two different theories in similar terms.

In multimode optical fibers which are commonly used in FOCS, there are literally thousands of allowed modes in which light can propagate. The total number of bound modes, $M_t$, in an optical fiber is given by

$$M_t = \frac{V^2}{2} \tag{4}$$

where V is called the waveguide parameter and is defined as

$$V = \frac{2\pi\rho NA}{\lambda} \tag{5}$$

Here, $\rho$ is the radius of the fiber core and $\lambda$ is the wavelength. For a step index fiber with $\rho = 100$ $\mu$m, NA = 0.5, and $\lambda = 0.5$ $\mu$m, there are approximately $2 \times 10^5$ bound modes.

When an optical fiber is placed against a light source emitting over a continuum of angles, those rays which are within the numerical aperture of the optical fiber are coupled into the fiber and excite the bound modes of the fiber. Conservation of energy requires that, at the fiber endface, the summation of the transverse components of the electric and magnetic field components of the source minus the energy lost due to surface reflection equals that of the bound fiber modes.

Since any given energy distribution impinging on the fiber endface can be decomposed into the spatial and angular plane wave components (rays) required to excite a full complement of modes, the optical fiber functions as a "light pipe" whose output is equal to the total energy transmitted in all of the excited modes. For most practical cases, where the fiber lengths are relatively short, the optical power distribution at the output is strongly dependent on the power distribution coupled to the fiber from the source. For long lengths of fiber, however, perturbations of the light propagating in the fiber caused by scattering, bends, connectors, etc., result in intermodal coupling and gradually alter the power distribution in the core.

### c. Origin of the Evanescent Field in Optical Fibers

The evanescent field in the cladding of an optical fiber arises because Maxwell's equations require that the tangential and normal components of the electric and magnetic field and flux densities, respectively, be continuous across the core/cladding interface. It is not possible to have finite electric and magnetic fields within the core and none whatsoever in the cladding. The magnitude of the evanescent field at the interface of a planar waveguide is given by

$$E_e = E_{0 \exp} - \left( \frac{x}{L} \right) \tag{6}$$

where $E_o$ is the field at the core/cladding interface and x is the distance into the cladding. The penetration depth, L, of the evanescent field is given by

$$L = \frac{\lambda}{2\pi N_1 \left( \sin^2 \theta_c - \sin^2 \beta \right)^{1/2}} \tag{7}$$

where $\beta$ is the angle of the ray within the waveguide and $\theta_c$ is the critical angle as defined in Equation 2. It is clear that the penetration depth of the evanescent field becomes very large as the ray angle approaches the critical angle.

### d. Bending Losses in Optical Fibers

The dependence of evanescent field penetration depth on the ray angle gives some insight into the losses associated with fiber bending. As an optical fiber is bent, the evanescent field penetration depth increases due to the increased ray angle caused by the bend. When the evanescent field reaches the edge of the cladding, energy will begin to radiate from the fiber. As the evanescent field penetration depth becomes infinite, the situation is equivalent to that where a ray of light impinges on the core/cladding interface at an angle exceeding the critical angle. In this case, a fraction of the energy propagating in the fiber will be transmitted into the cladding and radiated.

The relationship between the ray angle and the total evanescent field penetration depth for waveguides with numerical apertures of 0.512 and 0.122 is graphically illustrated in Figure 2. The total penetration depth is defined as five times the penetration depth given by Equation 7. At this point, the evanescent field has decayed to 0.67% of its value at the core/cladding interface and can be neglected. The value of $\beta$ used to plot Equation 7 is calculated by subtracting the percentage underfill from the critical angle for a given numerical aperture (found using Equation 2). From Figure 2, it can be inferred that optical fibers with high numerical aperture are intrinsically less sensitive to bending than are those of low numerical aperture since for a given percentage underfill, the evanescent field penetration is much lower.

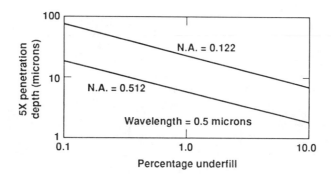

FIGURE 2. Plot of effective evanescent field penetration depth as a function of the numerical aperture underfill. The effective penetration depth is defined as five times the 1/e penetration depth. The underfill is defined as the percentage difference between the angle of a ray traveling within the optical fiber and the critical angle for a given numerical aperture. The core index is assumed to be 1.5.

Unfortunately, bending induced power attenuation is affected by many factors and quantitative prediction requires complex mathematical analysis. In general, the power attenuation due to a constant radius bend through an angle $\phi$ can be expressed by the integral over all ray angles and over the core cross section of

$$P(\phi) = P(0)\exp(-\gamma\phi) \tag{8}$$

where $P(0)$ is the power in the fiber preceeding the bend, $P(\phi)$ is the power remaining at an angle $\phi$, and $\gamma$ is the attenuation coefficient.[32] Equation 8 must be integrated because the attenuation coefficient is a complex function of the ray angle and the cross-sectional distribution. The attenuation coefficient can be expressed in the following functional form:

$$\gamma = f(\beta, NA, \lambda, Rb, \rho) \tag{9}$$

Here, $R_b$ is the radius of the bend and $\rho$ is the radius of the optical fiber and the other parameters are as previously defined.

In practice, it is difficult to accurately predict the effects of bending, and an empirical determination is usually necessary. Since it is normally desired to minimize the effects of bending, the numerical aperture of the source and/or the detector must be reduced below that of the fiber to the point that a bend of the minimum expected radius induces an acceptable output signal level change.

### e. Modal Noise in Optical Fibers

An additional phenomenon which can affect the performance of fiber optic sensors is modal noise. Modal noise is caused by perturbations that cause the light traveling in an optical fiber to change the mode in which it is propagating and can be exacerbated by the interference of the light traveling in different modes. Mode changing is caused by scattering from structural defects and fluctuations within the fiber itself, misaligned connectors, and time varying physical perturbation of the fiber, i.e., bending, vibration, thermal expansion, etc. The result is a fluctuating power density over the fiber cross section at the output of the fiber which can induce noise into the signal collected at a photodetector. Modal noise is most problematic when using laser light sources because of their high degree of coherence but can occur with incoherent sources as well.

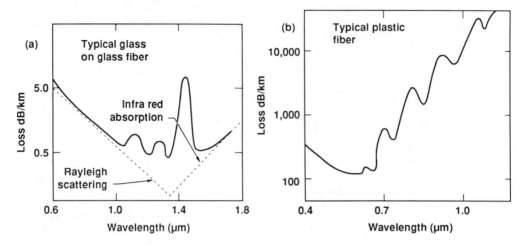

FIGURE 3. Typical spectral attenuation curves for (a) glass on glass and (b) plastic optical fibers.

In many cases, the modal composition of the light traveling in the optical fiber and modal noise is of little consequence in FOCS applications since most photodetectors are designed to be spatially underfilled and are therefore sensitive only to the total number of photons per unit time and not their spatial distribution. On the other hand, underfilling the detector does not guarantee freedom from the effects of modal noise. For example, modal noise can still be a significant problem in systems with underfilled detectors when it is caused by lossy optical fiber connectors upstream from the detector.

### f. Sources of Attenuation and Luminescence in Optical Fibers

The various sources of light loss in both silica- (or $SiO_2$) and plastic-based optical fibers combine to give rise to the typical transmission spectra shown in Figures 3a and 3b, respectively. For short wavelengths, the loss curves are dominated by ultraviolet (UV) absorption and Rayleigh scattering whereas the effects of infrared (IR) absorption become dominant at long wavelengths. As indicated in the figure, absorption peaks are present in the spectra of both glass and plastic optical fiber. The absorption peaks are caused by factors which are both intrinsic to the set of materials and manufacturing process used to fabricate each type of optical fiber and by the presence of unwanted impurities. The reader is referred to References 27 and 34 to 37 for more information on optical fiber fabrication technology and transmission characteristics.

As is evident from Figure 3, plastic optical fiber is much more lossy in the visible portion of the spectrum than optical fiber fabricated from silica. The attenuation at 0.65 μm is <10 dB/km for high quality silica fibers and 100 to 200 dB/km for plastic fiber. Attenuation is generally measured in decibels which is defined as 10 log $[P_{out}/P_{in}]$.

The total attenuation in either plastic or silica based optical fiber is generally attributed to a combination of intrinsic and extrinsic sources. Intrinsic absorption is associated with the compounds used in the fabrication of the optical fiber itself whereas extrinsic absorption is associated with the presence of impurities. For example, intrinsic absorption from compounds such as PMMA (polymethyl-methacrylate) or polystyrene used in the fabrication of plastic optical fibers results in much higher attenuation than that associated with high purity $SiO_2$. In addition extrinsic absorption, caused by the presence of unwanted impurities, is lower in silica-based optical fibers since silica can be manufactured with extremely high purity.

Intrinsic losses are caused by Rayleigh or Mie scattering as well as by absorption. Rayleigh scattering is caused by micro-fluctuations (small compared with the light wavelength) in the index of refraction of the fiber. The attenuation coefficient for Rayleigh scattering increases as the inverse of the forth power of the wavelength. For example, at 0.4 μm, the loss due to Rayleigh scattering in high quality glass fiber is about 23 dB/km whereas it is as much as an order of magnitude higher in plastic fiber. In many FOCS applications, the fiber lengths are less than 10 m and Rayleigh scattering can be neglected if glass fiber is used. Mie scattering occurs when the size of the structural fluctuations approach the wavelength of the light in the fiber. These fluctuations are generally caused by imperfections induced during the manufacturing process. Losses of this type can be very significant in plastic fibers but have been reduced to virtually negligible levels in high quality silica based fibers.

Intrinsic absorption also depends on the composition and manufacture of the materials used to construct an optical fiber. Generally speaking, as the wavelength decreases into the UV the loss increases as the absorption edge of a given material is approached. In UV-grade silica fibers, transmission losses of less than 1 dB/m at 0.2 μm are commercially available, whereas the losses are much higher for plastic fiber which becomes highly absorbing below about 0.4 μm.

As the wavelength approaches the IR, interaction of light with the vibrational modes of the fiber materials gives rise to absorption peaks and an effective cutoff wavelength for each material. Although silica fibers are a good choice in the 0.2 to 2 μm range, operation in the 4 to 10 μm range requires the use of fibers composed of more exotic materials such as fluoride, chalcogenide, $ZnCl_2$, KCl, and KRS-5 glasses. In plastic fibers, absorption peaks caused by C-H stretching vibrations and their harmonics are present throughout the visible spectrum and their presence in increasing numbers in the IR region of the spectrum leads to the steep increase in attenuation shown in Figure 3b.

Extrinsic absorption refers to absorption caused by substances which are added, either intentionally or unintentionally, to the fiber material during or subsequent to its manufacture. In silica core fibers, for example, metals such as chromium, copper, and vanadium give rise to absorption on the order of 1 to 3 dB/km when present at one ppb. When present at similar levels, hydroxl groups (OH) give rise to a series of absorption peaks as high as 1000 to 2000 dB/km (for the first overtone) which occur between 4.2 and 0.72 μm . High levels of (OH) are sometimes desirable because of the beneficial effects on radiation resistance and UV transmission. Fiber drawn from silica with low (OH) content, on the other hand, has a strong absorption peak at 630 nm but lower attenuation in the 0.85 to 1.0 μm wavelength range.

Extrinsic absorption processes generally result in photon-induced elevation of electrons from the ground state to excited states. Once excited, the atoms or molecules lose their energy by either radiative or nonradiative processes. Radiative processes can be particularly troublesome to FOCS since the radiated light is usually shifted slightly toward longer wavelengths and can interfere with the performance of sensors that are based on wavelength shifted emissions such as in fluorescence or phosphorescence. Care must also be taken in the selection and processing of optical fiber materials. For example, the evanescent field in the cladding is capable of exciting radiative molecules which can, in turn, couple wavelength shifted light into the fiber core. This can be particularly troublesome when using optical fiber with a polymer cladding since many polymers will fluoresce at short enough wavelengths. In single fiber sensors the return fluorescence signal is a small fraction of the incoming excitation and can be overpowered by even relatively low level spectrally overlapping fiber luminescence.

Other sources of potentially interfering emissions in optical fibers are those associated with Raman and Brillouin scattering. Of the two processes, Raman scattering is the most significant and should be considered in applications where low level signals are involved. Both Raman and Brillouin scattering produce scattered photons with wavelength shifts both above and below the excitation wavelength.[38,39]

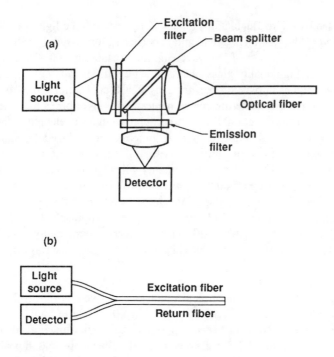

FIGURE 4. Illustration of (a) single and (b) multiple optical fiber sensor configurations.

## 2. Fiber Optic Sensor Configurations

There are numerous ways in which optical fibers can be configured to implement FOCS. Figure 4 shows examples of single fiber and dual fiber implementations. In the single fiber approach (see Figure 4a), the excitation and return light travel in the same fiber. This configuration offers the most compact and potentially least expensive implementation. These sensors work by direct illumination of a fluorophore through the endface or walls of the fiber or by excitation of a coating on the fiber surface through evanescent wave coupling. The disadvantages of the single fiber implementation include high sensitivity to interferences caused by internal fiber luminescence, limited utility for absorbance measurements, and the requirement of an optical splitter.

Figure 4b illustrates the dual fiber configuration. Here, separate fibers are used for the source and detector. Since the excitation and return light travel in separate fibers, the sensitivity to internal fiber luminescence is low and two fibers can readily be configured to measure absorbance or scattering. This configuration lends itself well to the use of inexpensive light sources such as LEDs and solid state photodiodes. Additionally, the throughput can be increased by replacing the individual fibers used for excitation and emission with fiber bundles. The disadvantages of this configuration include lower signal return efficiency, increased size, complexity, and potentially higher cost.

## 3. Basic Instrumentation Requirements for FOCS

The actual implementation details for each specific case will vary greatly depending on the type of light source, method of detection, referencing scheme, and sensor geometry. In Section III, physical and/or engineering models of the various building blocks will be presented in order to provide the basis for analysis and, ultimately, performance simulation of many possible FOCS implementations. Section IV provides an introduction to the various components used to form the basic instrumentation building blocks with the goal of introducing the

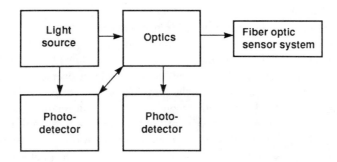

FIGURE 5. Block diagram of a generic instrument for a fiber optic sensor.

reader to the great variety of available options.

A block diagram for a generic instrument for FOCS is shown in Figure 5. The basic building blocks of all FOCS instrumentation are the light source, optics, and photodetector. Radiation from the light source is filtered and otherwise conditioned by the optical system and then launched into an optical fiber to provide excitation for the transducer system. Light returning from the sensor is steered back to the instrument and through the optical system where it is conditioned and directed to a photodetector.

Two photodetectors are shown in Figure 5 because it is often necessary to include a reference detector to track fluctuations in source intensity and/or color temperature (see Section III.A.2) as well as other parameters relating to the sensor such as fluorophore concentration. The small amount of energy required for the reference photodetector can be tapped directly from the light source or from some point downstream in the optical system.

In most real-world applications, some form of referencing scheme is needed to insure accurate and/or stable operation. This is particularly important in situations where the selectivity of a fiber optic sensor is an important factor. A sensor with perfect selectivity will sense only the analyte and be completely insensitive to all other forms of interference. Examples of such interferences are: (1) sensitivity to chemical species other than the analyte; (2) sensitivity to physical parameters such as temperature, fluorophore concentration variation (e.g., due to photo-bleaching), pressure, vibration, etc.; (3) sensitivity to drift in the light source intensity and color temperature; (4) fiber bending losses; (5) connector losses; and (6) periodically repeating transfer functions in interferometric sensors. The reader is referred to Chapter 3 of this volume for more information on referencing schemes.

## III. INSTRUMENTATION AND SENSOR PERFORMANCE MODELING

In the last section, the basic concepts of optical fibers were discussed in detail and the overall architecture of FOCS instrumentation was described at the block diagram level. In this section, the basic building blocks of the instrument will be described in more detail and models will be presented which allow each block to be simulated. By combining the results of the simulations for each block with a knowledge of the sensor transfer function, the complete sensor-instrument system can be simulated and its performance predicted.

In this section, models are presented for light sources, photodetectors, the optical system, and the sensor transfer function. As an example, these models are used to derive the complete system-level model for a prototype pH sensor. The performance of the prototype pH sensor is simulated for a specific set of physical characteristics, i.e., fiber size and numerical aperture, dye concentration, etc.

## A. LIGHT SOURCES

The optical excitation sources employed in instrumentation for FOCS are either continuous, pulsed, or some combination thereof. Continuous optical excitation can be intensity or frequency modulated with a carrier frequency between dc and many hundreds of megahertz, depending on what is actually being sensed and the mode of operation of the sensor. The optical excitation can also be delivered in the form of a narrow pulse or pulse train with pulse widths ranging from seconds to femto seconds ($10^{-12}$ sec) depending on the source. The characteristics of the most important types of light sources are presented in Section IV.A.

Regardless of the detection scheme, it is important to know the intensity of light coupled into a fiber from a given light source since the response of the photodetector is directly proportional to it. The first part of this section introduces and provides an analysis of techniques for coupling light into optical fibers from surface emitting sources such as incandescent lamps and LEDs as well as from collimated sources such as lasers.

As an aid to the comparison of various types of light sources, the concept of the blackbody radiator as a generic light source model for FOCS instrumentation is introduced. Once the equivalent blackbody temperature of a light source is known, the radiance and the power coupled into the sensor fiber can be calculated.

### 1. Coupling Light To and From Optical Fibers

In order to couple light into an optical fiber, the power emanating from a light source must ultimately impinge on the endface of an optical fiber either by direct contact with the source, projection of an image of the source, or focusing a precollimated beam onto the fiber endface.

Light sources such as an LED or the surface of an incandescent lamp filament emit light in all directions from the surface and are said to approximate a Lambertian source. The radiance in watts per square meter per steradian (see Section IV.A.2) as a function of the angle of observation from the normal to the surface, is given by

$$L(\theta) = L_0 \cos \theta \tag{10}$$

where $L_o$ is the source radiance at $\theta = 0$. The assumption that the source is Lambertian implies that $L_o$ is independent of $\theta$.

The power coupled into an optical fiber can be found by integrating Equation 10 over the allowable solid angle of entrance and area of the fiber core. The solid angle, $\Omega$, is defined as

$$\Omega = \frac{A}{d^2} \tag{11}$$

where A is the area on the surface of a sphere of radius d circumscribed by a cone of half-angle $\theta$. A differential element of solid angle, $d\Omega$, is shown in Figure 6a and is derived from an annular ring on the surface of the sphere and is expressed as

$$d\Omega = \frac{dA}{d^2} = \frac{2\pi d^2 \sin \theta d\theta}{d^2} = 2\pi \sin \theta d\theta \tag{12}$$

The differential element of area on the fiber endface shown in Figure 6b is simply

$$dA = rdrd\phi \tag{13}$$

The differential power, dP, coupled into an optical fiber is found by taking the product of Equations 10, 12, and 13 to be given by

$$dP = L_0 \cos \theta dA d\Omega \tag{14}$$

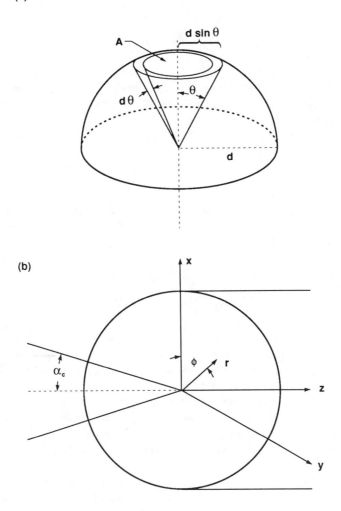

FIGURE 6. Illustration of the coordinate systems used to calculate the differential elements of (a) solid angle and (b) area at the endface of an optical fiber.

The total power is then obtained by integrating Equation 14 over the appropriate limits. To select out only the energy coupled into bound modes, the integration is carried out for angles less than the maximum angle of acceptance of the optical fiber or to the limit of $\theta = \alpha_c$. Solving Equation 3 for $\alpha_c$ yields the relation

$$\alpha_c = \arcsin\left[\frac{NA}{N_0}\right] \tag{15}$$

where NA is the numerical aperture of the optical fiber and $N_0$ is the index or refraction of the medium surrounding the optical fiber. For an optical fiber of radius $\rho$, Equation 14 is evaluated as follows:

$$P_{bound} = \int_0^\rho \int_0^{2\pi} \int_0^{\alpha_c} L_0 \cos\theta 2\pi \sin\theta d\theta d\phi r dr \tag{16}$$

Substituting for from Equation 15 and integrating, one obtains

$$P_{bound} = \pi L_0 A_f \frac{NA^2}{N_0^2} \qquad (17)$$

where the area of the fiber is $A_f = \pi\rho^2$.

When light impinges on the endface of the fiber, a small amount is reflected if the index of refraction of the medium surrounding the fiber differs from that of the core. This phenomenon is commonly referred to as Fresnel reflection and its magnitude at normal incidence is given by the relation

$$R = \left[ \frac{N_1 - N_o}{N_1 + N_o} \right]^2 \qquad (18)$$

where R is the Fresnel reflection coefficient, $N_1$ is the core index, and $N_o$ is the index of the medium surrounding the fiber.[40,41] For a typical optical fiber with a core index of 1.5 surrounded by air, is approximately 4%, whereas for a similar fiber immersed in water ($N_o = 1.33$), R is approximately 0.4%.

If the source output is relatively uniform over the operational wavelength band, then the total bound optical power coupled into a fiber, $P_f$, can be calculated using the relation

$$P_f \approx \frac{\pi R L_\lambda A_f (NA)^2 \Delta\lambda}{N_0^2} \qquad (19)$$

where $L_1$ is the radiance per unit wavelength and $\Delta\lambda$ is the wavelength band in microns.

Equation 19, has several important implications. First, the power coupled into an optical fiber is proportional to the square of the fiber NA, assuming that the NA of the source ($\sin \alpha$) is equal to or greater than that of the fiber. Second, if the NA of the source is less than that of the optical fiber, then the power coupled into the fiber is proportional to the square of the source NA. Lastly, the power is directly proportional to the optical bandwidth of the source, which can range from less than 0.001 μm for lasers to as much as 0.05 μm or more for filtered wideband sources.

Light coupled into the cladding will also be propagated if the index of refraction of the material surrounding the cladding is less than that of the cladding. Light coupled into the cladding can be an important factor in instrument design and should not be ignored.

In most cases, it is not practical to place the optical fiber directly in contact with the source. However, the source is readily imaged onto the surface of the fiber using optical elements such as lenses or mirrors. Figure 7a illustrates how a biconvex lens can be used to image a light source onto the endface of an optical fiber. The standard thin lens equation[40,41] expresses the relationship between the distances from the lens principal points to the fiber endface and source, $d_f$ and $d_s$, respectively, and the lens focal length $f_L$. For the configuration of Figure 7a, the thin lens equation is written as:

$$\frac{1}{f_L} = \frac{1}{d_r} + \frac{1}{d_s} . \qquad (20)$$

The magnification of the system, m, is defined as:

$$m = \frac{\rho_f}{\rho_s} = \frac{d_f}{d_s} \qquad (21)$$

(a)

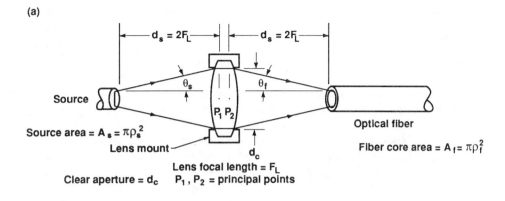

(b)

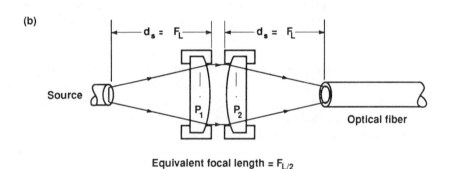

**Equivalent focal length = F$_{L/2}$**

FIGURE 7. Illustration of the use of lenses to couple energy from a light source to an optical fiber. In (a), a single bi-convex lens is used to project the image of an optical source onto an optical fiber. In (b), the single lens is replaced by a pair of plano-convex lenses.

where $\rho_f$ and $\rho_s$ are the fiber and source radii respectively. For unit magnification, $d_f = d_s = 2f_L$. The numerical apertures at the source, $NA_s$, and at the optical fiber, $NA_f$, are given by

$$NA_s = \sin\theta_s < \sin\left[\tan^{-1}\frac{d_c}{2d_s}\right] \tag{22}$$

and

$$NA_f = \sin\theta_f < \sin\left[\tan^{-1}\frac{d_c}{2d_f}\right] \tag{23}$$

where $\theta_s$ and $\theta_f$ are the source collection and fiber illumination angles, respectively. The use of the clear aperture, $d_c$, in Equations 22 and 23, as opposed to the lens diameter, accounts for the restriction in numerical aperture due to lens mounts and other real-world constraints such as ground edges on the lens. The source and fiber numerical apertures must always be less than that predicted by the right-hand side of Equations 22 and 23 because of the refraction that occurs at the lens surfaces. This effect can be precisely accounted for by tracing the path of the outermost ray through the optical system under consideration.

An improved method of imaging a source onto an optical fiber is shown in Figure 7b. Here, the single biconvex lens is replaced by a pair of plano-convex lenses with the plano sides facing outward. It is assumed that the system operates at unity magnification as in Figure 7a. This configuration has several advantages over that illustrated in Figure 7a. First, because the lenses are positioned at a distance of one focal length from the source and fiber, the light passing between them is collimated and can be conveniently processed by angle sensitive devices such as interference filters or diffraction gratings. The second advantage of this system is that if plano-convex lenses with the same focal length as the biconvex lens used in Figure 7a are used in Figure 7b, then $d_f$ and $d_s$ are cut in half. This results in a smaller optical system and a numerical aperture increase of about 26%. According to Equation 19, this corresponds to a 60% increase in power coupled into the fiber provided that the numerical aperture of the source does not exceed that of the optical fiber. It should be noted that the advantage of the plano-convex lens system may not be as great as implied above since biconvex lenses are available with shorter focal lengths than plano-convex lenses.

An additional benefit of the dual lens configuration of Figure 7b is that light loss due to spherical aberration is minimized, and therefore operation at higher numerical aperture is more efficient. The effect of spherical aberration is a reduction in the focal length of an optical system as the distance from a ray traveling parallel to the optical axis increases, i.e., between the lenses in Figure 7b. In other words the ability to focus the system for all ray angles is compromised by the presence of spherical aberration. Another approach to the minimization of spherical aberration is through the use of aspheric lenses.

Equation 17 represents the maximum possible energy coupling for the case when the output of the optical system is operated at the NA of the optical fiber and the source and fiber core areas are equal. If the imaging system is designed to reduce or enlarge the size of the source while maintaining optimal power transfer, then certain restrictions apply. By analogy to Equation 17, it can be shown that the power collected from a surface emitting source of area $A_s$ by an optical system operating with numerical aperture, $NA_s$, (see Equation 22) is given by

$$P_{source} = \pi L_0 A_s (NA)_s^2 \tag{24}$$

Ideally, the optical system should recreate $L_0$ at the fiber endface and thereby couple the power collected from the source to the optical fiber so that $P_{source} = P_{fiber}$. Assuming that $N_0 = 1$, Equations 17 and 24 can be equated and simplified to give the relationship, $\rho_f NA_f = \rho_s NA_s$, which when rearranged and combined with Equation 21 yields

$$m = \frac{\rho_f}{\rho_s} = \frac{NA_s}{NA_f} \tag{25}$$

The consequence of Equation 25 is that, for a given fiber core size and numerical aperture, the source numerical aperture must change in direct proportion to any change in magnification. For example, when coupling power from a large source to a small fiber, $m < 1$ and $NA_s < NA_f$. Although the above is theoretically possible, it is often preferable to operate at unity magnification when the source area is larger than that of the fiber to allow for greater misalignment tolerance between the source and fiber.

## 2. The Blackbody Radiator Model

Planck's law expresses the relationship between the temperature of a radiant body and its radiance per unit wavelength, $L\lambda$. The radiance of the source, $L_0$, in watts per square meter per steradian is found by integrating Planck's law over the appropriate interval as in

$$L_0 = \int_{\lambda_1}^{\lambda_2} \frac{\varepsilon C_1}{\pi \lambda^5 \left( \exp\left( C_2 / \lambda T \right) - 1 \right)} d\lambda \approx \frac{\varepsilon C_1 \Delta\lambda}{\pi \lambda^5 \left( \exp\left( C_2 / \lambda T \right) - 1 \right)} \quad (26)$$

where $\varepsilon$ is the emissivity of tungsten at the center wavelength, $\lambda$ ($\mu$m), and filament temperature, T (degrees Kelvin), $C_1 = 3.748 \times 10^{-4}$ (watt-$\mu$m$^4$/$\mu$m$^2$), $\Delta\lambda$ is the wavelength interval in $\mu$m, and $C_2 = 1.4388 \times 10^4$ $\mu$m-K.[42]

Application of Equation 26 to incandescent filaments is straightforward if the filament temperature is known since the emissivity of tungsten as a function of temperature has been well characterized.[42] In order to compare the radiance or brightness of various types of sources, it is convenient to define an equivalent tungsten filament temperature which can be calculated for any source. The temperature to which a tungsten filament lamp must be raised in order to produce a given radiance, $L_0$, can be found using the expression

$$T = \frac{C_2}{\lambda \ln\left( \frac{\varepsilon C_1 \Delta\lambda}{\pi \lambda^5 L_0} + 1 \right)} \quad (27)$$

where $\varepsilon$ is the emissivity of tungsten at the center wavelength and filament temperature. If $L_0$ is determined either by measurement or calculation from data sheet information for a given source, then an equivalent tungsten filament temperature can be calculated by solving Equation 27 for temperature.

Tungsten-halogen lamps are generally operated in the range of 2500 to 3200 K. Arc lamps, flash lamps, LEDs, and lasers, depending on the particular situation, can operate at equivalent tungsten filament temperatures as high as 10,000 to 30,000 K between 0.4 and 0.7 $\mu$m. Modlin[43] has correlated measurements of the power coupled into a 120 $\mu$m optical fiber from an E.G.&G. FY-904 flashlamp with an applied voltage of 750 V and a 0.05 $\mu$F discharge capacitor to an equivalent filament temperature of about 10,000 K. Modlin has also observed that the predictions of Equations 19 and 26 at wavelengths of 0.405, 0.485, 0.530, and 0.605 $\mu$m agree within ±20% with actual measurements of power coupled into a 120 $\mu$m optical fiber with NA = 0.49.

## B. PHOTODETECTOR MODELS

As was previously mentioned, the optical excitation sources employed in instrumentation for FOCS are generally operated in either the continuous or pulsed mode. Given the mode of operation of the light source and other important criteria, it is possible to choose an appropriate type of photodetector. Since most applications are satisfied by either a solid state photodiode or a photomultiplier tube (PMT), the discussion in this section is restricted to the noise performance of solid state photodiodes and PMTs, whereas Section IV.C presents an overview of the characteristics of other types of detectors as well. Before discussing the photodetectors themselves, an introduction to the various types of noise sources is presented.

### 1. Introduction to Sources of Noise in Photodetectors

The random fluctuations of currents and voltages in electronic circuits are referred to as "noise". In order to calculate the signal-to-noise ratio of a photodetector, and ultimately the resolution of a FOCS, it is crucial to understand the important sources of noise and how to calculate the total noise for a given photodetector circuit. Since an in-depth analysis of noise sources is beyond the scope of this discussion, the approach taken here will be to focus on results of relevance to FOCS instrumentation and to draw on the conventional literature on noise theory for background material as necessary.[44-46]

Noise can be represented as either a voltage, current, or power source. Noise quantities are usually measured or specified in terms of their root mean squared (rms) values. The rms value of a noise signal represents the corresponding dc signal required to generate a power equivalent to that generated by the noise source. For example, the rms value of a noise voltage source placed across a resistor is the value to which a dc voltage source placed across the resistor must be raised in order to dissipate the same power in the resistor as the noise voltage source.

The rms noise power from multiple noise sources dissipated in a single resistor can be directly added, whereas noise currents and voltages must be squared first and then added. This follows since $P = i^2R = v^2/R$. The lower case letters v and i are conventionally used to denote noise voltages and currents, respectively, whereas upper case letters are used for dc signals. The total noise power dissipated in a resistor, R, by n independent noise sources is given by

$$P_{total} = \sum_{j=1}^{n} P_J = R \sum_{j=1}^{n} i_j^2 = \frac{1}{R_j} \sum_{j=1}^{n} v_j^2 \tag{28}$$

where $i_j^2$ and $v_j^2$ are the square of the rms noise currents and voltages, respectively, of the j th noise source. Another way of looking at this is that the square of the rms values are added in order to avoid the need to account for the unknown phase relationships of the independent noise sources.

It is also important to correlate the peak-to-peak noise signal fluctuation to the rms value. The peak-to-peak fluctuation is found by multiplying the rms signal by a factor, $\kappa$, which is generally between 6.6 and 8.8. The actual peak to peak fluctuation will exceed that calculated using multiplicative factors of 6.6, 7.8, and 8.8 less than 0.1, 0.01, and 0.001% of the time, respectively.[44] For conservative estimations, larger values of $\kappa$ should be used. In the examples presented here, a value of 8.8 is used for $\kappa$.

### a. Johnson Noise

Johnson noise is produced by the random thermal motions of electrons in a resistor. An ideal Johnson noise source has a uniform output at all frequencies and is referred to as a "white noise" source. The rms power produced by a Johnson noise source, $P_j$, is proportional to the rms thermal energy of the electrons, the bandwidth over which the measurement is made, and is given by the relation

$$P_j = 4kTB \tag{29}$$

where k is Boltzman's constant ($1.3806 \times 10^{-23}$ J/K), T is the temperature in degrees Kelvin, and B is the equivalent noise bandwidth in hertz of the measurement device. The Johnson noise of the resistor in Figure 8 is represented by a parallel noise current source and is given by

$$i_j = \sqrt{\frac{P_j}{R}} = \sqrt{\frac{4kTB}{R}} \tag{30}$$

where R is the value of the resistor in ohms. A capacitor is included in parallel with the resistor since all real resistors have some shunt capacitance and it is the shunt capacitance that usually determines the equivalent noise bandwidth. Thus the power dissipated in the resistor decreases monotonically with frequency due to presence of the shunt capacitor. In this case, the equivalent noise bandwidth is the bandwidth in hertz which results in the same power dissipated in the resistor as in a noise current source having a constant power within its bandwidth and zero power elsewhere. The measurement bandwidth can be greater than that of the source without affecting the accuracy of the measurement.

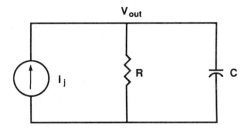

FIGURE 8. Parallel equivalent circuit for a Johnson noise source with an external shunt capacitance.

The equivalent noise bandwidth can be calculated by integrating the power transfer function, $T_p(f)$ of the circuit of Figure 8, over the frequency range of $f = 0$ to $f = \infty$. For the simple parallel R-C circuit shown in Figure 8, $T_p(f)$ is given by

$$T_p(f) = \frac{P_{out}(f)}{P_{in}(f)} = \frac{V^2_{out}(f)}{i^2_j(f)R^2} = \frac{1}{1+(2\pi fRC)^2} \tag{31}$$

The corner frequency of the circuit of Figure 8 is $F(1,2\pi RC)$ which is defined as the frequency at which the power has dropped by a factor of two or by 3 dB (the decibel is defined in Section II.B.1.f). The equivalent noise bandwidth, B in hertz, is thus given by

$$B = \int_0^\infty T_p(f) = \int_0^\infty \frac{df}{1+(2\pi fRC)^2} = \frac{1}{4RC} \tag{32}$$

It is interesting to note that the noise bandwidth exceeds the corner frequency by a factor of $\pi/2$ or 1.57.

### b. Shot Noise and "Photon Noise"

Shot noise, $i_s$ (measured in amps rms), is generated by random fluctuations in the number of current carriers (electrons or holes) flowing through an electronic circuit or device. The random fluctuation in the number of electrons surmounting the energy barrier of a PMT photocathode and carrier generation and recombination in the depletion region of a photodiode are two of the most important shot noise sources in instrumentation for FOCS. Shot noise can be expressed in terms of a shot noise current, $i_s$, which is given by

$$i_s = \sqrt{2qIB} \tag{33}$$

where q is the electronic charge ($1.602 \times 10^{-19}$ C), I is the dc current in amps, and B is the bandwidth in hertz.

In either solid state photodiodes or PMTs the "photon noise", which is inherent in the input light signal, is converted into shot noise when the photon to electron conversion takes place within the device. The noise current due to photon noise, $i_p$, is thus given by

$$i_p = \sqrt{2qBI_p} = \sqrt{2qBP_{in}\overline{S_d}} \tag{34}$$

where $_{IP}$ is the photon induced diode current, $P_{in}$ is the power incident on the photodiode in watts, and $S_d$— is the diode sensitivity in amps per watt (see Section IV.C.2.b for definition).

### 2. Noise Models for Solid State Photodiodes

Fundamental analysis of the physics of solid state photodiodes leads to the well known expression for the photodiode current-voltage (I-V) characteristic.[47] The I-V characteristic relates the current flowing through the leads of the device, $I_d$, the applied voltage, $V_d$, and the photon induced current, $I_p$, and is given by

$$I_d = I_s\left(\exp\left(\frac{qV_d}{kT}\right) - 1\right) - I_p \tag{35}$$

where $I_s$ is the reverse saturation current, q is the electronic charge (1.602 E-19 C), k is Boltzmann's constant (1.3806 E-23 J/ K), and T is the temperature in degrees Kelvin. The dc model of a photodiode is schematically illustrated in Figure 9a. $R_s$ is the equivalent series resistance of the diode.

The dc model must be expanded to account for the noise performance of the diode since noise is fundamentally an ac phenomenon. The ac model, which includes both the diode shunt and series resistances is illustrated in Figure 9b. The photocurrent noise is represented by an ac current source, $i_p$, which is found using Equation 34.

The Johnson noise associated with the diode shunt resistance, $i_{Rsh}$, is represented by an ac current source in parallel with, $R_{sh}$, and the diode shunt capacitance, $C_{sh}$. Differentiation of Equation 35 with respect to $V_d$ and inversion leads to the following expression for the diode shunt resistance:

$$R_{sh} = \frac{kT}{qI_s}\exp\left(\frac{-qV_d}{kT}\right) \tag{36}$$

Typical and worst case values for $R_{sh}$ and $I_s$ are usually specified in manufacturer's data sheets. The Johnson noise current is found by substituting $R_{sh}$ into Equation 30. For operation with zero reverse bias voltage, the Johnson noise current is given by

$$i_{R_{sh}} = \sqrt{\frac{4kTB}{R_{sh}}} \tag{37}$$

The depletion region of the diode gives rise to a capacitance, $C_{sh}$, which appears across $R_{sh}$ and in series with $R_s$. This is because $R_s$ results primarily from the resistivity of the bulk semiconductor material which surrounds the depletion region. Since $R_s$ is very small compared to $R_{sh}$, $R_s$ has virtually no effect on the low frequency noise performance of the diode. The exact value of the shunt capacitance, $C_{sh}$, is strongly dependent on the area and doping profiles of the photodiode and is therefore specified in manufacturers data sheets. However, as a rule of thumb, for reverse voltages greater than about 1 V, $C_{sh}$ is inversely proportional to the one third to one half power of the applied voltage.

Since the frequency response of a photodetector will generally increase as $C_{sh}$ decreases, it is sometimes desirable to operate the diode under conditions of reverse bias, i.e., $V_d \leq 0$. The noise current is then due to the shot noise resulting from the current through $R_{sh}$ caused by the applied voltage (Equation 33) and can be expressed as

$$i_{R_{sh}} = \sqrt{2qBI_d} = \sqrt{\frac{2qB|V_d|}{R_{sh}}} \tag{38}$$

where $R_{sh}$ is found using Equation 36.

The noise current for the photodiode can be determined with or without the application of reverse bias by using either Equations 37 or 39, respectively, to find $i_{Rsh}$. Equation 39 will underestimate the value of because the effect of the applied voltage on the width of the

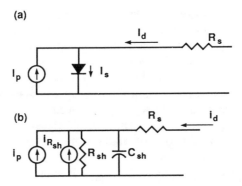

FIGURE 9. Photodiode circuit models shown for (a) dc operation and (b) ac operation and noise analysis.

depletion region (and the resultant increased carrier generation) have not been accounted for. In practice, it is best to use manufacturer's data sheets to find the dark current, $I_d$, as a function of reverse bias voltage and find $i_{R_{sh}}$ using the relation

$$i_{R_{sh}} = \sqrt{2qI_dB} \tag{39}$$

With the application of light the total noise current flowing from the diode, $i_d$, is just the sum of the squares of the photon induced noise current, $i_p$, and the noise due to the presence of $R_{sh}$,

$$i_d = \sqrt{\left(i_p^2 + i_{sh}^2\right)} \tag{40}$$

### 3. Noise Models for PMTs

The total noise current at the cathode of a PMT (see Section IV.C) is equal to the shot noise due to the sum of the dark current, $i_{dark}$, and the current stimulated by incident photons, $i_p$. The dark current is thermionically emitted from the photocathode, multiplied by the dynode stage gain, G, and appears at the anode as $I_{dark} = G\, i_{dark}$ where G is on the order $10^3$ to $10^7$ and $I_{dark}$ is on the order of 1 to 100 nA at room temperature. The noise current at the anode of a PMT due to the dark current is simply the shot noise at its cathode multiplied by the product of G and the noise deterioration factor, $\delta/_{(d-1)}$, due to secondary emission noise at the dynodes.[45] The total PMT noise at the anode, $i_a$, is thus given by

$$i_a = \sqrt{\frac{\delta 2qG^2B\left(i_p + i_{dark}\right)}{(\delta - 1)}} \tag{41}$$

where $\delta$ is the gain per dynode stage. The quantity, $\delta/_{(d-1)}$, is included for completeness but is generally on the order of 4/3 and therefore has little effect. The noise can also be expressed in terms of the PMT average anode sensitivity, $\overline{S}_a$ at a given wavelength (in amps/watt), the input power $P_{in}$ in watts, and the anode dark current, $I_d$, as

$$i_d = \sqrt{\frac{\delta 2qGB\left(P_{in}\overline{S}_a + I_{dark}\right)}{(\delta - 1)}} \tag{42}$$

Equation 42 is the most useful since the PMT anode sensitivity and dark current are usually specified in the manufacturer's data sheets.

PMT dark current can pose a significant problem in low-frequency, low-light-level applications because the dark current produces a potentially overwhelming temperature-sensitive offset current. Although the dark current can be greatly reduced or eliminated by cooling the PMT and operating in the photon-counting mode (see Section IV.C.1), constraints on size and cost and less stringent performance requirements often dictate that the effects of PMT dark current and other dc offset sources be minimized using synchronous detection techniques.

With synchronous detection, the light source is chopped at a steady frequency and the detector is synchronously switched to recover the original signal.[48] Nonsynchronous signals such as the dark current and amplifier offset voltages are converted into ac signals by the synchronous switch following the detector and are filtered out by a low pass filter. Using this technique, it is possible to virtually eliminate the effects of PMT dark current and many other sources of DC offsets.

## 4. Signal-to-Noise Ratio Models for Photodetector Circuits

In this section, photodetector circuits compatible with both continuous and pulsed light sources will be discussed. Both solid state photodiodes and PMTs can be used with either of the circuits described below, and the noise performance of either type of photodetector can be predicted by inserting the expressions for the noise currents for solid state photodiodes or PMTs presented in Sections III.B.2 and III.B.3, respectively, into the overall circuit models presented in this section. While numerous other circuits are also used for photodetection, the two described in the following sections are representative examples and therefore serve as a good place to start.

### a. Photodetectors for Continuous Light Sources

The op amp current to voltage converter shown in Figure 10 is probably the most commonly used photodetector circuit. Given the combination of its extremely high gain ($\approx 10^5$) and negative feedback through $R_f$ and $C_f$, the op amp maintains its inverting and noninverting terminals at virtually the same potential by adjustment of its output voltage. Since the positive terminal is connected to ground (zero volts), the op amp will attempt to maintain its inverting terminal at zero volts as well. It is this regulating effect coupled with the very high input impedance looking into the op amps inverting terminal ($\approx 10^{12}$ ohms) that allows the circuit of Figure 10 to function as a nearly ideal current to voltage converter.

When a positive current flows from the photodiode to the inverting terminal of the op amp, the op amp adjusts its output to a negative voltage of precisely the amount required to shunt the input current away from the inverting terminal and through the parallel combination of $R_f$ and $C_f$. The potential of the inverting terminal is thus maintained at approximately zero volts and the output voltage, $V_o$, is given by

$$V_0 = -P_{in} S_d Z_f \tag{43}$$

where $P_{in}$ is the input power (in watts), $S_d$ is the diode sensitivity (in amps per watt, see Section IV.C.2.b), and $Z_f$ is the impedance (in ohms) of $R_f$ in parallel with $C_f$. If the circuit is operated at dc, then $Z_f$ equivalent to $R_f$.

It is also crucial to consider the stability of the circuit since high gain photodetectors are prone to oscillation. Oscillation in any electronic amplifier generally occurs at the frequency where the total phase shift around the feedback loop (including the op amp) reaches 360° and the input and feedback signals constructively interfere with each other. A detailed analysis of the stability criteria of the circuit in Figure 10 is beyond the scope of this discussion and can be found eleswhere.[49,50]

It is relatively straightforward to use an empirical method to stabilize the circuit. As the size

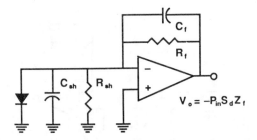

FIGURE 10. Photodetector based on a current to voltage (I to V) converter circuit.

of the feedback capacitor, $C_f$, is increased, the tendency for the circuit to oscillate decreases because the component of the loop phase shift due to the series combination of the feedback resistor, $R_f$, and the diode shunt capacitance, $C_s$, also decreases. As $C_f$ is increased above some threshold value, the circuit will no longer oscillate. This phenomenon can be observed on an oscilloscope by applying a repetitive light signal having a 50% duty cycle to the photodiode. For small values of $C_f$, the output of the circuit will ring or oscillate. As $C_f$ is increased, the amplitude of the ringing will decrease until it disappears altogether. Unfortunately, the bandwidth of the circuit will decrease as $C_f$ increases since it is inversely proportional to the product of $R_f$ and $C_f$.

The reality that must be faced when designing photodetector circuits employing solid-state photodiodes is that the product of the gain and the bandwidth is relatively constant with the consequence that high gain is invariably accompanied by low bandwidth. On the other hand, PMT-based circuits suffer much less from this problem because the high internal gain of the PMT requires a lower value of $R_f$ to achieve a given overall gain. Hence, PMT-based circuits achieve higher bandwidth than solid state photodiodes for the same overall gain.

The noise performance of the circuit in Figure 10 can be modeled if one inserts values for Johnson noise, shot noise, and op amp voltage and current noises in the appropriate places as shown in Figure 11. The values for the current and voltage noise, $i_n$ and $v_n$, respectively, of a given op amp are provided in the manufacturer's data sheet and are usually referred to as the input of the op amp over a specified frequency range.

Many factors must be taken into account when choosing an op amp. Field effect transistor (FET) input op amps usually give superior noise performance as compared to bipolar op amps in photodetector applications because the high impedance FET input stage provides a large reduction of the current noise over that obtainable with a bipolar input device. However, depending on the application, other factors such as offset voltage and temperature stability may also be important to consider. For example, the OPA 111 BM FET input op amp specifies $I_A = 0.4 \text{ fA}/\sqrt{Hz}$, $v_A = 100 \text{ nV}/\sqrt{Hz}$ at 1 Hz, and $v_A = 7 \text{ nV}/\sqrt{RHz}$ at 1 KHz.[50]

Since the noise performance of the circuit shown in Figure 11 has been analyzed in detail elsewhere,[49-51] the general expression for the total noise appearing at the output will be presented here without derivation. The form of the expression follows directly from the discussion on noise sources presented in Sections III.B.1 and III.B.2. The total rms noise output at a frequency, f, over a bandwidth, B, of the circuit in Figure 11 is equal to the sum of the contributions from the diode noise current, $i_d$ (from Equation 40 for the solid state diode or Equation 42 for the PMT), the op amp noise current, $i_A$ (from the data sheet), the feedback resistor noise current, $i_{R_f}$ (substitute $R_f$ into Equation 30), and the op amp voltage noise, $v_A$ (from the data sheet), and is given by

$$\text{rms noise output} = \sqrt{Z_f^2\left[i_d^2 + i_A^2 + i_{R_f}^2\right] + V_A^2\left[1 + \frac{Z_f}{Z_d}\right]^2} \qquad (44)$$

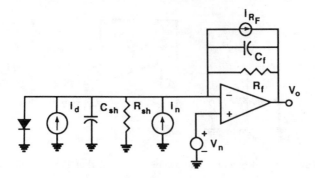

FIGURE 11.   Current to voltage photodetector including the relevant noise sources.

Here, the data sheet values for the amplifier noise components have each been multiplied by and the ac impedances of the feedback network and diode equivalent circuit, $Z_f$ and $Z_d$, respectively, are given by

$$Z_f = \frac{R_f}{1+\left(2\pi f R_f C_f\right)} \tag{45}$$

and

$$Z_d = \frac{R_{sh}}{1+\left(2\pi f R_{sh} C_{sh}\right)} \tag{46}$$

In both Equations 45 and 46, it is important to include the effects of parasitic resistances and capacitances from cables and circuit board traces, etc., when calculating $R_{sh}$ and $C_{sh}$ since the parasitic elements can sometimes be dominant. If the circuit is to be operated at dc, then $Z_f = R_f$ and $Z_d = R_{sh}$.

The quantity $(1 + Z_f/Z_d)$ is often referred to as the "noise gain" of the circuit since it is the factor by which the noise is amplified. It should be noted that Equation 44 applies only over a limited operational bandwidth. Operation over a wide bandwidth requires integrating each noise term over the entire bandwidth to insure accuracy. It is common practice to use Equation 44 in a piecewise fashion to approximate noise performance over a wide bandwidth.

The signal-to-noise ratio (SNR) can now be calculated given a value for $\kappa$ which expresses the SNR in terms of the peak-to-peak noise fluctuation (see Section III.B.1). For the solid state diode, the signal (Equation 43) and noise (Equation 44) at the output are divided to give

$$SNR = \frac{P_{in} S_d Z_f}{\kappa\sqrt{Z_f^2\left[i_d^2 + i_A^2 + i_{R_f}^2\right] + V_A^2\left[1 + Z_f\big/Z_d\right]^2}} \tag{47}$$

It is important to evaluate Equation 47 for each case since it is difficult to know in advance which of the terms in the denominator will dominate. In the case of the PMT, its extremely high gain virtually assures that the anode noise current of the PMT will be dominant. Thus for the PMT:

$$SNR = \frac{P_{in} S_a Z_f}{\kappa\sqrt{Z_f^2\left[i_a^2 + i_A^2 + i_{R_f}^2\right] + V_A^2\left[1 + Z_f\big/Z_a\right]^2}} \approx \frac{P_{in} S_a}{\kappa i_a} \tag{48}$$

where $i_a$ is the anode noise current from Equation 41, $Z_a$ is the impedance looking into the PMT and is analogous to Equation 46, $S_a$ is the anode sensitivity, and all other parameters are the same as in Equation 47.

In practice, the SNR for a PMT-based photodetector is generally limited by the noise from the PMT and is therefore one or more orders of magnitude better than what can be achieved with a solid state photodiode (see Section IV.C.3). This is important when dealing with very low light levels. In addition, the transition from electronics limited SNR to photon noise limited SNR occurs at a lower level for the PMT. This can be very important when the signal has a wide dynamic range.

### b. Photodetectors for Pulsed Light Sources

When the output of the light source is a high-amplitude pulse with a low duty cycle, such as that from a flash lamp or pulsed LED, the gated integrator is usually the detector circuit of choice. The gated integrator (shown in Figure 12) works by integrating the output from the photodiode using a capacitor when the gate (or switch) is open for a time $\Delta t$. The switch is closed at other times. Since capacitors are essentially noiseless, it is possible to achieve a signal-to-noise ratio which is dominated by the photon noise of the source or the intrinsic detection limit of the detector, whichever is larger.

After being closed for a long time, the switch is opened for a short time $\Delta t$ where $\Delta t \ll R_{sh}C_t$. When a light pulse strikes the photodiode, the output current pulse, $I(t)$, flows into the parallel combination of $C_t$ and $R_{sh}$ where $C_t = C_{sh} + C_{in}$. Virtually all of the current will flow into $C_t$ and none into $R_{sh}$ since the voltage across $C_t$ is initially zero. The output from the unity gain buffer amplifier, $V_o$, is therefore approximately given by

$$V_0 \approx \frac{1}{C_t} \int_{t_0}^{t_1} i_{R_{sh}}(t)dt \qquad (49)$$

Each time a light pulse strikes the photodiode, the output will increase in accordance with Equation 49. The switch, which is normally closed to discharge the capacitor and allow a path for the op amp bias current and diode reverse current, is opened just prior to the arrival of the light pulse. As soon as the light pulse is over, the switch is again closed. This circuit has the advantage that RF is not present and therefore does not contribute to the total noise at the output.

Equation 49 can also be used to understand the effect of gated integration on the noise appearing at the output. Neglecting, for the moment, the contribution from the op amp voltage and current noises, the voltage at the output of the op amp due to current noise from $R_{sh}$ can be calculated to be approximately

$$V_0 \approx \frac{1}{C_t} \int_{t_0}^{t_1} I(t)dt \qquad (50)$$

where $\Delta t = t_1 - t_0 \ll R_{sh}C_t$ as would usually be the case for a gated integrator. If many samples are collected, the rms value of the output voltage distribution will be given by

$$V_{0(rms)} \approx \frac{1}{C_t} \int_{t_0}^{t_1} i_{R_{sh}}(rms)dt \approx \frac{\Delta t}{C_t} i_{R_{sh}} \qquad (51)$$

As described in Section III.B.1, the peak-to-peak fluctuation can be estimated by multiplying the rms value of the diode voltage by the factor, $\kappa$, which is between 6.6 and 8.8 depending on the confidence level desired.

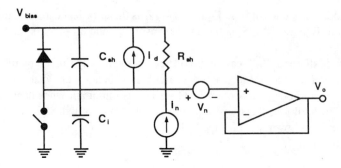

FIGURE 12.    Photodetector based on a gated integrator.

As in Equation 44, the total noise appearing at the output of the op amp is just the sum of the contributions from the noise sources $i_d$, $i_A$, and $v_A$:

$$\text{rms noise output} = \sqrt{\left[\frac{\Delta t}{C_t}\right]^2 \left[i_d^2 + i_A^2\right] + V_A^2} \tag{52}$$

It is assumed that the bandwidth of the unity gain op amp and other downstream measurement equipment is greater than the noise bandwidth of the input circuit ($B = F(1/4R_{sh}C_t)$). In order to meet the rise time requirements of high frequency pulsed sources, the diode is often reverse biased.

As was the case for the current to voltage converter (Equation 47), the signal-to-noise ratio for the gated integrator employing a solid state diode is found by dividing the signal (Equation 49) by the noise (Equation 52) at the output which gives

$$\text{SNR} = \frac{P_{in} S_d \Delta t}{\kappa C_t \sqrt{\left[\frac{\Delta t}{C_t}\right]^2 \left[i_d^2 + i_A^2\right] + V_A^2}} \tag{53}$$

Similarly, the SNR for the PMT is given by

$$\text{SNR} = \frac{P_{in} S_a \Delta t}{\kappa C_t \sqrt{\left[\frac{\Delta t}{C_t}\right]^2 \left[i_a^2 + i_A^2\right] + V_A^2}} \approx \frac{P_{in} S_a}{\kappa i_a} \tag{54}$$

The SNR for the gated integrator with a PMT is limited by the intrinsic PMT noise, whereas the SNR for the solid state diode can be limited by other noise sources if the average power from the light pulse is too low or the circuit is not designed correctly. Because the peak power is usually higher with pulsed sources than continuous sources, the SNR is more often limited by photon noise, and consequently the performance of the gated integrator can exceed that of the current to voltage converter.

## C. OPTICAL SYSTEM MODELING

Although there are many types of components used in optical systems employed in instrumentation for FOCS, the throughput of the system can be represented by the product of the transmission of each element integrated over the operational wavelength band. For an optical system containing n elements, the overall transmission of the system between $\lambda_1$ and $\lambda_2$ is expressed as

$$\text{transmission} = \prod_{i=1}^{n} \int_{\lambda_1}^{\lambda_2} T_i(d\lambda) \tag{55}$$

where $T_i$ is the transmission of the ith element per unit wavelength. Although this model is simple in concept, it often takes considerable effort to accurately estimate the transmission for each element. This is especially true for components such as interference filters and diffraction gratings which are highly wavelength sensitive. The transmission spectrum of these components must be either measured or graphically integrated since analytical prediction is usually quite difficult. The reader is referred to Section IV.B for more discussion on the transmission properties of optical components.

## D. SYSTEM PERFORMANCE MODELING

In order to predict the performance of a fiber optic sensor, it is necessary to have a combined overall model for the sensor, optical system, and the photodetector. In this section, the concepts of the system throughput model and sensor transfer function are introduced and their application to an amplitude modulated pH sensor is given as an example.

### 1. System Throughput Modeling

In the development of FOCS, it is crucial to determine that the throughput of the sensor-instrument system is adequate to provide the desired signal to noise ratio or sensor resolution. Even for the case of non-amplitude-modulated sensors, the performance will be degraded if the throughput of the system is too low. This is true since the signal-to-noise ratio of all electro-optical systems will become limited by the noise from the electronic system when the received optical power drops below a critical threshold.

Given an optical system with n attenuating elements, the output of the photodetector can be calculated using the relation

$$\text{output in volts} = P_f H_t G_A \int_{\lambda_1}^{\lambda_2} S_d d\lambda \prod_{i=1}^{n} \int_{\lambda_1}^{\lambda_2} T_i d\lambda \tag{56}$$

where $P_f$ is the source power coupled into the fiber, $H_t$ is the sensor transfer function (which varies in value between 0 and 1), $G_A$ is the transconductance of the photodetector input amplifier in volts per amp, $S_d$ is the detector sensitivity in amps per watt per unit wavelength, and $T_i$ is the transmission of the ith element per unit wavelength. The sensor transfer function, $H_t$, is discussed in detail in the following section.

For amplitude-modulated sensors, the resolution of the sensor (in measurement units) is equal to the change in value of the measured quantity required to cause the output of the photodetector (given by Equation 56) to change by an amount equal to the peak-to-peak noise of the photodetector. The peak-to-peak noise is found by multiplying the rms noise given by Equation 44 for continuous sources or Equation 52 for pulsed sources by the factor $\kappa$ given in Section III.B.1. For the case of non-amplitude-modulated sensors, the resolution is also related to the photodetector noise but not as directly. For example, in sensors based on time decay measurement, the minimum resolvable change in decay time can be limited by the noise at the output of the photodetector, since the decay time must ultimately be derived from measurements of the photodetector output. If the amplitude of the returning light is very high, then the noise at the output of the photodetector will be dominated by photon noise or by some noise source other than that of the photodetector. If the amplitude of the returning light is low, then the noise at the output of the photodetector will be dominant and therefore will determine the resolution of the sensor.

## 2. Sensor Transfer Function

In order to convert the output from a fiber optic sensor to a usable output in analytic measurement units (e.g., pH units, ppm, concentration, etc.), an overall transfer function for the sensor-instrument system must be determined. The overall transfer function relates the output from the measurement circuit, (e.g., in volts or analog to digital converter counts) to the desired analytic measurement units, and can either be derived from the basic physical principles of operation of the transducer or estimated by statistical curve-fitting to experimental data.

The sensor transfer function, $H_t$, in conjunction with Equation 56 determines the relationship between the measured quantity and the output from the photodetector. The sensor transfer function is defined by the relation

$$H_t \equiv \frac{\text{power returned}}{\text{input power}} = \eta_s M \tag{57}$$

where $H_t$ is a dimensionless quantity which can range in value from zero to one, $\eta_s$ is the sensor efficiency, and M is the modulation function. The sensor efficiency, $\eta_s$, accounts for losses due to geometrical factors and other nonidealities in the sensor and the modulation function, M, expresses the relationship between the measured quantity and sensor output. While the units of $\eta_s$ and M can vary depending on the type of sensor system under consideration, their product must be dimensionless to be consistent with Equation 57.

## E. AMPLITUDE-MODULATED pH SENSOR EXAMPLE

In this section, the basic concepts presented thus far will be applied to the measurement of pH using a pH-sensitive fluorescent dye. In this example, the sensor system shown in Figure 13 in conjunction with the instrument in Figure 4 will be analyzed. An optical fiber of radius $\rho$ and NA = 0.5 is immersed in an aqueous solution ($N_o$ = 1.33) which contains the pH sensitive dye. Excitation light coming from the instrument is emitted into the solution from the fiber endface and forms a light cone with a divergence angle of 22° (arcsin [0.5/1.33]). After absorbing incident photons, fluorescent dye molecules within the excitation light cone emit isotropically, i.e., into a solid angle of $4\pi$ sr. The fraction of the emitted light that is incident on (and transmitted through) the endface and within the NA of the fiber is returned to the instrument for detection.

## 1. Derivation and Calculation of the Sensor Transfer Function

It is clear that the amount of emitted light which is recaptured by the optical fiber is a function of many factors which include the NA and radius, $\rho$, of the optical fiber, the quantum efficiency, $\phi$, extinction coefficient, $\varepsilon'$ (see Appendix I, Equation 92), and molar concentration of the fluorescing dye molecules, $C_F$, and the index of refraction of the solution surrounding the fiber, $N_o$. The sensor transfer function, $H_t$, which accounts for these factors, has been derived for the geometrical arrangement of Figure 14 and the reader is directed to Appendix I for a more complete presentation of the basic concepts and details of the analysis. Under the set of simplifing assumptions also given in Appendix I (see Equation 100), $H_t$ is found to be closely approximated by the following relation

$$H_t \approx \frac{\Phi \varepsilon' C_F}{N_0^2} \sin^2\left[\frac{\alpha_c}{2}\right] \int_0^{Z_m} \frac{\exp(-\varepsilon' C_F z)}{\left[1 + \dfrac{2z \tan \alpha_c}{\rho}\right]^2} dz \tag{58}$$

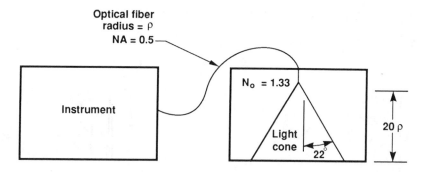

FIGURE 13.  Fiber optic pH sensor based on a cleaved optical fiber immersed in a solution containing a pH sensitive fluorescent dye.

where z is the distance from the fiber endface and $\alpha_c$ is the angle of a ray of light emerging from the fiber at the limit of the numerical aperture of the fiber. For the case where $\rho = 100$ µm, NA = 0.5, and $N_o = 1.33$, numerical integration of Equation 58 shows that about 79% of the light collected from within a distance of $200\rho$ from the fiber endface is collected from within a distance of $4\rho$. From this result, it is clear that the region within a few fiber diameters of the fiber endface is responsible for most of the returned light.

Although numerical integration of Equation 58 provides the most accurate estimation for the sensor transfer function, a set of simplifing assumptions based on practical constraints for real sensors are given in Appendix I with the result that Equation 58 can be expressed in a more tractable and intuitive form. Given that, as mentioned above, most of the returned light is collected from within a few fiber diameters of the fiber endface (or from within a distance z $= z_m$), and that $\varepsilon' C_F z_m \ll 1$, the exponential term in Equation 58 can omitted. Under these assumptions, Equation 58 can be integrated from $z = 0$ to $z = z_m$ to give

$$H_t \approx \frac{\Phi\varepsilon' C_F \rho}{2N_0^2 \tan\alpha_c} \sin^2\left[\frac{\alpha_c}{2}\right] \left[\frac{1}{\dfrac{\rho}{2z_m \tan\alpha_c} + 1}\right] \qquad (59)$$

Assuming that $2z_m\tan\alpha_c \gg \rho$ and that for small $\theta$, $\sin\theta \approx \theta$ and $\tan\theta \approx \theta$, Equation 59 can be further simplified to yield the following expression which gives an intuitive feel for the functional dependencies of the various parameters:

$$H_t \approx \frac{\Phi\varepsilon' C_F \rho NA}{8N_0^3} \qquad (60)$$

Equation 60 has been evaluated for several practical cases and has been found to agree with the results of numerical integration of Equation 58 to within 10 to 15%. It is therefore reasonable to use Equation 60 in the remainder of the analysis of the amplitude modulated pH sensor since the errors associated with its use are within reasonable bounds. It is much more important for the reader to acquire an intuitive feel for the behavior of the pH sensor than to pursue accuracy at the expense of simplicity.

The concentration of fluorescing dye molecules, $C_F$, is the only factor in Equation 60 which is not a constant. Additionally, $C_F$ varies with the pH in a manner that is characteristic of each pH-sensitive fluorescent dye. In any case Equation 60 can be expressed in the form of Equation 57 wherein the sensor effieiency is given by

$$\eta_s \approx \frac{\Phi\varepsilon' \rho NA}{8N_0^3} \qquad (61)$$

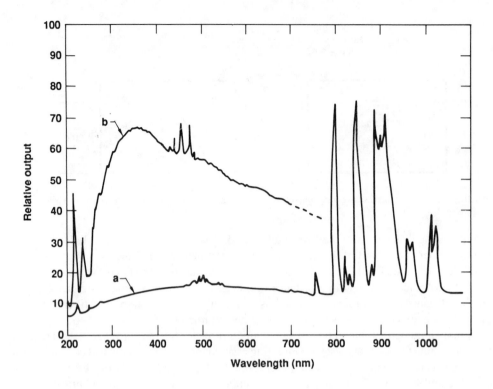

FIGURE 14.  Typical output of a xenon flashlamp for (a) low energy discharge and (b) moderate to high energy discharge.

and the modulation function is simply

$$M = N_F \qquad (62)$$

Assuming that the fluorescent dye is an acid-base indicator such as fluorescien, conservation of matter requires that the total molar concentration, $C_t$, must equal the sum of the molar concentrations of the acid form, [A], and the base form, [B], i.e.,

$$C_t = [A] + [B] \qquad (63)$$

For this example, it is assumed that only the base form absorbs the excitation light and participates in the fluorescense so that

$$C_F = [B] \qquad (64)$$

By definition, acid-base indicators equilibrate with the ambient hydrogen ion concentration in accordance with the reaction

$$[A] \rightleftharpoons [B] + \left[H^+\right] \qquad (65)$$

Using the law of mass action to express the equilibrium and substituting for [A] and [B] from Equations 63 and 64, respectively, one obtains

$$K_a = \frac{C_F[H^+]}{[A]} = \frac{C_F[H^+]}{C_t - C_F} \tag{66}$$

where $K_a$ is the equilibrium constant. The modulation function is now found, by solving Equation 66 for $C_F$, and substituting $pH = -\log[H^+]$ and $pK_a = -\log K_a$, to be given by

$$M = N_F = \frac{C_t}{[H^+] + K_a} = \frac{C_t}{10^{pK_a - pH} + 1} \tag{67}$$

The complete expression for the sensor transfer function, $H_t$, is then obtained by substituting Equation 61 and Equation 67 into Equation 57:

$$H_t = \eta_s M \approx \frac{\Phi\varepsilon'\rho NA}{8N_0^3} \frac{C_t}{\left(10^{pK_a - pH} + 1\right)} \tag{68}$$

Evaluating Equation 68 with $NA = 0.5$, $\rho = 100 \ \mu m$, $N_o = 1.33$, and $\phi\varepsilon' C_t = 10^{-4} \ \mu m^{-1}$ (which is typical for a pH sensitive fluorescent dye such as fluorescein), $H_t$ is found to be

$$H_t \approx \frac{2.66 \times 10^{-4}}{\left(10^{pK_a - pH} + 1\right)} \tag{69}$$

At $pH = pK_a$, Equation 69 yields

$$H_t \approx \frac{2.66 \times 10^{-4}}{2} = 1.33 \times 10^{-4} \tag{70}$$

It is thus observed that $H_t$ is a very small number and that FOCS with similar geometries often return only about 0.01% of the excitation light to the instrument!

## 2. Calculation of pH Sensor Resolution

The first step in determining the resolution of the pH sensor is to differentiate Equation 69 with respect to pH:

$$\frac{\partial H_t}{\partial pH} \approx \frac{2.66 \times 10^{-4} \times 2.302 \times \left(10^{pK_a - pH}\right)}{\left(10^{pK_a - pH + 1}\right)^2} \tag{71}$$

For simplicity, it is assumed that the pH to be measured is equal to the $pK_a$ of the dye, and therefore Equation 71 is evaluated at $pH = pK_a$ giving

$$\left[\frac{\partial H_t}{\partial pH}\right]_{pH = pK_a} \approx \frac{2.66 \times 10^{-4} \times 2.302}{4} = 1.53 \times 10^{-4} \ pH \ unit^{-1} \tag{72}$$

The next step is to calculate the excitation power coupled into the optical fiber. Assuming that the light source in Figure 4a is a tungsten filament operating at 3000 K with $\varepsilon = 0.455$ at $\lambda = 0.467 \ \mu m$,[42] the source radiance is found using Equation 26 to be $2.054 \times 10^{-9} \ W/\mu m^2/$

sr over the wavelength band of 0.475 to 0.495 μm. A correction factor of 0.5 is then applied to account for the gap between the filament coils. The effective source radiance is thus found to be $1.027 \times 10^{-9}$ W/μm²/sr when the source radiance is multiplied by the ratio of the filament diameter to the pitch of the filament (0.5 in this example). Assuming that the excitation filter and dichroic beam splitter each have a transmission of 70% in their passbands and that 4% of the incident light is reflected at each of the seven air-glass interfaces (two from the lamp envelope, four from the lenses, and one from the fiber), the effective percent transmission in the excitation path is calculated to be equal to 0.37. The power coupled into the optical fiber (NA = 0.5, and ρ = 100 μm) from the source is then found by taking the product of Equations 17 and 55:

$$P_f = \pi A_f \frac{NA^2 L_0}{N_0^2} \prod_{i=1}^{n} \int_{\lambda_1}^{\lambda_2} T_i(\lambda) d\lambda = \pi A_f (NA)^2 L_0 \overline{T}_{ex} = 9.38 \ \mu \text{ watts} \tag{73}$$

Here, $\overline{T}_{ex}$ is the percent transmission in the excitation light path and it is assumed that $N_0 = 1$ at the instrument.

The output of the photodetector, $V_0$, is then found by evaluating Equation 56 using the results of Equations 70 and 73 (assuming pH = pK$_a$). The expression below is evaluated assuming that $P_f = 9.38 \ \mu$W, $H_t = 1.33 \times 10^{-4}$, $G_A = 10^9$ V/amp, $\overline{S}_d = 0.25$ amps/watt, and $\overline{T}_{em} = 0.521$. The photodetector output is then calculated to be

$$V_0 = P_f H_t G_A \int_{\lambda_1}^{\lambda_2} S_d \, d\lambda \prod_{i=1}^{n} \int_{\lambda_1}^{\lambda_2} T_i d\lambda = P_f H_t G_A \overline{S}_A \overline{T}_{em} = 0.162 \text{ V} \tag{74}$$

where $\overline{S}_d$ and $\overline{T}_{em}$ are the values of the detector sensitivity and transmission of the emission light path, respectively, integrated over the operational wavelength band. The value for $T_{em}$ of 0.521 used in the above calculation is based on the assumption of 4% reflection from each of the six air-glass interfaces (one fiber endface, four lens surfaces, and one detector window), as well as 95% transmission through the dichroic beam splitter and 70% through the emission filter.

The resolution of the pH sensor can now be calculated if it is assumed that the circuit of Figure 10 is used for the photodetector and the op amp is an OPA 111 BM, whose noise properties were given in Section III.B.4.a. At pH = pK$_a$, the sensor resolution can be estimated using the relation

$$\text{sensor resolution} = \frac{(\text{rms amplifier noise}) \times (\kappa)}{\left[\frac{\partial V}{\partial pH}\right]_{pH=pK_a}} \tag{75}$$

where it is assumed that $\kappa = 8.8$ for a 99.999% confidence level (see Section II.B.1). The rms amplifier noise is found from Equation 4 to be 5.77 μV rms, and the term in the denominator is found to be equal to 0.187 V per pH unit by substituting the result of Equation 72 for $H_t$ into Equation 74. It is assumed that $R_s = 10^9 \ \Omega$, $C_s = 10^{-10}$ F, $R_f = 10^9 \ \Omega$, $C_f = 2.5 \times 10^{-10}$ F (for a noise bandwidth of 1 Hz, see Equation 32), T = 25 C, and that the circuit is operated at dc. Using these values, the resolution of the pH sensor is now calculated from Equation 75 to be $2.72 \times 10^{-4}$ pH units. This is indeed a very impressive resolution for any pH sensor and can be considered to be a theoretical limit since other factors may limit the resolution achieved in practice. Additionally, it is evident from Equation 71 that the resolution of the pH sensor will be degraded by one order of magnitude for each pH unit that the operating point is displaced from the $pK_a$ of the fluorescent dye.

An examination of the magnitude of the noise components which make up the total photodetector output noise for this example reveals that the current and voltage noise components of the op amp are relatively insignificant and that the diode shunt resistance and feedback resistor are the dominant noise sources. Therefore, the noise at the output will increase or decrease if either the diode shunt resistance or feedback resistor are decreased or increased in value, respectively.

It should also be pointed out that the output from the photodetector at $pH = pK_a$ is a relatively low level signal (162 mV), which could prove problematic if the circuit is operated at DC. This is because the offset voltage drift of the op amp with temperature or other extraneous factors sources such as power supply noise could introduce significant errors. It therefore may be desirable to AC couple the circuit, further amplify the signal, and make use of synchronous or narrow band detection to reduce the sensitivity to ambient noise and convert the AC signal into a form suitable for acquisition.[48]

At this point it is useful to combine the major results of this Section to obtain a complete model for the pH sensor. This allows one to readily comprehend the impact of all of the important factors on the performance of the sensor. Equation 68 is solved for pH and combined with Equations 73 and 74 to give the complete model for the pH sensor:

$$pH = pK_a - \log\left[ \frac{L_0 \pi^2 \rho^3 (NA)^3 G_A \overline{S}_d \overline{T}_{em} \overline{T}_{ex} \Phi \varepsilon' C_t}{8 V_0 N_0^5} - 1 \right] \qquad (76)$$

Relationships such as Equation 76 are indispensible when evaluating and/or designing instrumentation for FOCS and provide the basis for quantitative prediction of sensor performance.

## IV. COMPONENTS

### A. LIGHT SOURCES

In this section, an introduction to several laser and lamp sources that have potential application to FOCS instrumentation is presented. These sources can be succinctly classified as either laser or lamp. However, within these two classifications there exists a myriad of options. The purpose of this section is to present a broad overview of the various types of light sources relevant to FOCS instrumentation, whereas system-level analytical models for generic light sources are presented in Section III.A.1.

### 1. Lasers

The laser is nearing the end of its third decade of existence. In this short time it has grown from a laboratory curiosity to a having a presence, either directly or indirectly, in every home and business in modern society. Lasing action has been demonstrated in matter in all its phases and nearly all the elements have participated in the lasing process. Presently, there are in excess of 400 commercially cataloged lasers (comprising over 40 laser types) available for purchase. The following sections provide a brief introduction to the subset of commercially available lasers which are potentially useful in FOCS instrumentation. Readers are directed to Reference 52 as well as the various laser application handbooks (published in conjunction with the trade journals in the laser field) for additional background and manufacturer's information.

### a. Laser Characteristics

Laser light is routinely generated in an optical cavity composed of one total and one partially reflecting mirrored surface. This cavity is called an optical resonator and its design, which is not unique, is responsible for the output characteristics of the laser. One of the more important laser characteristics in FOCS applications is the cross-sectional (or transverse) structure of the light intensity. Most laser resonators are designed to operate in the lowest order transverse electromagnetic mode (often called $TEM_{00}$). This mode generates a circular beam with an intensity profile that decreases radially from a central maximum (a Gaussian profile). Since the boundary of the beam cannot be precisely located, it is customary to define the beam diameter as the distance between the $1/e^2$ points, i.e., the points on opposite edges of the beam where the intensity has fallen to $1/e^2$ of its maximum value. It is common for laser manufacturers to specify the transverse mode and beam size of their products.

$TEM_{00}$ is the desired mode for fiber instrument applications because of its relation to two very important and interrelated parameters — beam divergence and focal spot size. Divergence is a measure of the increase in beam spot size with distance from the source and is typically given in milliradians. It is generally specified by laser manufacturers for the far field; a distance approximated by $100\ w^2/\lambda$, where w is the laser minimum waist diameter and is its wavelength. The near field divergence is a complicated function of resonator design and is seldom specified. For a laser operating in the $TEM_{00}$ mode the full angle, far-field beam divergence is given by

$$\theta = 1.27\ \lambda\!\big/\!_w \tag{77}$$

Here, w is again the minimum waist of the laser beam which can be approximated as the output diameter to the $1/e^2$ points. By comparison, the full angle, far-field divergence created by a circular aperture is almost twice as large as the divergence of the $TEM_{00}$ mode of a laser.

The diameter of a focused spot is related to the divergence through the following approximate expression

$$s = f\theta \tag{78}$$

where s is the diameter of the focussed spot at the $1/e^2$ points, f is the focal length of the lens, and $\theta$ is the full angle, far-field divergence (even if the lens is in the near-field). For example, a laser beam with full angle divergence of 0.5 mrad can be focussed to a diameter of approximately 25 μm with a 25 mm focal length lens. Since multimode fiber applications generally utilize fibers with diameters of 100 μm or more, this light can be efficiently coupled into the fiber.

Lasers not operating in the $TEM_{00}$ mode are referred to as multimode and are characterized by higher beam divergence and nonuniform transverse intensity patterns. This results in a larger minimum focal diameter and in "hot spots". Consequently, lasers operating in higher order modes are undesirable for fiber instrumentation applications.

Lasers can be operated in either the pulsed or continuous wave (CW) mode. In the pulse mode, lasers can output a single pulse or a train of pulses with pulse widths as narrow as 1 psec or less. Lasers of this type can be extremely useful in applications such as time decay spectroscopy or when observing lifetimes in the nanosecond or subnanosecond range.

### b. Laser Types
#### i. Argon Ion Lasers

The argon ion laser[53] is a gas phase laser that has experienced enormous popularity in the scientific research community for the past two decades. It generates its lasing action from singly ionized atomic argon that is excited through electron collision. The energy for this collision is provided by a low pressure, high current, continuous plasma discharge. This discharge generates considerable waste heat and its removal requires a significant amount of environmental control. As a consequence, these lasers generally have a very large power supply and require water cooling. This has, until recently, restricted their application to somewhat sophisticated laboratory environments. However, the advent of the air-cooled version of the argon ion laser, for which there are now numerous manufacturers, has greatly increased its potential applications, and it is now a component in a variety of commercially available measurement and test equipment.

Regardless of the above restrictions, the water cooled version of the argon laser has made a deep penetration into the scientific research community for the following reasons: (1) its large number of available output wavelengths, (2) its low beam divergence (typically <1 mrad) and high pointing stability (within 10 μrad), and (3) its stable output (often < 0.3% RMS variation). These are highly desirable characteristics for fiber sensor research and account for its use in early laboratory scale bench-top instrumentation.[12]

### ii. He-Ne Lasers

The He-Ne laser[54] is also a plasma discharge gas laser. It differs from the argon ion laser in that it generates its lasing state through atomic collision. This process is somewhat more efficient resulting in less waste heat and, consequently, a smaller overall laser system. The He-Ne laser shares many of the advantages of the argon ion laser: low beam divergence, good pointing stability, and fair output power stability. In addition, it is low cost, small in size, and requires only minor environmental control. However, He-Ne lasers are typically low power devices that lack internal power regulation. Consequently, they require external optical support such as variable attenuators to adjust the power delivered to the final destination.

For many years the He-Ne laser was available in only one visible output wavelength — 632.8 nm. This put considerable limitations on its applications for fiber fluorescence sensor research because of the greatly reduced number of chromophores available at this wavelength. Recently, however, several manufacturers have developed a "green" version of this laser with an output wavelength as short as 543 nm that may find application in FOCS instrumentation.

### iii. Diode Lasers

The diode laser is a complex semiconducting device composed of an active layer that is sandwiched between material of lower refractive index. Current flowing through the device generates electron-hole pairs in the active medium which spontaneously emit radiation upon recombination. The surrounding lower index material creates a waveguide that confines the emitted light to a narrow region. At sufficient current levels the light intensity becomes high enough to produce stimulated emission and hence lasing action. The output characteristics of the diode laser are almost solely attributed to the active material.

The rapid commercialization of the diode laser is one of the most exciting current developments in laser research.[55] The original motivation for this development was the need for a compact, inexpensive, and efficient source for long haul fiber optic telephony. The transmission peak of silica optical fibers in the near infrared (IR) led to the development of the communications industry standard 1300 nm diode laser. More recently, the laser printing and optical disk industry has generated a need for a shorter wavelength diode laser. This has resulted in the commercial availability of several new diode laser types[56] with wavelengths as short as 670 nm. The development of shorter wavelength diode lasers is a materials research problem. Indeed, there is no fundamental theoretical limitation to the eventual development of diode lasers with output throughout the visible region of the spectrum. Short wavelength (400 to 600 nm) diode lasers would have a major impact on fiber optic sensor development. Table 1 lists available diode lasers and their significant operating characteristics.

### iv. Excimer Lasers

These are rare gas-halogen pulsed lasers that are characterized by moderately high average power ultraviolet (UV) output. Emission frequencies are available at 193 nm (argon fluoride), 248 nm (krypton fluoride), 308 nm (xenon chloride), and 351 nm (xenon fluoride). Excimers are relatively expensive lasers and require significant support utilities; nevertheless, they are becoming increasingly popular in medical research with several companies using them in laser/catheter hot tip angioplasty applications.

Owing to their high cost, compexity, and short output wavelength (poor fiber transmission and increased fluorescence background), eximer lasers have found limited use in the fiber sensor field. Despite the drawbacks of excimer lasers, their UV output provides the unique capability to directly probe for target compounds through their natural fluorescence, and hence there is an important niche for these lasers in certain fiber sensor applications.

### v. Nd-YAG Lasers

This is a flashlamp or diode pumped solid state laser with the neodymium ion lasing within a crystal host. Besides being a mainstay R&D laser for many years, it is also a popular choice for laser surgery and hot tip angioplasty. Its principal output wavelength is 1064 nm with harmonics of this wavelength (532, 355, and 266 nm) being achievable at reduced energy with non-linear optical devices. It is also available with 1320 nm output for fiber communications applications.

Although flashlamp pumped Nd-YAG lasers are expensive and require significant utilities support, diode pumped versions have recently become commercially available. The diode laser pumped version takes advantage of the high electrical to optical power conversion efficiency of the diode laser and the excellent frequency match for Nd absorption to produce a compact, air cooled laser operating at 532 nm. These lasers are currently quite expensive, but the price should fall dramatically as they are produced in larger quantities.

### vi. Dye Lasers

The dye laser is composed of an organic dye that has been dissolved in a suitable solvent and sequestered in an optical cavity. It typically receives its excited state energy by the direct absorption of the output of a second laser (the pump). This light is absorbed by an imprecisely defined (as contrasted with an atomic laser scheme) energy level, thus making a band of emission frequencies possible. By careful design of the dye laser resonator cavity it is possible to generate a tunable laser output. Both pulsed and CW dye lasers are available with outputs ranging from the UV to the near IR.

At first glance this appears to be the laser of choice for laboratory bench-top sensor research. Indeed, commercially available Nd-YAG pumped dye laser systems can generate continuously tunable radiation from 0.26 to about 4 μm. However, this impressive capability does not come without its cost. In addition to now having two lasers to maintain, the dyes have a limited tuning range (10s of nanometers) and lifetime, and both of these factors necessitate frequent dye changes. Also, the pump laser wavelength often must be changed to match the dye absorption band. As a consequence, a sophisticated dye laser system can consume nearly as much resource as the research that it is supporting.

### 2. Lamps

In contrast to the laser, lamp sources are distinguished by their relatively broad spectral emission and incoherent output. They are also generally less expensive and often require less utilities support for routine applications. Because of the large variety of existing lamps and their associated unique characteristics, the potential user must have considerable background in order to select the appropriate lamp for a given application. In this section, a brief overview of lamp technologies and characteristics that have specific application to FOCS instrumentation is presented. Before proceeding, however, it is useful to present a consistent set of measurement units since there is much confusion in the literature in this regard.

## TABLE 1
### Commercially Available Diode Lasers

| Type | Wavelength (nm) | Applications |
|------|-----------------|--------------|
| InGaAsP | 1100—1600 | Communication standard |
| GaAlAs | 750—905 | Optical storage, laser printing |
| GaInP | 670 | Bar-code reader |

There are two basic, and distinctly different, sets of units commonly used in the measurement of light: radiometric and photometric units. Radiometric measurement units are used to characterize the total power of radiation, while photometric measurement units are used to characterize visible radiation, i.e., how light is perceived by the human eye. In the latter, characterization of radiation over the visible portion of the spectrum requires correction by a weighting factor that is a measure of effectiveness in stimulating a human visual response. The weighting factor is dependent on the wavelength and whether the eye is light or dark adapted; the photopic or scotopic responces, respectively. The photopic curve is bell shaped with its peak value of 673 lm/W at 555 nm and its tails at the edges of the visual spectrum, i.e., 400 and 700 nm. This curve is commonly referred to as the standard "CIE" curve (short for Commision Internationale de l'Eclairage) and is reproduced throughout the commercial and scientific literature.[30,40] Photometric units are not discussed further here since they are rarely used when working with FOCS instrumentation. One exception is in converting manufacturer's data, which is sometimes expressed in photometric units, to radiometric units.

Terms used to describe radiometric quantities are often modified by the adjective "radiant" whereas similar terms in photometry frequently use the modifier "luminous". The reader is cautioned, though, that analogous quantities in radiometry and photometry are often denoted by the same symbol. Furthermore, it is not uncommon to find these terms misused in the open literature.

The fundamental unit of radiometry is the radiant power crossing an arbitrary surface. This is called the radiant flux, $\phi$, and is expressed in watts. The radiant flux impinging on a unit area of surface is called irradiation, E, and is measured in watts per square meter. Conversely, the flux per unit area emitted by a surface is called radiant emittance, $\varepsilon$, and also has units of watts per square meter. Both terms are defined in terms of the radiant flux by the equations

$$E = \frac{d\Phi}{ds} \quad \text{and} \quad \varepsilon = \frac{d\Phi}{ds} \tag{79}$$

where ds is the infinitesimal area over which $d\phi$ is measured.

At distances large compared to the size of the source, the irradiation decreases inversely with the square of the distance, r, as in

$$E = \frac{d\Phi}{ds_n} = \frac{I}{r^2} \tag{80}$$

where $ds_n$ is an area element normal to the propagation direction and I is a proportionality factor called the radiant intensity. Since the normal area element $ds_n$ at distance r subtends the solid angle $ds_n/r^2$, Equation 80 can be rewritten as

$$I = \frac{d\Phi}{d\Omega} \tag{81}$$

Thus the radiant intensity is the amount of flux radiated per steradian and has units of watts per steradian.

An important quantity, defined for those cases in which the area of the radiating source is not negligible, is radiance (or photometrically, brightness or luminance). Radiance is commonly denoted by the symbol L and is defined in terms of intensity as follows:

$$L = \frac{dI}{ds_n} \qquad (82)$$

Here, $ds_n$ is a differential area element of the emitting surface. Radiance has the units of watts per steradian per square meter.

Since the radiation emitted by lamps is distributed over a wide range of wavelengths, the quantity emitted in any particular wavelength region depends on the width of the region and on the functional form of the distribution. This distribution, often called the spectral radiance and denoted by, $L\lambda$, is the radiance per unit wavelength and is defined as

$$L_\lambda = \quad (\lim) \quad \frac{L(\lambda)}{\Delta\lambda} \qquad (83)$$
$$\Delta\lambda \to 0$$

There are two basic lamp types available — incandescent (or filament) and gas discharge. Although incandescent and gas discharge lamps are quite different in design, they share one common feature. For the purpose of our application to fiber instrumentation, they are all extended, or finite (as opposed to a point), light sources. As such, the important characteristic in comparing one lamp to another (within an application group) is its radiance (or photometrically its brightness or luminance). The reader is referred to Section III.A.2 for a discussion of the blackbody radiator model and its use in comparing different types of light sources on the basis of their equivalent tungsten filament temperature (Equation 27).

### a. Incandescent Lamps

The incandescent lamp consists of a current carrying filament enclosed in a glass or quartz envelope (bulb). Electric current is directed through the filament to resistively heat it to incandescence, thus forming a light source. The operating characteristics of the lamp are primarily determined by the length, diameter, and surface area of the filament. Tungsten wire or ribbon is by far the most popular filament material. There are dozens of filament geometric designs for the many commercial applications of incandescent lamps.

The spectral radiance (radiance/unit wavelength) of any source can be determined by multiplying the emittance of a blackbody at the same temperature by the emissivity of the radiator under consideration (see Section III.A.2). Since the emissivity of tungsten wire has been measured over the range of 0.25 to 2.6 μm,[42] it is relatively straightforward to predict the radiance of a tungsten filament lamp if its operating temperature is known.

The life of an incandescent lamp is primarily determined by the rate at which metal evaporates from the filament. This evaporation has a twofold detrimental effect, it ultimately weakens the filament to the point of mechanical failure and free metal deposits on the bulb causing a steady decrease in emittance. Filling the bulb with an inert gas slows the rate of evaporation and subsequent darkening of the lamp. However, the gas causes convective heat loss from the filament, thus lowering the apparent color temperature. To mitigate this side effect, tungsten is wound into a tight coil. This has been shown empirically to reduce heat losses both through convection and conduction to the filament supports.

A modern development of the incandescent lamp is the tungsten-halogen lamp. Here, a halogen gas, such as iodine, is added to the bulb atmosphere. The iodine is free to combine with the tungsten that has deposited on the bulb surface. In doing so it forms a volatile tungsten-iodine compound. This compound diffuses to the hot filament where it redeposits the tungsten metal. This process quickly achieves an equilibrium and provides a long-lived and stable lamp source, provided that the lamp is operated at a temperature high enough to activate the tungsten-halogen clean-up cycle. Consequently, the miniature tungsten-halogen lamp is a highly desirable incandescent source for FOCS applications.

### b. Gas Discharge Lamps

The gas discharge lamp is composed of two electrodes (anode and cathode) and a gas of neutral atoms contained by an envelope of suitable material (glass, quartz, or synthetic quartz). In its basic operation, free electrons in the gas are accelerated to high energy by a potential difference across the electrodes. These free electrons collide with the atoms of the gas and elevate bound electrons to excited states. The resulting luminosity originates from the gas and is due to a number of different light emission mechanisms. The most important are discrete radiation arising from spontaneous emission of excited atoms and nondiscrete continuum radiation from the heated plasma (ionized gas). With the proper conditions of gas pressure, voltage, and current, a steady-state light emission can be achieved. This emission can be either the glow discharge that we are familiar with in neon signs and fluorescent lighting or the high radiance of an arc source such as the carbon arc searchlight. Because of this significant difference in radiance, the pulsed (flashlamp) and continuous (CW) arc lamp are primary candidates for FOCS instrumentation.

#### i. Flashlamps

The flashlamp, primarily due to its commercially significant application as a laser pumping source, has experienced considerably more innovation in recent years than other light sources. As a consequence, the electrical, optical, and mechanical design parameters for their efficient use have been extensively investigated.[57] This has, in turn, led to routine commercial availability of a large variety of flashlamps and optimized power supplies. Therefore, a detailed physical knowledge of arc discharges is often not required for the researcher to obtain an appropriate lamp for a specified application. In this section we provide an overview of flashlamp technology that has particular relevance to FOCS instrumentation.

The flashlamp is composed of the electrodes, the fill gas, and an envelope (bulb). The output characteristics of the flashlamp are more strongly dependent on envelope material and gas type than on electrode design. The electrodes are designed to withstand high temperature and high current operation. As such, the material of choice is tungsten or an alloy thereof. The anode is always maintained at a positive potential relative to the cathode. The cathode is typically a composite material designed to freely provide electrons to the discharge. The size and shape of the electrodes depends very strongly on the lamp operating parameters. The higher the lamp output, the larger the electrodes must be in order to prevent damage.

There are a variety of envelope materials available. These include: fused silica, synthetic fused silica, and doped versions of the above. Fused silica is made by fusing natural quartz rock crystals. It is relatively inexpensive and is the primary flashlamp envelope material. However, impurities in the natural rock absorb UV radiation and tend to darken the envelope. This darkening (solarization) shortens the lamp lifetime as well as decreases its luminosity.

Synthetic fused silica is composed of highly processed, man-made quartz. As such, it is an ultra-pure material that is immune to solarization. Consequently, synthetic fused silica flashlamps have a higher radiance and last longer although they are considerably more expensive. Synthetic fused silica has a UV cutoff of approximately 170 nm while that of natural fused silica is approximately 260 nm. It is possible to alter this cutoff by the addition of suitable dopants. This is particularly desirable if the UV radiation is unnecessary for the application, since shortwave UV (below 243 nm) produces ozone by dissociation of oxygen molecules and also can solarize other components of an apparatus. The most popular dopant is cerium oxide which has a UV cutoff of approximately 375 nm.

The most common fill gases are xenon and krypton. This is due to their chemically inert nature and high conversion efficiency. Xenon lamps typically convert 60% of the input electrical energy to usable light in the wavelength region of 200 to 1200 nm. High radiance, compact xenon arc lamps typically have fill pressures greater than one atmosphere and as such they should be handled as a potential safety hazard.

The most important operating characteristics of a flashlamp are its spectral output, lifetime, and radiance. The spectral distribution (intensity vs. wavelength) of a flashlamp varies for the duration of the discharge. Consequently, the precise characterization of the lamp output is a three-dimensional function of time, intensity, and wavelength.[58] This level of detail is not readily available for most lamp types. In general, the output of a flashlamp is characterized by spectra that represent the total light intensity emitted at each wavelength integrated over the duration of the pulse discharge.

Figure 14a is an example of the output of a typical xenon flashlamp.[59] It is characterized by discrete emission lines in the region of 800 to 1000 nm, which are due to well characterized atomic state transitions, and a broad continuum that arises from the near blackbody emission of the plasma. For applications such as laser pumping, the discrete transitions are particularly useful whereas the broadband emission in the UV and visible portions of the spectrum is often used for a general fluorescence excitation source. It is possible, however (Figure 14b), to obtain a proportionally larger fraction of emission in the visible and near UV by operating the lamp at higher energy. Although this reduces lamp lifetime, as discussed below, it provides a more nearly optimum light source for fluorescence applications.

Due to the wide range of operating conditions for flashlamps, there is no universally applicable means to estimate lamp life. Reasonable empirical formulas exist for predicting lamp life when operation is in the low to moderate energy regime. Here, lifetime is primarily governed by material that is sputtered from the electrodes, particularly the cathode. This material deposits on the walls of the lamp, thereby reducing its output and increasing the thermal load on the envelope which, in turn, can lead to premature mechanical failure. The empirical equation below has been successfully used to predict the useful life of a variety of flashlamps:

$$\# \text{ useful pulses} = \left(\frac{E_0}{E_x}\right)^{-8.5} \tag{84}$$

Here $E_0$ is the operating energy and $E_x$ is the lamp explosion limit. This equation predicts that a lamp operated at 50% of its explosion limit will have a life of 350 pulses whereas operation at 20% of $E_x$ yields 1 million useful pulses.

Unlike the surface emitting tungsten filament lamp, the flashlamp is a partially transparent volume emitter. As such, the radiated power that can be collected is a function of the viewing depth of the optical system. In addition, there is a pronounced spatial distribution of lamp radiance, which can change with time, over the cross-section of the plasma. An optical system that observes a small region of the arc can experience intensity variations in excess of 20%. Consequently, the arc discharge lamp presents much more of a challenge to the optical designer than does the incandescent lamp.

Flashlamps have been designed for two basic applications, high radiance, such as the strobe light, or high aspect ratio, as found in many laser designs. The short arc (< 2 mm), compact xenon arc flashlamp, which closely approximates a point source, is the primary candidate for FOCS instrumentation. Operation at moderate energies will typically provide in excess of 10 million pulses. Versions of this lamp can be operated at rep rates up to several kilohertz.

*ii. Continuous Arc Lamps*

In general, many of the characteristics of flashlamps that were discussed above also apply to the continous wave (CW) arc lamp. The most significant difference is the gas fill pressure. CW lamps are typically operated at much higher pressure. This enables them to dissipate more heat between the electrodes, thereby generating high radiance with reasonable lifetimes.

As discussed previously, in optical applications that contain small limiting apertures such as a single optical fiber or the entrance slit of a monochromator, the highest radiance source provides the largest transmitted power. There are three arc lamp types that are readily commercially available: deuterium, mercury, and xenon. With rare exception, the highest radiance is available from the shortest arc version of these lamps.

The deuterium arc lamp is available as a moderately radiant source with source diameters as small as 1 mm. Its distinguishing characteristic is a continuum emission from 180 to 400 nm. Consequently, its principal application is as a source for UV spectroscopy.

Xenon gas and mercury vapor lamps are in a class referred to as short arc lamps. Arcs as small as 250 μm are available and are the highest radiance commercial lamp products. The mercury lamp emission is characterized by numerous discrete lines in the UV and visible portion of the spectrum. Xenon, on the other hand, has a relatively smooth and continuous emission throughout the visible region of the spectrum. As such, it is the preferred source for most visible spectroscopy applications. The short arc mercury lamp, however, is the highest radiance source available. The 100 W version of this lamp has five times the radiance of a 1000 W xenon arc lamp.

### c. Light Emitting Diodes

Light emitting diodes (LEDs) generate photons when electron-hole pairs recombine inside a pn-junction. In LEDs, the photons are generated via a spontaneous emission process as contrasted to stimulated emission in diode lasers. Electron-hole recombination results in the emission of photons with energy equal to the bandgap of the host material. For this reason, the wavelength the emitted light is inversely proportional to the bandgap. LEDs fabricated from a wide variety of materials are commercially available and operate at wavelengths ranging from 480 nm (silicon carbide)[60] to the near IR portion of the spectrum.

Although a wide variety of LEDs are available, it is often quite challenging to utilize them in FOCS applications. This is because (1) LEDs are available at relatively few wavelengths and are rather narrow band sources (20 to 40 nm), (2) the amplitude and frequency spectrum of LEDs exhibit drift with both temperature and aging, and (3) aside from the IR devices used in communications, most LEDs are packaged for use as indicator lamps rather than as optical sources. LEDs with visible output wavelengths are likely to find increasing application in FOCS instrumentation as their packaging is optimized for coupling to optical fibers and compatible sensing chemistries are developed. For more information on LEDs, the reader is directed to References 28, 30, 34-36, 47, 61, and 62.

### d. Summary

There are three major lamp types available for use in FOCS instrumentation: the incandescent (filament), gas discharge, and LED. In Figure 15, overlapping spectra typical of several important lamp types for FOCS instrumentation are shown for reference. The tungsten-halogen lamp offers a low price, relatively low power consumption, long lifetime, moderate radiance (2800 to 3200 K), stable output, and broad spectral output. The short arc versions of the xenon gas and mercury vapor lamps offer higher UV radiance than the tungsten-halogen lamp (5000 to 9000 K), but are shorter lived, less stable, expensive, and can be hazardous to operate. Flashlamps, on the other hand, have low average power dissipation, the highest UV radiance of all (10,000 to 15,000 K) and potentially long lifetimes provided that they are operated at low repetition rates and at low discharge energies. Unfortunately, flashlamps are relatively expensive, large in size, and require elaborate high voltage power supplies. LEDs have great, but as yet unrealized, potential as they are small in size and relatively inexpensive. While the radiance of long wavelength LEDs is quite high, the output power drops off dramatically as the wavelength decreases.

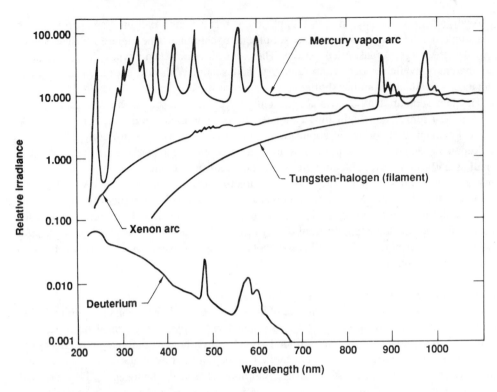

FIGURE 15.   Relative spectral output of important lamp sources for fiber optic sensor instrumentation.

## B. OPTICS
### 1. Refracting and Reflecting Elements

Refracting and/or reflecting optical elements are routinely used to manipulate light in FOCS instrumentation. Spherical, aspheric, and gradient index lenses fall into the category of refracting optics whereas elliptical, spherical, and parabolic mirrors are examples of reflecting optical elements. Each different optical element has its advantages and disadvantages in a given application, and it is beyond the scope of this chapter to provide more than a very brief introduction to this subject.

Section III.A.1 describes two standard methods of using simple lenses to couple light from an optical source to a fiber. This method is particularly suitable when intimate contact with the source is not possible and a relatively long focal length is needed for physical isolation, e.g., incandescent lamps, arc lamps, etc. In order to operate at high numerical apertures and long focal lengths, optical elements with diameters on the order of the focal length are required (since for lenses: lens diameter = 2 × focal length × NA).

Gradient index lenses (available under the trade name SELFOC)[63] and miniature spherical lenses are routinely used to couple light from solid state sources to fibers or from one fiber to another because of their small size and relatively low cost. Unfortunately, for the reason mentioned above, miniature lenses are usually not usable in coupling light from sources such as incandescent and arc lamps because of the need for physical isolation and long focal length.

It is also possible to use reflecting elements for any of the above applications. Reflecting elements have many potential advantages such as high coupling efficiency and a lack of chromatic aberration; however, reflecting elements are generally relatively large and expensive. For more information on refracting and reflecting optical elements, the reader is referred to References 30, 40, and 41 as well as the general literature.

## 2. Optical Couplers/Splitters

One of the principal advantages of the optical fiber based sensor is its small size. This advantage is ultimately realized in a sensor delivery and measurement system that is composed of a single fiber. This places constraints on the instrument designer since for most fluoresence-based sensors there are at least two wavelengths of interest propagating in the fiber: the excitation wavelength and the Stokes shifted fluoresence emission. A system based on a single fiber has the unavoidable characteristic of having the excitation and emission light sharing the same optical path. However, several innovative techniques that allow for the efficient injection of the excitation light while also efficiently separating the returning emission light have been developed. These techniques are either geometric or interferometric and they are described in more detail below.

### a. Geometric

The geometric coupler shown schematically in Figure 16 is the most versatile of the optical coupling schemes. Its principal optical element is simply a mirror with a small hole. The incident radiation is directed through the hole and into the fiber while the returning emission radiation, which exits the fiber at the numerical aperture of the fiber, is predominantly reflected by the mirror. The main advantages of this coupler are its small number of optical components and its achromatic response.

The efficiency of collection of emission light is determined by the ratio of the hole diameter to the diameter of the returning light at the mirror surface. On the other hand, the efficiency of rejection of excitation light is determined by the diameter of the incident light beam at the hole. Consequently, this configuration has its greatest applicability with laser excitation and high NA fibers.

A modified version of the above coupler is shown in Figure 17. Here the incident light is directed from off-axis by a small diagonal mirror into the focussing and viewing optic. The diagonal mirror occludes a fraction of the returning emission light that is directed through it and towards the detection elements. The principal disadvantage of this scheme is the mechanically fragile nature of the suspended diagonal mirror.

### b. Interferometric

The interferometric coupler relies on precision optical coatings to distinguish and separate incident and emission radiation (see Section IV.B.3.a). Schematic versions of this coupler are shown in Figure 18 and also in Figure 4b. Here, parallel incident and emission radiation that share the same optical axis are separated by a dichroic mirror. The dichroic mirror is an interference filter that is nominally designed to operate at 45° incidence (operation at other angles is possible) and to transmit excitation light and reflect emission light with high efficiency. The principal advantages of this technique are its potential for high efficiency and its ease of optical alignment.

In Figure 18, the dichroic is positioned in the light cone emerging from the optical fiber whereas in Figure 4b, only parallel light passes through the dichroic. Placement of the dichroic in the diverging light cone introduces a small but often acceptable blurring of the dichroic response curve. This is due to the fact that light near the edges of the cone travels through the dichroic at a different incident angles than light traveling along the optical axis.

A less sophisticated version of this coupler is obtained by replacing the dichroic mirror with a simple beamsplitter. The most common beamsplitter is a partially silvered mirror that partially transmits and partially reflects incident light. Mirrored beamsplitters can be made with almost any ratio of transmission to reflection. A 50% beamsplitter used in place of the dichroic mirror in Figure 18 would transmit only half of the available excitation light to the sensor and, likewise, it would only direct half of the returning emission light to the detector. This yields a maximum efficiency for the coupler of only 25%.

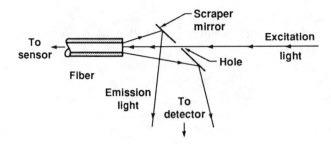

FIGURE 16.   Schematic diagram of a purely geometric fiber optic splitter.

The principal advantages of the silvered beamsplitter are its low cost and achromatic response throughout the visible portion of the spectrum. Metallic coatings, however, can have significant absorption which can produce thermally induced signal drift over short time periods. Beamsplitters formed with thin film interference coatings can operate at higher powers with considerably less drift. The principles of the design of reflectors based on thin film dielectric coatings is presented in Section IV.B.3.a.

### 3. Wavelength Selection

There are three basic requirements for wavelength selecting optics in fiber sensor instrumentation: (1) isolating an excitation band from a broadband optical source, (2) isolating an optical detector from stray light, and (3) separating excitation and emission light when they share the same optical path. The optical components that are readily available to achieve these results are the optical filter, the prism, and the diffraction grating. Advances in grating and filter technology have greatly diminished the utility of the prism in wavelength discrimination, and therefore its characteristics will not be further discussed here.

Although it is quite common to find comparisons of filters and gratings, they are actually quite dissimilar optical components. The filter is designed to pass a range (bandpass) or region (long or short pass) of wavelengths while the grating is designed to geometrically disperse multiple wavelength light. Consequently, the motivation for their use can often be quite different. The filter is most often incorporated in a dedicated end product of fixed application while the grating is most often found in an intermediate optical component such as the grating spectrometer or spectrograph. For the purpose of this discussion it is sufficient to make the observation that for a given optical configuration the filter offers the designer higher throughput whereas the diffraction grating offers higher resolution and tunability. As a consequence, one would expect to see the filter most often incorporated in dedicated or portable FOCS instrumentation and the grating in bench-top research level instrumentation. The characteristics of filters and gratings are discussed in more detail below.

### a. Optical Filters

As mentioned above, optical filters are generally constructed to pass a narrow band of wavelength (bandpass) or a broad region (long or short pass). The properties of these filters are best understood by referring to Figure 19a and b. Bandpass filters are characterized by their maximum transmission, center wavelength, bandwidth, and blocking (a concept that will be discussed in more detail later); whereas, long and short pass filters are characterized by their wavelength of 50% transmission, the steepness of cut-on or cut-off, and their maximum transmission. Two of the most popular and cost effective types of filters which have the properties described above are absorption and optical interference filters and are discussed in the sections that follow.

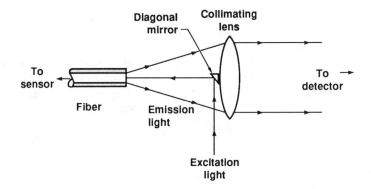

FIGURE 17. Schematic diagram of a diagonal injection fiber optic splitter.

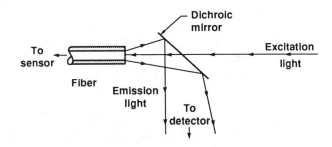

FIGURE 18. Schematic of a dichroic fiber optic splitter.

### i. Absorption Filters

Absorbing glass filters are available from the majority of optical supply companies and are typically made of high quality Schott or Hoya optical glass. Their light absorption properties are obtained through the incorporation of absorbing materials in a glass substrate. Since glass is an amorphous (noncrystalline) substrate these materials exhibit broad spectral features. As a consequence, absorption filters are characterized by large bandwidth (50 to 150 nm) and by slow cut-on or cut-off. The light absorbing component and its naturally occurring impurities often yield a nontrivial level of fluoresence.

There are several other characteristics of the absorbing glass filter that merit discussion. Since their spectral properties are achieved through absorption their transmission is a function of pathlength. Tilting the filter increases the pathlength thereby increasing absorption. They also exhibit a small thermal dependence which generally results in shifting their spectral curves to longer wavelength. Since these filters are predominantly glass, their transmission is limited to 2.5 μm in the IR and 290 nm in the UV.

Absorption filters are of limited utility in FOCS instrumentation because of their broad spectral features and potential for unwanted fluorescence emission. However, they are excellent for reducing stray light as a prefilter to a monochromator or as a blocker (see next section) for the more sophisticated interference filter.

### ii. Interference Filters

Figure 20 is a highly schematic drawing of a basic spectroscopic tool: the Fabry-Perot interferometer (or etalon). Its operation is based on two fundamental principles of electromagnetic theory: reflection and superposition. Reflection occurs at the interface or boundary of materials of different refractive index and its magnitude depends on the magnitude of the index difference. Superposition is manifest for light waves that occupy the same region of space and results in the observable phenomenon of constructive and destructive interference.

In Figure 20, two parallel and reflective surfaces are separated by an arbitrary distance. If this separation is equal to 1/2 the wavelength (or an integral multiple) of the light introduced into the cavity, then light reflected back to the first surface again reflects, but also constructively interferes with light traveling in the original direction. This results in the transmission spectrum of Figure 21 with transmission peaks at $\lambda/n$ where n = 1,2, .... Fabry-Perot interferometers are described in terms of their finesse, which is the ratio of the width of a transmission and the separation between successive peaks. The higher the reflectance of the mirrors, the higher the finesse and consequently the greater the resolution. These devices represent the basic technology underlying optical interference filters.

In contrast to the metallic mirrored surfaces of the Fabry-Perot, modern interference filters are composed almost entirely of highly transparent dielectric materials. Reflection is produced by depositing alternating layers of transparent materials of high and low refractive index onto a suitable substrate. If the thickness of these layers is equal to 1/4 the wavelength of the incident light, then a pronounced reflection occurs (it is possible to achieve reflectance greater than 99% by stacking many alternating layers). Such a device is called a quarter wave stack and is a fundamental building block of the interference filter.

In basic filter design, two quarter wave stacks are separated with a 1/2 wave thickness of low index material. This forms what is called the cavity or period of the filter. By layering several cavities in series, it is possible to produce a filter with the transmission spectrum shown in Figure 22. Here a well-defined central transmission, the passband, occurs at the design wavelength, with secondary transmission to both sides. Increasing the number of cavities results in a passband with steeper sides, a flatter top, and deeper (more effective) blocking. Filters made from as many as six cavities are routinely available while filters with ten or more cavities have been produced for special applications.

In most applications, a bandpass filter with the transmission spectrum shown in Figure 22 would allow an unacceptable amount of light to leak through its stopband. This leakage surpressed through the use of blockers. Blockers are usually laminated to the basic multiple-cavity interference filter and are chosen to optimize the performance of the filter over its operational wavelength band. Unwanted transmission to the short wavelength side is usually removed with long pass colored glass filters (see Figure 19b), whereas transmission to the long wavelength side is removed with appropriate quarter wave stacks and absorption filters.

Passband interference filters are usually specified by the width and location of the transmission band, angle of incidence, physical dimensions, and the degree of blocking. Since blockers reduce the passband transmission of the filter, it is not necessary or desirable to provide blocking in spectral regions where the source has no emission or the detector has limited or no sensitivity.

In addition to bandpass filters, interference filters are available as steep cut-off long and short pass filters (edge filters), as nonnormal incident long or short pass filters such as the dichroic filter, and as rejection band filters. Because of the custom nature of their spectral properties, these filters are excellent candidates for dedicated FOCS instruments. For more information, the reader is directed to References 40, 41, and 64 and to the literature provided by the various manufacturers and suppliers of optical filtering components.

### b. Diffraction Gratings

The diffraction grating works on the same principle as the interference filter; the constructive and destructive interference of light waves. However, the grating achieves the ultimate wavelength separation of incident light in quite a dissimilar fashion. Figure 23 shows a highly schematic drawing of a diffraction grating. Here a reflecting surface is ruled with a high

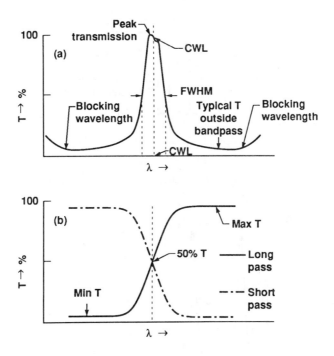

FIGURE 19.  Transmission characteristics of (a) bandpass and (b) long and short pass optical filters.

density of precision grooves. Parallel light incident on this surface sees a series of reflecting surfaces separated by equal spacing. In the example of Figure 23, parallel rays of light that are in phase as they pass through a plane normal to the plane of the figure and containing line a-a′ will remain in phase after reflection and hence constructively interfere if they travel an equal distance in arriving at line b-b′. In the example of Figure 23, ray 2 travels a distance (d sin α - d sin β) farther than ray 1. If the extra difference traveled is an integral multiple of the wavelength, $\lambda$, as in

$$n\lambda = d\sin\alpha - d\sin\beta \tag{85}$$

then the rays constructively interfere and the grating exhibits a strong reflection at an angle β from the normal. If the wavelength is shifted slightly while maintaining the angle of incidence constant, the condition for a strong reflection still exists but at a slightly different reflection angle. The grating is, therefore, a dispersing optic that reflects light of different wavelengths in different angular directions.

The significance of this can best be understood with the example shown in Figure 24. Here, white light from a nearly point source (such as an optical fiber) is collimated and reflected from a diffraction grating and then refocused at $s_2$. Rather than recover the point image of the fiber at $s_2$, the dispersion of the grating yields, to first order, a discrete line of light with wavelength increasing linearly from one end to the other. This illustrates the principle of linear dispersion of a grating, and that along with its ability to reject stray light are the fundamental measures of its performance.

It is rare for even the most sophisticated researcher to use a bare grating in an experimental design. The need to condition the light before and after incidence on the grating has given rise to devices such as the monochromator or spectrometer and the spectrograph. However, the ability of the researcher to specify the optimum spectrograph or spectrometer for his work requires a moderate understanding of grating and spectrometer design.

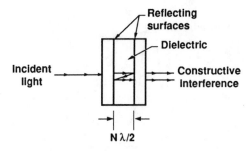

FIGURE 20. Schematic diagram of a typical optical coating cavity used in the design of many types of interference filters.

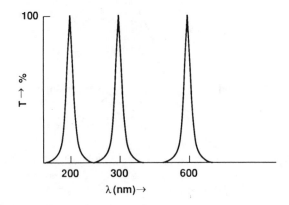

FIGURE 21. Transmission of a cavity tuned for 600 nm.

Figure 25 is a schematic representation of one of the most popular monochromator configurations, the Czerny-Turner (CZ). The most significant parameter of this and most spectrometers is its effective f-number (or f/#). The f/# is defined as the ratio of the focal length to the diameter of the limiting aperture and is a measure of the throughput of the monochromator. In the example of Figure 24 the limiting aperture is mirror M1. Due to purely physical limitations, most CZ spectrometers are designed to an f/# of approximately 4 to 5. A second important characteristic is stray light rejection which is a measure of the fraction of light that appears at the output incorrectly and is caused by scattering from the grating and other optical components within the spectrometer. Stray light rejection is typically specified by monochromator manufacturers for monochromatic incident light. For low signal and high background level applications, the stray light rejection can be greatly improved using a double monochromator. This is simply two identical monochromators separated by a central slit.

The principal specifications of the diffraction grating are its size, groove density, and blaze angle. The size of the grating is specified by the spectrometer manufacturer and is a measure of the limiting f/# of the spectrometer, the groove density is a measure of maximum resolution achievable, and the blaze angle specifies the most efficient wavelength region of operation.

The influence of the grating parameters on fiber instrument design are illustrated in the following example. Figure 25 shows a schematic of an experimental arrangement designed to spectrally analyze the light emitted from a 200 μm diameter fiber (NA = 0.4) with an f/5 CZ grating monochromator. Here, the fiber emits light into a cone of approximately f/1.25. This light is collimated by a lens (L1) of appropriate focal length and focussed into the spectrometer by the appropriate f/# matching lens (L2). The image of the fiber at the entrance slit of the

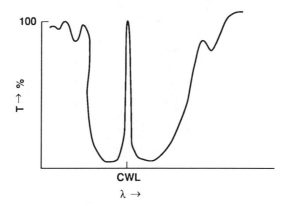

FIGURE 22. Typical transmission of an optical filter composed only of tuned cavities.

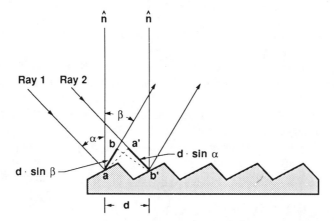

FIGURE 23. Schematic diagram illustrating the principles of the optical diffraction grating.

monochromator is thus magnified by the ratio 5/1.25 = 4. This yields an image size at the entrance slit of the monochromator of 800 μm which represents the minimum slit width for efficient coupling of the fiber into the monochromator. This slit width also sets the maximum resolution obtainable for this configuration without loss of signal. Since many fiber sensor applications are signal limited, it is undesirable to increase resolution by narrowing the entrance slit.

In addition to a changes in slit width, it is also possible to alter the resolution by a simple modification of the monochromator. Gratings are routinely available in groove densities of from 300 to 2400 grooves per millimeter. The higher the density the greater the resolution. For example, the resolution of the system shown in Figure 25 could be varied by a factor of 8 using such gratings. Consequently, groove density is an important specification for a monochromator.

Diffraction gratings are available as replicates of precision ruled gratings or as original or replicates of a holographic process. The latter process is a modern development that combines photosensitive polymer and laser technology. The nearly perfect interference fringes that can be formed with intersecting laser beams are used to expose a layer of photoresist that has been deposited on a glass substrate. Subsequent processing and coating of this surface with a suitable reflecting material yields an original grating of very high stray light rejection.

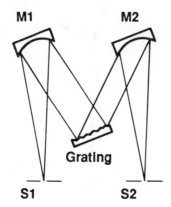

FIGURE 24.   Schematic representation for the use of a diffraction grating in a Czerny Turner monochromator. Here, diverging light from slit or pinhole S1 is collected by mirror M1 and directed parallel onto the grating surface. Wavelength separated light is then redirected by mirror M2 to an exit opening at S2.

Holographic gratings are capable of rejecting stray light as much as 10 to 100 times better than the convential ruled grating.

Plane holographic gratings are more expensive and less efficient than conventional ruled gratings and are often not the grating of choice for many routine optical applications. On the other hand, super-abberation corrected concave holographic gratings[65] are nearly ideal for FOCS applications. These gratings are both a dispersing and focusing element, consequently, they permit the design of monochromators that have no other optical components. Concave holographic gratings are small, relatively inexpensive, have high throughput and stray light rejection, and are ideally suited for use with array detectors. For additional information on the design of optical systems utilizing diffraction gratings, the reader is directed to Reference 66.

## C. OPTICAL RADIATION DETECTORS

The ideal optical radiation detection system would unambiguously detect and record all photons directed toward it. Such a system would require a perfect photodetector; one that generated a readily identifiable signal for each incident photon. Unfortunately, perfect photon detectors do not exist, and phenomena such as dark current, shot noise, and background radiation provide fundamental limitations to detection. Modern technology has, however, provided many effective options for the detection of light. Choosing the proper one for each particular application requires a comprehensive knowledge of the devices available and their operating characteristics.

Modern detectors are commonly divided into three classes: thermal, photoemissive, and quantum. In thermal detectors, the energy carried by incident photons causes a temperature increase which is measured using standard calorimetric techniques. Since thermal detectors are most often used as power meters, they will not be further discussed here. Nonetheless, it should be pointed out that thermal detectors are very useful because of their high accuracy and insensitivity to the wavelength of the incident radiation.

Photoemissive and quantum detectors are based on the emission of electrons from a photocathode and the generation of electron-hole pairs in a semiconducting element, respectively. In choosing a photodetector for use in FOCS instrumentation, the designer must consider trade-offs between a number of factors. These include: detectivity, sensitivity, noise level, longevity, accuracy, spectral response, and response time. In addition, cost must also be considered since the most sophisticated photodetectors can become prohibitively expensive in certain applications, particularly when the detector is cooled. Lowering the operating temperature of either PMTs of quantum detectors generally lowers the noise and thereby improves the performance.

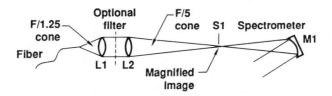

FIGURE 25.    Schematic diagram of an optical system that matches the output of an f/1.25 fiber to an f/5 monochromator with provisions for an optional filter.

Comparisons of the different types of detectors is often a complex and painstaking task. The figure of merit that is most often used to compare quantum detectors is D-star (D*). This is the signal-to-noise ratio obtained when 1 W of incident power illuminates a detector with 1 cm active area in a 1 H bandwidth at a given wavelength. D-star is directly proportional to the responsivity of the detector and inversely proportional to its inherent noise. Unfortunately, D-star is not a very useful concept with PMTs; however, it is possible to compare the performance of PMTs and quantum detectors, but only on a case-by-case basis due to the many factors that are involved. A comparison of PMTs and solid state photodiodes is presented in Section IV.C.3.

There are currently over 40 detector types (and literally hundreds of manufacturer-specific models) currently available. The spectral range covered by several of these detectors in the 0.2 to 20 μm region of the electromagnetic spectrum is given in Figure 26. An overview of the major detector types that have potential application to FOCS instrumentation is presented below. For more detail than presented here, the reader is referred to Section III.B.2 and References 44 to 47.

### 1. Photoemissive Detectors

The photomultiplier tube (PMT) is by far the most popular photoemissive detector in use today. A schematic diagram of a typical PMT is shown in Figure 27. It comprises a vacuum tube, a photoemissive surface (photocathode), and an electron multiplier.

When a photon of sufficient energy strikes the photocathode, it dislodges an electron via the photoelectric effect. This electron is accelerated by a voltage drop across a gap to a secondary emitting electron surface (dynode) where its numbers are multiplied. This process is continued for several stages and is the concept of gain in the PMT. The resulting pulse is then collected at the anode where it can be readily detected. Gains of $10^6$, which are routinely achieved in PMTs, are the primary reason for their uniquely high sensitivity.

The photocathode is chiefly responsible for the operating characteristics of the PMT. It consists of a thin (semitransparent) or thick (opaque) layer of predominantly alkalai metals with low work functions. The exact composition of these metals give the photocathode its unique spectral characteristics. For many years, manufacturers designated the most common cathodes with the nomenclature S1 to S20, although recent advances in number and type of cathode have resulted in a straying from this convention. The typical spectral response of a number of the more commonly used photocathodes (measured in milliamps of cathode current per watt of incident radiation) is shown in Figure 28. From the figure, it can be seen that the quantum efficiency ranges from 0.1 to about 25% depending on the photocathode material and the wavelength.

The PMT is uniquely suited to operation in extremely low light level situations. This is primarily due to the fast rise and fall times of the photo-pulses that arrive at the anode which allow the tube to be operated with a signal acquisition technique known as photon-counting. The principal components required for photon-counting are a PMT with suitable housing and high-voltage power supply, an amplifier, a discriminator, and a ratemeter or photon counter. Many available PMTs are specifically designed for photon-counting applications. The PMT should be selected for high quantum efficiency in the wavelength range of the application. It should be powered with a low noise, high stability supply, and maintained in a magnetically shielded housing. Since the pulses arriving at the anode are the result of a single photon event at the photo-cathode, they are low in amplitude and require amplification. A high speed amplifier is needed so as not to degrade the pulse rise time. This amplifier should be placed as close to the anode as possible to prevent pulse degradation due to the capacitance of the interconnecting cables. It is common practice to use a small (low gain) pre-amp directly attached to the anode followed by a variable gain post-amplifier. The resulting amplified pulse train is then passed through a discriminator. The descriminator is designed to perform two main functions. Its output is a conditioned TTL or NIM pulse train whose frequency is equal to the input train. This allows for very accurate recording of the signal with either a ratemeter or digital photon counter. It also has a variable threshold level whereby pulses less than the threshold level are rejected. This allows the user to discriminate against dark noise since most dark pulses are low energy pulses originating from random thermionic emission in the dynode chain. By proper selection of discriminator threshold (an empirical process) it is possible to virtually eliminate the effects of the PMT dark current.

Figure 29 is a schematic diagram of a common photon-counting system. Here, the output of the discriminator, which is proportional to the incident light level, is directed simultaneously into two data acquisition systems. One, a ratemeter, which typically responds to analog signals, generates an output voltage that is proportional to the input pulse frequency. This voltage is readily monitored with a chart recorder. The other, a digital photon counter, records the actual pulse frequency. It is typically interfaced to a computer for data acquisition and reduction.

### 2. Quantum Photodetectors

As stated previously, quantum detectors are semiconductors that either change their conductivity or generate a voltage upon illumination with light in material-specific wavelength ranges. Combinations of semiconducting materials along with various dopants have led to a large variety of detectors with sensitivities from the near UV to, and throughout, the visible and near IR portions of the spectrum. A small subset of these devices, the photoconductor and the solid state photodiode, have significant application to fiber sensor instrumentation and are discussed in more detail below.

#### a. Photoconductive Detectors

The photoconductive detector, often synonymously referred to as a photoconductor, is a semiconductor that changes its conductivity when exposed to light of the appropriate wavelength. The incoming light generates an increased number of charge carriers thereby increasing the conductivity of the device. The fundamental photoconductive principle is illustrated in Figure 30a. Here, a photon of sufficient energy elevates an electron from the valence band to the conduction band of a pure (or intrinsic) semiconductor.

The long wavelength cut-off limit of a photoconductor is directly related to the energy required to transit the bandgap, although is possible to extend the infrared response of a given semiconducting material by the addition of a suitable dopant. The effect of adding a dopant which acts as an electron donor and whose energy level is within the bandgap is shown in Figure 30b. Now, donor electrons can be elevated to the conduction band with photons of significantly lower energy or longer wavelength. In an analogus fashion, dopants that act as acceptors can be added to the semiconductor with similar results. Devices with donor or acceptor levels lying within the bandgap are called extrinsic photoconductors and generally are operated at liquid helium temperatures due to their undesirably high levels of thermally induced carrier generation.

Figure 31 shows the relative spectral response of several of the more popular photoconduc-

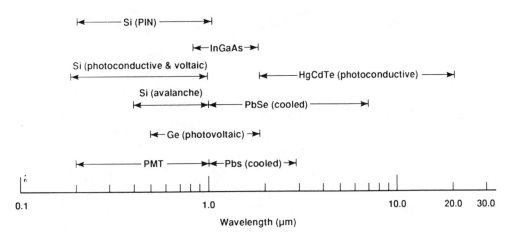

FIGURE 26. Wavelength response of popular detector types in the visible and near infrared portion of the spectrum.

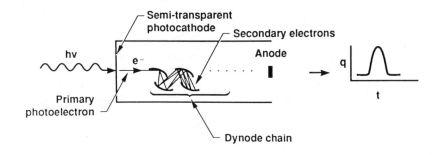

FIGURE 27. Schematic diagram of a photomultiplier tube.

tive detectors — lead sulfide (PbS), lead selenide (PbSe), and indium antinomide (InSb). Here, log D-star (see Section IV.C for definition) is plotted against log wavelength. As is evident from Figure 31, photoconductors have their greatest application in the region of 1 to 5 μm. The mercury-cadmium-telluride (MCT) photoconductor has excellent response in the 5 to 20 μm region and is used predominantly in infrared imaging applications. Because of their relatively high levels of generation-recombination and Johnson noise, photoconductors are most often cooled to improve their noise performance and many types are available with built-in thermo-electric cooling.

The photoconductor is a passive device that requires an external power source to establish a current flow. Figure 32 shows an abbreviated circuit diagram for the operation of a photoconductor. A bias voltage is applied to the detector, and the current flowing through it is forced, by the op amp, to flow through the feedback resistor thereby generating a voltage at the output which is proportional to the resistance of the photoconductor.

Photoconductors are not often used in FOCS applications because solid state photodiodes provide higher performance at a lower cost in the wavelength region where FOCS are traditionally operated. It is for this reason that a more detailed discussion of the noise performance of photoconducting detectors is not presented here. Readers interested in more information are referred to References 46 to 48.

### b. Solid State Photodiodes

As opposed to the photoconductor, which changes its resistance when exposed to light, the solid state photodiode generates a current. As shown in Figure 33, a depletion zone is formed at the junction of the p and n materials. The depletion zone is a region within a few microns

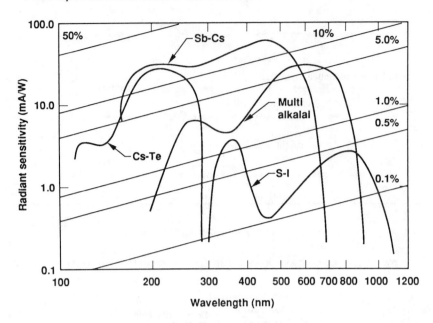

FIGURE 28.   Spectral response of the photocathodes of some of the more popular photomultiplier tubes.

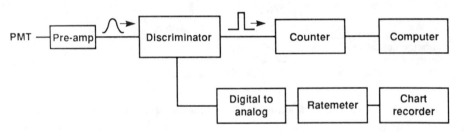

FIGURE 29.   Block diagram of a commonly used photon-counting scheme.

of the junction where the electrons and holes have diffused away from their respective donors and their concentrations are very low. The concentration difference of electrons and holes across the p-n junction (or doping levels) causes the electrons and holes to diffuse to the opposite side of the junction where their respective concentrations in the bulk material are much lower. Since the diffusion of charge leaves ionized donor and acceptor atoms behind, a potential difference (and its attendant electric field) between the p and n region builds up until the number of electrons and holes crossing the junction due to the electric field is exactly equal to the number crossing the junction due to diffusion and a steady state is achieved.

Incident photons having an energy greater than the bandgap generate electron-hole pairs in the semiconducting material. If these pairs are generated in the depletion zone, then the electrons in the p region are driven into the n region by the junction potential while the opposite occurs for holes. It is this separation of charge that causes a current to flow in an external circuit when the diode voltage is $\leq 0$ V and is the most popular mode of operation of solid state photodiodes.

The diode sensitivity $S_d$ (in amps/watt) is defined after van der Zeil[45] as

$$S_d = \frac{(1-R)\eta q \lambda}{hc} \tag{86}$$

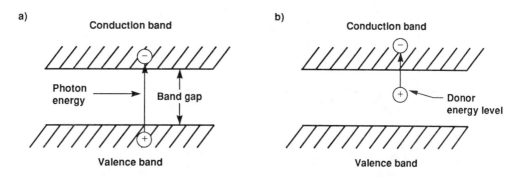

FIGURE 30. Illustration of the operating principles of a photoconductor. In (a), a photon of sufficient energy elevates an electron from the valence to conduction band, while in (b) a photon of lower energy achieves the same result when a donor atom is present whose energy level lies within the bandgap.

where R is the surface reflection coefficient, $\eta$ is the quantum efficiency of the diode ($\eta \approx 0.8$ to 0.9 for silicon), q is the electronic charge ($1.602 \times 10^{-19}$ C), $\lambda$ is the wavelength of the incident radiation in meters, h is Planck's constant ($6.6262 \times 10^{-34}$ W sec$^2$), and c is the speed of light in vacuum ($2.9979 \times 10^8$ m/sec). Assuming no special antireflective coating, i.e., 1-R = 0.96, Equation 86 can be expressed more simply as

$$S_d = 0.774\eta\lambda \qquad (87)$$

It should be noted that the diode sensitivity is generally referred to as "responsivity" in the open literature. The nomenclature used in this chapter departs from convention in order to be consistent with the symbol for the anode sensitivity of the PMT, $S_a$.

The diode sensitivity drops off dramatically from that predicted by Equations 86 and 87 at extremely short and long wavelengths as is evident from Figure 34. For short wavelengths, the photon absorption coefficient is very high. This causes the carriers generated by the incoming photons to recombine at or near the surface without reaching the depletion zone. For long wavelengths (corresponding to photon energies below the bandgap), the semiconductor becomes transparent and the photons pass through unabsorbed. An effectively designed photodiode can deliver a linear response to incident radiation over as many as eight decades of current.

The most common materials used in photodiodes are silicon (Si), gallium arsenide (GaAs), germanium (Ge), indium arsenide (InAs), and indium antinomide (InSb). Figure 34 shows the relative responsivity of these photodiode types. In general, photodiodes offer shorter wavelength response and higher D-stars at room temperature than their photoconducting counterparts, however, photoconductors are the only choice for wavelengths longer than 2 μm.

The capacitance of the p-n junction is primarily responsible for its response time. It is possible to significantly lower this capacitance by the application of a reverse bias voltage and/or with a modified p-n junction. One of the most successful modifications is the insertion of a high resistance layer (low doping level) between the p and n layers. This structure, called the pin diode, offers ultra-fast response times.

It is also possible to greatly increase the sensitivity of the photodiode with a structural and operational modification. In the avalanche photodiode, the p-n junction is separated by a weak electron donor. Free electrons in this region are accelerated by an external bias voltage and dislodge additional electrons in collisions with atoms in the lattice. This creates a multiplication effect very similar to that of a PMT and results in an internal gain. Because of their internal gain, avalanche photodiodes require less external gain and hence can outperform

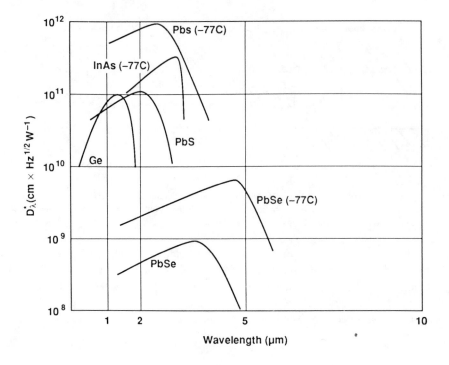

FIGURE 31.   Relative spectral response of several popular photoconductors.

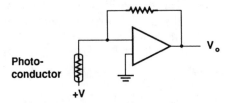

FIGURE 32.   Schematic diagram of a basic photoconductor circuit which produces an output voltage that is directly proportional to the resistance of the photoconductor.

standard photodiodes in certain applications. Unfortunately, avalanche photodiodes, like PMTs, require highly regulated power supplies, and the gain of avalanche photodiodes vary with temperature and must be compensated if a constant gain is required.

### 3. Comparison of Minimum Detectable Power of PMTs and Photodiodes

The minimum detectable power, $P_m$, of a PMT is defined as the amount of optical power required to produce a signal at the anode whose magnitude is equal to the PMT anode rms noise current, $i_a$, and is given by

$$i_a = P_m \overline{S}_a \tag{88}$$

where $\overline{S}_a$ is the average anode sensitivity (in amps/watt) over the operational wavelength band. An expression for the minimum detectable power is then found by equating Equation 88 to Equation 42 and substituting $P_m$ for $P_{in}$ as in

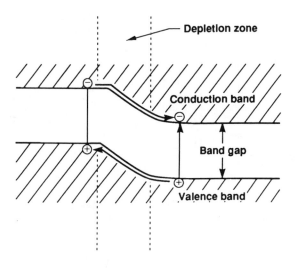

FIGURE 33. Energy band diagram of a p-n junction photodiode.

$$P_m \overline{S}_a = i_a = \sqrt{\frac{\delta 2qGB\left(P_m \overline{S}_a + I_{dark}\right)}{(\delta - 1)}} \tag{89}$$

Here, all of the terms are as defined in Section II.B.3. In most cases, the term $P_m \overline{S}_a$ is only a few percent of $I_{dark}$ and can be neglected.

For a bandwidth of 1 Hz, typical values for $P_m$ are in the range of $10^{-16}$ to $10^{-14}$ W for most PMTs depending on the wavelength. $P_m$ can be much higher if the PMT is operated in a spectral region of low sensitivity. This is one reason that PMTs are available with a wide variety of photocathode materials designed for operation in different parts of the optical spectrum.

The minimum detectable power for a solid state photodiode is simply the noise current due to the Johnson noise associated with its shunt resistance, $R_s$, (with 0 V across the device) divided by the average diode sensitivity, $S_d$, over a given wavelength interval:

$$P_m = \frac{\left(\sqrt{\dfrac{4kTB}{R_{sh}}}\right)}{\overline{S}_d} \tag{90}$$

The minimum detectable power for silicon photodiodes can range from $10^{-13}$ to $10^{-15}$ W.

It is, in general, much more difficult to realize the theoretical value for $P_m$ in practical circuits for solid state photodiodes than for PMTs. This is because the large and almost noise free built-in gain of the PMT permits operation in circuits with lower gain and hence the noise at the output is dominated by that of the PMT. Solid state photodiodes, on the other hand, require very high gains in order to boost the very small output current from low level light inputs to measureable voltage levels. Depending on the demands of each specific application, e.g., response time, bandwidth, etc., the total noise may not be dominated by the Johnson noise generated by the diode shunt resistance but rather by some other electronic component in the circuit.

In conclusion, if it is required to detect light levels less than $10^{-15}$ W, then PMTs present the only solution. Assuming that the associated circuit can be designed to allow the Johnson

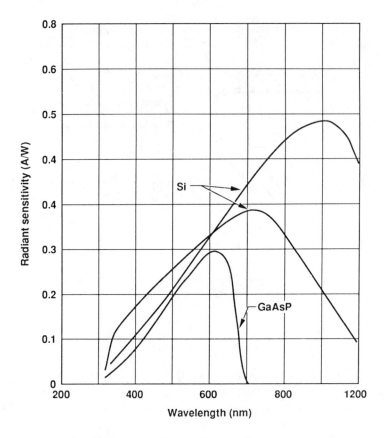

FIGURE 34.   Relative spectral response of several photodiode types.

noise of the solid state photodiode to be dominant, it is possible for the minimum detectable power of a silicon photodiode to come within a factor of 10 of that for a PMT. In most cases, however, photodetectors based on PMTs will outperform those based on solid state photodiodes by two to three orders of magnitude, and potentially even more when cooling and photon-counting detection schemes are employed.

## APPENDIX I

## DERIVATION OF THE SENSOR TRANSFER FUNCTION FOR A CLEAVED FIBER IN SOLUTION

In Section III.E.1, the resolution of a fiber optic pH sensor with the geometry shown in Figure 13 is calculated in order to provide an example of the method of analysis presented in this chapter. This appendix provides an introduction to some of the basic concepts as well as the details of the analysis leading to the calculation of the sensor transfer function for the case where a cleaved optical fiber is immersed in a solution containing a fluorescent dye having an index of refraction $N_o$. The volume of dye solution lying within the light cone emerging from the optical fiber is excited by the incident radiation and subsequently emits Stokes shifted radiation into the surrounding medium over a solid angle of $2\pi$ sr. Only that portion of the emitted light which is within the numerical aperture and incident on the endface of the optical fiber core is recovered.

The important parameters of the fluorescent dye are its quantum efficiency, $\phi$ (dimensionless), the extinction coefficient, $\varepsilon$ ($\mu m^{-1} M^{-1}$ where $M$ is molar concentration), the total dye concentration, $C_t$ (moles per liter), and the concentration of fluorescing dye molecules $C_F$ (moles per liter). As implied above and in Equations 63 and 64, only the fluorescent dye molecules are assumed to absorb and emit light and the total dye concentration is equal to the sum of that of the fluorescent and nonfluorescent molecules. The amount of light absorbed by a slab of fluorescent dye molecules of thickness u is proportional to the quantity $\varepsilon C_F u$ which is commonly referred to as the absorbance or optical density of the solution. The amount of light emitted by the same slab is proportional to the product $\phi\varepsilon C_F u$ .

The ratio of the incident to the final radiance, $L_u/L_0$, is given by the Beer-Lambert Law and can be expressed as

$$\frac{L_u}{L_0} = \exp\left(-\varepsilon' C_F u\right) = 10\left(-\varepsilon C_F u\right) \tag{91}$$

where $\varepsilon'$ is defined as

$$\varepsilon' = 2.303\varepsilon \tag{92}$$

Figure 35 illustrates how the radiation emitted from a disc shaped differential element of dye solution of thickness dz located at a distance u from the endface of the optical fiber returns to the endface of the optical fiber. The coordinate system of Figure 6 is assumed with the difference that the x and y axis are transposed so that z is increasingly positive as the distance into the solution and away from the fiber endface increases. The area of the fluorescing disc at z = u is

$$A(u) = \pi\left(\rho + u\tan\alpha_c\right)^2 \tag{93}$$

where $\rho$ is the radius of the fiber and $\alpha_c$ is the angle of a ray of light emerging from the fiber at the limit of the numerical aperture of the fiber (see Equation 6 and Figure 1). The radiance of the light emitted from the disc, $dL_{em}(u)$, is given by

$$dL_{em}(u) = L_{ex}(u)\Phi\varepsilon' C_F dZ \tag{94}$$

where $L_{ex}(u)$ is the radiance of the excitation light at z = u. The radiance of the excitation light along the z axis at a point z = u decreases due to the increased area over which the power emerging from the fiber is distributed in conjunction with absorption by the dye and is given by

$$L_{ex}(u) = L_{ex}(0)\frac{A_f}{A(u)}\exp\left(-\varepsilon' C_F u\right) \tag{95}$$

where $A_f$ is the area of the fiber ($\pi\rho^2$) and A(u) is given by Equation 93.

The next step in the analysis is to relate the radiance of the light emitted from the disc at z = u, $L_{em}(u)$, to that at the endface of the fiber $L_{em}(0)$ . As indicated in Figure 14, by symmetry, the area of the disc of emitted light at z = 0 is equivalent to the area of disc of the excitation light at z = 2u or A(2u). The conversion of excitation light to emission light can be thought of as a "virtual reflection" at z = u. Unlike ideal reflection, much of the energy of the returning light is lost in the conversion from excitation to emission light at the disc at z = u. This is because the light at each point on the disc emits into $4\pi$ sr. whereas the fiber collects only the light within the solid angle, $\Omega$, defined by its NA. The radiance at the fiber endface is thus equal to that of the disc at z = u reduced by the product of the ratio of the area of the light cone at z = 2u to the area at z = u and the ratio of the solid angle of acceptance of the fiber to $4\pi$ sr. This is expressed mathematically as

$$dL_{em}(0) = \frac{A(u)}{A(2u)} \frac{\Omega}{4\pi} dL_{em}(u) \tag{96}$$

In order to simplify the analysis, at this point it is assumed that the radiance of the disc of emitted light is approximately uniform. This implies that the radial decrease in $L_{ex}$ with respect to the axis of the fiber due to increased path length and absorption by the dye is negligible. This is a reasonable assumption because (1) with NA = 0.5 and $N_o$ = 1.33, the extra distance traveled by light leaving the fiber at an angle of $\alpha_c$ results in only 8% of additional path length ($u/_{cosac}$ = 1.08 u); (2) the amount of light collected from the periphery of the emitting disc is small compared to the total; and (3) for dye concentrations used in practice, the total attenuation of the excitation light is generally only a few percent in the volume effectively sensed by the optical fiber. The assumption of radial uniformity allows the transfer function to be computed using a one dimensional integral in z.

The complete expression for the radiance of the emitted light at z = o is now found, by combining the results of Equations 93 to 96, to be given by

$$dL_{em}(0) = L_{ex}(0)\Phi\varepsilon'C_F \frac{\Omega}{4\pi} \frac{A_f}{A(2u)} \exp(-\varepsilon'C_F u)dz \tag{97}$$

where the solid angle, $\Omega$, is found by integrating Equation 3 from $\theta = 0$ to $\theta = \alpha_c$:

$$\Omega = \int_0^{\alpha_c} 2\pi \sin\theta \, d\theta = 2\pi(1 - \cos\alpha_c) = 4\pi \sin^2 \frac{\alpha_c}{2} \tag{98}$$

The complete expression for the sensor transfer function, $H_t$, can now be obtained using Equation 8 to convert $L_{ex}(0)$ and $L_{em}(0)$ to incident and returned power, respectively, and by substituting $\Omega = 4\pi\sin^2\alpha_c/2$, $A(2u) = \pi(\rho + 2u\tan\alpha_c)^2$, and z for u into Equation 97. After rearranging and setting up the integration from z = 0 to z = $\infty$, one obtains the relation

$$H_t = \frac{P_{em}}{P_{ex}} \approx \frac{\left(NA^f\right)^2 \left(N_0^s\right)^2}{\left(NA^s\right)^2 \left(N_0\right)^2} \Phi\varepsilon'C_F \sin^2\left[\frac{\alpha_c}{2}\right] \int_0^\infty \frac{\exp(-\varepsilon C_f z)}{\left[1 + \dfrac{2z\tan\alpha_c}{\rho}\right]^2} dz \tag{99}$$

where $NA^s$ and $NA^f$ are the numerical apertures of the source and fiber, respectively, and $N_0^s$ is the index of refraction of the medium surrounding the endface of the optical fiber where the excitation light is coupled in. For the case where $NA^s = NA^f$, $N_0^s = 1$, and all of the light is collected from within a distance z= $z_m$, Equation 99 simplifies to

$$H_t \approx \frac{\Phi\varepsilon'C_F}{N_0^2} \sin^2\left[\frac{\alpha_c}{2}\right] \int_0^{z_m} \frac{\exp(-\varepsilon'C_F z)}{\left[1 + \dfrac{2z\tan\alpha_c}{\rho}\right]^2} dz \tag{100}$$

With the additional assumption that $\varepsilon'C_t z_m \le 0.2$, the exponential inside the integral can be approximated by $(1 - \varepsilon'C_F z)$ to an accuracy of better than 5%. After integrating between z = 0 and z = $z_m$, the expression for $H_t$ becomes

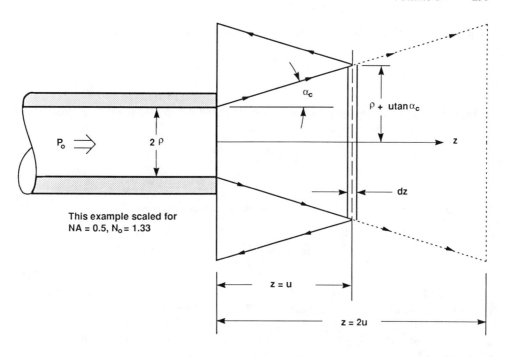

FIGURE 35. Close-up view of the tip of the fiber optic pH sensor shown in Figure 13. The outermost ray of the excitation light emerges in a cone defined by the critical angle (22° for NA = 0.5). Emission light from a disc shaped element of differential thickness at a distance z = u from the fiber endface results in a secondary light cone directed back at the fiber endface. The total collected emission light is composed of the sum of all such discs lying between z = o and infinity.

$$H_t \approx \frac{\Phi\left(\varepsilon'C_F\rho\right)^2}{4\pi N_0^2 \tan\alpha_c} \sin^2\left[\frac{\alpha_c}{2}\right]\left[1 + \frac{2\tan\alpha_c}{\varepsilon'C_F\rho} - \frac{1 + \dfrac{2\tan\alpha_c}{\varepsilon'C_F\rho}}{1 + \dfrac{2z_m\tan\alpha_c}{\rho}} - \log\left[1 + \frac{2z_m\tan\alpha_c}{\rho}\right]\right] \quad (101)$$

For the case where absorption by the dye is negligible and the exponential term can be ignored, Equation 100 becomes

$$H_t \approx \frac{\Phi\varepsilon'C_F}{N_0^2} \sin^2\left[\frac{\alpha_c}{2}\right]\int_o^{z_m} \frac{1}{\left[1 + \dfrac{2z\tan\alpha_c}{\rho}\right]^2} dz \quad (102)$$

which when evaluated yields

$$H_t \approx \frac{\Phi\varepsilon'C_F\rho}{2N_0^2 \tan\alpha_c} \sin^2\left[\frac{\alpha_c}{2}\right]\left[\frac{1}{\dfrac{\rho}{2z_m\tan\alpha_c} + 1}\right] \quad (103)$$

Assuming that $2 z_m \tan \alpha_c \gg \rho$, Equation 103 simplifies to

$$H_t \approx \frac{\Phi \varepsilon' C_F \rho}{2 N_0^2 \tan \alpha_c} \sin^2 \left[ \frac{\alpha_c}{2} \right] \tag{104}$$

With the additional substitutions that for small $\theta$, $\sin \theta \approx \theta$ and $\tan \theta \approx \theta$, Equation 104 can be further simplified to yield

$$H_t \approx \frac{\Phi \varepsilon' C_F \rho NA}{8 N_0^3} \tag{105}$$

If $2 z_m \tan \alpha_c \ll \rho$, then Equation 103 can be expressed in simplified form as

$$H_t \approx \frac{\Phi \varepsilon' C_F z_m (NA)^2}{4 N_0^4} \tag{106}$$

Equation 105 should be used when the dye solution extends a long distance from the fiber endface, whereas Equation 106 applies when the dye solution extends only a short distance from the fiber endface. It is useful to note the differences for the two cases in the functional dependencies of Equations 105 and 106 on the fiber radius, the NA of the optical fiber, and the thickness of the dye solution in front of the fiber endface.

# REFERENCES

1. **Bright, F. V., Monnig, C. A., and Hieftje, G. M.,** Rapid frequency-scanned fiber-optic fluorometer capable of subnanosecond lifetime determinations, *Anal. Chem.*, 58, 3139, 1986.
2. **King, P. R., Driver, I., Dawson, J. B., Ellis, D. J., and Feather, J. W.,** Fiber optic probe for in-situ fluorescence measurements, *Proc. SPIE*, 906, 150, 1988.
3. **Newby, K., Reichert, W. M., Andrade, J. D., and Benner, R. E.,** Remote spectroscopic sensing of chemical absorption using a single mode optical fiber, *Appl. Opt.*, 23, 1812, 1984.
4. **Ratzlaff, E. H., Harfmann, R. G., and Crouch, S. R.,** Absorption-corrected fiber optic fluorometer, *Anal. Chem.*, 56, 342, 1984.
5. **Schwab, S. D. and McCreery, R. L.,** Versatile, efficient Raman sampling with fiber optics., *Anal. Chem.*, 56, 2199, 1984.
6. **Trott, G. R. and Furtak, T. E.,** Angular resolved Raman scattering using fiber optic probes, *Rev. Sci. Instrum.*, 51, 1493, 1980.
7. **Zhang, J. D.,** The Feasibility of Constructing a Fiber-Optic Based Multi-Spectra Spectrometric System, Masters thesis, University of Hawaii, Honolulu, 1987.
8. **Anick, H.,** Optical fiber sensors broaden temperature measurement, *Res. Dev.*, 28 ,64 , 1986.
9. **Gehrich, J. L., Lübbers, D. W., Opitz, N. O., Hansmann, D. R., Miller, W. W., Tusa, J. K., and Yafuso, M.,** Optical fluorescence and its application to an intravascular blood gas monitoring system, *IEEE Trans. Biomed. Eng.*, BME-33, 117, 1986.
10. **Gammage, R. B., and Vo-Dinh, T.,** Luminescence monitoring of oil and tar contamination, *Nucl. Instrum. Methods*, 175, 236, 1980.
11. **Herron, N. R., Simon, S. J., and Eccles, L.,** Remote detection of organochlorides with a fiber optic based sensor III. Calibration and field evaluation of an improved chloroform fiber optic chemical sensor with a dedicated portable fluorimeter, submitted to *Anal. Instrum.*, 1988.
12. **Hirschfeld, T., Deaton, T., Milanovich, F., and Klainer, S.,** Feasibility of using fiber optics for monitoring groundwater contaminants, *Opt. Eng.*, 22, 527, 1983.

13. **Milanovich, F. P., Daley, P. F., Klainer, S. M., and Eccles, L.,** Remote detection of organochlorides with a fiber optic based sensor II. A dedicated portable fluorimeter, *Anal. Instrum.*, 15, 347, 1986.
14. **Wolfbeis, O. S.,** Analytical chemistry with optical sensors, *Fresenius. Z. Anal. Chem.*, 325, 387, 1986.
15. **Guilbault, G.,** Practical Fluorescence, Marcel Dekker, New York, 1973.
16. **Haugland, R. P.,** *Handbook of Fluorescent Probes and Research Chemicals*, Molecular Probes, Eugene, OR, 1989.
17. **Wickersheim, K. A. and Sun, M. H.,** Fiberoptic thermometry and its applications, *J. Microwave Power Electromagn. Energy*, 22, 85, 1987.
18. **Wickersheim, K. A. and Jensen, E. M.,** Optical technique for measurement of currents induced by microwave frequency radiation. I. Basic technology and instrument design, Paper T15, in EMC Int. Conf. Electromagnetic Compatibility, Washington, D.C., 1988.
19. **Krohn, D. A.,** Fiber optic sensors: phase modulation, *Photonics Spectra*, 22, 61, 1987.
20. **Kieli, M. and Herzfield, P. R.,** Novel fiber optic sensor based on modal power distribution (MPD) modulation, *Proc. SPIE*, 798, 331, 1987.
21. **Angel, S. M.,** Optrodes: chemically selective fiber-optic sensors, *Spectroscopy*, 2, 38, 1987.
22. **Culshaw, B.,** *Optical Fibre Sensing and Signal Processing*, Peter Peregrinus, London, 1984.
23. **DePaula, R. P., More, E. L., and Jamieson, R. S.,** Overview of fiber-optical sensors, JPL Invention Rep. NPO-16817/6329, JPL, Pasadena, CA, 1987.
24. **Dils, R. R.,** High-temperature optical fiber thermometer, *J. Appl. Phys.*, 54, 1198, 1983.
25. **Giallorenzi, T. G., Bucaro, J. A., Dandridge, A., and Cole, J. H.,** Optical fiber sensors challenge the competition, *IEEE Spectrum*, 23, 44, 1986.
26. **Krohn, D. A.,** Fiber optic sensors: intensity modulation, *Photonics Spectra*, 21, 59, 1987.
27. **Basch, E. E.,** *Optical Fiber Transmission*, Howard Sams & Co., Indianapolis, 1986.
28. **Wilson, J. and Hawkes, J. F. B.,** *Optoelectronics: An Introduction*, Prentice-Hall, Englewood Cliffs, NJ, 1983.
29. **Cockrum, R. H. and Wang, K. T.,** *NASA Tech. Briefs*, 12, 19, 1988.
30. **Hentschel, C.,** *Fiber Optics Handbook*, Hewlett Packard GmbH, HP product #5952-9654, Germany, 1988.
31. **AMP Inc.,** *Designers Guide to Fiber Optics*, AMP Inc., Harrisburg, PA, 1982.
32. **Snyder, A. W. and Love, J. D.,** *Optical Waveguide Theory*, Chapman and Hall, New York, 1983.
33. **Kapany, N. S. and Burke, J. J.,** *Optical Waveguides*, Academic Press, New York, 1972.
34. **Senior, J. M.,** *Optical Fiber Communications, Principles and Practice*, Prentice-Hall, London, 1985.
35. **Cheo, P., K.,** *Fiber Optics: Devices and Systems*, Prentice-Hall, Englewood Cliffs, NJ, 1985.
36. **Li, T., Ed.,** *Optical-Fiber Communications, Vol.1, Fiber Fabrication*, Academic Press, Orlando, FL, 1985.
37. **Kaino, T., Fujiki, M., and Nara, S,** Low-loss polystyrene core-optical fibers, *J. Appl. Phys.*, 52, 7061, 1981.
38. **Kittel, C.,** *Introduction to Solid State Physics,* 6th Ed., John Wiley & Sons, New York, 1986.
39. **Walrafen, G. E., Krishnan, P. N., and Freiman, S. W.,** Raman investigation of optical fibers under high tensile stress, *J. Appl. Phys.*, 52, 2832, 1981.
40. **O'Shea, D .C.,** *Elements of Modern Optical Design*, John Wiley & Sons, New York, 1986, 23.
41. **Smith, W. J.,** *Modern Optical Engineering*, Mc Graw Hill, New York, 1966, 21.
42. **Weast, R. C. Ed.,** *Handbook of Chemistry and Physics*, 66th ed., CRC Press, Boca Raton, FL, 1988, radiation constants from page F-194, Tungsten emissivity data from page E-377.
43. **Modlin, D. N.,** Optimized photodetection and light sources for multimode fiber optic sensors, *Proc. SPIE*, 906, 11, 1988.
44. **Ott, H. W.,** *Noise Reduction Techniques in Electronic Systems*, John Wiley & Sons, New York, 1976.
45. **van der Ziel, A.,** *Noise in Measurements*, John Wiley & Sons, New York, 1967.
46. **Dereniak, E. L. and Crowe, D. G.,** *Optical Radiation Detectors*, John Wiley & Sons, New York, 1984.
47. **Sze, S. M.,** *Physics of Semiconductor Devices*, 2nd ed., John Wiley & Sons, New York, 1981.
48. **Meade, M. L.,** *Lock-in Amplifiers: Principles and Applications*, Peter Peregrinus, London, 1983.
49. **Graeme, J. G., Tobey, G. E., and Huelsman, L. P.,** *Operational Amplifiers, Design and Applications*, McGraw-Hill, New York, 1971.
50. **Graeme, J.,** FET op amps convert photodiode outputs to usable signals, *EDN*, 205, 1987.
51. *Burr-Brown Integrated Circuits Data Book*, Burr Brown Corp., Tuscon, AZ, 1989.
52. **Wilson, J. and Hawkes, J.F.B.,** *Lasers, Principles and Applications*, Prentice Hall, Englewood Cliffs, NJ, 1987.
53. **Wiedemann, R.,** Green light from ion lasers, *Lasers Optronics*, 6 , 55, 1987.
54. **Boscha, B.,** Designing with the He-Ne laser, in *Photonics Design and Applications Handbook*, Laurin Publishing, Pittsfield, MA, 1988, 193.
55. **Dreyfus, J.,** Visible wavelength diode lasers, *Lasers Optronics*, 7, 53, 1988.

56.  NEC Electronics, Inc., P.O. Box 7241, 410 Ellis Street, Mountain View, California, 94039, USA, product availability announced in 1988.
57.  **Dishington, R. H., Hook, W. R., and Hilberg, R., P.,** Flashlamp discharge and laser efficiency, *Appl. Opt.*, 13, 2300, 1974.
58.  **Kelley, J. H., Brown, D. C., and Teegarden, K.,** Time resolved spectroscopy of large bore Xe flashlamps, *Appl. Opt.*, 19, 3817, 1980.
59.  **Powell, H.,** Unpublished data, Lawrence Livermore National Laboratory, 1988.
60.  **Gabardi, D. R. and Shealy, D. L.,** Coupling of domed light-emitting diodes with a multimode step-index optical fiber, *Appl. Opt.*, 25, 3435, 1986.
61.  **Gillessen, K. and Schairer, W.,** *Light Emitting Diodes, An Introduction*, Prentice Hall, Englewood Cliffs, NJ, 1987.
62.  **Siemens Components, Inc.,** product #LDB5410, Optoelectronics Division, 19000 Homestead Road, Cupertino, California, 95014, USA.
63.  **NSG America, Inc.,** 28 World's Fair Dr., Somerset, New Jersey, 08873, USA.
64.  **Macleod, H.A.,** *Thin Flim Optical Filters*, Macmillan, New York, 1986.
65.  **Lerner, J. M., Thevenon, A., and Touzet, B.,** Advances in holographic diffraction gratings point to aberration-free spectrometry, *Laser Focus*, 24, 90, 1988.
66.  **Francis, J. J. and Sternberg, R. S.,** *The Design of Optical Spectrometers*, Chapman and Hall, London, 1969.

Chapter 7

# SENSOR CHEMISTRY

**Ernest Koller and Otto S. Wolfbeis**

## TABLE OF CONTENTS

# I. INTRODUCTION

This chapter covers the chemical procedures and materials that are frequently employed in the fabrication of optrodes. In the overwhelming majority of cases, the sensing material is composed of an indicator (and/or an enzyme) placed in, or on, a polymeric support, using a proper immobilization technique. The physical chemistry of indicators and their spectral properties are treated in Section II herein. The next step in sensor design (which is the choice of a proper polymeric support and the immobilization of the indicators) will be described in Sections III and IV. In immunosensing, monitoring of the intrinsic fluorescence of antigens and antibodies is possible, but fluorescence labeling is the preferred method. Labels and labeling methods are treated in Sections V and VI.

# II. INDICATORS

## A. GENERAL DISCUSSION

All kinds of extrinsic second generation optical sensors require some sort of indicator chemistry which converts the analyte concentration into an optical signal. In other words, an indicator acts as a transducer for the chemical species that cannot be determined directly by optical means. Consequently, it is the concentration of the indicator species that is determined rather than the analyte itself. (The measurement of surface potentials is a notable exception.) This situation has far-reaching implications which will be discussed later. The chemistry of indicators is fairly established[1,2] but not optimized for chemical sensing purposes. Many indicators cannot be used in fiber optic chemical sensors (FOCSs) because of unfavorable analytical wavelengths, poor photostability, low molar absorbance, additional reagents (such as strong acid or alkali) that are needed in conventional spectrometry to adjust optimal conditions, or simply because they are not available in a purity required for FOCS application (e.g., xylenol orange).

Indicators can be categorized as either absorbance- and fluorescence-based. Most absorbance-based indicators undergo a color change (with one band appearing as the other disappears) rather than an intensity change of one band only. Usually, both the complexed and uncomplexed species have an absorption of comparable intensity in the visible. These are referred to as two-color indicators and are mostly compatible with LED or filament lamp light sources. They easily lend themselves to two-wavelength internal referencing methods.

Fluorescent indicators, in contrast, are frequently of the yes/no type in that only one of the species (the complexed or the uncomplexed one) is fluorescent. Hence, absolute fluorescence intensity can be measured with no background from a second species. However, internal two-wavelength referencing is sometimes more difficult. Another disadvantage of fluorescent indicators results from the fact that they are prone to quenching by species other than the analyte. Finally, most fluorophores have lower molar absorbance (as compared to color indicators) and are not excitable by green, yellow, or red LEDs, or by semiconductor lasers. On the other hand, fluorescence indicators provide distinctly improved sensitivity (which is important in case of minute sensor size) and selectivity because it is unlikely that an interfering species has the same absorption and emission as the indicator. Finally, fluorimetry offers a broad variety of additional techniques including dynamic quenching, lifetime measurements, energy transfer, and combinations thereof.

Absorbance- or fluorescence-based indicators for use in FOCS are known to undergo either ion combination reactions (such as with protons, cations or anions) or oxidation-reduction reactions. Neglecting the charges on the various species, the general reaction in ion-combination reaction is

$$xB + y \, Ind \rightleftharpoons B_x Ind_y \tag{1}$$

where B is an ion and Ind the indicator.

Virtually all ion-combination indicators (including those for pH) are weak acids or, less commonly, weak bases, or salts thereof, and are acid-base indicators so that even when used for purposes other than hydrogen ion indication the equilibria involve hydrogen ion. When indication of metal ion concentration is required, it is necessary to maintain the hydrogen ion concentration nearly constant by using an excess of strong acid or strong base or by buffering the solution. Under these conditions, the hydrogen ion concentration is either treated as a separate factor or included in the conditional stability constant, which is given, for cationic indicators, by

$$K_s = \frac{\left[B_x Ind_y\right]}{[B]^x [Ind]^y} \tag{2}$$

The indicator constant is $K_s$, the conditional formation constant of the product $B_x Ind_y$. This constant applies only to the particular experimental conditions under which it has been determined, and will have a different value if conditions such as the concentration of hydrogen ion or any ligands competing for B or Ind are changed.

In case of pH indicators and some metal chelators, x and y are 1. This situation is sufficiently common to warrant separate treatment, particularly as it reveals certain facts without the added complication of the stoichiometric effect which will be considered separately. Even when $x \neq y$, it is often sufficient to consider only the single final step leading to the complex product, or, since such complexes often have widely differing stepwise formation constants, another single step which gives the color change, and in these cases the treatment simplifies to the case where $x = y = 1$, with the appropriate species substituted in the equilibria.

In general, for a two-color indicator the ratio of two species becomes difficult to detect in the presence of the other when the intensity of the one is less than one tenth of the intensity of the other. The indicator appears therefore to be completely combined when

$$\varepsilon_{BInd}[BInd] = 10\ \varepsilon_{Ind}[Ind] \tag{3}$$

and to be completely uncombined when

$$\varepsilon_{BInd}[BInd] = \frac{\varepsilon_{Ind}[Ind]}{10} \tag{4}$$

so that the color changes completely over the range,

$$pB = {}^{10}\log K_s + {}^{10}\log\left(\varepsilon_{BInd}/\varepsilon_{Ind}\right) \pm 1 \tag{5}$$

The transition interval (i.e., the analytically useful range), is therefore 2 pB units.

For a *one-color indicator* the transition range becomes

$$pB = {}^{10}\log K_s \pm 1 \tag{6}$$

and the transition interval remains at 2 pB units. (See also Chapter 9 for the special case of metal cation indicators.)

## B. pH INDICATORS
### 1. Indicator Theory
According to Ostwald, indicators are weak acids (HI) or bases (IOH) whose color is different from that of the indicator ion formed by their dissociation. The relation between pH, $pK_a$, and absorbance or fluorescence intensity of the two species is given by

$$pH = pK_a + \log\frac{[I^-]}{[HI]} \tag{7}$$

In this equation, [HI] represents the *concentration* of the undissociated indicator molecule whose color is called acid color, while [I$^-$] denotes the concentration of the indicator anion. $pK_a$ represents the negative logarithm of the dissociation constant $K_a$. At the transition point of the titration curve, $pH = pK_a$.

Taking the ion product of water ($K_w$) into consideration, defined as

$$K_w = [H^+][OH^-] = 10^{-14} \tag{8}$$

in pure water, it follows that

$$pK_a = 14 + \log \frac{\left[I^-\right]}{[HI]} - \log\left[OH^-\right] \tag{9}$$

Because, however, $\log K_w$ is $-14$ only in case of purely aqueous solution, it becomes obvious that addition of organic solvent can considerably change the $pK_a$ of an immobilized indicator as used in a pH sensor.

The indicator bases may be characterized similarly to the indicator acids in that

$$\frac{\left[I^+\right]\left[OH^-\right]}{[IOH]} = K_b \tag{10}$$

Taking again the ion product of water into consideration, it can be derived that

$$pH = 14 - pK_b + \log \frac{[IOH]}{\left[I^+\right]} \tag{11}$$

where $pK_b$ represents the negative logarithm of the dissociation constant, and IOH means the undissociated indicator base, the color of which is the alkaline color. The color change effected by pH may be interpreted similarly to the color change of indicator acids. At the transition point,

$$\left[H^+\right] = \frac{K_w}{K_b} \tag{12}$$

and

$$pH = 14 - pK_b \tag{13}$$

Indicator bases which contain amino groups are able to bind protons due to the unshared electron pair of nitrogen atoms, and so dye cations can be produced of different charges depending upon the actual pH value. On the other hand, indicator acids possessing hydroxy groups release in alkaline medium the hydrogens of the hydroxy groups and dye anions are produced.

$pK_a$ values can be determined by spectrophotometry or fluorimetry with the help of Equation 14:

$$pK_a = pH - \log \frac{\left(E_x - E_A\right)}{\left(E_B - E_x\right)} \tag{14}$$

where $E_x$ is the absorbance at the analytical wavelength at a certain pH, and $E_A$ and $E_B$ are the absorbances at this wavelength for the pure acid and base forms, respectively. This equation assumes the Lambert-Beer law to be valid.

In a more general sense than treated so far, indicator acids are not only phenols and carboxylic acids, but also protonated azo compounds and amines. Some indicators such as phenolphthalein, however, change color as a result of a complete rearrangement of the molecule and cannot be treated this way. Aside from this, this rearrangement requires some time and involves various intermediates. It should be noted, too, that even simple acid-base

## TABLE 1
### Spectral Data (in nm) as well as pK$_a$ Values of the Common Absorbance-Based Indicators for the pH 5—9 Range

| Indicator | Absorbance maximum or color of acid/base forms | pKa value or pH range |
|---|---|---|
| Bromothymol blue | 430/617 | 6.8 |
| Calcichrome | — | 7.2 |
| o-Chlorophenol-indophenol | 555/625 | 7.1 |
| Chlorophenol red | 460/530 | 6.3 |
| o-Cresoltetrachloro-sulfonephthalein | 430/590 | 7.5 |
| Dibromo-xylenol blue | ca. 420/614 | 7.6 |
| 4',6'-Dinitro-4-hydroxy-azobenzene | 383/524 | 6.4—8.2 |
| 4-Ethoxy-4'-hydroxy-azobenzol | 390/476 | 6.0—8.0 |
| α-Naphthol-phthalein | 428/661 | 6.7 and 7.9 |
| Neutral red | 527/453 | 7.4 and 5.9 |
| Nitrazine yellow | 460/590 | 6.5 |
| 4-Nitrophenol | 360/415 | 7.2 |
| p-Nitrophenyl-acetylhydrazin | 360/465 | 7.6, 11.44 |
| Palatine chrome black | 520/643 | 7.4 |
| Phenolbenzein | Yellow/red | 6.0—7.6 |
| Phenol red | 432/576 | 7.6 |
| Phenoltetrachloro-sulfonephthalein | 435/575 | 7.0 |
| Pinachrome | 360/575 (EtOH) | 7.34 |
| Quinoline blue | 339/595 | 7.0—8.0 |
| Rosolic acid | Brown/red | 6.9—8.0 |
| Solochrome violet RS | 515/562 | 4.35, 7.4, 9.35 |
| Tropaeolin 000-1 | 470/508 | 8.1 |

equilibria such as the nitrophenol-nitrophenolate equilibrium in fact can involve more than two species, in the given example the quinoid aci-form of nitrophenol.

Consequently, the variation of the pH causes not only a change in the electrolytic dissociation equilibrium of the indicator base or acid, but also an inner rearrangement of the molecules. The colorless normal form, the so-called pseudo-form, changes into the ionogenic colored, mostly quinonoid aci- or baso-form. Generally, aci-acids are those strong acids which are produced through molecular rearrangement of pseudo-acids, i.e., of weak acids or of nonacidic compounds. The pseudo-ionogenic transition generally requires a finite time.

Since sensors for physiological pHs are of greatest importance, we first focus on the respective indicators. They may be classified according to their chemical structures in order to demonstrate the influence of substituents on pK$_a$ values and spectral properties. But although a great variety of both colored and fluorescent pH indicators are known,[1,3-5] only a few meet the requirements of an ideal pH sensor for measurement in the physiological pH range. These include (1) an appropriate pK$_a$ (7 to 8), (2) analytical wavelengths for both absorption or excitation at above 400 nm to allow the use of inexpensive waveguide optics and light sources, (3) a large Stokes' shift to minimize optical interferences in case of fluorescent indicators, (4) photostability and chemical stability, (5) lack of toxicity, and (6) the availability of functional groups suitable for chemical immobilization.

## 2. Absorbance-Based Indicators

Among the great variety of organic chromophores such as azo dyes, nitrophenols, phthaleins, sulfophthaleins, aniline-sulfophthaleins, triphenylmethane dyes, polymethines, and others, only a few have so far been considered to be useful and applied to sensor technology.

These dyes are phenol red,[6,7] chlorophenol red and bromothymol blue,[8,9] and alizarin.[7] Table 1 lists a variety of absorbance-based pH probes of potential utility in optical sensors, along with their $pK_a$ values and spectral data. It is obvious that many of them can be excited with a blue, green, or yellow LED, or with a frequency-doubled diode laser. The lack of directly diode laser-excitable dyes is obvious, though. It is important to note that a number of nonfluorescent indicators display fluorescence when they are bound to a rigid solid support which is frequently the case in optical sensor chemistry.

Various other indicators covering the range outside pH 6 to 8 are compiled in Table 2. Some of the dyes can be immobilized covalently in this form because they possess suitable functional groups, while many others lack such a function. Also given are the absorption maxima or colors of the acid or base form. Preference was given to LED-excitable dyes. However, the ideal pH indicators for absorbance or reflectance based optrodes seem not to exist yet for many pH ranges.

## 3. Fluorescent Indicators

Fluorescent indicators have been applied more often than absorbance based indicators in optrodes and shall be discussed in more detail according to their chemical structures.

### a. Coumarins

7-Hydroxycoumarines such as umbelliferone and 4-methylumbelliferone (4-MU) play an important role in measuring physiological pH values by fluorimetry (Table 3). 4-MU has been used in a pH optrode (Chapter 8) and in $CO_2$ sensors. Its $pK_a$ of 7.8 is well suited for measuring physiological pHs, but its spectral properties are disadvantageous: fluorescence is maximally excited between 325 and 355 nm, which is a spectral region where background fluorescence from biological material such as serum or intracellular liquids is fairly expressed, and most optical components are not well suited for. Introducing a trifluoromethyl group in place of the methyl group in 4-MU results in a probe (4-TMU) having distinctly longwave-shifted of the spectra (Table 3) along with an increase in fluorescence efficiency and a decrease in the $pK_a$.

Certain coumarins possessing an electron-withdrawing substituent in position 3 have also been considered to be useful as physiological pH indicators.[10-14] Typical examples are summarized in Table 3. They have intense fluorescence and both longwave absorption and emission. In aqueous solution both the absorption and fluorescence spectra are highly pH dependent in near neutral pH range. Among these, 7-hydroxycoumarin-3-carboxylic acid (HCC) has found application in a fluorescence sensor for monitoring physiological pH values because it can be easily immobilized.[15]

An interesting property of 7-hydroxycoumarins (and many other phenols) is their capability of undergoing an excited state dissociation, resulting in an anion (phenolate) fluorescence emission, even when the phenol species is excited.[16] This is due to the protolytic dissociation of a photoexcited molecule during the lifetime of its excited state because the $pK_a$ of the $S_1$ state is lower than that of the ground state. Thus, by exciting the acid and base form of the indicator, respectively, and ratioing the emission intensity as obtained at two excitation wavelengths at a fixed emission wavelength, a signal for pH is obtained which is independent of dye concentration and therefore makes dye bleaching and leaching effects less serious a problem. The technique also can account for light source and detector fluctuations.

7-Hydroxy-3-pyridylcoumarins[17] have two $pK_a$s and can be used as two-step indicators for titrations and the determination of pHs over a wide range (typically 2 to 9). The presence of a 4-cyano group gives rise to a dramatic longwave shift in absorption and emission ($R_1$ = CN in Table 3). In aqueous solution, the fluorescence maxima are at around 570 to 600 nm, with excitation maxima lying between 410 and 510 nm, depending on whether the phenol or

## TABLE 2
### Selected Absorption Indicators for Measurement of Acidic or Basic pHs, and Respective pK$_a$ Values (or Transition Ranges) at Room Temperature

| Indicator | Absorbance maximum or color of acid/base forms | pK$_a$ value or pH range |
|---|---|---|
| Acidic range | | |
| N-Benzylaniline sulfophthalein | Yellow/blue | 0.30 |
| Methyl violet | Yellow/blue | 0.0—1.6 |
| Methyl green | Yellow/blue | 0.2—1.8 |
| Crystal violet | Yellow/blue | 0.2—2.0 |
| Ethyl violet | Yellow/blue | 0.2—2.4 |
| Malachite green | Yellow/blue-green | 0.2—1.8 |
| 2-(p-Dimethylaminophenyl-azo)pyridine | Yellow/red | 0.4—1.8 |
| Cresol red | Red/yellow | 1.0—2.0 |
| Aniline sulfonphthalein | Yellow/blue | 1.6 |
| Quinaldine red | Colorless/red | 1.0—2.2 |
| Thymol blue | Red/yellow | 1.2 - 2.8 |
| Meta-cresol purple | Red/yellow | 1.2—2.8 |
| Orange IV | Red/yellow | 1.4 - 2.8 |
| Erythrosin, disodium salt | Orange/red | 2.2—3.6 |
| Benzopurpurin 4B | Violet/red | 2.2—4.2 |
| N,N-Dimethyl-p-(tolylazo)aniline | Red/yellow | 2.6—4.8 |
| p-Dimethylaminoazobenzene | Red/yellow | 2.8—4.4 |
| Bromophenol blue | Yellow/blue | 2.8—4.8 |
| Methyl orange | Orange/yellow | 3.0—4.4 |
| 4,4'-Bis(2-amino-1-naphthylazo-2,2'-stilbene-disulfonate | Purple/red | 3.0—4.0 |
| Congo red | Blue/red | 3.0—5.0 |
| Tetrabromophenolphthalein | Yellow/blue | 3.2—4.8 |
| Bromocresol green | Yellow/blue | 4.6 |
| Resazurin | Orange/violet | 3.8—6.4 |
| 4-Phenylazo-1-naphthylamine | Red/yellow | 4.0—5.6 |
| Ethyl red | Colorless/red | 4.0—5.8 |
| 2-(p-Dimethylaminophenylazo)-pyridine | Red/yellow | 4.4—5.6 |
| Lacmoid | Red/blue | 4.4—6.2 |
| Methyl red | Red/yellow | 4.6—6.0 |
| Alizarin red S | Yellow/red | 4.6—6.0 |
| Propyl red | Red/yellow | 4.8—6.6 |
| Bromocresol purple | Yellow/purple | 6.3 |
| Alizarin | Yellow/red | 5.6—7.2 |
| 2-(2,4-Dinitrophenylazo)-1-naphthol-3,6-disulfonate | Yellow/blue | 6.0—7.0 |
| | | |
| Alkaline range | | |
| Brilliant yellow | Yellow/orange | 6.6—8.0 |
| Cresol red | Yellow/red | 7.0—8.8 |
| Curcumin | Yellow/red | 7.4—8.6 |
| Meta-cresol purple | Yellow/purple | 7.4—9.0 |
| α-Naphthol-sulfophthaleine | Yellow/blue | 7.5—9.0 |
| Tetrabromo anilinesulfophthalein | Violet/yellow | 8.5 |
| 4,4'-Bis(4-amino-1-naphthylazo)-2,2'-stilbenedisulfonate | Blue/red | 8.0—9.0 |
| Thymol blue | Yellow/blue | 8.0—9.6 |
| Cresolphthalein | Colorless/red | 8.2—9.8 |
| Naphtholbenzein | Orange/blue | 8.2—10.0 |
| Phenolphthalein | Colorless/purple | 8.4—10.0 |
| Ethyl bis(2,4-dinitrophenyl)-acetate | Colorless/blue | 8.6—9.6 |

**TABLE 2 (continued)**
**Selected Absorption Indicators for Measurement of Acidic or Basic pHs, and Respective pK$_a$ Values (or Transition Ranges) at Room Temperature**

| Indicator | Absorbance maximum or color of acid/base forms | pK$_a$ value or pH range |
|---|---|---|
| Thymolphthalein | Colorless/blue | 9.2—10.8 |
| N-(p-Aminophenyl)aniline-sulfophthalein | Blue/yellow | 10.5 |
| Alizarin yellow R | Yellow/red | 10.0—12.0 |
| Curcumin | Red/orange | 10.2—11.8 |
| Azoviolet | Yellow/violet | 11.0—13.0 |
| Alizarin | Red/purple | 11.0—12.4 |
| Tetraethyl anilinesulfophthalein | Blue/yellow | 13.2 |
| Indigocarmine | Blue/yellow | 11.4—13.0 |
| 2,4,6-Trinitrotoluene | Colorless/orange | 11.4—13.0 |
| Clayton yellow | Yellow/amber | 12.2—13.2 |

*Note:* Indicators with color changes from colorless to yellow (or vice versa) not included. For more dyes, see Table 6.

phenolate species is excited. Thus, some of these are compatible with blue or green LEDs. Due to their low pK$_a$ values of around 6.0, the 4-cyano-7-hydroxycoumarins are, however, more suitable for measuring pHs of weakly acidic liquids such as urine rather than for blood pHs.

Aside from 7-hydroxycoumarins, certain 7-arylsulfonylamino-coumarins also show spectral changes in the near neutral pH range which have been described to be useful for the measurement of physiological pHs.[18] These dyes complement the 7-hydroxycoumarins and possibly have improved photostability. Delisser-Matthews and Kauffman[11] evaluated various 3-aryl-7-aminocoumarins as indicators and found them useful for a wide spectrum of applications such as adsorption indicators for precipitation titrations, and as indicators for acid-base titrations in aqueous medium (pH transition range 3.4 to 5.8) and in nonaqueous solvents.

### b. Fluoresceins

Fluorescein is a widely used pH probe[19,20] and has been immobilized[21] to give pH sensors (see Chapter 8 on pH). The popularity of the fluoresceins results from the close match of its absorption with the 488-nm line of the argon laser, and its availability as the isothiocyanate (FITC). However, fluoresceins are poor pH probes in having small Stokes' losses, overlapping pK$_a$s, and limited photostability. The spectral properties of fluoresceins are similar to bilirubin and flavin nucleotides which therefore can interfere unless lifetime discriminations are made. The influence of substituents upon pK$_a$ and spectra of fluoresceins is demonstrated by the data given in Table 4. Aside from FITC, the 5(6)-carboxyl derivatives are possible candidates for covalent attachment to polymer supports.

A variety of other fluorescein-type pH probes is commercially available in activated form ready for covalent attachment to solid supports.[22] 2′,7′-Bis-carboxyethyl-5(6)-carboxyfluorescein is frequently used for estimation of intracellular pH[12,23] since it possesses a pK$_a$ near 7 and is better retained in cells and vesicles than fluorescein due to its many negative charges. Unfortunately, most fluoresceins can be excited at one band only so that no two-wavelength (ratio) spectroscopy is possible.

Recently, a new class of fluorescent benzo[c]xanthene pH indicators has been developed by Whitaker et al.[24] They exhibit long wavelength emissions in acidic and basic media with

## TABLE 3
### Spectral Properties and pK$_a$ Values of Coumarin-Based Fluorescent Indicators as a Function of their Substituents in Position 4 (R$_1$) and Position 3 (R$_2$) of the Coumarin Ring

| Indicator | | pK$_a$ (at 22°C) | Excitation maximum at pH 4/pH 10 (nm) | Fluorescence maximum (nm) |
|---|---|---|---|---|
| R$_1$ | R$_2$ | | | |
| H | H | 7.80 | 330/358 | 450 |
| CH$_3$ | H (4-MU) | 7.78 | 325/355 | 445 |
| CF$_3$ | H (4-TMU) | 7.26 | 355/410 | 503 |
| H | Cyano | ca. 7.0 | 355/410 | 453 |
| H | Carboxy | 7.04 | 360/400 | 448 |
| H | Carboxamido | ca. 7.1 | 350/398 | 445 |
| H | Phenyl | 7.80 | 340/386 | 463 |
| H | 2-Benzothiazolyl | 7.02; 6.79[a] | 385/439 | 490 |
| H | 4-Pyridyl | 7.64; 4.74[b] | 377/400 | 470 |
| H | 5-Methyl-7-sulfo-2-benzoxazolyl | 6.80 | 378/431 | 470 |
| CN | 2-Benzothiazolyl | 6.38 | 432/505 | 595 |
| CN | 5-Methyl-7-sulfo-2-benzoxazolyl | 6.08 | 416/494 | 577 |
| Calcein blue | | 3.0, 6.9, 11.3 | 320/365 | 455 |

[a]　At 37°C.
[b]　Two pK$_a$ values.

well-defined isosbestic points and pK$_a$s within the physiological range. Their major advantage is their excitability by green and yellow LEDs (Table 4). Their pKas are around 7.6 in bulk solution. Unlike many other phenols, the dyes do not undergo photodissociation and therefore display two distinct pH-dependent emission maxima at about 570 and 630 nm in acidic and alkaline solution, respectively. This makes them the first fluoresceins having two distinct absorption and emission bands.

When excitation is performed at the isosbestic wavelength of the *excitation* spectrum, emission intensity is pH-independent. Another isosbestic (better: isoemissive) point exists in the emission spectra. Given the variety of absorption and emission bands, various spectroscopic methods can be applied to measure pH independent of dye concentration: one-wavelength excitation and two-wavelength emission, two-wavelength excitation and one-wavelength emission, and two-wavelength excitation and two-wavelength emission seem feasible. While one-wavelength excitation and one-wavelength emission appears to be the most simple method, the advantages of ratio measurements are lost under these circumstances. Properties similar to those displayed by the SNARF dyes are displayed by a dye called Vita Blue (Table 4).

In an approach to cover the pH 0 to 7 range with fluorescent pH probes,[25] fluorescein was immobilized on fibrous cellulose and partially converted (by gentle bromination) into eosine. The two indicator materials combined have a pH dependence that covers the whole gastric pH range. Measurements can be performed by exciting the fluorescein molecule at 490 nm and detecting the emission of eosin at 550 nm.

### c. *Naphthyl and Pyrene Derivatives*

Naphtholsulfonates[10,12] possess pK$_a$ values in the weakly alkaline pH range and are of

**TABLE 4**
**Absorption and Emission Maxima (in nm) as well as pK$_a$ Values of Fluorescein and Other Xanthenes**

| Indicator | pKa | Excitation/emission maximum at pH 10 |
|---|---|---|
| Fluorescein | 2.2, 4.4, 6.7 | 490/520 |
| Eosin | 3.25, 3.80 | 518/550 |
| 2',7'-Dichlorofluorescein | 0.5, 3.5, 5.0 | 502/526 |
| Dimethylrhodole[a] | ca. 6.0 | 510/545 |
| 5(6)-Carboxyfluorescein | ca. 6.4 | 490/520 |
| 5(6)-Carboxyeosin | ca. 3.6 | 525/560 |
| 5(6)-Carboxy-2',7'-dichlorofluorescein | 5.0–5.3 | 505/530 |
| 2',7'-Bis-carboxyethyl-5(6)-carboxyfluorescein | 6.98 | 500/530 |
| SNARF[b] | ca. 7.6, 7.3[c] | 560/625[d] |
| SNAFL[b] | ca. 7.6, 7.3[c] | 550/620[e] |
| Vita blue[f] | ca. 7.5 | 610/665[g] |

[a]  The chemical name is 9-(2-carboxyphenyl)-6-dimethylamino-xanthen-3-one; it is a chemical hybride between fluorescein and rhodamine.
[b]  Registered tradename of Molecular Probes, Inc., Portland, OR.
[d]  530/575 nm at pH 4.
[e]  515/540 nm at pH 4.
[f]  Several isomers known: Lee, L. G., Berry, G. M., and Chen, C. H., Vita Blue: a new 633-nm excitable fluorescent dye for cell analysis, *Cytometry,* 10, 151, 1989.
[g]  524/570 nm at pH 5.2.

limited utility for physiological pHs because they also lack visible excitation. 8-Hydroxypyrene-1,3,6-trisulfonic acid trisodium salt (HPTS), in contrast, was found to be the best of a series of indicators examined.[10] Its major advantages are high water solubility, high fluorescence quantum yield, visible excitation (455 nm) and emission (515 nm), large Stokes' shift, two-wavelength excitability, and the presence of sulfonato groups which facilitate immobilization. Due to its excellent properties, HPTS has found application in pH and pCO$_2$ sensor technology (see Chapters 8 and 19). However, due to its triple or quadruple charge, HPTS is rather sensitive to changes in ionic strength. Like the coumarins, naphthalenes and pyrenes display two distinct absorption maxima, but one emission band only, which makes them well suited for two-wavelength excitation.

### d. Lipophilic pH Indicators

Various pH probes have been made lipophilic[22,26,27] by linking them to lipid chains to make them suitable for incorporation into lipids and membranes. It was noted, however, that they tend to undergo apparent shifts in their pK$_a$ when placed at the interface between lipid and aqueous phase. Typical examples for lipophilic indicators[22] include 3-hexadecanoyl-7-hydroxy-coumarin and several fluorescein and eosin derivatives. Thus, eosin esters have been incorporated into Langmuir-Blodgett films to give pH-sensitive lipid bilayer membranes,[28a] and lipophilic nile blue was incorporated into ion sensors using neutral ion carriers and applying pH transduction.[28b] An anionic lipophilic dye was used in another ion sensor based on ion-pair formation.[28c]

### 4. Ionic Strength Effects

The mass action law is valid for highly dilute solutions only. In practice, the working concentrations (C) are such that it is no longer identical with the actual activity (a) of the ion.

The relation between a and C is given by

$$a = C \cdot f \tag{15}$$

where f is the activity coefficient which has been shown by Debye and Hückel to be related to ionic strength I and ion charge z by

$$\log f = -0.512 \, z^2 - I^{1/2} \tag{16}$$

at 25°C. I can be calculated from the concentrations and charges of the various ions being present in the solution according to

$$I = 0.5 \sum c \cdot z^2 \tag{17}$$

The relation between f and I is somewhat more complicated when the total ionic strength exceeds 10 mM, now being

$$\log f = \frac{-0.512 z^2 I^{1/2}}{\left(1 + 1.6 I^{1/2}\right)} \tag{18}$$

The effect of *foreign* neutral electrolytes, that is, the salt-effect, manifests itself first of all by altering the indicator equilibrium. The phenomenon may be easily explained especially for media of small ionic strength by the theory of Debye and Hückel. Secondly, foreign electrolytes change the activity of the analyte.

For the three most common types of charged indicators (that is, cationic, neutral, and mono-anionic), the $pK_i$ alters with ionic strength (I) in the following way:

$$HInd^+ \rightleftharpoons H^+ + Ind \qquad pK_a' = pK_a + 0.5 \cdot I^{1/2} \tag{19}$$

$$HInd \rightleftharpoons H^+ + Ind^- \qquad pK_a' = pK_a - 0.5 \cdot I^{1/2} \tag{20}$$

$$HInd^- \rightleftharpoons H^+ + Ind^{2-} \qquad pK_a' = pK_a - 1.5 \cdot I^{1/2} \tag{21}$$

Consequently, the salt error of di- and tri-sulfonate indicators is relatively great, since the alkaline forms of these indicators are ions of several negative charges. In solutions of small or medium ionic strength those indicators have a small salt error, which exhibit a dipolar ion structure (methyl orange, methyl red, etc.), because the dipolar ion behaves like a neutral molecule. In solutions of great ionic strength, dipolar ions possess two separate charges. Hence, the salt error increases (see also Table 1 in Chapter 8). The difference in the ionic strength dependence of indicators has been exploited to sense the ionic strength of an unknown solution by making use of two pH sensors with different ionic strength sensitivity.[29]

Aside from foreign electrolytes, the *indicator itself* contributes to a slight shift in $pK_a$ if the activities of acid and conjugate base form are different. By definition,

$$K_a = \frac{[H^+][B]}{[A]} \tag{22}$$

where [H$^+$], [B], and [A] are the *concentrations* of protons, indicator base, and conjugate acid, respectively. Applying the law of mass action to the above equilibrium and taking *activity coefficients* into account, it follows that

$$K_a = \frac{a_H \cdot a_B}{a_A} = \frac{a_H \cdot [B] \cdot f_B}{[A] \cdot f_A} \tag{23}$$

where $a_H$, $a_B$, and $a_A$, respectively, are the activities of the protons and the acid and conjugate form of the indicator, and $f_A$ and $f_B$ represent the activity coefficients of the alkaline and acid form of the indicator. Taking the logarithm of the above equation and ordinating it, we see that

$$-\log a_H = pH = pK_a + \log \frac{[B]}{[A]} + \log \frac{f_B}{f_A} \tag{24}$$

The apparent indicator exponent ($pK_a$) depending upon the ionic strength of the indicator is defined by the expression:

$$pK_a' = pK_a + \log \frac{f_B}{f_A} \tag{25}$$

According to the above definition of the apparent dissociation exponent, the salt error is already reflected in the variation of $pK_a'$, due to the variation of the ratio $f_A/f_B$.

Summarized, it may be said that in the presence of foreign neutral salts the transition interval of the indicator acids will be shifted towards lower pH values, whereas that of the indicator bases will be shifted in the direction of higher pHs. Beside the alteration of indicator equilibria, the presence of foreign salts also changes the optical absorption intensity of the indicator colors. The color of solutions containing neutral salts is in general less intensive than that of diluted acidic or alkaline solutions.

The ionic strength dependence is considered to present a major obstacle in precise determination of pH using optrodes.[30,31] It has been concluded[31] that direct use of a fiber optic sensor for the determination of pH in an unknown solution with an accuracy of ±0.1 pH units is unrealistic, and this may well be true. Janata[30] gives an excellent discussion of the sources of errors in optical pH determination, some of which have been thoroughly discussed decades ago for dissolved indicators but also hold for immobilized reagents. The main sources are the effects of ionic strength and dissolved polyelectrolytes (that is, proteins), of added solvent, and of surface structural effects of optrodes. It has been followed that for thermodynamic reasons neither optical nor electrochemical sensors can measure pH precisely, but that on grounds of error minimization in electrodes, the electrochemical measurements of ion activities are superior to the optical sensors.

### 5. Effect of Solvents

Different solvents exercise different effects upon indicator dyes, so color changes as well as indicator exponents of the indicators vary with the solvent. In aqueous methanolic or ethanolic solutions the alteration is relatively not so significant; in anhydrous alcohol, however, it becomes greater, while in other solvents one can meet quite new phenomena.[1,32]

It often happens in analytical practice that the aqueous solution contains alcohol. Alcohol alters the equilibrium of the indicator system, but the observed effect depends not only upon the indicator but also on the acid-base system present in the solution. The dissociation of weak acids and bases varies in the presence of alcohol also on account of the decrease of the dielectric constant of the solution. If alcohol is added to strong acid solutions, the color of indicator acids is shifted in the direction of the acid color. This effect is much smaller in solutions of weak acids, whereas in buffer solution there is no change at all.

**TABLE 5**
**Dissociation Constants of the Acidic Forms of Some**
**Indicators in Ethyl Alcohol[a]**

| Indicator | pK (EtOH) | pK (H$_2$O) | $\Delta$ pK |
|---|---|---|---|
| Thymol blue | 5.7 | 1.65 | 4.05 |
| Dimethylaminoazobenzene | 5.2 | 3.25 | 1.95 |
| Methyl orange | 2.9—3.5 | 3.0 | –0.5 |
| Bromphenol blue | 9.1 | 4.1 | 5.0 |
| Bromcresol green | 10.3 | 4.9 | 5.4 |
| Bromcresol purple | 11.5 | 6.4 | 5.1 |
| Bromthymol blue | 12.8 | 7.3 | 5.5 |
| Phenol red | 13.4 | 8.0 | 5.4 |
| α-Naphtholphthalein | 13.8 | 8.3 | 5.5 |
| Thymol blue (acid region) | 15.1 | 9.2 | 5.9 |
| Phenolphthalein | 15.3 | 9.3 | 6.0 |

[a]    Data from Reference 32.

The behavior of indicator bases is just the reverse. On the effect of alcohol in strong acid solutions the color of indicator bases is shifted in the alkaline direction. This shift is even greater in weak acid solutions and is greatest in buffer solutions. Methyl orange, for instance, shows its transition color in 0.01 $M$ aqueous solution of acetic acid, whereas in the presence of 40% alcohol the color is definitely yellow.

Table 5 summarizes pK$_a$ values for a set of indicators in water and ethanol solution. It will be noted that the transition ranges of K$_a$ values of indicators in general follows the same relative order as in water. Thus, as a first approximation, the transition intervals of indicators in water can serve as a useful guide in selecting an indicator for a particular acid-base titration in a "waterlike" organic solvent or in an aqueous-organic solvent mixture. The final selection of the indicator may have to be partly trial-and-error, but at least the concepts discussed should permit some intelligent guesses.

In the presence of alcohol, the color intensity and shade of indicators are also different. Phenolphthalein, for instance, in aqueous solutions of sodium hydroxide is cherry colored, whereas this color is blended in the presence of alcohol more and more with a shade of violet. The color intensity is also less.

The effect of added solvent results from the involvement of water molecules in the dissociation step of indicators:

$$A + H_2O \rightleftharpoons H_3O^+ + B^- \tag{26}$$

H$_3$O$^+$ is usually meant when H$^+$ is written. The equilibrium is governed by the usual dissociation constant. Taking activities into account and logarithms, we get

$$pH = pK_a + \log \frac{B}{A} + \log \frac{f_B}{f_A} - \log a_{H_2O} \tag{27}$$

It is the last term in Equation 27 that cannot be ignored if (1) the ionic strength of the solution exceeds about 4 $M$, and (2) a mixed organic/aqueous solvent is employed, so that the activity term of free, uncoordinated water decreases with an increasing fraction of organic solvent.

## 6. Influence of Proteins and Colloids

Proteins and substances consisting of macromolecules may adsorb the indicators, through which the color change will become completely different. Proteins bind the indicator acids through their basic group and indicator bases through their acid group. The charge of the particles plays an important role in the phenomena taking place on surfaces. The result of such binding processes is a shift in the apparent dissociation or binding constant. Among these effects, the apparent shift in the $pK_a$ of pH indicators upon addition of protein is most important. This so-called protein error is defined as the difference between the colorimetrically determined pH value in the presence and absence of the protein in solution whose pH has been adjusted (electrometrically) to its original value. The error can be substantial.[30] In practice, pH optrodes are covered with a protein-impermeable but proton-permeable membrane such as cuprophane, so to minimize interaction of protein with the surface-bound indicator. It has been shown more recently[33] that the specific interaction theory may be extended to cover the interaction of polyelectrolytes with acid-base interactions.

## 7. Influence of Temperature

The color of many indicators depends on the temperature. When heated up in solution to the boiling point, the color of alkali-sensitive indicators is shifted in the direction of the alkaline, whereas that of acid-sensitive indicators is shifted in the direction of the acid side, which of course means the displacement of the transition intervals. The alkali-sensitive methyl orange changes, for instance, at room temperature between pH 3.1 and 4.4, whereas at 100°C the pH interval is between 2.5 and 3.7. This is due first of all to the fact that the ion product of water changes significantly with the temperature. It is 14.2 at 18°C, but 12.2 at 100°C. Data on the effect of temperature are given in Table 6 and, in Chapter 8, Table 2.

The second source for the observed temperature coefficient (tempco) is the shift in the $pK_a$ of the indicator itself. It is difficult to separate the tempco of the ion product of water and that of the dissociation constant. Various indicator pK values and the respective tempcos at defined ionic strength are given in Table 6., showing that temperature has a considerably varying effect on the $pK_a$ and, hence, on the precision of measurement.

## 8. Acid-Base Indicator Resins

The indicator resins provide an interesting form of acid-base indicators.[1] Known for quite a time, they are the precursors of the immobilized indicators materials now used in optrodes. The indicators change their absorption (or fluorescence) at about the known pH, but even small $pK_a$ shifts due to immobilization can adversely affect the precision of a pH sensor that is intended for use in blood pH measurement. The preparation of the material is very simple in that the resin beads are shaken for a fixed period of time with an alcoholic solution of the indicator and the material is washed until no more leaching is observed. Thymol blue, bromocresol green, and phenolphthalein have been bound to Amberlite cation ion exchangers, while others studied the binding of dyes to anion exchange resins of the Dowex type. The use of resin-bound indicators for pH sensing is described in more detail in Chapter 8.

## C. METAL INDICATORS (CHELATORS)

There are many types of compounds that form colored complexes with metal ions[1,2] and therefore can be employed as indicators in optical sensors provided that the experimental conditions are suitably chosen. However, the color reaction must be sufficiently selective and the value of the stability constant of the complex formed should be such as to make the reaction reversible in order to make the device a sensor rather than a single-shot probe (see Chapter 3.III.A and Chapter 8.).

**TABLE 6**
**pK_a Values and Temperature Coefficients of**
**Acid-Base Indicators at 20°C and an Ionic Strength *I***
**of Zero, Unless Otherwise[a]**

| Indicator | $pK_a$ |
|---|---|
| m-Cresol purple | 1.51 (I = 0.1) |
| Thymol blue | 1.65 (15—30°C) |
| Quinaldine red | 2.63—0.007 (t—20°C) |
| Dimethyl yellow | 3.25 (18°C) |
| Hexamethoxytriphenylcarbinol | 3.32 + 0.007 (t—20°C) |
| Methyl orange | 3.46—0.014 (t—20°C) |
| 2,6-Dinitrophenol | 3.70—0.006 (t—20°C) |
| 2,4-Dinitrophenol | 4.10—0.006 (t—20°C) |
| Bromphenol blue | 4.10 (15—25°C) |
| Chlorophenol blue | 4.43 (25°C) |
| Iodophenol blue | 4.57 (25°C) |
| Bromocresol green | 4.90 (15—30°C) |
| Methyl red | 5.00—0.006 (t—20°C) |
| 2,5-Dinitrophenol | 5.20—0.0045 (t—20°C) |
| Heptamethoxytriphenylcarbinol | 5.90 (I = 0.05) |
| Chlorophenol red | 6.25—0.005 (t—20°C) |
| Bromcresol purple | 6.40—0.005 (t—20°C) |
| p-Nitrophenol | 7.15—0.011 (t—20°C) |
| Bromothymol blue | 7.30 (15—30°C) |
| Pinachrome | 7.34—0.013 (t—20°C) |
| Phenol red | 8.00—0.007 (t—20°C) |
| m-Nitrophenol | 8.35—0.01 (t—20°C) |
| Cresol red | 8.46 (30°C) |
| m-Cresol purple | 8.32 (I = 0.1) (30°C) |
| Thymol blue | 9.20 (15—30°C) |

[a]   Data from Reference 1.

Metal indicators are usually salts of polybasic acids which change color when the acidity of the solution is varied. It is therefore necessary to add a buffer mixture to the solution when an indicator of this type is used. The theoretical basis of the use of metal indicators in complexometry can be discussed in terms of the so-called *conditional constant* (Equation 2). When a metal ion *M* reacts with an indicator in a molar ratio of 1:1, it is

$$K_s = \frac{[MI]}{([M'][I'])} \tag{28}$$

where [I'] denotes the concentration of the indicator which is not bound in the complex MI, and [M'] the concentration of the metal ion that is not bound to the indicator as (MI).

The effects of other species such as hydrogen and hydroxyl ion and competitors of M and I can be separately evaluated, or incorporated in K which is then the conditional formation constant. The formation constant is the stability constant of the product of reaction and is the inverse of the instability or dissociation or ionization constant. The two are sometimes confused. It is common practice to deal with equilibrium constants in logarithmic form, particularly in tabulations.[2] These are normally given to the base 10 and are [10]log K for formation constants, but −[10]log K (= pK) for dissociation or ionization constants. Chapter 9 in this book gives a discussion on the possibilities provided by immobilizing metal chelators and how the conditional formation constants affect response and dynamic range.

Metal indicator chemistry is fairly well established,[1,2,4] and a number of textbooks and reviews are available for long-known indicators so that there is no need to go into detail. However, there is an obvious lack in optical indicators for alkali and earth alkali ions. In view of the interest in sensing these species in clinical samples (and for which no indicators useful at pH 7 have been known for long times), we confine this chapter to an introduction into the state of the art in alkali ion indicators, with a few more recent other chelators being included.

Crown ether dyes[34-36] have found particular attention. These compounds are designed to bring about specific color changes on the interaction with metal cations such as alkali and alkaline earth metals, thus serving as probes or photometric reagents selective for these metal ions. The chromophoric groups can bear one or more dissociable protons or can be nonionic. In the former, the ion exchange between the proton and appropriate metal cations causes the color change, while in the latter the coordination of the metal ion to the chromophoric donor or acceptor of the dye molecule induces a change of the charge transfer band of the dye. So far, the proton-dissociable crown ether dyes especially have been used for extraction photometric determination of alkali and alkaline earth metal ions. Similarly, the neutral crown ether dyes are potentially useful for such determinations in lipophilic homogeneous media.

Monoprotonic crown ether dyes consist of a monoprotonic chromophore that is introduced into a crown ether skeleton in such a proximity to the ethereal function that the dissociation of the chromophoric proton is assisted by the complexation of the positively charged metal ion with the crown ether macrocycle. The first crown ether dyes designed according to this principle[34,35] are 4′-picrylamino-substituted derivatives of benzo[15]crown-5. To enhance the metal extraction efficiency one can introduce a chromophoric side arm into a crown ether skeleton in such a manner that the deprontonated anionic group can interact directly with or coordinate to the metal ion which is bound in the crown ether cavity.

Synthetically, such side-armed crown ethers are most readily attained by using azacrown ethers. Typical examples include p-nitrophenyl type and umbelliferone type azacrown ethers. Crowned phenols are interesting crown ether dyes in which a phenolic hydroxyl group constitutes an integrated part of the crown ether skeleton. Such chromogenic crown ethers consisting of a phenolic crown ether and a chromophore, i.e., a merocyanine dye,[36] a carbocyanine dye,[37] and an azo dye,[38] have found application in optical alkali sensors. The introduction of two of the proton-dissociable chromophores into the crown ether skeleton leads to dyes which are selective for divalent metal ions. Typical examples include p-nitrophenyl and 4-methylumbelliferyl derived crown ethers.[34]

A great variety of uncharged, neutral crown ether dyes has also been synthesized, e.g., azo dyes[34] or triphenylmethane dyes.[35] Because of the higher sensitivity of fluorescence, a great variety of fluorescent crown ether derivatives has been developed. Typical fluorophores that have been introduced into various crown ether skeletons include 4-methylumbelliferone,[39] benzo- and naphthothiazolylphenols,[40] p-(1,8-naphthalenedicarboximido)phenol,[41] 7-(dimethylamino)-3-(styryl)-1,4-benzoxazin-2-one,[42] and 4-nitrobenzofurazan.[43]

Fluorescent chelators have found widespread application in the fluorimetric measurement of intracellular ion concentrations. The recently developed indicators SBFI and PBFI are fluorescent benzofurane-derived crown ethers.[44] These polycarboxylic acids can be loaded into cells as their acetoxymethyl ester forms. The polycarboxylic acids are regenerated after immobilization by mild hydolysis at alkaline pH. SBFI has found to be a selective sodium chelator that is unaffected by physiological concentrations of potassium and protons. Its excitation/emission maxima are at 335/505 nm. The potassium indicator PBFI effectively selects potassium in the presence of sodium and protons. FCryp-2, another fluorescent crown ether, has also found to be selective for sodium and is unaffected by potassium, calcium or magnesium in the normal physiological concentration ranges.[45]

Quin-2 probably was the first $Ca^{2+}$ probe useful at near-neutral pH. The dye has been widely used to determine $Ca^{2+}$ in cells.[46] The more recent calcium chelators fura-2 and indo-

1, however, represent significant improvements over quin-2. Fura-2 shifts to shorter wavelength upon $Ca^{2+}$ complexation (from 375 to 350 nm in excitation), while emission remains at 510 nm. The fluorescence emission of indo-1 shifts from about 490 nm to about 405 nm when complexed by $Ca^{2+}$. Since the intensity at selected wavelengths can be measured and formed into a ratio, it is not necessary to measure changes in absolute fluorescence intensities.[47]

The fluorescence emission maximum of indo-1 shifts from about 490 nm in $Ca^{2+}$-free medium to about 405 nm when saturated with $Ca^{2+}$. Even more long-wavelength calcium indicators have been developed recently.[48] Fluo-3 and rhod-2, derived from fluorescein and rhodamine, respectively, do not undergo significant emission or excitation wavelength shifts on $Ca^{2+}$ binding, but a fluorescence enhancement. Similar to fura-2 on calcium binding, a dye named furaptra[49] undergoes excitation wavelength shifts on $Mg^{2+}$ binding. It is structurally related to the $Ca^{2+}$ chelator fura-2, but displays an excellent selectivity for $Mg^{2+}$ over $Ca^{2+}$.

A color indicator for potassium is MEDPIN whose anion forms a stable ternary complex with valinomycin/potassium.[50] The dye is 8(n-decyl)-2-methyl-4-(3',5'-dichloro-4'-phenone)-indonaphth-1-ol. Detection of potassium with MEDPIN is based on liquid-liquid solvent extraction between octanol and water containing the ion on a millimolar level. The water-immiscible octanol was combined with valinomycin and MEDPIN. The ternary complex of valinomycin/$K^+$/MEDPIN that forms has an absorption maximum at 610 nm (at pH 7.45) while MEDPIN alone absorbs at 472 nm. No shift was observed when sodium replaced potassium.[50] In summary, it may be said that a broad spectrum of alkali indicators is available by now. If these can be immobilized without compromising their binding capability and selectivity, useful optrodes for these clinically important species should be obtained.

An interesting new and fairly specific fluorescent zinc chelator has been described recently.[51] The fluorescence emission intensity of 9,10-bis(TMEDA)-anthracene increases over 1000-fold in the presence of zinc chloride. N-Butyl-N'-dansylthiourea in contrast, is a stable, fluorescent complex agent for a broad spectrum of metal ions.[52] A new class of chelating agents has been synthesized by condensation of Kemp's triacid with aromatic diamines such as m-xylidendiamine.[53] The obtained compounds are excellent chelators for bivalent cations such as $Ca^{2+}$ and $Mg^{2+}$. Carbonate-selective ion carriers based on trifluoroacetylazobenzene have been synthesized and the influence of carbonate on their UV-spectra has been studied previously.[54]

## D. REDOX INDICATORS

The classes of indicators to be described are all organic dyestuffs, exhibiting reversible redox reactions. If a redox half reaction is represented by Ox + n e $\rightleftharpoons$ Red, where Ox as usual represents the oxidized species and Red the reduced species, and n is the number of electrons, e, transferred, then the electrode potential of the redox couple will be given by the Nernst equation

$$E_h = E^0 + \frac{RT}{nF} \cdot \ln \frac{a_{ox}}{a_{red}} \qquad (29)$$

where $a_{ox}$ and $a_{red}$ are the activities of the oxidized and reduced forms respectively, T is the absolute temperature, R and F have their usual meaning, $E_h$ is the potential referred to the normal hydrogen electrode, and $E^0$ is a constant, characteristic of the redox couple. $E^0$ corresponds to the value of $E_h$ when the oxidized and reduced forms are present at unit activity, and is called the standard potential of the redox couple.[55]

Most frequently, however, the term $E_o'$, called the formal potential, is used. This is the potential of the solution containing equal *concentrations* (not necessarily 1 *M*) of the oxidized and reduced species, under specified conditions in terms of other electrolytes. For any redox indicator system it will therefore be necessary to know the formal potential under a variety of experimental conditions. Unfortunately, while this information is available for many inorganic redox systems,[56] it is not so readily available for the organic systems of interest in this chapter, Clark[55] being the only major source of reference.

The most important application of formal potentials which must be considered here is that of the variation of $E_o'$ with the pH value of the solution. Both $E_o$ and $E_o'$ apply to pH = 0, that is a hydrogen ion activity of 1. For values of $E_o'$ at other pH values the term $E_m$, first introduced by Michaelis, is used, with a superscript indicating the pH value at which the measurement is relevant. For example, $E_m^7$ means the value of the formal potential at pH 7. The term $E_o'$ will always refer to pH 0.

The distinction between $E_o'$ and $E_m$ is very important in the consideration of redox indicators as their formal potentials are dependent on the pH of the solution. Measurement of $E_m$ values over a wide range of pH affords a fairly simple method of establishing the pK values of many reversible oxidation/reduction systems.

Fluorescent redox dyes where color depends on oxidation state have been useful in biochemical studies in conjunction with microscopic or flow cytofluorometric techniques, but await their application to optical sensors for the redox state of systems such as biofermenters. The change from the oxidized to the reduced form should be accompanied by a marked color change and fully reversible. The redox potential of the indicator may then be chosen appropriately so to provide some selectivity for the redox process to be monitored in the system under investigation. A list of colored redox dyes with their respective standard potentials is given in Table 7.

A promising approach to fluorescent molecular sensors for cations (including the proton) employs the principle of photo-induced electron transfer.[57] Here, excess energy of a photoexcited state is used to drive a one-electron reduction/oxidation of a species A and a one-electron oxidation/reduction of species B. Assuming, for instance, an indicator molecule whose electronic excited state at least partially is deactivated by electron transfer in the deprotonated state, but not in the protonated state, then fluorescence intensity would be distinctly lower in the former case. Protonation thus would modulate fluorescence intensity and lifetime. Such a dye was found in 9-diethylamino-anthracene whose fluorescence quenching dramatically changes at the $pK_a$ of the amino group, although the absorption spectrum remains the same, obviously because the amino group is not part of the chromophore. An indicator for potassium ion was found in N-(anthracene-9-yl-methyl)monoaza-18-crown-6. The use of such redox indicators involving electron transfer and subsequent fluorescence quenching holds great promise for future sensing schemes because of straightforward chemical techniques that allow the design of a number of new indicators.

A few other redox systems seem worth mentioning. Thus, the reduction of tetrazolium salts to formazans has been long used in histo-, cyto-, and biochemical determinations especially of oxidases and dehydrogenases. This redox system meets most of the preconditions mentioned above. However, only very weak fluorescence intensities have been reported.[58] Certain 3-cyano-1,5-diaryl-formazans have been synthesized and were found to be redox-active (like their long-known isomers).[58] Only the crystals are fluorescent (bright red) when observed under fluorescence microscopy, whereas no fluorescence can be detected of an ethanolic solution of the dye. After incubation of Ehrlich ascites cells with colorless, watersoluble 5-cyano-2,3-ditolyltetrazolium salts, bright red water-insoluble formazan crystals on the cell surface can be observed under fluorescence microscopy. Possible application of these monotetrazolium salts are fiber optic cytochemical investigations of oxidative metabolic reactions. In an elegant approach that utilizes the change in fluorescence intensity as a result of radical reactions, Blough and Simpson[59] have successfully sensed various types of redox reactions.

**TABLE 7**
**Selected Reduction-Oxidation Indicators[a]**

| Indicator | Redox potential $E_o$ (at pH 0 and 20°C) | $E_m$ (at pH 7 and 30°C) | Color change |
|---|---|---|---|
| Amidoblack 10B | +0.57 | +0.84 | Yellow/blue |
| Anilic acid | +0.89 | — | Purple/colorless |
| Benzidine | +0.873 | — | Blue/colorless |
| 2,2'-Bipyridine | +1.03 | — | Blue/red |
| Brillant cresylblue BB | +0.58 | +0.05 | Blue/colorless |
| o-Dianisidine | +0.849 | — | Red/colorless |
| 2,6-Dichlorphenol-indophenol | +0.67 | +0.23 | Blue/colorless |
| 3,3'-Dimethyl-naphthidine | −0.78 | — | Colorless/purple |
| Diphenylamine | +0.7 | — | Colorless/purple |
| Diphenylamine-4-sulfonate | +0.84 | — | Colorless/violet |
| Erioglaucine A | +1.00 | — | Red/yellow |
| Eriogreen B | +1.01 | — | Orange/yellow |
| Ferroin | +1.06 | — | Blue/orange |
| Indigocarmine | +0.29 | −0.11 | Blue/yellow |
| Indophenol, Na salt | +0.65 | +0.23 | Purple/colorless |
| Cacotheline | +0.525 | — | Yellow/purple |
| m-Cresol-indophenol | +0.21 | — | Blue/colorless |
| Methylene blue | +0.53 | +0.01 | Blue/colorless |
| Methylviologene | — | −0.45 | Colorless/violet |
| Neutral red | +0.24 | −0.29 | Purple/colorless |
| Nile blue | +0.41 | −0.12 | Blue (red)/colorless |
| Nitroferroin | +1.25 | — | Red/blue |
| 1,10-Phenanthroline | +1.14 | +1.06 | Blue/red |
| N-Phenyl-anthranilic acid | +0.89 | — | Purple/colorless |
| Resazurine | +0.38 | −0.051 | Blue/colorless |
| Safranin T | +0.24 | −0.29 | Violet/colorless |
| o-Tolidin | +0.873 | — | Blue/colorless |
| Toluylene blue | +0.60 | +0.11 | Violet/colorless |
| 1,10-Phenanthroline-$V^{2+}$ | +0.15 | — | Blue/green |
| Variamin Bluesalt B | +0.712 | +0.31 | Violet(yellow)/ colorless |
| Xylenecyanol FF | +1.00 | — | Blue/pink |

[a]   Data from Reference 1.

## E. POTENTIAL-SENSITIVE DYES

These comprise a quite different class of dyes which respond to an electrical potential (V/cm) created at an interface. Thus, they do not directly report the concentration or an activity of a chemical species, but rather an electromagnetic parameter. They provide an interesting alternative to sensors based on conventional chromogenic chelators. Most of the work with potential-sensitive dyes (PSDs) so far is on measurement of membrane potential of cells and tissue.[60-62] PSDs are usually placed directly at the site where an electrical voltage is applied or where such a potential is created in a fashion similar to ion-selective electrodes, using an ion carrier incorporated into a lipid membrane (see Chapters 9.I. and 21)

Various reasons have been given for the response of PSDs towards electrolytes.[61] Changes in membrane potential can occur as a result of (1) charge separation across the membrane (caused, for instance, by an added ionophore); (2) streaming potentials at the membrane surface caused by the sample solution passing by; and (3) possibly, the formation of diffusional potentials. These changes in potential result in optical changes which, in turn, may be due to a variety of effects:[60]

1.    The Stark effect (i.e., a change in the absorption and emission spectrum of a fluorophore when an external field is applied);
2.    Changes in the otherwise homogeneous distribution of the fluorophore within the lipid membrane when an electric field is created, leading, for example, to aggregation and self-quenching;
3.    A field-dependent distribution of the dye between regions of different polarity within the lipid membrane, resulting in a solvatochromic effect.

The Stark effect is best observed with chromophores that have a large difference in the permanent dipole moments of ground and excited state. It can be shown that, in a first approximation, the following equation is valid:

$$\Delta \upsilon \cdot hc = \Delta \mu \cdot F_e \qquad (30)$$

where $\Delta \nu$ is the resulting wavenumber shift ($cm^{-1}$), $\Delta \mu$ the difference in the permanent dipole moments of $S_o$ and $S_1$ states, $F_e$ the electric field strength, h Planck's constant, and c the velocity of light. The signal change will be most expressed, if $\Delta \mu$ is large, $F_e$ exceeds $10^5$ V/cm, and the transition moment of the dye fits the direction of the electric field.

It is known for many dyes that the fluorescence intensity is changed by about 5% when an external field of $10^6$ V/cm is applied. The signal changes found in most work are in the order of 2 to 8% per logarithmic concentration unit. One way to create a high field strength (V/cm) is to minimize the diameter of the membrane or layer over which the potential is effective or the voltage is established. This has been achieved by using the Langmuir-Blodgett technique for fabrication of thin lipid layers.[28,63] Assuming a lipid bilayer to be approximately 5 nm thick, one can calculate the actual electric field strength to be in the order of 250 mV/5 nm when a -10% signal change is observed and the Stark effect is solely responsible for the signal change.

It has been assumed,[64] however, that the selective response for a given ion not only is caused by the effects listed above, but also by streaming potentials and, possibly, diffusional potentials. The unselective response is caused by the limited selectivity of the ionophore together with other kinds of streaming and diffusional potentials. The latter two can be compensated for by using a reference optrode, but the unselectivity of the ionophore cannot.

Empirically it was found[65] that the relation between relative signal changes (after correction for nonspecific response by a two-sensor technique) and the negative logarithm of the ion concentration is linear over a concentration range of at least two decades and can be described by a equation of the type

$$\frac{\Delta I}{I} = \frac{\Delta I_o}{I} + S \log[Me] \qquad (31)$$

where $\Delta I/I$ is the relative signal change, and $\Delta I_o/I$ is a membrane-specific parameter which corresponds to the signal change caused by a 1 $M$ ion solution, and [Me] is the metal ion concentration. S is an analyte- and material-dependent constant. Various optrodes based on optical measurement of membrane potential have been reported[63-67] (see Chapter 9). The corresponding dyes have the advantage of possessing longwave excitation and emission, but some cyanines display cell toxicity and photolability. They also are difficult to covalently immobilize.

## 1. Cationic Potential-Sensitive Dyes

Positively charged carbocyanines are membrane-permeant and distribute across the membrane according to the membrane potential. Fluorescence changes as large as 80% associated with a redistribution of the dyes between cells and the extracellular medium during hyperpolarization have been monitored.[61] During hyperpolarization carbocyanines accumulate in the cell and form aggregates that show altered absorption and no fluorescence emission. The fluorescence changes are relatively slow (1 to 20 s) compared to fast responding probes such as styryl dyes. Carbocyanines have longwave absorption and emission, strong molar absorbance, but relatively small Stokes' shifts. They are weakly fluorescent in water, but are highly fluorescent in organic solvents or when bound to membranes. The most widely used carbocyanines are $DiSC_3(5)$ and $DiOC_5(3)$ and their close homologs. Cationic styryl dyes are also potential-sensitive, but styryls have found wider use as their neutral zwitterionic derivatives.

The rhodamine dyes comprise another class of cationic PSDs. Among these, rhodamine 6G,[69,70] rhodamine 123, and rhodamine B octadecyl ester[64,66,68] have been found to be useful for measurements of membrane potential in sensors. The methyl and ethyl ester of tetramethyl rhodamine perchlorate have been synthesized recently and investigated as membrane potential probes.[71] The rhodamine esters are nontoxic, highly fluorescent dyes which do not form aggregates or display binding-dependent changes in fluorescence efficiency. Thus, their reversible accumulation is quantitatively related to the contrast between intracellular and extracellular fluorescence and allows membrane potentials in individual cells to be continuously monitored.

## 2. Anionic Potential-Sensitive Dyes

Historically one of the first PSDs, merocyanine 540 continues to find use, also in optrodes.[66] Oxonol dyes undergo slow potential dependent absorption changes without alteration in the state of dye aggregation. The voltage-dependent partitioning between water and the membrane is usually measured by differential absorption changes rather than fluorescence. Of the oxonols studied by Smith and Chance,[72] oxonol VI gave the largest spectral shifts. Symmetric bis-oxonols[73] are slow-responding fluorescent anions, whose distribution across the membrane is potential dependent. The dianionic merocyanine oxonol dye WW 781 is retained by (bio)membranes due to its high anionic charge and, hence, needs no covalent immobilization. Spectral response of WW 781 is much faster than the carbocyanines,[74] although the magnitude of the response is smaller. The lipophilic fluorescent pH indicator, 7-hydroxy-4-tridecylcoumarin, also has found use as a PSD.[75]

## 3. Neutral Zwitterionic Potential-Sensitive Dyes

Styryl dyes containing an (aminostyryl)pyridinium chromophore are fast-responding PSDs. The fast-response is the result of a direct potential-sensitive change in the electronic distribution within the dye that results in spectral shifts. The properties of a series of zwitterionic styryl dyes have been explored,[76] and spectral properties were discussed in terms of the excitation-induced charge shift from the pyridine to the aniline. This charge shift provides the basis for the voltage dependence of the spectra according to an electrochromic mechanism.

The spectral responses to a membrane potential are quite similar for all probes tested. The more subtle variations from dye to dye can be partially rationalized by consideration of binding parameters, the depth within the membrane, and structural factors. The most responsive PSD in this collection has been designated di-4-ANEPPS and has a 6-amino-2-naphthyl group in place of the p-anilino group on the parent chromophore. The probe has been reported to have a fairly uniform 10%/100 mV response in several types of cells. Among the styryl dyes, a great number has been developed by Grinvald et al.[77] and found nomination as RH dyes (from R. Hildesheim).

## F. QUENCHABLE FLUOROPHORES

A most promising class of indicators are the quenchable fluorophores. In dynamic quenching, interaction between analyte and fluorophore is in the excited state only and leads to reduction of both fluorescence emission intensity and lifetime. The relation between the two parameters and analyte concentration is described by the Stern-Volmer equation (Equation 16 of Chapter 3). The process is fully and rapidly reversible. Quenchable fluorophores have therefore been preferably applied where possible. Typical examples for oxygen optrodes are presented in Chapter 10.

The most significant parameter for judging the utility of a quenchable fluorophore is its Stern-Volmer quenching constant. The larger it is, the more sensitive the sensor will be. Methods for determination of static and dynamic quenching constants are described in Chapters 2 and 3. The following is a presentation and discussion of dynamically quenchable fluorophores according to the charge of the quencher. Because of their great importance, oxygen and biological quenchers are listed separately. However, given the tremendous literature on fluorescence quenching, only those quenchers are discussed that are considered to hold some promise in optical sensor technology.

## 1. Fluorophores Quenchable by Neutral Species

The quenching of fluorescence by aliphatic and aromatic amines has been reported for anthracene[78] and phenylated anthracenes.[79] The fluorescence of naphthalene and fluorinated naphthalenes is quenched by triethylamine.[80] These findings offer the possibility of designing the respective optrodes, but cross-sensitivity to oxygen may be a problem. The rate constants for the fluorescence quenching increase in the order of naphthalene, 1-fluoronaphthalene, and octafluoronaphthalene and depend on the solvent used. Typical quenching parameters of triethylamine, other amines, and anilines are summarized in Table 8. Quenching efficiency is usually expressed in terms of $K_d$ (the overall dynamic quenching constant in the Stern-Volmer equation), or $K_q$ (the bimolecular quenching constant, with $K_d = K_q \cdot \tau$).

The strong fluorescence of 6-methoxy-4′-methyl-2-phenylquinoline-4-methanol is reported to readily be quenched by organic sulfides.[81] The mechanism of quenching involves the formation of an exciplex between the quencher and the quinoline compound. Typical examples for sulfide quencher include isopropyl disulfide (log $k_q$ 9.04), isopropyl sulfide (log $k_q$ 8.19), thiomorpholine (log $k_q$ 9.56), and 4-methyl-thiazole (log $k_q$ 8.30). Obviously, this could be exploited for continuous sensing of these species.

The fluorescence of 9,10-dicyanoanthracene is quenched by a variety of acyclic and cyclic dienes and alkenes.[84] These are important precursors in polymer fabrication, and respective sensors are likely to be of great interest in monitoring polymerization. The values of the Stern-Volmer constants for representative olefins are listed in Table 9. Olefins also quench the fluorescence of other aromatic hydrocarbons such as pyrene, naphthalene, and various methylnaphthalenes, the rate of which is is extremely solvent dependent.[85]

Fluorescence quenching of 2,5-diphenyloxazole[86] and aromatic hydrocarbons[87] by the toxic solvent tetrachloromethane has been reported. Carbon disulfide quenches the fluorescence of anthracene groups that have been covalently bound to poly(methyl methacrylate) particles.[88] The pesticide lindane (hexachlorocyclohexane) quenches the fluorescence of the lipid probe methyl 11-(N-carbazolyl)-undecanoate.[89] A moderate quenching rate constant of $7.0 \times 10^8$ $M^{-1}s^{-1}$ in methanol has been determined. All these species are of interest in environmental analysis and could be determined with the corresponding sensors, although the quenching constants are such that sensitivity is unlikely to be very high.

A series of fluorescent dyes have been investigated for quenching by sulfur dioxide.[90] Fluorescence quenching by sulfur dioxide has been found extremely efficient, particularly for fluoranthene ($K_d$ 284 $M^{-1}$), pyrene ($K_d$ 238 $M^{-1}$), and benzo(b)fluoranthene ($K_d$ 225 $M^{-1}$), and exclusively dynamic for all aromatic hydrocarbons studied so far. Rhodamine 6G is quenched as well, albeit with much smaller efficiency ($K_d$ 12.4 $M^{-1}$), and has more favorable analytical wavelengths. These findings led to the design of a fiber optic sulfur dioxide sensor.[91]

The paramagnetic nitroxides are also known to be efficient quenchers of excited single

**TABLE 8**
**Typical Fluorophores Quenched by Amines, and the Respective**
**Quenching Constants**

| Fluorophore | Quencher | $K_q$[a] | Solvent | Ref. |
|---|---|---|---|---|
| Naphthalene | Triethylamine | 5.5 | Acetonitrile | 80 |
| 1-Fluoronaphthalene | Triethylamine | 8.7 | Acetonitrile | 80 |
| Octafluoronaphthalene | Triethylamine | 10.8 | Acetonitrile | 80 |
| 6-Methoxy-2-phenyl-quinoline | N,N-Dimethylaniline[b] | | Various | 81 |
| Naphthoquinone-SO$_3^-$ | Primary amines | — | — | 82 |
| Pyrene | Dialkylanilines | — | DPPC liposomes | 83 |

[a]    $K_q$ is the bimolecular quenching constant (see Chapter 3) and is given in $10^9$ $M^{-1}$ $s^{-1}$ units.
[b]    Various other amines also investigated.

**TABLE 9**
**Stern-Volmer Quenching Constants ($K_d$)**
**for the Quenching of the Fluorescence of**
**9,10-Dicyanoanthracene by Olefins[a]**

| Quencher | $K_d$ $(M^{-1})$ |
|---|---|
| 2,5-Dimethyl-2,4-hexadiene | 286 |
| cis-trans-2,4-hexadiene | 365 |
| 1,3-Cyclohexadiene | 327 |
| cis,cis-2,4-hexadiene | 592 |
| 2,3-Dimethyl-2-butene | 309 |
| 1,3-Cycloheptadiene | 220 |
| trans-1,3-pentadiene | 227 |
| cis,cis-1,3-cyclooctadiene | 206 |
| 2-Methyl-2-butene | 217 |

[a]    Data from Reference 85.

states of aromatic hydrocarbons,[59] presumably through an intermolecular electron-exchange interaction between the ground-state nitroxide and excited state compound within a collision complex. Nitroaromatics, some of which are potent explosives, statically quench the fluorescence of rhodamines[78] and electron-rich aromatics including tetrapyrrols like chlorophyll and porphyrins.

## 2. Fluorophores Quenchable by Anionic Species

The halide ions are quenchers of several fluorophores with the increasing order of chloride, bromide, and iodide. The rate constants for the fluorescence quenching of a series of aromatic hydrocarbons by inorganic anions in acetonitrile have been reported.[92] The rate constant for the fluorescence quenching by iodide decreases in the order of p-terphenyl, biphenyl, fluorene, naphthalene, fluoranthene, anthracene, pyrene, perylene, phenanthrene, and coronene. The fluorescence quenching of aromatic hydrocarbons has also been studied[103] quantitatively in the solvent system water-ethanol (1:1). 7-Ethoxycoumarin and 7-ethoxy-4-methylcoumarin have also been found to be quenched by halide ions.[93]

Fluorescence quenching of cationic fluorophores (acridinium and acridizinium ions) by halide ions in methanol has been investigated.[94] The system shows the largest Stern-Volmer constants that have been determined for quenching by halides so far. Other highly sensitive

**TABLE 10**
**Spectral Properties and Quenching Constants ($K_d$s) for Typical Halide-Sensitive Quenchable Fluorophores[a]**

| Indicator | Solvent | Excitation/emission maximum (nm) | $K_d$ ($M^{-1}$) | | |
|---|---|---|---|---|---|
| | | | $Cl^-$ | $Br^-$ | $I^-$ |
| Quinine cation | 0.1 N Sulfuric acid | 347/450 | 115 | 185 | 255 |
| 6-Methoxyquinoline | 0.1 N Sulfuric acid | 347/450 | 115 | 185 | 255 |
| SPQ[b] | Water | 347/450 | 120 | 193 | 258 |
| SPA[c] | Water | 358/488 | 9.3 | 296 | 390 |

[a]  Data from Reference 107.
[b]  1-Sulfopropyl-6-methoxyquinolinium inner salt.
[c]  10-Sulfopropyl-acridinium inner salt.

halide indicators were found in cationic 6-methoxyquinolines and acridines.[95] Data on the most sensitive halide ion indicators are summarized in Table 10. In the first optical sensor for halides and pseudohalides, 3-(10-methylacridinium-9-yl)propionic acid and 6-methoxy-N-(3-sulfonatopropyl)quinolinium (SPQ) were used as fluorescent probes and immobilized, via spacer groups, onto a glass surface.[96] The sensors are able to indicate the concentration of halides in solution by virtue of the decrease in fluorescence intensity due to the dynamic quenching process. Other sensing optrodes for iodide ion[97] are based on dynamic quenching of rhodamine 6G. The lipophilic quinolinium probe 6-methoxy-N-hexadecyl-quinolinium perchlorate has been incorporated in a Langmuir-Blodgett type bilayer to form an optrode for halide anions.[28]

The chloride indicator SPQ, first synthesized by Wolfbeis and Urbano,[95] is now widely used in the measurement of membrane transport of chloride in biological systems. To understand the structure-activity relationships of compounds with chloride-sensitive fluorescence properties, structural analogs of SPQ were synthesized.[98] The effect of variations in ring structure, length of sulfoalkyl chain, position of ring substituent, and nature of ring substituent were examined. It was concluded that a high chloride sensitivity ($K_d > 50\ M^{-1}$) requires the presence of a quinoline backbone substituted with electron-donating groups such as methyl and methoxy, but does not depend on length of the sulfoalkyl chain or on the position of ring substituents.

Other anions such as thiocyanate,[93] nitrate,[92] and azide[93,99] are weak quenchers of certain aromatic hydrocarbons. The fluorescence of SPQ ($K_d$ 211 $M^{-1}$) and other quinolinium compounds is quenched by thiocyanate with high efficiency.[98] Some quinolinium compounds are quenched by bicarbonate ($K_d$ 8 to 12 $M^{-1}$) and citrate ($K_d$ 9 to 28 $M^{-1}$) with moderate efficiency,[98] and by hydrogen sulfide ($K_d$ 2.4 $M^{-1}$),[100] the latter probably by a static process (the reversible addition of hydrogen sulfide anion). Sulfate and phosphate are practically not capable of quenching any of the investigated fluorphores to a remarkable extent. An exception is the quenching of tyrosyl residues in proteins by phosphate.[101] This may be exploited to design a phosphate sensor, although the fluorescence excitation/emission maxima of tyrosine (280/303 nm) are at rather unfavorable wavelengths.

## 3. Fluorophores Quenchable by Cationic Species

Among the cationic quenchers, particular attention has obviously been paid to the heavy metal ions. Among the popular fluorophores, cationic derivatives of 6-methoxyquinoline and acridine, rhodamine B, cresyl violet, aromatic hydrocarbons such as perylene, and carbazole have been reported in numerous papers to be quenched by metal ions including $Cu^{2+}$, $Ag^+$,

$Fe^{3+}$, $W^{6+}$, $Pb^{2+}$, $Co^{2+}$, and also by cationic organic molecules. Thus, sulfopropyl-6-methoxyquinolinium is quenched by $Cu^{2+}$, $Ag^+$, $Pb^{2+}$, or $Hg^{2+}$ in methanol with Stern-Volmer quenching constants ($K_d$ of 64.7, 20.7, 10.9, and 35 $M^{-1}$, respectively). For sulfopropylacridinium, the corresponding data are 8.4, 42.6, <1, and <1 $M^{-1}$, respectively. $Fe^{3+}$ is a moderate quencher of the fluorescence of cresyl violet and rhodamine.

Trace levels of lanthanide ions efficiently quench the fluorescence of calcein blue.[102] The Stern-Volmer constants obtained for the obviously static quenching of calcein blue by trivalent lanthanide-metal ions range from 90,000 to 120,000 $M^{-1}$. The analytical method based on this quenching is by far more sensitive than any other. Concentrations as low as 0.01 to 0.02 μg/ml can be detected. Other coumarin derivatives such as methyl calcein blue and 4-methylumbelliferone are also subject to lanthanide quenching,[123] but not to such an extreme extent. Stern-Volmer constants range from 241 to 590 $M^{-1}$ depending on the identity of the lanthanide ion. Dynamic fluorescence quenching of the aromatic amino acids tyrosine is an elegant means to study the accessibility of Tyr and Trp in proteins. Static quenching of indols by $Hg^{2+}$, respectively, was exploited to continuously sense mercury ion which acts as a static quencher of the fluorescence of indole.[103] Indole, however, has to be excited at around 290 nm, which is compatible with quartz fibers only.

The blue fluorescence of anionic fluorescent whiteners is efficiently quenched by cationic aromatic nitro compounds.[104] The fluorescence quenching of sodium pyrene-3-sulfonate by poly(dimethyl diallylammonium chloride)[105] and aromatic compounds by diazonium cations[106] in aqueous solutions has been investigated. Efficient static and dynamic fluorescence quenching has been reported for 1-amino- and 1-alkoxypyrene-trisulfonates by cationic detergents such as cetyl trimethyl ammonium bromide and quaternized thiazolium and pyridinium ions.[107,108]

### 4. Fluorophores Quenchable by Oxygen

By virtue of its triplet ground state configuration, molecular oxygen is a notorious (and almost ubiquitous) quencher of luminescence. Because of the distinctly longer lifetime of phosphorescence, this kind of luminescence is practically never observed in fluid solution at room temperature. Fluorescence lifetimes, in contrast, are comparable to the inverse bimolecular quenching rates so that quenching strongly can modulate fluorescence intensity and lifetime (Chapter 3.III.A.). As a matter of fact, quenching has always considered a nuisance until it was recognized that the effect can be a blessing with regard to the old problem of sensing oxygen. Later, many other procedures were introduced when the potential of fluorescence quenching analysis became evident. Nowaday, fluorescence quenching forms the basis of almost all kinds of oxygen optrodes, and numerous other quenching-based optical sensor schemes have been described.

The polycyclic aromatic hydrocarbons (PAHs) form a class of fluorophores that are known to be very efficiently quenched by oxygen. Table 11 summarizes spectral data and quenchabilities of a number of PAHs. Among these, decacyclene and benzoperylene have been used most often in oxygen optrodes. Decacyclene has fairly longwave excitation and emission maxima (which shift to even longer wavelengths when t-butylated in order to make it better soluble in hydrophobic polymers), high quenchability, and stability. The limitations of PAHs with respect to sensor application are their mutagenicity (decacyclene, however, being an exception) and their incompatibility with solid-state optoelectronics. Although certain PAHs are known which are excitable at or above 488 nm, these suffer from rapid photodecomposition.

A detailed discussion of fluorescent oxygen indicators listed in Table 11 can be omitted since Chapter 10 deals, in some detail, with the indicators that have been used in practice. If oxygen indicators are assessed according to longwave absorption/fluorescence and quenching

**TABLE 11**
**Spectral Characteristics (in nm) and Quenchability of**
**Oxygen-Sensitive Fluorophores[a]**

| Fluorophore | Solvent | Excitation/emission | Quenching (%)[a] |
|---|---|---|---|
| PAHs | | | |
| Decacyclene | Silicone | 390-420/510 | −55 |
| Solubilized decacyclene | Silicone | 400-440/520 | −50 |
| Diphenylanthracene | Xylene | 394/435 | −25 |
| Benzo(ghi)perylene | Silicone | 380-410/430 | −60 |
| Anthracene | Silicone | 430/460 | −10 |
| Coronene | Silicone | 340/446 | −70 |
| Carbazole | Silicone | 345/360 | −28 |
| Indenopyrene | Xylene | 410-430/480, 510 | −15 |
| Pyrenebutyric acid | Silicone | 345/400 | −68 |
| Dibenzoanthracene | Xylene | 350/420 | −58 |
| Fluoranthene | Xylene | 360/425-525 | −30 |
| Chrysene | Xylene | 320/430 | −68 |
| Benzo(a)anthracene | xylene | 360/436 | −62 |
| Others | | | |
| Various heterocycles[c] | Toluene | 520—561/570—612 | −15 to −28 |
| Ru(bipy)₃(II) | Silica-adsorbed | 460/610 | −47[d] |

<small>
[a]  Data from References 109 and 113.
[b]  Expressed in the reduction of fluorescence intensity in going from nitrogen to air at 750 Torr.
[c]  From Reference 112.
[d]  From References 110 and 111.
</small>

efficiency, the top five of the PAH-type indicators in Table 11 are probably most suitable for sensor application.

Obviously, there is a considerable need for more longwave absorbing quenchable fluorophores. One such dye is the ruthenium(II)tris(bipyridyl)$^{2+}$ cation which can be excited with the blue LED at around 430 to 480 nm and may be physically immobilized on hydrated silica gel particles.[110,111] Notwithstanding this, the complex has a poor molar absorption and quantum yield, but a lifetime of several hundred nanoseconds so that it is amenable to strong quenching. Even more longwave absorbing transition metal complexes are likely to exist.

A series of heterocyclic fluorophores with excitation maxima ranging from 460 to 565 nm, and emission maxima from 511 to 612 nm was investigated with respect to quenching by oxygen.[112] In going from pure nitrogen to pure oxygen in toluene solution at room temperature, the dyes were found to undergo fluorescence quenching by oxygen by −10 to −28%. Oxygen has also been reported to quench the fluorescence of other heterocycles such as acridine, fluorescein, quinine, rhodamine B, and eosine, but with poor efficiency. Certain dyes undergo more efficient quenching when adsorbed on polar surfaces (see Chapter 10.II).

Minute amounts of oxygen can quench the phosphorescence of adsorbed luminophores such as chlorophyll, acridine orange, or certain coumarins. This forms the basis for some phosphorescence-based oxygen sensors which are described in more detail in Chapter 10.III. The main driving force behind the design of phosphorescent oxygen indicators is the need for long-lived emitters in lifetime-based sensors. Pt(II) and Pd(II) porphyrins were shown[114a-e] to exhibit intense and long-lived emission as well as good solubility in polymers such as polystyrenes.[114b,c] Typical lifetimes for platinum porphyrins are in the order of 50 to 100 μs in polystyrene solvent under nitrogen, but around 20 μs under air. Quantum yields are as high as 0.4 to 0.6 at room temperature.[114a] Such properties make metalloporphyrins most attractive for lifetime-based sensors and for monitoring processes during which oxygen is consumed or produced.[114c]

Generally it is observed that fluorescence quenching of dyes dissolved in a polymer matrix or bound to polymer surfaces is much more complicated than in fluid homogeneous solution. Stern-Volmer plots are frequently nonlinear at higher oxygen tension. An unusual effect is the *enhancement* of fluorescence of certain dyes by molecular oxygen when the dye is incorporated into polymer matrices. Instead of the expected reversible quenching of fluorescence and phosphorescence, Geacintov et al.[116] observed that dibenzanthracene in poly(vinyl acetate) showed a 10% fluorescence enhancement in air compared to evacuated samples. Similar results were obtained for other PAHs in PVA and other polymers.[117] Explanations for the effect were given.[118]

### 5. Fluorophores Quenchable By Biomolecules

Fluorescence quenching along with decay measurements and energy transfer experiments is a very popular technique in bioanalytical studies. Important biological quenchers include pyruvate,[119] nicotinamide,[120] thiamine[107] (vitamin $B_1$), and nucleosides and nucleotides.[121] Pyruvate, which participates in many metabolic reactions, quenches the fluorescence of ε-ATP, while phospho-enolpyruvate does not. Pyruvate also quenches neutral, cationic, and anionic pyrenes. Thiamine is a mainly static quencher of 1-amino- and 1-alkoxypyrene-trisulfonates, but the mechanism is rather complex.[107] Nucleotides like AMP, GMP, TMP, and CMP quench the fluorescence of acridinium chlorides, and fluorescence decay follows a single exponential decay law. Both static and dynamic quenching processes seem to be involved. The nucleosides are also known to quench proflavine, acridine orange, and 9-aminoacridone.

Near-infrared semiconductor laser fluorimetry has been applied to assays of xanthine and xanthine oxidase.[122] In this reaction, xanthine is converted into uric acid by xanthine oxidase, in a reaction that also produces hydrogen peroxide. The fluorescence of indocyanine green in the near-infrared region is said to be quenched by hydrogen peroxide but more likely the effect is due to chemical decomposition. Xanthine can be determined by measuring the decrease in fluorescence intensity of the dye.

# III. POLYMERIC SUPPORTS

## A. GENERAL ASPECTS

Everybody ever having been active in the field of optical sensors will probably agree that polymer chemistry (as a part of the broad field of material sciences) is an extremely important part of this technology. Both the light guide (including its cladding and coating) and the sensing chemistry of indicator-mediated sensors are made from organic or inorganic polymer. Within this book, only the chemistry of polymeric supports for use in chemical sensors will be treated, rather than the materials used for fabrication of cores and claddings. The knowledge of the respective materials is, however, of considerable importance when it comes to make the material of the working chemistry compatible with the material of the fiber.

The polymers used in optrodes can have one or more of the following functions: (1) It acts as a rigid support onto which the dye (or receptive element) is immobilized. (2) It may act as a solvent or cage for the material to be immobilized. (3) It can provide selectivity for certain species by virtue of the permselectivity of most polymers. (4) Polymeric covers are frequently used as protective covers for sensitive working chemistries. (5) They can serve as optical isolation so to avoid ambient light to enter the optical system of the optrode.

The choice of polymer is governed by the permeability of the polymer for the analyte, its stability and availability, its suitability for dye immobilization, its compatibility with other materials used in the fabrication of optrodes, and its compatibility with the sample to be investigated. Quite a number of reviews and books on various polymers and their applications is available.[123-128] A useful compilation of new polymeric materials and blends, along with their use in various fields (except optrodes), is found in a book edited by Seymour and Mark[125].

The choice of polymer material has a pronounced effect on the sensor performance. The response time, for example, will be governed by the diffusion coefficients of gases or liquids, and the quenching efficiency by the solubility of the gas in the polymer. The solubility and diffusion coefficients for various gas/polymer combinations have been compiled.[129-131] However, although these authors have compiled a considerable amount of data on various polymers, numerous new materials are available for which no data are known. It is also known that copolymers and polymer mixtures do not necessarily display the properties that may be expected from averaging the data of the pure components.[125]

Among the variety of polymers tested, silicones have unique properties[131,132] in having a higher permeability for most gases than any other polymer, and also a high solubility for oxygen and $CO_2$. Even water vapor (unlike liquid water!) readily dissolves in, and passes through, silicone membranes. Typical examples for sensors in which the analyte diffuses into the polymer to quench a dissolved indicator are the oxygen fluorosensors. In certain cases, the inner polymer is covered with an outer polymer that inparts some selectivity. Halothane, for instance, interferes in many oxygen optrodes because it acts as a quencher as well. By covering the working chemistry with a 6-μm PTFE membrane, the cross sensitivity can be eliminated.[133] Similarly, the selectivity of sensors for $CO_2$ results from the fact that interfering ions such as the proton do not pass hydrophobic membranes and therefore cannot interact with a dissolved indicator. Most hydrophilic polymer membranes, on the other side, are penetrated by low molecular-weight analytes, but not by proteins.

Another group of polymers functions in a quite different way in that it acts as a rigid support. Only its surface (or sample-accessible sites) are covered with indicator molecules. Hence, the polymer is not a solvent for the indicator, and the analyte does not equilibrate between sample phase and polymer phase, but rather with its surface only. Here, the response time is governed by the rate of equilibration between sensor surface and sample. Typical examples are given in the Chapters on pH and electrolyte sensors.

A final function of the polymer material is to act as an optical insulator. A variety of sensors have been described which are covered with a light-impermeable but analyte-permeable membrane so to prevent sample fluorescence and ambient light to interfere with the optical system of the sensor. A typical material for use in a gas sensor would be a 6-μm black PTFE membrane, while for electrolyte sensors or biosensors a layer of black hydrogel is a good choice.

Some polymers such as polystyrenes and polyethylenes display an intrinsic fluorescence under UV excitation. On the other side, pure poly(vinyl chlorides), poly(vinyl alcohols), and polysiloxanes are fairly clean under >300 nm excitation. However, most organic polymers have added plasticizers to make them softer and more permeable. Among these, esters of phthalic acid are fluorescent by themselves under UV excitation and can give rise to considerable background signal.

## B. SILICONES

Silicones are polymers with excellent optical and mechanical properties and excellent gas permeability.[131,132] In case of oxygen it exceeds all other polymers. Numerous silicone prepolymers are available commercially and allow easy manufacturing of membranes, emulsions, suspension, or other kinds of sensing chemistries. One may differentiate between one-component and two-component silicone prepolymers. The former cure in the presence of moisture (e.g., from air) by splitting off acetic acid, methanol, or amines. In two-component prepolymers, a catalyst accelerates the addition reaction of component A to component B to give a long-chain polymer. The catalyst is usually contained in one of the prepolymers. Some catalysts have been found to act as quenchers of the fluorescences of charged indicators. Many silicones are of the room-temperature vulcanizing (RTV) type, and the prepolymers can be dissolved in aprotic solvents such as toluene. This greatly facilitates their handling.

There are certain disadvantages of silicones, too. First, they do not easily lend themselves to surface modification. Hence, the covalent immobilization of indicators on cured silicone membranes is extremely difficult. It is probably for that reason that silicones have mainly served as solvents for indicators or as gas-permeable covers, but never as rigid supports with surface-immobilized indicators. Second, silicones have limited compatibility with other polymers and are difficult to glue onto other materials. As a matter of fact, certain sensor types described in the literature and based on silicone rubber materials in combination with other materials have extremely poor long-term stability because of material incompatibility. Third, most silicones are very good solvents for most gases including oxygen. This may lead to a depletion of gas when the sample volume comes to lie below the 100-fold volume of the sensing layer. Finally, silicones are poor solvents for most quenchable fluorophores, and various attempts have been made to improve solubility without compromising quenchability. One such way is to provide the indicator with tertiary butyl groups,[133] another is to covalently link the indicator to a terminal functional group of the polymer chain.[134]

Fluorinated silicones offer a quite promising alternative.[134] In these, the C-H bonds of the Si-alkyl groups are partially or totally replaced by C-F bonds. Polyfluorosilanes display extreme chemical resistance to most solvents and better solubility for indicators, but gas diffusion is somewhat slower when compared to their conventional counterparts (see Section III.G).

Block copolymers of poly(dimethyl siloxanes) (the soft component) and polycarbonate (as the hard component) have been used in both $CO_2$ and ammonia electrodes and optrodes as pinhole-free covering membranes that are permeable to these gases but act as a barrier to protons. However, gas permeability is reduced when compared with silicone, but mechanical strength is substantially increased.

Silicones and their prepolymers are commercially available in various forms from several companies. Excellent data handbooks are available too,[135-137] and the reader is referred to these for a general discussion of their chemistry. The chemistry of silanizing agents is compiled in the books written by Leyden and Collins[138] and by Plueddeman.[139] The main application of silicone materials is in sensors for oxygen and other quenching species, such as $SO_2$, and as gas-permeable covers in sensors for $CO_2$ or $NH_3$ based on pH changes in an internal buffer.

## C. OTHER HYDROPHOBIC MATERIALS

Poly(vinyl chloride) (PVC), polyethylene, poly(tetrafluoroethylene) (PTFE), and polystyrene are other hydrophobic materials that efficiently retain ionic species. Except for polystyrene, they are difficult to chemically modify so that their function is confined to a solvent for indicators that penetrate the polymer, or as a gas-permeable cover. However, the diffusion of analytes through, and the solubility of gases in such membranes is distinctly smaller and results in drastically limited quenching constants. Zhang et al.[140] have prepared clear PVC membranes by dissolving, in tetrahydrofurane, PVC, plasticizer, and poly(oxyethylene imine), and evaporating the solvent. The membranes remain clear indefinitely in water, although some amine is washed out. The material was used to sense trinitrotoluene which causes the membrane to turn to red.

More often than have membranes, PVC and polystyrene have been applied as small beads soaked with an indicator solution. PVC and polystyrene are much better solvents for uncharged indicators than is silicone rubber. PTFE, in contrast, is a poor solvent as are silicones, but also has poor permeability for larger molecules. It has been used in the CDI oxygen sensor as a halothane-impermeable, but oxygen-permeable barrier in front of the actual sensor chemistry (Chapter 19).

## D. SILICA MATERIALS

Given the mechanical stability and favorable optical properties of glass, it has been the material of choice for many sensing purposes. Because glass is impermeable to practically every analyte, its function has always been that of a rigid support whose surface was modified by chemical means. The popularity of glass is also due to the fact that its surface can be made both hydrophilic and hydrophobic, simply by treating it with the proper reagent.[138,139] Glass does not measurably swell and is fairly easy to handle. It has been used in planar form which has the disadvantage of having a small specific surface. To overcome this situation, glass beads may be sintered onto planar glass supports to give a material of high specific surface.[15] Further enlargement of the specific surface is achieved by using porous glass beads. They are available in various forms, even with chemically modified surfaces, so that immobilization is particularly simple. However, porous glass is fragile and is easily ground to fine powder. The ease of surface modification of glass[138] has two beneficial consequences. First, all kinds of hydrophilic or hydrophobic indicators may be covalently bonded to its surface. Second, glass can be made compatible with all kinds of other materials including polymers as different as silicone rubber and hydrogel by proper modification of the surface. The major disadvantage of glass is its brittleness.

An interesting material was obtained[141] by controlled hydrolytic polycondensation of $Si(OEt)_4$ to give a fairly inert inorganic glassy matrix whose porosity and size of pore network can be varied, to a certain degree, by polymerization conditions. Organic dye molecules can be incorporated into these glassy matrices having a low-density network structure. These so-called sol-gel glasses are formed at room temperature and capable of supporting the transport of small molecules, but also of entrapping large molecules such as enzymes. Because the material has no absorption in the near UV and visible, it is well suited for fabrication of dyed materials in the form of films, fibers, or monoliths.

## E. MIXED HYDROPHILIC/HYDROPHOBIC MATERIALS

When the surface of a hydrophobic polymer is modified with charged functions, or when a hydrophilic polymer is surface-modified with apolar groups, new materials with promising properties are obtained. They can serve a double purpose in that they act as a solid support for the indicator to be immobilized, and sometimes provide some selectivity by retaining undesired components of the analyte to reach the indicator. On the other hand, these "mixed polarity" materials sometimes have limited compatibility with other polymers.

Typical examples are ion-exchange materials.[142] Ion-exchange membranes with high chemical resistance are usually of the block copolymer type such as the polystyrene/polyisoprene/polystyrene (SIS) type, onto whose surface $SO_3^-$ groups are attached in a second step. These materials have found widespread application in electrodialysis and are commercially available (e.g., from Asahi). Other large-scale cation exchange materials include the poly(acrylonitrile/methallylsulfonate) copolymer for use in hemodialysis and ultrafiltration (Rhone-Poulenc), polysulfones for hyperfiltration (Rhone-Poulenc), and Nafion™ (a sulfonated and partially fluorinated polyethylene from Dupont). A typical weak anion exchange material would be the (methacrylic acid on polyethylene) block polymer (from RAI Corporation). Anion exchangers, for example, are obtained by bonding vinylpyridine onto PTFE (RAI Corporation) or by introducing trimethylammonium groups onto the surface of polystyrenes (Amincon).

Charged indicators can easily be immobilized, e.g., by soaking ion-exchanging polystyrene beads (with their charges positioned on or near the surface of the polymer) with a solution of an oppositely charged organic indicator.[9,143] Because of the small exclusion size of ion-exchange beads, small indicator molecules and analytes easily diffuse into the interior regions of the beads. Large molecules, namely proteins or polysaccharides, or even cells, do not have access to the binding sites and hence cannot produce adverse interferences. For instance, no "protein error" is observed when the pH of serum is measured using a sensing material prepared from an anionic pH indicator that was electrostatically immobilized onto cation-exchanging polystyrene beads.[143]

Aside from polystyrene which is available in various physical forms (beads, membranes, etc.) and with various surface modifications (cationic and anionic), several other hydrophobic polymers have been shown to candidates for surface modification with hydrophilic groups. Among these, Nafion™ is widely used. It is a perfluorinated ion-exchange membrane containing sulfonic acid groups and good for immobilization of cationic indicators such as rhodamines.[97] It is available in bead form as well. Its excellent mechanical and chemical stability make it a most useful material for sensing purposes. Other mixed polarity polymers include chemically modified cellulose, copolymers of hydrophilic and hydrophobic monomers (e.g., styrene/ acrylamide), and carboxy-modified Latex™ particles.[144a] The latter are available in preactivated form. Cellulose triacetate plasticized with isodecyl diphenylphosphate was used in a sensor membrane for detection of volatile explosives.[144b]

## F. HYDROPHILIC SUPPORTS

Hydrophilic supports are characterized by a large number of hydrogen-bridging functions such as OH or $NH_2$, or by large numbers of charged groups such as $COO^-$ or $SO_3^-$ on the polymer chain. Typical examples are the polysaccharides (celluloses), polyacrylates, polyacrylamides, polyimines, polyglycols, and the variety of so-called hydrogels. Depending on the degree of polymerization and cross-linking, they are water-soluble or water-insoluble but water-swelling. Generally, they are easily penetrated by aqueous solutions and have limited compatibility with hydrophobic polymers.

Cellulose[145,146] in either bead or membrane form has found widespread application as a support for indicators,[21] chelating agents,[148] and proteins.[149-151] The ease of penetration by water results in distinctly shorter response times. Membranes as thin as 6 μm are commercially available. We found the cuprophane membrane type as used in artificial kidneys to be an excellent support for pH and metal ion indicators. However, thin cellulose membranes require careful handling and are easily populated by bacteria or algae. Aside from the plain membranes, cellulose bound to glass or on polyester is also available. Aside from cellulose, dextrans and agarose have been used for dye and enzyme immobilization to produce a sensing chemistry for water-soluble analytes, but with no obvious advantages over cellulose.

Chemical modification of the cellulose hydroxy groups is easily accomplished by various methods[21,145,146,148] and results in celluloses with either hydrophilic functional groups such as COOH or $NH_2$ or lipophilic groups such as long chain fatty acids on their surface. They also may be equipped with charged groups to make them an ion-exchange material. The latter are offered by various manufacturers, albeit usually optimized for chromatography purposes. Cellulose is soluble in strongly alkaline solutions of copper(II) or zinc(II) ions containing a complexing amine, and can be precipitated from these solutions by acidification. The solubility in strong alkali only is sometimes disadvantageous and contrasts the good solubility of many hydrogels in alcohols.

Cellulose membranes have been placed in front of another sensor material to prevent large molecules to reach the sensor because most cellulose membranes have exclusion molecular weights between 10.000 and 30.000 D. When cellulose dries out, it becomes very brittle and then is difficult to handle. On the other hand, dried cellulose requires a considerable time to hydrate again and thereby undergoes considerable swelling. Both the swelling rate and the hydration number are pH-dependent. Aside from swelling, such porous membranes generally have permselective properties, and their contraction or expansion may change the pore size and can lead to a marked change in water permeability.

Polyacrylamides (PAAs),[124,152] poly(vinyl alcohols),[124] and polyglycols[124,153] are good solvents for a number of indicators but are water-soluble unless crosslinked. These polymers can be retained at the sensing site using a cellulose membrane but dissolve quite an amount of water when in contact with aqueous samples. Cross-linked PAAs form mechanically stable and water-insoluble supports which are easily handled and chemically modified, but lack the good permeability of cellulose. They tend to swell in water, thereby giving rise to slow drifts when used in sensors that previously were stored in dry state. PAAs are also applied in bead form, and their surface can easily be fitted with functional groups such as COOH or primary amine. However, an excess of these functions can introduce a considerable buffer capacity, resulting in very long response time at the respective pH range when used as pH sensing material.

Poly(ethylene glycol) was used as a support in a $CO_2$ sensor, with fluorescein particles dissolved in it.[154] Zhujun et al.[155] have evaluated poly(vinyl alcohol) (PVA) as a substrate for immobilizing indicators used in fiber optic chemical sensors. Cross-linking was implemented by adding glutaraldehyde and HCl to a 5% (w/w) aqueous PVA solution. The resulting gel is clear and transparent in the VIS and UV region down to 230 nm. Swelling properties depend on the amounts of glutaraldehyde and HCl. Immobilization and sensor fabrication involve the following steps: PVA is reacted with cyanuric chloride, the cyanuric chloride/PVA conjugate is reacted with an indicator, HCl and glutaraldhyde are added to initiate cross-linking, and before the gel starts to form, a precise volume of indicator/PVA conjugate is transferred to the end of the fiber and allowed to solidify *in situ*. The ability of this procedure to control both the amount of indicator and the amount of substrate was illustrated by using fluorescein as a pH indicator, chelators such as eriochrome black or morin, and the polarity probe dansyl chloride. The polymer is permeable to proteins up to a molecular weight of about 19,000 Da.

A final group of polymers that have been used in chemical sensors for water-soluble species are the hydrogels.[156-158] Chemically, it is a rather inhomogeneous group of polymers comprising poly(hydroxyethyl acrylate) (PHEMA), polyacrylamides, polyvinylpyrrolidones, poly(vinyl alcohols), polyurethanes, and the like. Hydrogels are cross-linked macromolecular networks swollen in water or biological fluids and possess excellent biocompatibility, probably due to their high water content and special surface properties. Low water content hydrogels with more hydrophobic properties can be synthesized by copolymerization of hydrophilic and hydrophobic vinyl monomers and are of interest in possible blood compatibles. The application and biocompatibility of PHEMA-based materials have been reviewed.[157,158]

Typical applications of hydrogels are in the fabrication of thin lipid membranes, e.g., composed of phosphatidyl choline and cholesterol which can be interfaced to PAA hydrogel via Langmuir-Blodgett film deposition to form electrochemical biosensors,[159] and in the fabrication of disposable planar optical sensors that can be mass-produced.[160]

## G. DIFFUSION AND PERMEATION OF GASES THROUGH POLYMERS

In order to select the proper material for use in optrodes it is frequently necessary to have knowledge of the solubility of gases in, and the permeability of gases through, the polymer. Proper choice of materials can provide substantial selectivity. Quantitative data also form the basis for any kinetic treatment of the diffusional processes involved, both in gas sensors and in sensors where gas sensors are used as transducers for enzyme-based assays.

The most important parameters for characterization of gas diffusion and permeation are the diffusion coefficient D, the gas solubility S, and the permeation coefficient P. The permeation of small molecules through flawless and pinhole-free polymers occurs through consecutive steps of solution of a permeant in the polymer, and diffusion of the dissolved permeant through the inner free volume of the polymer, so that

$$P = D \cdot S \qquad (32)$$

The respective units are $cm^3(STP) \cdot cm/s \cdot cm^2(Hg)$ for P (which corresponds to $3.36 \cdot 10^{-6} mol/N \cdot m$ in SI units), and $cm^3(STP)/cm^3(cm\ Hg)$ for S (which is $33.9 \cdot 10^{-3}\ mol/N \cdot m$ in SI units).

The temperature coefficients of P, D, and S can be represented in Arrhenius-type equations

$$P = P_0 \cdot \exp\left(\frac{-E_p}{RT}\right) \tag{33}$$

$$D = D_0 \cdot \exp\left(\frac{-E_D}{RT}\right) \tag{34}$$

and

$$S = S_0 \cdot \exp\left(\frac{-\Delta H_S}{RT}\right) \tag{35}$$

Consequently,

$$E_p = E_D + \Delta H_S \tag{36}$$

where $E_p$ and $E_D$ are the respective activation energies, and $\Delta H_S$ is the heat of solution.

Permeability P generally decreases with increasing density of the polymer, its crystallinity, and orientation. Cross-linking a polymer reduces P, as do added fillers, while adding plasticizers can increase it. Humidity increases P of some hydrophilic polymers.

Table 12 gives the permeation coefficients for oxygen, $CO_2$, nitrogen, and water vapor of various polymers, demonstrating the unique properties of poly(dimethyl siloxane). Tables 13 and 14 give the respective diffusional coefficients and solubilities. Additional data can be found in the compilations given by Yasuda and Stannett.[129]

# IV. IMMOBILIZATION TECHNIQUES

Following the choice of indicator and polymeric support, the next step in sensor design frequently will involve immobilization of the dye on the support to give the so-called sensing chemistry or working chemistry. Three methods are important for the preparation of sensing chemistries, namely mechanical, electrostatic, and covalent immobilization. Excellent reviews cover all aspects of the chemistry and physics of immobilized reagents and dyes, proteins, and even whole cells.[147-151] The methods known from protein chemistry (probably the most thoroughly studied field of immobilization techniques) may, of course, also be adapted to problems associated with dye immobilization.

Immobilization of indicators and reagents usually can be performed under more drastic conditions than those required for enzyme immobilization. In addition, the chemistry is not confined to reactions in aqueous solutions, so that a broader range of coupling procedures becomes available. However, as a general rule it may be said that once the indicator is immobilized, the chemistry should be kept as limited as possible and confined to mild reaction conditions and procedures that give very high yields. Ideally, no further step is necessary after the indicator has been reacted with the modified sensor layer.

Fluorescent indicators should not be covalently bound to polymers by azo-coupling because most azo dyes do not fluoresce, except for certain metal complexes of *ortho*-hydroxy azo compounds. However, certain indicators nonfluorescent in bulk solution exhibit fluorescence when adsorbed on a solid support or contained in a rigid polymer. Coupling of indicators or proteins via phenylazo groups can result in the formation of a light-absorbing chromophore that acts as an inner filter. Attention should also be paid to not use a chelating functional group for the immobilization chemistry which would render the reagent useless.

Immobilization of most dyes results in a small change of their spectral characteristics. In

**TABLE 12**
**Permeation Coefficients P (in $10^{15}$ mol·m·s$^{-1}$·N$^{-1}$) for Various Polymers at 25°C[a]**

| Polymer | $CO_2$ | $N_2$ | $O_2$ | $H_2O$ |
|---------|--------|-------|-------|--------|
| Polyisoprene | 51 | 3.2 | 7.7 | 874 |
| Poly(1,3-butadiene) | 46 | 2.2 | 6.4 | 1,700 |
| Poly(dimethyl siloxane) | 1,090 | 94 | 203 | 13,400 |
| Fluorosilicone | — | — | 37 | — |
| Polyethylene (0.964 g.cm$_1$) | 0.6 | — | 0.134 | 4 |
| Polyethylene (0.922 g.cm$_1$) | 9.4 | — | 2.3 | 30 |
| Polypropylene | 3.1 | 0.15 | 0.77 | 22 |
| Poly(vinyl chloride) | 0.053 | 0.004 | 0.015 | 92.4 |
| Polystyrene | 3.5 | 0.27 | 0.88 | 403 |
| Polycarbonate | 2.4 | 0.084 | 0.5 | 470 |
| Poly(ethyl methacrylate) | 1.7 | 0.074 | 0.39 | — |
| Polyacrylonitrile | 0.0003 | — | 0.00007 | 10 |
| Polytetrafluoroethylene | 4.3 | — | 1.65 | 11 |
| Nylon 66 | 0.054 | — | 0.013 | 92 |
| Cellulose acetate | 0.81 | — | 0.27 | 2,280 |

[a] Data from Reference 130.

**TABLE 13**
**Diffusion Coefficients D (in $10^{-12}$·m$^2$·s$^{-1}$) of Gases at Room Temperature[a]**

| Polymer | $H_2$ | $CH_4$ | $N_2$ | $O_2$ | $CO_2$ |
|---------|-------|--------|-------|-------|--------|
| Poly(vinyl chloride) | 50 | 0.13 | 0.38 | 1.2 | 0.25 |
| Polycarbonate | 64 | — | 1.5 | 2.1 | 0.48 |
| Polystyrene | 436 | — | 6 | 11 | 5.8 |
| Poly(ethyl methacrylate) | — | — | 2.2 | 10 | 3.3 |
| Polyethylene (0.914 g·cm$^{-1}$) | — | 19 | 32 | 46 | 37 |
| Polyethylene (0.964 g·cm$^{-3}$) | — | 5.7 | 9.3 | 17 | 12 |
| PTFE | — | — | 8.8 | 15 | 9.5 |
| Poly(dimethyl siloxane)[b] | 4700 | — | 850 | 1200 | 1500 |
| Natural rubber | 1020 | 89 | 117 | 173 | 125 |

[a] Data from Reference 130.
[b] Containing 10% w/w filler.

addition, there are frequently considerably shifts in $pK_a$ values, binding constants, and, in particular, dynamic quenching constants. The changes reflect the various interactions between neighbor dye molecules on the polymer surface, interactions between dyes and polymer surface, and also electronic effects of covalent bonds. In each instance it is therefore required to determine the respective data of the immobilized dye rather than to use the values determined in bulk solution. The large differences in the quenching constants of dissolved and immobilized indicators also explain why reservoir sensors (with an indicator in fluid solution) will more sensitively respond to a quencher than an indicator phase sensor (with an immobilized indicator), despite the same indicator being used.

## A. MECHANICAL IMMOBILIZATION
Mechanical (physical) immobilization involves (1) adsorption (which plays a minor role in

**TABLE 14**
**Solubility S (in $10^{-6}$ mol·N$^{-1}$·m$^{-1}$) of Simple Gases in Polymers at Room Temperature**

| Polymer | H$_2$ | CH$_4$ | N$_2$ | O$_2$ | CO$_2$ | H$_2$O |
|---|---|---|---|---|---|---|
| Poly(vinyl chloride) | 12 | 76 | 11 | 13 | 214 | 39,200 |
| Polycarbonate | 0.26 | — | — | 2.3 | 58 | 758 |
| Polystyrene | — | — | — | 25 | 290 | — |
| Poly(ethyl methacrylate) | — | — | 34 | 38 | 500 | 110,000 |
| Polyethylene (d 0.914) | — | 49 | 10 | 21 | 116 | — |
| Polyethylene (d 0.946) | — | 23 | 5.4 | 8 | 9.4 | — |
| PTFE | — | — | 0.7 | 1.3 | 5.4 | — |
| Natural rubber | 17 | — | 25 | 50 | 401 | — |
| Silicone rubber[a] | 33 | — | 89 | 138 | — | — |

   [a]   containing 10% (w/w) filler.

sensor chemistry) and (2) inclusion of molecules in a sphere which they cannot leave. Thus, proteins can be included in the interior of polyurethane, and indicators may be entrapped in capsules of only a few nanometers in diameter. The method is usually confined to high mol-weight species which react with low mol-weight analytes that can permeate the membrane. In enzyme optrodes, encapsulation has the advantage of high enzyme activities and unchanged enzyme properties. Enzymes may also be entrapped in a polymer network[161] which also can result in high enzyme loading, but usually poor flow properties. Peterson et al.[6] have shown that copolymerization of acrylamide with methylene bis(acrylamide) in the presence of phenol red leads to microspheres with the dye firmly bound to the polymer. Such nondiffusible forms of pH indicating dyes are obtained by emulsion copolymerization of phenol red with aqueous acrylamide in the presence of emulsifier and toluene under nitrogen to give microspheres that are useful in optical pH sensing.

For certain purposes (in particular for disposable probes) it may suffice to immobilize the dye or protein by adsorption on a polar surface. Binding results from weak electrostatic or van der Waals interaction. This is the simplest way of immobilization and can give high yields at low costs. However, the binding strength to solid surfaces is not very high.

Though not a physical method in its original sense, lipophilic indicators dissolved in a lipophilic polymer practically can be considered to be immobilized and are not washed out by aqueous samples because of a Nernst distribution that strongly favors the presence of the indicator in the lipophilic phase. Typical examples include oxygen-sensitive polymers with dissolved lipophilic dyes,[133] and lipophilic indicators dissolved in lipid bilayers.[28] Dissolving the dye in a polymer and casting it on the surface of the core of the fiber is another easy way of fabricating sensing chemistries on fibers. Shaksher and Seitz[162] have studied the binding of a cationic fluorescent probe to the surfaces of unmodified silica and hydrocarbon-modified silicas. For the latter case, it was concluded that the organic fluorophore is excluded from the solvent and experiences a primarily surface environment whose polarity is different from that of the sample solution passing by. Surface adsorption or deposition is the favored way in case of gaseous samples which do not present the risk of washing out the indicator dye.

## B. ELECTROSTATIC IMMOBILIZATION

When the surface of a rigid support is fitted with charged groups such as sulfo groups or quaternized ammonium groups, the material is capable of binding ions having opposite charge. Sulfonated polystyrene, for instance, binds cations with varying binding strength. This effect is widely used in order to separate anions or cations from a solution, and for enrichment of traces of ions. They may be displaced from the solid phase with strong acid or base. Like metal ions, the charged functional groups at the polymer surface are also capable of binding oppositely charged indicators. Binding to the surface occurs as a result of electrostatic interaction between the charged dye and the oppositely charged functional group and can result in considerable binding strength.

Numerous ion exchange materials are commercially available and may be classified into "strong" and "weak" ion exchangers. This refers to the binding strength of the material to the respective cation or anion. Both membrane and bead type ion exchange materials are available. In order to firmly bind organic dye ions, the use of strong ion exchangers is advisable so to prevent washout of the dye with time. In optical sensor technology, ion exchangers have mainly been used to immobilize pH indicators. Typical examples include bromothymol blue (on anionic polystyrene)[8,9] and hydroxypyrene-trisulfonate (on cationic quaternized aminostyrene).[143] Further examples are given in the section on acid-base indicator resins (II.A.8), in Chapter 8 (on pH sensors), and in Bishop's book.[1]

The major advantages of electrostatic immobilization are the ease of the procedure and its reproducibility. Loading can be easily governed by the time of immobilization. The fabrication is very simple in that the charged polymer is immersed, for a defined period of time, into a methanolic solution of the dye. Because the indicator molecules are situated at sites on the surface of the polymer that is easily accessible to protons, but not easily accessible to proteins, the corresponding pH sensors are said to display no protein error.[143]

## C. CHEMICAL (COVALENT) IMMOBILIZATION

Chemical immobilization is accomplished by creating a covalent bond between indicator and a polymer surface. Covalent immobilization of enzymes, antigens, antibodies, and even whole bacteria can be achieved by both surface immobilization on a solid support and by cross-linking the material with proteins like albumin. Numerous methods of surface modification of polymers exist[163] and can yield materials capable of covalently binding indicators via their reactive groups. At best, however, a few monolayers are produced on the surface. Therefore, methods of surface amplification have been introduced and will be discussed later. The final decision on how to immobilize any species should be governed by the criteria of ease of operation, reproducibility, costs, performance of the immobilized species, and the number of required reaction steps.

In order to covalently bind an indicator to a polymer surface, one or two activation steps are usually required to make the reagents undergo a facile room temperature reaction. The first step usually involves the modification of the *polymer* to provide it with a sufficiently reactive function. A similar procedure may be required when the *indicator* does not possess chemical functions suitable for immobilization. Polycyclic aromatic hydrocarbons and many metal chelates, for instance, are devoid of functional groups suitable for covalent binding. Therefore, they have to be furnished with sufficiently reactive functional groups. One of the few useful approaches is to introduce a vinyl group into the chromophore and to copolymerize it with vinylic monomers such as styrene[164] or methyl methacrylate.[88]

Covalent surface modification of quartz, kieselgel, silica gel, conventional glass, and even metals such as iron and platinum, and elemental carbon, is almost exclusively achieved with reagents of the type $(RO)_3Si-R'$, with R being ethyl or methyl, and R' being 3-aminopropyl, 3-chloropropyl, 3-glycidyloxy, vinyl, or a long-chain amine.[135-139] The resulting materials are easily reacted with the indicator or peptide to be immobilized. Because of the widespread use of immobilization techniques, porous glass with various types of organo-functional extension arms is commercially available and has been widely used for the design of biosensors of the planar waveguide type. Antibodies have been bound to silica surfaces this way and spectroscopically characterized.[165]

Cellulose (linear chains of 1,4-linked β-glucose organized in fibers of a high degree of crystallinity) can be surface-modified by reaction with cyanogen bromide, followed by treatment with a long-chain diamine.[145,146] The latter can be omitted, and the nasty cyanogen bromide may be replaced by cyanuric chloride or epichlorohydrine. The products thus obtained can be reacted with acid chlorides in an aprotic solvent, or with carboxylic acids or even proteins using a carbodiimide as a coupling agent. A review on the immobilization of indicators on various materials including celluloses has been presented by Seitz.[147]

In a typical procedure, 500 mg of cellulose beads or discs are activated at pH 10.5 for 10 min with 30 ml of a 4% solution of cyanogen bromide in water. After filtering and careful washing, the material is suspended in a solution 5 mg of the protein to be immobilized (or another amine) in 50 ml 0.12 $M$ NaHCO$_3$ of pH 8.4, and incubated overnight at room temperature under gentle mixing or stirring. The material is washed with saline and water, air-dried, and stored dry. Residual binding sites on the cellulose first may be blocked by treatment with ethanolamine.

Carboxymethylcellulose (CMC) is obtained from cellulose and chloroacetic acid and can serve as a support for amines using similar immobilization methods. CMC and aminoethyl cellulose are commercially available in various modifications. After several years of experience with covalent immobilization it is recommended to prefer standardized, i.e., commercially available support material over self-prepared material whenever possible. The former material is available in continuous quality, whereas the material produced in a laboratory on a small scale varies from preparation to preparation. Agarose (alternating 1,3-linked β-D-galactose and 1,4-linked 3,6-anhydro-α-L-galactose) is similar in its properties to cellulose. It requires the same surface chemistry and does not offer particular advantages over cellulose.

A simple method to immobilize indicators on cellulose and related material is based on the creation of an interpenetrating network within the cellulose backbone.[166] The cellulose membrane is simply soaked in an aqueous solution of poly(ethylene imine) or a long-chain diaminoalkane. Within the cellulosic material the amine forms a network, to which an indicator can be attached by reaction with its chloride or sulfochloride. As a result, the indicator-loaded amine chain can no longer be washed out from the cellulose material, since it is anchored in the cellulose network.

Polyacrylamide (PAA) is of low reactivity, but can be activated by surface saponification with strong alkali hydroxide.[124,152] Subsequent acidification yields a material having surface carboxy groups which are capable of covalent binding to amines. Carboxy-modified PAA, crosslinked poly(acrylic acid), carboxy-modified PVC, and carboxy-modified polystyrene are commercially available and are capable of coupling to amine-type indicators. Also, long-chain diamines acting as spacer groups may be linked to the surface.

PAA may be covalently linked to silica surfaces by first derivatizing the glass with aminosilane reagents, then attaching a polymerizable group such as the acryloyl group, and copolymerizing support and acrylamide (in solution) in the presence of a crosslinker.[167] If this reaction sequence is performed in the presence of an acryloyl-derivatized enzyme, the enzyme will be incorporated into the layer. A simple way to introduce a defined amount of amino groups is by copolymerization of acrylamides with, e.g., N-(β-aminoethyl)acrylamide, resulting in a copolymer with free amino groups suited for dye coupling.[168] We were able to couple numerous isothiocyanates and sulfochlorides to such a material which later may be crosslinked if required. The material should also be useful for coupling free carboxy groups to its surface. Like cellulose, polyacrylamides have a large specific surface which allows heavy dye loading.

Polymers such as poly(vinyl alcohols), polyglycols, or the strongly basic poly(ethylene imines) offer attractive alternatives but tend to swell with, or dissolve in, water. This limits their utility in case of sensing in an environment of varying humidity. Klainer and Harries[169] have used an interesting material that was obtained by copolymerization of ethylene oxide with glycidol. A branched poly(ethylene glycol) (BPEG) was obtained suitable for controlled immobilization of enzymes. It can also be used as a dye carrier and for the surface amplification of fibers by attaching a three-dimensional active surface on the core. BPEG was first activated by reaction with cyanuric chloride and then reacted with aminopropyl groups at the glass surface. The activated BPEG is then amenable to the introduction of sensor chemistries by reaction with the third chloride group.

Poly(vinyl alcohol) lacks direct functions which can be used for direct coupling to many dyes and to enzymes. Therefore, it has been reacted first with cyanuric chloride to form a C–O–C link. The remaining active chloride group may be reacted with the amino group of an indicator.[155] This works well with samples of near neutral pH, but the C–O–C bond is labile at both high and low pH.

Polystyrene is the material that, aside from polysaccharides, undergoes the broadest variety of chemical modification reactions.[163] It can be easily benzylated (e.g., with hydroxybenzyl groups to with chromophores can be bound via the OH group), chloromethylated, carboxylated, sulfonated, or halogenated. Metalation with lithium and subsequent reaction with an electrophilic reagent provides further possibilities for introducing functional groups. Polyolefins and poly(dimethyl siloxanes), in contrast, can be modified by halogenation only. PVC also is a material whose surface is difficult to modify.

### D. PREACTIVATED MATERIALS FOR IMMOBILIZATION

Given the widespread use of immobilized compounds in chemical sciences, various preactivated polymers have become commercially available in the past years. The surfaces of these materials are equipped with a reactive group which usually, under mild conditions, reacts with amines and related nucleophiles. Thus, the Millipore Company (Bedford, ME) offers various types of nylon, fluorocarbon, or nitrocellulose membranes which require a very simple immobilization protocol: immerse the membrane into the protein solution for a defined time and wash after.

Excellent results have also been obtained in the author's laboratory with the Biodyne™ immunoaffinity membrane (Pall Company, Glen Glove, NY) which is a hydrophilic microporous membrane with activated COOH grous at its surface. Riedel de Haen (Seelze, FRG) offers beads of copolymerized vinyl acetate and divinyl ethylene urea with surface epoxy groups that can be utilized for protein immobilization. A new ion exchange membrane, composed of resin beads permanently enmeshed in a web of PTFE, is offered by the BioRad Company (Richmond, CA). It is available in strong anion, strong cation, and chelating functional group form. A general method for preparing enzyme-based fiber optic biochemical sensors was described.[170] The glass surface of the fiber, after being furnished with amino groups, was reacted with preactivated biotin, and after being furnished with amino groups, was reacted with preactivated biotin, and labeled enzymes were introduced into the fiber through a sandwich biotin-avidin-biotin interaction.

Several reagents are offered that allow simple conversion of polymer surfaces into reactive surfaces. The surface chemistry of glass has been mentioned before. Polystyrene surfaces can be converted to hydrophilic (COOH-modified) surfaces by the CML tube coating offered by Seragene Diagnostics, Inc. (Indianapolis, IN). The COOH group can then be coupled to an amino group using a peptide coupling reagent like N,N-carbonyl-diimidazole or various carbodiimides. A useful reagent is the Polycup™ resin, a water-soluble, polyamide-epichlorohydrin-type cationic material effective as cross-linking agent for carboxylated polystyrene surfaces, carboxymethyl cellulose, poly(vinyl alcohols), or other water-soluble polymers. It also reacts with starch to impart water resistance.

## V. LABELS AND LABELING TECHNIQUES

The term chromogenic or fluorescent label is used for molecules consisting of a chromophore or fluorophore and a reactive group that can be attached to other molecules by covalent

**TABLE 15**
**Important Fluorescence Derivatization Reactions**

| Reagent | Functional group | Product |
|---|---|---|
| Sulfonyl chlorides | Amines | Sulfonamides |
| Sulfonic acid p-nitrophenyl-esters | Amines | Sulfonamides |
| Isothiocyanates | Amines | Thiourethanes |
| Aldehydes | Amines | Schiff bases |
| o-Dialdehydes | Amines[a] | Isoindoles |
| Carboxylic acids | Amines[b] | Carboxamides |
| Carboxylic acid succinimidyl esters | Amines | Carboxamides |
| 4-Halogeno-7-nitrobenzo-2-oxa-1,3-diazole (NBD) | Amines | 4-Amino-NBD |
| Dichlorotriazines | Amines | Monochloramino-triazines or di-aminotriazines |
| Carboxylic acid chlorides | Alcohols | Carboxylic esters |
| Carbonyl azides | Alcohols | Carbamic acid esters |
| Carbonyl nitriles | Alcohols | Carboxylic esters |
| 4-Halogeno-7-nitrobenzo-2-oxa-1,3-diazole (NBD) | Thiophenols | NBD thiols |
| Haloalkyls | Thiols | Thioethers |
|  | Carboxylic acids | Carboxylic esters |
| Diazoalkyls | Carboxylic acids | Carboxylic esters |
| Amines | Carboxylic acids[c] | Carboxamides |
|  | Carbonyl compounds | Schiff bases |
| Isocyanates | Amines | Urethanes |
|  | Alcohols | Carbamic acid esters |
| Hydrazines | Carbonyl compounds | Hydrazones |
| Maleimides | Thiols | Thioethers |
| Iodoacetamides | Thiols | Thioethers |

[a]   In the presence of a thiol.
[b]   In the presence of carbodiimides.
[c]   In the presence of carbodiimides.

binding. The term is also used for the chromophore or fluorophore of the labeled molecule itself. Noncovalent fluorescent probes, in contrast to covalent probes, form a reversible association with other molecules by a combination of hydrophobic, dipole-dipole, and ionic interactions. The two types of probes are complementary and each has advantages, limitations, and specific applications to fluorescence studies.

Labels are frequently used in precolumn or postcolumn derivatization procedures for fluorescence detection in HPLC.[171] They have gained increasing interest in the field of clinical immunology, not only as an additional alternative to radioimmunoassays, but also in producing cheap, stable, and safe reagents and rapid and sensitive assays.[172,173] It is not the aim of this chapter to give an exhaustive overview of the tremendous literature on fluoroprobes. Rather, representative candidates are discussed according to their spectral properties and labeling behavior. Labeling procedures of the most important dyes for biomolecules, e.g., proteins, will finally be given in some detail.

The choice of a label for a particular application is determined by the nature of the biomolecule and the reactive groups available, mainly $NH_2$, COOH, SH, and OH. Amino groups are mostly protonated at physiological pH, while carboxy groups are dissociated. Selective modification reactions of biomolecules have been reviewed by Haugland.[174a] The most important reactions leading to covalently bound fluorophores are summarized in Table 15. Spectral data for the various labels are summarized in Table 16. Diode laser compatible

## TABLE 16
**Spectroscopic Properties of Ovalbumin Labeled with Frequently Used Fluorescent Labels in 0.1 *M* Phosphate Buffer of pH 7.4**

| Label | Excitation/ emission Maxima (nm) | Decay time (ns)[a] | Solvent[b] |
|---|---|---|---|
| 4-Acetamido-4'-isothiocyanato-stilbene-2,2'-disulfonic acid | 314/406 | 0.2 | A |
| Anthracene-2-isocyanate | 351/466 | 17.5 | D |
| 4-Chloro-7-nitrobenzo-2-oxa-1,3--diazole | 470/554 | 2.3 | E |
| Dansyl chloride | 340/535 | 18.9 | Ac |
| Dansyl aziridine | 346/511 | 22.7 | Ac |
| Eosin-5-iodoacetamide | 529/556 | 3.0 | Ac |
| Eosin-5-isothiocyanate | 524/551 | 0.6 | Ac |
| Fluorescamine[c] (fluram) | 370/488 | 6.9 | Ac |
| Fluorescein-5-iodoacetamide | 490/522 | 2.6 | Ac |
| Fluorescein-5-isothiocyanate | 490/525 | 3.7 | Ac |
| 2-(4'-Iodoacetamidoanilino)--naphthalene-6-sulfonic acid | 316/454 | 8.0 | E |
| 5-[2-(Iodoacetyl)amino ethyl]-amino-naphthalene-1-sulfonic acid | 345/462 | 23.5 | A |
| Lucifer yellow CH | 419/535 | 9.3 | W |
| 3-(4-Maleimidyl-phenyl)--4-methyl-7-diethylaminocoumarin | 387/468 | 3.0 | Ac |
| 4-Maleimidyl salicylic acid | 306/424 | 10.2 | Ac |
| 5-Maleimidyl salicylic acid | 314/434 | 10.3 | A |
| Monobromobimane | 386/463 | 14.7 | E |
| Monobromotrimethylammoniobimane | 366/472 | 4.4 | E |
| o-Phthaldialdehyde (+ mercaptoethanol) | 326/434 | 19.8 | W |
| 1-Pyrenylmethyl iodoacetate | 332/391 | 56.5 | Ac |
| Rhodamine B isothiocyanate | 547/585 | 2.3 | Ac |
| Tetramethylrhodamine isothiocyanate | 546/582 | 1.9 | Ac |

[a]  Long-lived component only.
[b]  Abbreviations: A, 50% aqueous ethanol; E, ethanol, Ac, acetone; W, water; D, dioxane.
[c]  Not to be confused with fluoresceinamine.

dyes have been listed,[174b] and phosphorescent labels for immunoassay were discussed[174] as an alternative to rare earth labels (see later).

## A. AMINE MODIFICATION

Amine modification methods are employed in protein labeling and indicator immobilization on solid supports having surface amino groups. Probably the most widely used covalent fluorescent probes for amines are sulfonic acid chlorides and isothiocyanates. Sulfonic acid chlorides form highly stable sulfonamide derivatives with primary and secondary amines, but are also reactive to hydroxy and even sulfhydryl compounds. Horner et al.[175] have introduced sulfonic acid esters, e.g., the p-nitrophenylester, derived from dansylchloride, and found them more selective than the sulfonic acid chloride. Isothiocyanates form thiourethane derivatives upon reaction with amines and the adducts are somewhat unstable to hydrolysis, in particular in media of extreme acid or alkaline pH.

Aside from the probably most widely used fluorescent labels, namely dansyl chloride, sulforhodamine 101 sulfonyl chloride (Texas Red) and fluorescein-5-isothiocyanate (FITC isomer I), o-phthalaldehyde (OPA) and fluorescamine are also important labels. OPA reacts with amines in the presence of a mercaptane to give a fluorescent isoindole derivative that is not completely stable. Other o-dialdehydes such as naphthalene-2,3-dicarboxaldehyde[176] have been introduced recently to replace OPA. Fluorescamine forms highly fluorescent conjugates only with primary amines including amino acids, but a decisive disadvantage is its instability. The hydrolysis rate of fluorescamine in water is so fast that conjugation must be complete in minutes. It reacts also with other nucleophilic functional groups such as alcohols, secondary amines and water, but yielding weakly fluorescent or nonfluorescent products only.

Fluorescent carboxylic acids can be coupled to proteins by their activated compounds known from the methods used in synthetic peptide chemistry.[177] The highly reactive carboxylic acid chlorides are usually not used since high reactivity is always accompanied by a considerable loss in selectivity. Among the activated esters with p-nitrophenol or N-hydroxy-succinimide, the succinimidyl esters show high specificity for amine modification and aliphatic amines are readily derivatized by succinimidyl esters even at a near-neutral pH. Thus, labeling with succinimidyl esters is an excellent alternative to labeling with isothiocyanates, since in addition the formed carboxamides possess high chemical stability. Fluorescent carboxylic acids can also be coupled to proteins by conversion to mixed anhydrides. Typical examples include mixed anhydrides with ethyl chloroformate and trifluoroacetic anhydride. Other important peptide coupling reagents are the carbodiimides and particularly water-soluble carbodiimides such as 1-ethyl-3-(3-dimethylamino-propyl)-carbodiimide hydrochloride.[178]

Dichlorotriazines (the reactive acid chlorides of cyanuric acid) react with amines by replacement of one or even both chlorides which would result in cross-linking. But this reaction usually requires harsher conditions than those used to label proteins. However, dichlorotriazines also react with phenols and thiols. Dichlorotriazinylamino derivatives of fluorescein have been proposed[179] as alternatives to FITC for use in immunology due to its easier conjugation procedure with better reproducibility at a lower pH.

4-Halogeno-7-nitrobenzo-2-oxa-1,3-diazole derivatives such as NBD chloride react with either thiols or amines resulting in adducts with different spectral properties. The reactivity of NBD derivatives decreases[180] in the order NBD-F, NBD-Cl, and NBD-Br. Other reagents such as chloroformates and isocyanates are also reactive to amines. However, they have become of more importance in the derivatization of alcohols and will be discussed below.

Modification of the $\varepsilon$-lysine amino group by all types of reactions described above lead to a change of the net charge of proteins due to the loss of positive charge. Such a change of the net charge may have a negative effect on the stability, solubility, and biological activity of proteins. This alteration in the net charge can be prevented by reductive alkylation reactions which retain the basic character of the $\varepsilon$-amino group of lysine residues present in the protein. In this type of modification reaction, a fluorescent aldehyde such as 2-(7-nitrobenz-2-oxa-1,3-diazol-4-yl)methylaminoacetaldehyde forms Schiff bases with the amino group which is than reduced with sodium borohydride or sodium cyanoborohydride.[181]

## B. THIOL MODIFICATION

Thiol (sulfhydryl) groups are import modification sources of proteins which occur primarily as cysteine residues.[182] The three major classes of reactive groups for selective thiol modification are maleimides, haloacetyl probes, and aziridines. Maleimide reagents by themselves are not significantly fluorescent until after conjugation of a reactive nucleophile thiol across the maleimide double bond. The fluorescence properties of the conjugate are determined by the fluorophore attached to the maleimide. A decisive limitation of maleimide reagents is their chemical instability through ring-opening hydrolysis of the anhydride-like maleimide which occurs at an appreciable rate especially above pH 8. This maleimide ring opening can occur also after conjugation to the biomolecules which can result in different fluorescence properties of the closed-ring thiol adduct and the open-ring adduct. Dimaleimides have been used for cross-linking of spatially adjacent thiols.

Haloacetyl probes such as the iodoacetamides and iodoacetyl derivatives are widely used for thiol modification. The two major probes that have been used are 1,5-IAEDANS at shorter wavelengths[182] and 5-iodoacetamidofluorescein[183] at longer wavelengths. All iodoacetamides are rapidly photolyzed by ultraviolet light, whereas the covalent conjugates possess much greater photostability. Haloalkyl derivatives such as monobromobimane form fluorescent thioethers for analysis and quantitation of thiols in plasma.[184]

Activated aziridines are amides of carboxylic acids or sulfonic acids such as dansyl aziridine[185] which is by far the most widely used fluorescent aziridine. The unactivated aziridines are N-alkyl- or N-arylethyleneimine derivatives such as NBD-aziridine. Activated aziridines undergo ring-opening nucleophilic reactions with thiols or amines in basic media, while the unactivated aziridines are very stable in base and undergo ring-opening reactions only with acid catalysis.

Attempts to improve the selectivity of NBD derivatives towards thiols led to the introduction of 4-fluorobenzo-2-oxa-1,3-diazole-7-sulfonate (SBD-F)[186] and 4-(aminosulfonyl)-7-fluoro-benzo-2-oxa-1,3-diazole (ABD-F)[187] as derivatization reagents for sulfhydryl groups. Disulfides such as didansyl cystine are sulfhydryl-specific in that they can only react in an exchange reaction with thiols to form mixed disulfides. The reaction is reversible, and the label can be displaced by common thiols such as 2-mercaptoethanol, cysteine, or DTT. The unsymmetrical disulfide reagent probes are probably the most thiol-specific fluorescence labeling reagents available for macromolecules.[188] Sulfhydryl groups also can selectively be blocked and marked by fluorescent aryl vinyl sulfones.[189,190]

## C. CARBOXYLIC ACID MODIFICATION

Direct modification of carboxylic acids of polymeric supports and in biomolecules is usually difficult because of the low reactivity of the carboxylate anion which is the actual species at neutral pH. A popular way is to couple fluorescent amines to carbodiimide-activated carboxy groups. Since a large number of carboxy groups usually is present, selectivity of modification depends on the ability of the carbodiimide to reach the carboxy group for activation. For surface carboxy groups a polar water-soluble carbodiimide is probably most suitable, but for the rarer buried carboxy groups a nonpolar carbodiimide such as dicyclohexylcarbodiimide (DCC) may give better results. If the polymer material withstands the required aggressive reagents, transformation of the carboxy group into the carboxylic acid chloride results in a highly reactive group that can be coupled directly to amines. An interesting alternative to chemical-activated labeling procedures is enzyme-mediated fluorescent labeling, since enzymatic reactions are usually highly specific. A widely used system is transglutaminase-catalyzed incorporation of dansylcadaverine into proteins.[191]

Diazoalkanes are highly reactive to carboxylic acids leading to the formation of carboxylic acid esters. Thus, 9-diazomethylanthracene[192] and 4-diazomethyl-7-methoxycoumarin[193] have been introduced as fluorescent labeling reagents for carboxylic acids. Bromo- and chloroalkanes are frequently used to detect carboxy compounds in chromatographic procedures. Typical examples include 4-bromomethyl-7-methoxycoumarin,[194] 4-bromomethyl-6,7-dimethoxycoumarin,[195] 9-(chloromethyl)anthracene,[196] 1-chloromethylbenz[c,d]indol-2(1H)-one,[197] and 3-bromomethyl-6,7-methylenedioxy-1-methyl-2-(1H)-quinoxalinone.[198] All these labels require excitation in the UV or violet part of the spectrum.

## D. HYDROXYL MODIFICATION

Isocyanates such as anthracene isocyanate[199] are reactive to primary, secondary, and even tertiary alcoholic groups forming carbamic acid esters. 7-Methoxycoumarin-3(and 4)-carbonyl azides[200] and anthroyl nitriles[201] were synthesized as fluorescent labeling reagents, but only with primary and secondary alcohols reaction products were found. 3,4-Dihydro-6,7-dimethoxy-4-methyl-3-oxo-quinoxaline-2-carbonyl azide also reacts with tertiary alcoholic groups to form fluorescent derivatives[202]. Carboxylic acid chlorides such as 7-[(chlorocarbonyl)methoxy]-4-methylcoumarin,[203] benzocoumarin-3-carboxylic acid chloride,[204] or 3,4-dihydro-6,7-dimethoxy-4-methyl-3-oxo-quinoxaline-2-carbonyl chloride[205] are also important alcohol reagents. Diazomethanes are also reactive to alcohols in the presence of fluoboric acid as a catalyst giving highly fluorescent ethers.[193]

Dansyl phenylphosphinic acid p-nitrophenyl ester has been found to be selective to hydroxyl groups in the presence of amino and sulfhydryl groups.[175] This type of compounds[206] as well as phosphonofluoridates such as the respective pyrene derivative[207] have found application in the selective modification of serine residues at the active sites of chymotrypsin, trypsin, thrombin, and the cholinesterases. Dansyl fluoride, the fluoride analog of dansyl chloride, does not react with amines, but reacts with chymotrypsin specifically at the active-site serine.[208] Tyrosine in proteins can be modified by several fluorescent labels but with low selectivity in the presence of other common amino acids such as cysteine and lysine. Dansyl chloride forms stable sulfonate derivatives, FITC reacts to some degree, and the iodoacetamides react readily at elevated pH.

A common structural property of sugars, polysaccharides, glycoproteins, and RNA is the vicinal glycol group. These can be oxidized by periodate at neutral pH to give aldehydes, ketones, or mixtures thereof. After removal of the excess periodate salt, the carbonyl compounds can be modified as described below (Section V.E). 1,2- or 1,3-glycols form derivatives with boronic acids such as dansyl amino boronic acid[209] without periodate oxidation.

## E. CARBONYL MODIFICATION

Carbonyl compounds can be condensed with fluorescent amine, hydrazine, thiosemicarbazide, or hydroxylamine derivatives. Fluorescent amines, e.g., 7-amino-4-methylcoumarin,[210] form Schiff bases that usually are reduced with sodium borohydride or preferably with sodium cyanoboro-hydride in order to increase the stability of the linkage. Fluorescent hydrazides and hydrazines may be prepared from dansyl hydrazide,[211], N-(2-aminophenyl-6-methylbenzthiazol)-acetylhydrazide,[211] 7-hydrazino-4-nitrobenzo-2-oxa-1,3-diazole,[212] 1-pyrenebutyryl hyrazide,[213] and lucifer yellow carbohydrazide.[214] Fluorescent hydrazines have been used for the labeling of glycoproteins.[214] The method consists of oxidation of oligosaccharides present in the protein, reaction of the resulting aldehydes with the labels, and reduction of the hydrazones and unreacted aldehydes with $NaBH_4$.

## F. SPECTRAL PROPERTIES OF FLUORESCENT LABELS

Fluorescent labels for proteins are widely used in fluorescence immunoassay, histochemistry, cell sorting, study of macromolecular structure and function, fluorescence photobleaching recovery experiments, and other experimental systems. The choice of a label for a particular application is frequently determined by spectral and lifetime properties, and to some extent, the polarization spectrum and quantum yield. For optical sensing purposes, the stipulations of longwave excitation and emission are predominant. Long lifetimes are desired as well in many cases. A comparison of dyes from literature data is rather complicated, since spectra and lifetimes are affected by several parameters. The spectral properties depend on temperature, solvent, ionic components, pH, the labeling conditions, the degree of labeling, and even the properties of the particular protein studied. An atlas of fluorescence spectra and lifetimes of dyes attached to proteins has been published.[215] A series of fluorescent conjugates have been prepared with chicken egg ovalbumin which was chosen as a typical protein, since it contains many groups reactive with a variety of labeling reagents. The spectral properties of the most important labels are summarized in Table 16.

Anthracene, naphthalene, pyrene, and 4-methyl-coumarin derived labels are UV excitable only and fluoresce in a region where interferences from background fluorescence of biological material could be fairly high. Pyrenes have the longest lifetimes of the common organic fluorescent dyes, typically 40 to 160 ns. Absorbance and fluorescence is much higher for pyrenes than for either anthracene- or naphthalene-based labels. They are strongly subject to oxygen quenching. Dansyl-containing reagents result in conjugates having different spectra. The dansyl group suffers from high sensitivity to small environmental effect not related to the analyte. In addition, conjugates differing in degree of labeling and/or in types of sites labeled would be expected to have different spectra. Other fluorophores having strong environment-depending spectra are the 7-nitrobenzo-2-oxa-1,3-diazole and 7-dialkylamino-coumarin derivatives. They are not considered to be good labels for optical sensing purposes.

One of the most widely used label with longwave absorption and fluorescence is fluorescein isothiocyanate. Its advantages are high extinction coefficient, high quantum yield (that is almost invariant with environment, except pH), and water solubility. The spectra are far removed from the usually naturally fluorescent components of biomolecules although serum with high bilirubin interferes. The disadvantages are a small Stokes shift of about 25 nm and tendency toward photobleaching which is retarded by propyl gallate or p-phenylenediamine. The limiting factor in detecting fluorescein fluorescence is the Raman scattering peak buried under the emission. In addition, its absorption is highly pH dependent in the physiological pH range due to a $pK_a$ of 6.7. In order to obtain a pH-independent group of longwave labels, new labels derived from 3-substituted coumarins have been designed and synthesized.[216] The respective albumin-dye conjugates possess fluorescein-like fluorescence with emission maxima in the range of 510 to 540 nm, when excited at 440 to 470 nm.

Rhodamines have the advantage over fluorescein of higher photostability. They can be used together with fluorescein in dual labeling studies, since they absorb and fluoresce at longer wavelengths. Particularly, Texas Red[217] is useful in combination with fluorescein isothiocyanate, since the excitation and emission spectra of Texas Red conjugates are widely separated from those of molecules labeled with fluorescein isothiocyanate. In addition, Texas Red displays excellent water solubility due to its hydrophilic nature, whereas other rhodamines are difficult to dissolve in the conjugation medium.

Europium chelates[218] have found advantageous application in time-resolved fluoroimmunoassays, because of their long lifetimes of 200 to 800 μs and Stokes shift as large as 290 nm. The europium chelating reagent 4,7-bis(chlorosulfophenyl)-1,10-phenanthroline-2,9-dicarboxylic acid (BCPDA)[219] can be covalently bound to proteins via its sulfochloride groups.

## VI. LABELING PROTOCOLS

Rather than a lengthy discussion how to label proteins, we present, in the following, a series of typical procedures with proven reliability for most proteins. For a discussion of fluorescence labeling in immunosensing, see also Chapter 17.

### A. PROCEDURE FOR CONJUGATION OF ANTIBODIES WITH FLUORESCEIN ISOTHIOCYANATE (FITC)

1.  Decide on final volume of conjugate desired and/or amounts of protein solution necessary (final protein concentration in grams should equal 1% of final volume in milliliters, i.e., weight of protein in 10 mg equals final volume in milliliters).
2.  Calculate volume of 0.5 *M* carbonate buffer (10% of final volume).
3.  Determine volume of saline (unbuffered) necessary to bring protein solution and carbonate buffer to final volume.
4.  Constantly agitating at 4°C, add the following reagents in the order given: saline (step 3), carbonate buffer, and protein solution (step 1).
5.  Continue stirring, slowly sprinkle in fluorescein isothiocyanate (1/60 weight of protein used).

6.   Continue stirring for 18 h at 4°C.
7.   Dialyze against PBS until dialysate is free of green. For 500 mg protein, several changes of PBS are made in the morning and afternoon for 2 or 3 d, using a total of about 5 to 10 l PBS.
8.   Add 1:10,000 merthiolate (or a similar biozide).
9.   Divide into 0.5 ml portions and freeze.

Carbonate buffer: Stock A: 0.5 $M$ $Na_2CO_3$ (13.25 g/250 ml distilled water); Stock B: 0.5 $M$ $NaHCO_3$ (10.50 g/250 ml distilled water). To prepare the buffer of pH 9.2 to 9.5, mix 9 parts A with 1 part B, check pH, and adjust with A (the base) or B (the acid) as needed.

## B. CONJUGATION OF GLOBULINS WITH TETRAMETHYLRHODAMINE ISOTHIOCYANATE (TRITC)[220]

The temperature of the reagents is brought to 4°C and all procedures are carried out at this temperature. To 2 ml of the globulin fraction is added 0.2 ml of a 0.5 $M$ carbonate-bicarbonate buffer of pH 9.0. The TRITC solution is prepared by dissolving 0.02 mg of the dye per milligram protein to be conjugated in 2 ml freshly prepared 2% bicarbonate solution with a pH of 8.2. Prolonged shaking for several hours at room temperature is necessary. The fluorochrome solution is added dropwise to the globulin fraction under continuous stirring continued for 30 min after the dye has been added. The mixture is then left overnight. The free dye is removed by gel filtration with a Sephadex G-50 column. Merthiolate is added to a final concentration of 1:10,000. The sample is divided into aliquots of about 0.5 ml and stored at –70°C.

## C. ASSESSMENT OF DEGREE OF FITC/TRITC CONJUGATION[221]

The fluorescein to protein (F/P) ratio is estimated by measuring the extinction at 495 and 280 nm for FITC-labeled conjugates and at 550 and 280 nm for TRITC-labeled conjugates. With a nomogram reading, F/P ratios are expressed in micrograms of dye per milligram protein. Wells et al.[221] have provided a useful nomogram in which linearity of the extinction values is assumed over a wide range.

## D. PREPARATION OF TEXAS RED CONJUGATED IgG[173]

Place goat IgG solution (10 mg per milliliter in 0.175 $M$ sodium phosphate, pH 7.0) in a test tube over a magnetic stirrer at room temperature and raise the pH to 9.0 with 1 N NaOH. Dissolve 1 mg of Texas Red in 100 µl of anhydrous dimethylformamide (stored over molecular sieves). Add the fluorophore solution (4 µl per milligram of IgG each time) in two aliquots over a period of 1 h with constant stirring. Readjust the pH to 9.0 if necessary. At the end of reaction, apply the mixture to a long Sephadex G-50 column (approximately 20 times the volume of the reaction mixture is needed for good resolution) equilibrated with 0.0175 $M$ sodium phosphate, pH 6.3. The blue conjugate elutes in the void volume. The protein recovery is about 90%. Under these conditions (reagent: protein weight ratio = 0.08:1) the conjugate has an average ratio of absorbances at 595 and 280 nm of 0.8. A detailed description of the reaction conditions and physical properties of Texas Red conjugated IgGs, lectins, and avidin is given by Titus et al.[217]

## E. LABELING OF PROTEINS USING DYES ADSORBED ONTO DIATOMACEOUS EARTH

Some reactive dyes are not sufficiently soluble and stable for use in labeling procedures requiring long labeling periods of more than 12 h. To overcome these problems, dyes have been adsorbed onto diatomaceous earth such as Celite. The tremendously large reactive surface permits drastic reduction of the reaction time. A period of 30 min or less has been found adequate for satisfactory labeling of protein solutions compared with 12 to 24 h used in other labeling techniques.

**General procedure using Celite**[222] — Approximately 1.5 g of the dye is dissolved in 150 ml dry chloroform or another appropriate dry organic solvent. After filtration from insoluble material the dye solution is added to 15 g of Celite which had been dried by heating at 300°C for 10 min and cooled to room temperature in dry atmosphere. The solvent is then evaporated under reduced pressure and traces of solvent removed from the residual powder with a high vacuum pump. Such preparations, when stored at room temperature in a tightly closed container, retained satisfactory labeling activity for at least 1 year. For the labeling tests 10 to 40 mg of fluorescent Celite powder is added to 3 ml of rabbit serum or protein solution and 3 ml of 0.05 $M$ sodium carbonate-bicarbonate buffer (pH 9, see Section VI.A herein) in a centrifuge tube. The mixture is shaken mechanically for 30 min and centrifuged. The supernatant is dialyzed for 3 d against running water.

## F. PREPARATION OF A 4-METHYLUMBELLIFERONE-3-ACETIC ACID (4-MUAA) IgG₁ CONJUGATE[223]

The condensation of 4-MUAA with available active amino groups of the purified anti-E₂ antibody to form a peptide link was effected using cyclohexyl-3-(2-morpholino-ethyl)carbodiimide metho-p-toluene sulphonate (CMC). An active intermediate of 4-MUAA and CMC was first formed in 5 ml of 0.01 $M$ phosphate buffer, pH 5.0, by reacting 4 mg 4-MUAA and 120 mg CMC (high molar excess of CMC) at 10 to 12°C for 45 min. This intermediate was added slowly to 2 ml solution of the same buffer containing 165 μg of the purified IgG₁, whose antibody binding sites had been protected by adding 40 μg of estrone. The antibody site protection was considered necessary in case essential active amino acid residues in the active site were involved in 4-MUAA conjugation, resulting in possible antibody deactivation. The mixture was stirred at 10 to 12°C for 3 h and then dialyzed in 0.2 $M$ NaCl containing 0.1% sodium azide before extracting with ether three times to remove the estrone. The volume of the preparation after dialysis was reduced using poly(ethylene glycol) powder. It was then layered on a Sephadex G-25 column, equilibrated with 0.1 $M$ assay buffer of pH 7.1 (consists of 195 ml of 0.2 $M$ NaH₂PO₄, 305 ml of 0.2 $M$ Na₂HPO₄, 1 g of sodium azide, and 9 g of sodium chloride made up to 1 l with water) and 1 ml fractions collected. Aliquots were removed from each column fraction and these were used to determine both fluorescence and the ability of the antibody binding.

## G. PROCEDURE FOR PREPARING THE DRUG/DYE CONJUGATE, β-GALACTOSYL-UMBELLIFERONE-SISOMICIN[224]

This conjugate has been applied in a homogeneous reactant-labeled fluorescent immunoassay to the measurement of therapeutic drug concentrations in human serum, e.g. gentamicin.[220] β-Galactosyl-umbelliferone-sisomicin was prepared by mixing 50 mg (117 μmol) of the commercial β-galactosidase substrate, β-[7-(3-carboxycoumarinoxy)]-D-galactoside potassium salt, with 171 mg of sisomicin sulfate (223 μmol of sisomicin free base) in 2 ml of water. The pH was adjusted to 3.8 by dropwise addition of 1 $M$ HCl. The solution was cooled in an ice bath and 30 mg (150 μmol) of 1-ethyl-3-(3-dimethylaminopropyl)-carbodiimide hydrochloride was added. After 2 h the mixture was chromatographed at 25°C on a 2.5 × 50 cm column of CM-Sephadex C-25, 5.8 ml fractions were collected, and their absorbance was monitored at 345 nm. The column was washed with 200 ml of 50 m$M$ ammonium formate to elute unreacted β-galactosyl-umbelliferone-sisomicine. A linear gradient, formed with 400 ml of 50 m$M$ and 400 ml of 1.8 $M$ ammonium formate, was applied to the column. A peak of material absorbing at 345 nm eluted at approximately 1.4 $M$ ammonium formate. After the gradient, the column was washed with 600 ml of 1.8 $M$ ammonium formate. Three 345 nm absorbing peaks were eluted in this wash. Unreacted sisomicin was eluted well separated from the last 345 nm peak which represents the major peak of β-galactosyl-umbelliferone-sisomicin.

## H. LABELING OF MORPHINE-SPECIFIC ANTIBODIES WITH 4′,5′-DIMETHOXY-5- AND 6-CARBOXYFLUORESCEIN, N-HYDROXYSUCCINIMIDYL ESTER[225]

A solution of 0.07 mg of the label in 25 μl of dimethylformamide was added over 20 min to 7.5 mg of sheep antimorphine gamma-globulin in 0.5 ml of 0.05 *M* phosphate buffer, pH 8.0, at 0 to 5°C. After stirring overnight in the cold, the solution was centrifuged for 2 min at room temperature and the red solution gel filtered on a Sephadex G-25 column (1 × 15 cm) with the same buffer. A single high molecular weight fraction was obtained which was practically nonfluorescent and had visible absorption with a maximum at 517 nm. About 80 to 90% labeling efficiency was achieved.

## I. COUPLING OF 4-METHYLUMBELLIFERONE-3-ACETIC ACID, N-HYDROXYSUCCINIMIDYL ESTER TO CASEIN[226]

Approximately 1 g of casein was left to dissolve overnight in 20 ml of 50 m*M* sodium tetraborate buffer, pH 9.0. After removal of any undissolved particles by centrifugation, 99 mg of the label was added in three portions over a period of 1 h with vigorous stirring. The reaction mixture was kept at room temperature for a further 5 h. The solution was then centrifuged, and the soluble casein was precipitated by the addition of 0.5 *M* sodium formate/HCl buffer, pH 4.0. The precipitate was filtered off and washed, first with the above buffer and then with 10 m*M* phosphate-buffered saline (0.15 *M* NaCl per 10 m*M* sodium phosphate buffer), pH 4.0, until the washings were substantially free of fluorescence. Final traces of fluorescent contamination were removed by redissolving the 4-methylumbelliferyl-casein in 20 ml of 10 m*M* phosphate-buffered saline, pH 9.0, and dialyzing overnight against a large volume of 10 m*M* phosphate-buffered saline, pH 7.2, before freeze-drying the product.

## J. LABELING OF (NA,K)-ATPASE WITH LUCIFER YELLOW CARBOHYDRAZIDE[214]

Purified dog kidney (Na,K)-ATPase (1 ml of 2 to 4 mg per milliliter) was incubated with 60 to 120 units of galactose oxidase, 1.4 to 2.8 units of neuraminidase, and 2000 to 4000 units of catalase in 50 m*M* Hepes-NaOH per 1 m*M* EDTA buffer of pH 7.5, for 2 h at 0°C. The reaction was stopped by the addition of 5 m*M* dithiothreitol (DTT). The pH was decreased to 6.3 by addition of 1 *M* succinate per 2 m*M* EDTA, pH 5.9, lucifer yellow carbohydrazide was added at the final concentration of 4.5 m*M* and the incubation was continued at 0°C in the dark. After various lengths of time, the pH was raised to 7.4 with 1 *M* Hepes per 2 m*M* EDTA buffer of pH 8.0, 25 m*M* NaBH$_4$ was added, and the incubation was continued for 30 min at 0°C. Reduction with NaBH$_4$ results in reduction of both hydrazone and excess aldehyde groups and therefore eliminates the possibility of protein cross-linking. The labeled (Na,K)-ATPase was separated from excess probe by centrifugation through a discontinuous sucrose gradient (37, 28, and 15%) in an SW-41 rotor for 3 h at 35,000 rpm. The enzyme was removed from the 28 to 37% interface, washed twice by sedimentation and resuspended in 50 m*M* Mops-Tris per 1 m*M* EDTA, pH 7.0, and then resuspended in an identical buffer containing 30% (w/v) glycerol. The preparation was divided into aliquots, frozen in liquid nitrogen, and stored at -70°C until use.

Oxidation of (Na,K)-ATPase oligosaccarides with periodate, instead of the above enzyme system, resulted in 50 to 80% inhibition of the (Na,K)-ATPase activity with low or undetectable labeling.

# REFERENCES

1. **Bishop E., Ed.,** *Indicators*, Pergamon Press, Oxford, 1972.
2. **Cheng K. L., Ueno, K., and Imamura T., Eds.,** *Handbook of Organic Analytical Reagents*, CRC Press, Boca Raton, FL, 1982.
3. **Kotyk, A. and Slavik, J.,** Intracellular pH and its Measurement, CRC Press, Boca Raton, 1988.
4. **Fernandez-Guttierez, A. and Munoz de la Pena, A.,** Determination of Inorganic Substances by Luminescence Methods, in *Modern Luminescence Spectroscopy. Methods and Applications*, Vol. 1, Schulman, S. G., Ed., John Wiley & Sons, New York, 1985.
5. **De Ment, J.,** Fluorescence titration, *J. Chem. Educ.*, 30, 145, 1953.
6. **Peterson, J. I., Goldstein, S. R., Fitzgerald, R. V., and Buckhold, D. W.,** Fiber-optic pH probe for physiological use, *Anal. Chem.*, 52, 864, 1980.
7. **Suidan, J. S., Young, B. K., Hetzel, F. W., and Seal, H. R.,** pH measurement with a fiber-optic tissue-pH monitor and a standard blood pH meter, *Clin. Chem.*, 29, 1566, 1983.
8. **Bacci, M., Baldini, F., and Scheggi, A. M.,** Spectrophotometric investigations on immobilized acid-base indicators, *Anal. Chim. Acta*, 207, 343, 1988.
9. **Kirkbright, G. F., Narayanaswamy, R., and Welti, N. A.,** Fiber-optic pH probe based on the use of an immobilized colorimetric indicator, *Analyst*, 109, 1025, 1984, and references cited therein.
10. **Wolfbeis, O. S., Fürlinger, E., Kroneis, H., and Marsoner, H.,** A study on fluorescent indicators for measuring near neutral ("physiological") pH-values, *Fresenius Z. Anal. Chem.*, 314, 119, 1983.
11. **Delisser-Matthews, L. A. and Kauffman, J. M.,** 3-Arylcoumarins as fluorescent indicators, *Analyst*, 109, 1009, 1984.
12. **Graber, M. L., DiLillo, D. C., Friedman, B. L., and Pastoriza-Munoz, E.,** Characteristics of fluoroprobes for measuring intracellular pH, *Anal. Biochem.*, 156, 202, 1986.
13. **Wolfbeis, O. S., Koller, E., and Hochmuth, P.,** The unusually strong effect of a 4-cyano group upon electronic spectra and dissociation constants of 3-substituted 7-hydroxycoumarins, *Bull. Chem. Soc. Jpn.*, 58, 731, 1985.
14. **Wolfbeis, O. S.,** pH-dependent fluorescence spectra of 3-substituted umbelliferones, *Z. Naturforsch.*, 32A, 1065, 1977.
15. **Offenbacher, H., Wolfbeis, O. S., and Fürlinger, E.,** Fluorescence optical sensors for continuous determination of near-neutral pH values, *Sensors Actuators*, 9, 73, 1986.
16. **Schulman S. G., Ed.,** *Molecular Luminescence Spectroscopy: Methods and Applications*, Vol. 2., John Wiley & Sons, New York, 1988, chap. 6.
17. **Wolfbeis, O. S. and Marhold, H.,** A new group of fluorescent pH indicators for an extended pH range, *Fresenius Z. Anal. Chem.*, 327, 347, 1987.
18. **Wolfbeis, O. S. and Baustert, J. H.,** Synthesis and spectral properties of a new class of fluorescent pH indicators. *J. Heterocycl. Chem.*, 22, 1215, 1985.
19. **Leonhardt, H., Gordon, L., and Livingston, R.,** Acid-base equilibria of fluorescein and 2',7'-dichlorofluorescein in their ground and fluorescent states, *J. Phys. Chem.*, 75, 245, 1971.
20. **Diehl, H. and Horchak-Morris, N.,** Studies on fluorescein. V. The absorbance of fluorescein in the ultraviolet, as a function of pH, *Talanta*, 34, 739, 1987, and previous papers.
21. **Saari, L. and Seitz, W. R.,** pH sensor based on immobilized fluoresceinamine, *Anal. Chem.*, 54, 821, 1982.
22. *Handbook on pH Probes & Enzyme Substrates*, Lambda Probes & Diagnostics, A-8053 Graz, Austria, 1989.
23. **Rink, T. J., Tsien, R. Y., and Pozzan, T.,** Cytoplasmic pH and free $Mg^{2+}$ in lymphocytes, *J. Cell Biol.*, 95, 189, 1982.
24. **Whitaker, J. E., Haugland, R. P., and Prendergast, F. G.,** Seminaphtho-fluoresceins and seminaphtho-rhodafluors — dual fluorescence pH indicators, *Biophys. J.*, 53, 197, 1988.
25. **Posch, H. E. and Wolfbeis, O. S.,** Towards a gastric pH sensor: an optrode for the pH 0—7 range, *Fresenius Z. Anal. Chem.*, 334, 762, 1989.
26. **Fromherz, P. and Masters, B.,** Interfacial pH at electrically charged lipid monolayers investigated by the lipoid pH indicator method, *Biochim. Biophys. Acta*, 356, 270, 1974.
27. **Alpes, H. and Pohl, W. G.,** 3-Palmitoyl-7-hydroxy-coumarin — a new lipoid pH indicator, *Naturwissenschaften*, 65, 652, 1978.
28a. **Schaffar, B. P. H. and Wolfbeis, O. S.,** New optical sensors based on the Langmuir-Blodgett technique, *Proc. SPIE*, 990, 122, 1988.
28b. **Morf, W. E., Seiler, K., Rusterholz, B., and Simon, W.,** Design of a calcium-selective optode membrane based on neutral ionophores, *Anal. Chem.*, 62, 738, 1990.
28c. **Suzuki, K., Ohzora, H., Tohda, K., Miyazaki, K., Watanabe, K., Inoue, H., and Shirai, T.,** Fibre-optic potassium ion sensors based on a neutral ionophore and a novel lipophilic anionic dye, *Anal. Chim. Acta*, 237, 155, 1990.

29. **Wolfbeis, O. S. and Offenbacher, H.,** Fluorescence sensor for monitoring ionic strength and physiological pH values, *Sensors Actuators*, 9, 85, 1986.

30. **Janata, J.,** Do optrodes really measure pH?, *Anal. Chem.* 59, 1351, 1987, and references cited therein.

31. **Edmonds, T. E., Flatters, N. J., Jones, C. F., and Miller, J. N.,** Determination of pH with acid-base indicators: implications for optical fiber probes, *Talanta*, 35, 103, 1988.

32. **Kolthoff, I. M.,** The dissociation of acid-base indicators in ethanol with a discussion of the medium effect, *J. Phys. Chem.*, 35, 2632, 1931

33. **Salvatore, F., Ferri, D., and Palombari, R.,** Salt effect on the dissociation constant of acid-base indicators, *J. Solution Chem.*, 15, 423, 1986.

34. **Katayama, Y., Fukuda, R., Iwasaki, T., Nita, K., and Takagi, M.,** Synthesis of chromogenic crown ethers and liquid-liquid extraction of alkaline earth metal ions, *Anal. Chim. Acta*, 204, 113, 1988 and references cited therein.

35. **Dix, J. P. and Vögtle, F.,** Ion-selective crown ether dyes, *Angew. Chem. Int. Ed. Engl.*, 17, 857, 1978.

36. **Van Gent, J., Sudhölter, E. J. R., Lambeck, P. V., Popma, T. J. A., Gerritsma, G. J., and Reinhoudt, D. N.,** A chromophoric crown ether as a sensing molecule in optical sensors for the detection of alkali metal ions, *J. Chem. Soc. Chem. Commun.*, 893, 1988.

37. **Kreuwel, H. J. M., Lambeck, P. V., Van Gent, J., and Popma, T. J. A.,** Surface plasmon dispersion and luminescence quenching applied to planar waveguide sensors for the measurement of chemical concentrations, *Proc. SPIE*, 798, 218, 1987.

38. **Alder, J. F., Ashworth, D. C., Narayanaswamy, R., Moss, R. E., and Sutherland, I. O.,** An optical potassium ion sensor, *Analyst*, 112, 1191, 1987.

39. **Haugland, R. P.,** Handbook of Fluorescent Probes, Mol. Probes Inc., Eugene (OR), 1991.

40. **Tanigawa, I., Tsuemoto, K., Kaneda, T., and Misumi, S.,** Synthetic macrocyclic ligands. VI. Lithium ion-selective fluorescent emission with crowned benzo- and naphthothiazolylphenols, *Tetrahedron Lett.*, 25, 5327, 1984.

41. **Kimura, K., Iketani, S.-I., and Shono, T.,** Synthesis of a fluorescent 14-crown-4 derivative bearing a proton-dissociable moiety and its use for selective lithium-ion extraction, *Anal. Chim. Acta*, 203, 85, 1987.

42. **Fery-Forgues, S., Le Bris, M.-T., Guette, J.-P., and Valeur, B.,** Ion-responsive fluorescent compounds. 1. Effect of cation binding on photophysical properties of a benzoxazinone derivative linked to monoaza-15-crown-5, *J. Phys. Chem.*, 92, 6233, 1988.

43. **Street, K. W. and Krause, S. A.,** A new metal sensitive fluorescence reagent, *Anal. Lett.*, 19, 735, 1986.

44. **Tsien, R. J.,** Intracellular sodium regulation in rabbit gastric glands determined using a fluorescent sodium indicator, *J. Gen. Physiol.*, 92, Abstr. 53, 1988.

45. **Smith, G. A., Hesketh, T. R., and Metcalfe, J. C.,** Design and properties of a fluorescent indicator of intracellular free $Na^+$ concentration, *Biochem. J.*, 250, 227, 1988.

46. **Rink, T. J. and Pozzan, T.,** Using quin-2 in cell suspensions, *Cell Calcium*, 6, 133, 1985.

47. **Grynkiewicz, G., Poenie, M., and Tsien, R. Y.,** A new generation of $Ca^{2+}$ indicators with greatly improved fluorescence properties, *J. Biol. Chem.*, 260, 3440, 1985.

48. **Tsien, R. Y. and Minta, A.,** *J. Biol. Chem.*, in press (1989).

49. **Murphy, E., Freudenrich, C. C., Levy, L. A., London, R. E., Lieberman, M.,** *Proc. Natl. Acad. Sci. U.S.A.*, 86, 2981, 1989.

50. **Roe, J. N., Szoka, F. C., and Verkman, A. S.,** Optical indicator for alkaline earth metal ions, (to be published), 1989.

51. **Huston, M. E., Haider, K. W., and Czarnik, A. W.,** Chelation-enhanced fluorescence in 9,10-bis(TMEDA)anthracene, *J. Am. Chem. Soc.*, 110, 4460, 1988.

52. **König, K.-H., Bosslet, J., and Holzner,** N-Butyl-N'-dansylthiourea as fluorescent broad-band complexing agent, *Chem. Ber.*, 122, 59, 1989.

53. **Marshall, L.,** A new class of chelating agents, *J. Am. Chem. Soc.*, 110, 5192, 1988.

54. **Behringer, C., Lehmann, B., and Simon, W.,** Carbonate-selective chromoionophores, *Chimia*, 41, 397, 1987.

55. **Clarke, W. M.,** *Oxidation-Reduction Potentials of Organic Systems*, Williams and Wilkins, Baltimore, 1960.

56. **Charlot, G.,** *Selected Constants: Oxidation - Reduction Potentials*, Pergamon Press, Oxford, 1958.

57. **Bryan, A. J., De Silva, A. P., De Silva, S. A., Rupasinghe, R. A. D. D., and Sandanayake, K. R. A.,** Photo-induced electron transfer as a general design logic for fluorescent molecular sensors for cations, *Biosensors*, 4, 169, 1989.

58. **Glenner, G. G.,** Formazans and tetrazolium salts, in *H. J. Conn's Biological Stains*, Lillie, R. D., Ed., Williams and Wilkins, Baltimore, 1977, 225.

59. **Blough, N. V. and Simpson, D. J.,** Chemically mediated fluorescence yield switching in nitroxide-fluorophore adducts: optical sensors of radical/redox reactions, *J. Am. Chem. Soc.*, 110, 1915, 1988, and references cited therein.

60. **Waggoner, A. S.,** Dye indicators of membrane potential, *Annu. Rev. Biophys. Bioeng.*, 8, 47, 1979.

61. **Beeler, T. J., Farmen, R. H., and Martonosi, A. N.,** The mechanism of voltage-sensitive dye responses on sarcoplasmic reticulum, *J. Membr. Biol.*, 62, 113, 1981.

62. **Szöllösi, J., Damjanovich, S., Mulhern, S. A., and Tron, L.,** Fluorescence energy transfer and membrane potential measurements monitor dynamic properties of cell membranes: a critical review, *Prog. Biophys. Mol. Biol.*, 49, 65, 1987.

63. **Schaffar, B. P. H. and Wolfbeis, O. S.,** Effect of Langmuir-Blodgett layer composition on the response of ion-selective optrodes for potassium, based on the fluorimetric measurement of membrane potential, *Analyst*, 113, 693, 1988.

64. **Wolfbeis, O. S. and Schaffar, B. P. H.,** Optical sensors: An ion-selective optrode for potassium, *Anal. Chim. Acta*, 198, 1, 1987.

65. **Schaffar, B. P. H. and Wolfbeis, O. S.,** A calcium-selective optrode based on fluorimetric measurement of membrane potential, *Anal. Chim. Acta*, 217, 1, 1989.

66. **Zhujun, Z. and Seitz, W. R.,** Ion-selective sensing based on potential sensitive dyes, *Proc. SPIE*, 906, 74, 1988.

67. **Opitz, N. and Lübbers, D. W.,** Electrochromic dyes, enzyme reactions and hormone-protein interactions in fluorescence optic sensor (optode) technology, *Talanta*, 35, 123, 1988.

68. **Mokhova, E. N. and Rozovskaya,** The effects of mitochondrial energetics inhibitors on the fluorescence of potential-sensitive dyes rhodamine 123 and DiS-C3-(5) in lymphocyte suspensions, *J. Bioenerg. Biomembr.*, 18, 265, 1986.

69. **Aiuchi, T., Daimatsu, T., Nakaya, K., and Nakamura, Y.,** Fluorescence changes of rhodamine-6G associated with changes in membrane-potential in synaptosomes, *Biochim. Biophys. Acta*, 685, 289, 1982.

70. **Dietzmann, K., Letko, G., and Sokolowski, A.,** Mitochondrial membrane potential in living cells: evidence from studies with rhodamine 6G as fluorescent probe, *Exp. Pathol.*, 31, 47, 1987.

71. **Ehrenberg, B., Montana, V., Wei, M.-D., Wuskell, J. P., and Loew, L. M.,** Membrane potential can be determined in individual cells from the Nernstian distribution of cationic dyes, *Biophys. J.*, 53, 785, 1988.

72. **Smith, J. C. and Chance, B.,** Kinetics of the potential-sensitive extrinsic probe oxonol VI in beef heart submitochondrial particles, *J. Membr. Biol.*, 46, 255, 1979.

73. **Brauner, T., Hulser, D. F., and Strasser, R. J.,** Comparative measurements of membrane potentials with microelectrodes and voltage-sensitive dyes, *Biochim. Biophys. Acta*, 771, 208, 1984.

74. **Freedman, J. C. and Novak, T. S.,** Membrane potentials associated with Ca-induced K conductance in human red blood cells: studies with a fluorescent oxonol dye, WW 781, *J. Membr. Biol.*, 72, 59, 1983.

75. **Fromherz, P. and Kotulla, R.,** Fluorescent dye in soap lamella as a probe of the electrical potential, *Ber. Bunsenges. Phys. Chem.*, 88, 1106, 1984.

76. **Fluhler, E., Burnham, V. G., and Loew, L. M.,** Spectra, membrane binding, and potentiometric responses of new charge shift probes, *Biochemistry*, 24, 5749, 1985, and references cited therein.

77. **Grinvald, A., Hildesheim, R., Farber, I. C., and Anglister, L.,** Improved fluorescent probes for the measurement of rapid changes in membrane potential, *Biophys. J.*, 39, 301, 1982.

78. **Obyknovennaya, I. E. and Cherkasov, A. S.,** Quenching of anthracene and rhodamine fluorescence by aromatic nitro compounds and amines in aqueous-micellar solvent, *Opt. Spektrosk.*, 64, 325, 1988, and references therein.

79. **Lishan, D. G., Hammond, G. S., and Yee, W. A.,** Amine quenching of fluorescence of phenylated anthracenes, *J. Phys. Chem.*, 85, 3435, 1981.

80. **Kano, K., Yanagimoto, M., Uraki, H., Zhou, B., and Hashimoto, S.,** Fluorescence quenching of perfluoronaphthalene by triethylamine, *Bull. Chem. Soc. Jpn.*, 59, 993, 1986.

81. **Epling, G. A. and Lin, K.-Y.,** Quenching of the fluorescence of quinoline derivatives by exciplex formation with heteroatom-containing compounds, *J. Heterocycl. Chem.*, 21, 1205, 1984.

82. **De Moerloose, P. and Baeyens, W.,** Fluorescence quenching of sodium 1,2-naphthoquinonesulfonate by primary amines, *Verh. K. Vlaam. Acad. Geneeskd. Belg.*, 38, 252, 1976.

83. **Kano, K., Kawazumi, H., Ogawa, T., and Sunamoto, J.,** Fluorescence quenching of pyrene and pyrenedecanoic acid by various kinds of N,N-dialkylanilines in dipalmitoylphosphatidylcholine liposomes, *Chem. Phys. Lett.*, 74, 511, 1980.

84. **Chang, S. L. P. and Schuster, D. I.,** Fluorescence quenching of 9,10-dicyanoanthracene by dienes and alkenes, *J. Phys. Chem.*, 91, 3644, 1987.

85. **Encinas, M. V., Guzman, E., and Lissi, E. A.,** Intramicellar aromatic hydrocarbon fluorescence quenching by olefins, *J. Phys. Chem.*, 87, 4770, 1983.

86. **Takahashi, T., Kikuchi, K., and Kokubun, H.,** Quenching of excited 2,5-diphenyloxyzole by carbon tetrachloride, *J. Photochem.*, 14, 67, 1980.

87. **Wiczk, W. M. and Latowski, T.,** Photophysical and photochemical studies of polycyclic aromatic-hydrocarbons in solutions containing tetrachloromethane. 2. The solvent effect on the fluorescence quenching of aromatic hydrocarbons by tetrachloromethane, *Z. Naturforsch., Teil A*, 42, 1290, 1987, and references cited therein.

88. **Winnik, M. A., Egan, L. S., Owens, S. M., and Ottewill, R. H.,** Fluorescence quenching studies on poly(methyl methacrylate) particles, *Anal. Chim. Acta*, 189, 89, 1986.

89. **Daems, D., Boens, N., and De Schryver, F. C.,** Determination of partition and diffusion of lindane in unilamellar L-α-dimyristoylphosphatidylcholine vesicles by fluorescence quenching of the intramembrane probe methyl 11-(N-carbazolyl)-undecanoate, *Anal. Chim. Acta*, 205, 61, 1988.

90. **Sharma, A. and Wolfbeis, O. S.,** The quenching of the fluorescence of polycyclic aromatic hydrocarbons and rhodamine 6G by sulphur dioxide, *Spectrochim. Acta*, 43A, 1417, 1987.

91. **Wolfbeis, O. S. and Sharma, A.,** Fibre-optic fluorosensor for sulphur dioxide, *Anal. Chim. Acta*, 208, 53, 1988.

92. **Watkins, A. R.,** Kinetics of fluorescence quenching by inorganic anions, *J. Phys. Chem.*, 78, 2555, 1974.

93. **Moriya, T.,** Excited-state reactions of coumarins in aqueous solutions. III. The fluorescence quenching of 7-ethoxycoumarins by the chloride ion in acidic solutions, *Bull. Chem. Soc. Jpn.*, 59, 961, 1986, and references cited therein.

94. **Bendig, J., Kreysig, D., Gebert, H., and Regenstein, W.,** Deactivation behaviour of arenes and heteroarenes. XVII. Fluorescence quenching of cations by anions, *J. Prakt. Chem.*, 321, 420, 1979.

95. **Wolfbeis, O. S. and Urbano, E.,** A fluorimetric method for analysis of chlorine, bromine, and iodine in organic materials, *Fresenius Z. Anal. Chem.*, 314, 577, 1983.

96. **Urbano, E., Offenbacher, H., and Wolfbeis, O. S.,** Optical sensor for continuous determination of halides, *Anal. Chem.*, 56, 427, 1984.

97. **Wyatt, W. A., Bright, F. V., and Hieftje, G. M.,** Characterization and comparison of three fiber-optic sensors for iodide determination based on dynamic fluorescence quenching of rhodamine 6G, *Anal. Chem.*, 59, 2272, 1987.

98. **Krapf, R., Illsley, N. P., Tseng, H. C., and Verkman, A. S.,** Structure-activity relationships of chloride-sensitive fluorescent indicators for biological application, *Anal. Biochem.*, 169, 142, 1988.

99. **Reszka, K., Hall, R. D., and Chignell, C. F.,** Quenching of the excited states of 2-phenyl-benzoxazole by azide anion. A fluorescence and electron-spin-resonance study, *Photochem. Photobiol.*, 40, 707, 1984.

100. **Wolfbeis, O. S. and Trettnak, W.,** Fluorescence quenching of acridinium and 6-methoxyquinolinium ions by $Pb^{2+}$, $Hg^{2+}$, $Cu^{2+}$, $Ag^+$ and hydrogen sulphide, Spectrochim. Acta, 43A, 405, 1987.

101. **Homer, R. B. and Allsopp, S. R.,** An investigation of the electronic and steric environments of tyrosyl residues in ribonuclease A and Erwinia carotovora L-asparaginase through fluorescence quenching by cesium, iodide and phosphate ions, *Biochim. Biophys. Acta*, 434, 297, 1976.

102. **Brittain, H. G.,** Submicrogram determination of lanthanides through quenching of calcein blue fluorescence, *Anal. Chem.*, 59, 1122, 1987.

103. **Wolfbeis, O. S.,** Fiber optic fluorosensors in analytical and clinical chemistry, in *Molecular Luminescence Spectroscopy. Methods and Applications*, Vol. 2, Schulman S. G., Ed., John Wiley & Sons, New York, 1988, Chap. 3.

104. **Harnisch, H. and Siegel, E.,** Cationic aromatic nitro compounds and their use in fluorescence quenching, German Patent. DE 3,415,103, 1985.

105. **Becker, H. G. O., Schütz, R., Kuzmin, M. G., Sadovskii, N. A., and Soboleva, I. V.,** Quenching of fluorescence and excimer formation of sodium pyrene-3-sulfonate in aqueous solutions of poly(dimethyl diallylammonium chloride), *J. Prakt. Chem.*, 329, 87, 1987.

106. **Becker, H. G. O., Schuetz, R., Kuz'min, M. G., and Soboleva, I. V.,** Quenching of fluorescence of aromatic compounds by diazonium cations in aqueous solutions of polyelectrolytes, *J. Prakt. Chem.*, 329, 95, 1987.

107. **Koller, E., Kriechbaum, M., and Wolfbeis, O. S.,** 1-Aminopyrene-3,6,8-trisulfonate: a fluorescent probe for thiamine and pyridinium ion, *Spectroscopy*, 3, 37, 1988.

108. **Marhold, S., Koller, E., and Wolfbeis, O. S.,** Unpublished results, 1989.

109. **Kroneis, H. W.,** Ph.D. dissertation, Techn. University Graz, 1983.

110. **Wolfbeis, O. S., Leiner, M. J. P., and Posch, H. E.,** A new sensing material for optical oxygen measurement, with the indicator embedded in an aqueous phase, *Mikrochim. Acta (Vienna)*, III, 359, 1986.

111a. **Lippitsch, M. E., Pusterhofer, J., Leiner, M. J. P., and Wolfbeis, O. S.,** Fibre optic oxygen sensor with the decay time as the information carrier, *Anal. Chim. Acta*, 205, 1, 1988.

111b. **Demas, J. N. and DeGraff, B. A.,** Design of transition metal complexes as luminescence probes, *Proc. SPIE*, 1172, 216, 1989.

112. **Wolfbeis, O. S. and Carlini, F. M.,** Long wavelength fluorescent indicators for the determination of oxygen partial pressure, *Anal. Chim. Acta*, 160, 301, 1984.

113. **Olmsted J.,** Oxygen quenching of the fluorescence of organic dye molecules, *Chem. Phys. Lett.*, 26, 33, 1974.

114a. **Eastwood, D. and Gouterman, M.,** The luminescence of metalloporphyrins, *J. Mol. Spectrosc.*, 35, 359, 1970.

114b. U.S. Pat., 4,810,655, 1989.

114c. **Vanderkooi, J., Maniara, G., Green, J., and Wilson, D. F.,** An optical method for measurement of dioxygen concentration based up on quenching of phosporescence, *J. Biol. Chem.*, 262, 5476, 1987.

114d. **Papkovsky, D. B., Savitsky, A. P., and Yaropolar, A. T. I.,** Optical oxygen and glucose sensors based on phosphorescence quenching of metalloporphyrins, *Zhurn. Anal. Khim.,* 45, 1441, 1990; (in Russian).

114e. **Papkovsky, D. B.,** Phosphorescent polymer films for optical oxygen sensors, *Biosensors & Bioelectronics,* 1991, in press.

115. **Egan, L. S., Winnik, M. A., and Croucher, M. D.,** Oxygen quenching studies of nonaqueous dispersions of poly(vinyl acetate) labeled with phenanthrene groups, *Langmuir,* 4, 438, 1988.

116. **Geacintov, N., Oster, G., and Cassen, T.,** Polymeric matrices for organic phosphors, *J. Opt. Soc. Am.,* 58, 1217, 1968, and references cited therein.

117. **Bolton, P. H., Kenne,r R. D., and Khan, A. U.,** Molecular oxygen enhanced fluorescence of organic molecules in polymeric matrices, *J. Chem. Phys.,* 57, 5604, 1972.

118. **Jones, P. F. and Nesbitt, R. S.,** Molecular oxygen enhanced fluorescence of organic molecules in polymer matrices, *J. Chem. Phys.,* 59, 6185, 1973.

119. **Ando, T. and Miyata, H.,** Pyruvate as a fluorescence quencher: a new spectroscopic assay for pyruvate reactions, *Anal. Biochem.,* 129, 170, 1983.

120. **Hüttenrauch, R., Fricke, S., and Matthey, K.,** On the relation between water structure and fluorescence quenching, *Pharmazie,* 37, 225, 1982.

121. **Kubota, Y., Motoda, Y., Shigemune, Y., and Fujisaki, Y.,** Fluorescence quenching of 10-methylacridinium chloride by nucleotides, *Photochem. Photobiol.,* 29, 1099, 1979, and references cited therein.

122. **Imasaki, T., Okazaki, T., and Ishibashi, N.,** Semiconductor laser fluorimetry for enzyme and enzymatic assays, *Anal. Chim. Acta,* 208, 325, 1988.

123. **Bartl, H. and Falbe, J., Eds.,** *Houben Weyl's Methods of Organic Chemistry, Vol. E 20. Macromolecular Materials,* Thieme, Stuttgart, 1987.

124. **Davidson, R. L.,** *Handbook of Water-Soluble Gums and Resins,* McGraw-Hill, New York, 1980.

125. **Seymour, R. B. and Mark, H. F., Eds.,** *Applications of Polymers,* John Wiley & Sons, New York, 1988.

126. **Wall, L. A.,** *Fluoropolymers,* John Wiley & Sons, New York, 1972.

127. **Molyneux, R.,** *Water-Soluble Synthetic Polymers. Properties and Behaviors,* CRC Press, Boca Raton, FL, 1984.

128. **Goethals, E. J.,** *Polymeric Amines and Ammonium Salts,* Pergamon Press, London, 1980.

129. **Yasuda, H. and Stannett, V.,** Permeability coefficients, in *Polymer Handbook,* Brandrup, J. and Immergut, E. H., Eds., John Wiley & Sons, New York, 1981, III.

130. **Batzer, H., Ed.,** *Polymeric Materials,* Thieme, Stuttgart, 1985.

131. **Robb, W. L.,** Thin silicone membranes — their permeation properties and some applications, *Ann. N.Y. Acad. Sci.,* 146, 119, 1968.

132. **Stern, S. A., Shah, V. M., and Hardy, B. J.,** Structure-permeability relationships in silicone polymers, *J. Polym. Sci., Part B,* 25, 1263, 1987.

133. **Wolfbeis, O. S., Posch, H. E., and Kroneis, H. W.,** Fiber optic fluorosensor for determination of halothane and/or oxygen, *Anal. Chem.,* 57, 2556, 1985.

134. **Hsu, L. and Heitzmann, H.,** US Patent 4,712,865, 1987.

135. **N. N.,** Selection Guide to Dow Corning Organosilane Chemicals, The Dow Corning Co., Midland, MI, 1986.

136. **Anderson, R., Arkles, B., and Larson, G. L.,** Silicone Compounds. Register and Review, Petrarch Systems, Bristol, PA, 1987.

137. **N. N.,** The Pierce Handbook and General Catalog, Pierce Chem. Co., Rockford, IL, 1987.

138. **Leyden, D. E. and Collins, W. T., Eds.,** *Silylated Surfaces,* Gordon and Breach, New York, 1980.

139. **Plueddemann, E. P.,** *Silane Coupling Agents,* Plenum Press, New York, 1982.

140. **Zhang, Y., Seitz, W. R, Grant, C. L., and Sundberg, D. C.,** A clear amine-containing PVC membrane for in situ optical detection of 2,4,6-trinitrotoluene, *Anal. Chim. Acta,* 217, 217, 1989.

141a. **Knobbe, E. T., Dunn, B., and Gold, M.,** Organic molecules entrapped in a silica host for use as biosensor probe materials, *Proc. SPIE,* 906, 39, 1989.

141b. **Kaufman, V. R., Avnir, D., Pines-Rojanski, D., and Huppert, D.,** Water consumption during early stages of the sol-gel tetramethyl-orthosilicate polymerization, *J. Non-Crystall Solids,* 99, 379, 1988, and references cited therein.

142. **Pusch, W. and Walch, A.,** Synthetic membranes: Synthesis, structure and applications, *Angew. Chem. Int. Ed. Engl.,* 21, 660, 1982, and references 138 to 140 cited therein.

143. **Zhujun, Z. and Seitz, W. R.,** A fluorescence sensor for quantifying pH in the range from 6. 5 to 8. 5, *Anal. Chim. Acta,* 160, 47, 1984.

144a. **Bangs, L. B.,** Uniform Latex Particles, Seragen Diagnostics Inc., Indianapolis, IN, 1985.

144b. **Jian, C. and Seitz, W. R.,** Membrane for *in-situ* optical detection of organic nitro compounds based on fluorescence quenching, *Anal. Chim. Acta,* 237, 265, 1990.

145. **Rogowin, Z. A. and Galbraich, A.,** *Chemical Treatment and Modification of Cellulose,* Thieme, Stuttgart, 1983.

146. **Bartl, H. and Falbe, J., Eds.,** *Houben Weyl's Methods of Organic Chemicstry. Vol. E 20. Macromolecular Materials,* Thieme, Stuttgart, 1987, 2046.

147. **Seitz, W. R.,** Chemical sensors based on immobilized indicators and fiber optics, *CRC Crit. Rev. Anal. Chem.,* 19, 135, 1988.

148. **Wegscheider, W. and Knapp, G.,** Preparation of chemically modified cellulose exchangers and their use for the preconcentration of trace elements, *CRC Crit. Rev. Anal. Chem.,* 11, 79, 1981.

149. **Carr, P. W. and Bowers, L. D.,** *Immobilized Enzymes in Analytical and Clinical Chemistry,* John Wiley & Sons, New York, 1980.

150. **Sharma, B. P., Bailey, L. F., and Messing, R. A.,** Immobilized biomaterials: techniques and applications, *Angew. Chem. Int. Ed. Engl.,* 21, 837, 1982.

151. **Woodward, J., Ed.,** *Immobilized Cells and Enzymes,* IRL Press, Oxford, 1985.

152. **Bartl, H. and Falbe, J., Eds.,** *Houben Weyl's Methods of Organic Chemistry. Vol. E 20, Macromolecular Materials,* Thieme, Stuttgart, 1987, 1185.

153. **Bartl, H. and Falbe, J., Eds.,** *Houben Weyl's Methods of Organic Chemistry. Vol. E 20, Macromolecular Materials,* Thieme, Stuttgart, 1987, 1367.

154. **Kawabata, Y., Kamichika,T., Imasaka, T., and Ishibashi, N.,** Fiber optic sensor for $CO_2$ with a pH indicator dispersed in a poly(ethylene glycol) membrane, *Anal. Chim. Acta,* 219, 223, 1989.

155. **Zhujun, Z., Zhang, Y., Wangbai, M., Russell, R., Shakhsher, Z. M., Grant, C. L., Seitz, W. R., and Sundberg, D. C.,** Poly(vinyl alcohol) as a substrate for indicator immobilization to fiber-optic chemical sensors, *Anal. Chem.,* 61, 202, 1989.

156. **Peppas, N. A., Ed.,** *Preparation, Methods and Structures of Hydrogels,* CRC Press, Boca Raton, FL, 1986.

157. **Peppas, N. A., Ed.,** *Hydrogels in Medicine and Pharmacy,* CRC Press, Boca Raton, FL, 1986.

158. **Ratner, B. D.,** Biomedical applications of hydrogels: a critical appraisal, in *Biocompatibility of Clinical Implant Materials, CRC Series in Biocompatibility,* Vol. 2, Wiliams D. F., Ed., CRC Press, Boca Raton, FL, 1981, 145.

159. **Arya, A., Krull, U. J., Thompson, M., and Wong, H. E.,** Langmuir-Blodgett deposition of lipid film on hydrogel as a basis for biosensor development, *Anal. Chim. Acta,* 173, 331, 1985.

160. **Leiner, M. J. P. , Weis, L., and Wolfbeis, O. S.,** ur. Pat. Appl. 354, 204, 1990.

161. **Kulp, T. J., Camins, I., and Angel, S. M.,** Enzyme-based fiber optic sensors, *Proc. SPIE,* 906, 134, 1988.

162. **Shaksher, Z. M. and Seitz, W. R.,** Fluorescence study of organic cation binding to hydrocarbon-bonded silica surface, *Anal. Chem.,* 61, 590, 1989.

163. **Bartl, H. and Falbe, J., Eds.,** *Houben Weyl's Methods of Organic Chemistry. Vol. E 20, Macromolecular Materials,* Thieme, Stuttgart, 1987, 1994.

164. **Egan, L. S., Winnik, M. A., and Croucher, M. D.,** Oxygen quenching studies of non-aqueous dispersions of poly(vinyl acetate) labeled with phenanthrene groups, *Langmuir,* 4, 438, 1988.

165. **Lin, J. N., Herron, J., Andrade, J. D., and Brizgys, M.,** Characterization of immobilized antibodies on silica surfaces, *IEEE Trans. Biomed. Sci.,* 35, 466, 1988, and references cited therein.

166. **Wolfbeis, O. S., Kroneis, H., and Offenbacher, H.,** U.S. Patent 4,568,518, 1986.

167. **Munkholm, C., Walt, D., Milanovich, F., and Klainer, S.,** Polymer modification of fiber optic chemical sensors as a method for enhancement of signals for pH measurement, *Anal. Chem.,* 58, 1427, 1986.

168. **Wolfbeis, O. S. and Marhold, S.,** Unpublished results, 1989.

169. **Klainer, S. M. and Harris, J. M.,** The use of fiber optic chemical sensors in medical applications: enzyme-based systems, *Proc. SPIE,* 906, 139, 1988.

170. **Luo, S. and Walt, D. R.,** Avidin-biotin coupling as a general method for preparing enzyme fiber-optic sensors, *Anal. Chem.,* 61, 1069, 1989.

171. **Hülshoff, A. and Lingeman, H.,** Fluorescence detection in chromatography, in *Modern Luminescence Spectroscopy. Methods and Applications,* Vol. 1, Schulman, S. G., Ed., John Wiley & Sons, New York, 1985, chap. 7.

172. **Hemmilä, I.,** Fluoroimmunoassays and immunofluorometric assays, *Clin. Chem.,* 31, 359, 1985.

173. **Wang, K., Feramisco, J. R., and Ash, J. F.,** Fluorescent localization of contractile proteins in tissue culture cells, *Methods Enzymol.,* 85, 514, 1982.

174a. **Haugland, R. P.,** Covalent fluorescent probes, in *Excited States of Biopolymers,* Steiner, R. F., Ed., Plenum Press, New York, 1983, chap. 2.

174b. **Imasaka, T. and Ishibashi, N.,** Diode lasers, *Anal. Chem.,* 62, 383A, 1990.

174c. **Savitskii, A. P., Papkovskii, D. B., Ponomarev, G. V., and Berezin, I. V.,** Phosphorescent immunoassay: Are metalloporphyrins an alternative to rare earth fluorescent labels, *Dokl. Acad. Nauk SSSR,* 304, 1005, 1989; engl. ed. p. 48 (Plenum Publ. Corp.).

175. **Horner, L., Vogt, M., and Rocker, M.,** A new method for detection of functional groups OH, $NH_2$ and SH by induction of fluorescence, *Phosphorus Sulfur,* 32, 91, 1987.

176. **Carlson, R. G., Srinivasachar, K., Givens, R. S., and Matuszewski,** New derivatizing agents for amino acids and peptides. 1. Facile synthesis of N-substituted 1-Cyanobenz[f]isoindoles and their spectroscopic properties, *J. Org. Chem.*, 51, 3978, 1986.

177. **Bodanszky, M. and Bodanszky, A., Eds.,** *The Practice of Peptide Synthesis*, Springer-Verlag, Berlin, 1984.

178. **Sheehan, J. C., Preston, J., and Cruikshank, P. A.,** A rapid synthesis of oligopeptide derivatives without isolation of intermediates, *J. Am. Chem. Soc.*, 87, 2492, 1965.

179. **Blakeslee, D. and Baines, M. G.,** Immunofluorescence using dichlorotriazinylaminofluorescein (DTAF). Preparation and Fractionation of Labelled IgG, *J. Immunol. Methods*, 13, 305, 1976.

180. **Imai, K. and Watanabe, Y.,** Fluorimetric determination of secondary amino acids by 7-fluoro-4-nitrobenzo-2-oxa-1,3-diazole, *Anal. Chim. Acta*, 130, 377, 1981.

181. **Angelides, K. J. and Nutter, T. J.,** Preparation and characterization of fluorescent scorpion toxins from leiurus quinquestriatus quinquestriatus as probes of the sodium channel of excitable cells, *J. Biol. Chem.*, 258, 11948, 1983.

182. **Hudson, E. N. and Weber, G.,** Synthesis and characterization of two fluorescent sulfhydryl reagents, *Biochemistry*, 12, 4154, 1973.

183. **Taylor, D. L. and Wang, Y. L.,** Fluorescently labeled molecules as probes of the structure and function of living cells, *Nature (London)*, 284, 405, 1980.

184. **Velury, S. and Howell, S.,** Measurement of plasma thiols after derivatization with monobromobimane, *J. Chromatogr.*, 424, 141, 1988.

185. **Scouten, W. H., Lubcher, R., and Baughman, W.,** N-Dansylaziridine: a new fluorescent modification for cysteine thiols, *Biochim. Biophys. Acta*, 336, 421, 1974.

186. **Imai, K., Toyo'oka, T., and Watanabe, Y.,** A novel fluorogenic reagent for thiols: ammonium 7-fluorobenzo-2-oxa-1,3-diazole-4-sulfonate, *Anal. Biochem.*, 128, 471, 1983.

187. **Toyo'oka, T. and Imai, K.,** Isolation and characterization of cysteine-containing regions of proteins using 4-(aminosulfonyl)-7-fluoro-2,1,3-benzoxadiazole and high-performance liquid chromatography, *Anal. Chem.*, 57, 1931, 1985.

188. **Teale, F. W. J. and Constable, D.,** Intramolecularly-quenched thiol-specific fluorescence probes, in *Fluorescent Probes*, Beddard, G. S. and West, M. A., Eds., Academic Press, London, 1981, 1.

189. **Horner, L. and Lindel, H.,** Potential-governed cleavage of S-S bonds and fluorescence derivatization studies on bovine insuline, *Liebigs Ann. Chem.*, 1985, 40.

190. **Horner, L. and Lindel, H.,** Gezielte und reversible Fluoreszenzmarkierung von SH-Lysozym mit Vinylsulfonen, *Liebigs Ann. Chem.*, 1985, 34.

191. **Pober, J. S., Iwanij, V., Reich, E., and Stryer, L.,** Transglutaminase-catalyzed insertion of a fluorescent probe into the protease sensitive region of rhodopsin, *Biochemistry*, 17, 2163, 1978.

192. **Barker, S. A., Monti, J. A., Christian, S. T., Benington, F., and Morin, R. D.,** 9-Diazomethylanthracene as a new fluorescence and ultraviolet label for the spectrometric detection of picomole quantities of fatty acids by high-pressure liquid chromatography, *Anal. Biochem.*, 107, 116, 1980.

193. **Takadate, A., Tahara, T., Fujino, H., and Goya, S.,** Synthesis and properties of 4-diazomethyl-7-methoxycoumarin as a new fluorescent labeling reagent for alcohols and carboxylic acids, *Chem. Pharm. Bull.*, 30, 4120, 1982.

194. **Dünges, W.,** 4-Bromomethyl-7-methoxycoumarin as a new fluorescence label for fatty acids, *Anal. Chem.*, 49, 442, 1977.

195. **Farinotti, R., Siard, P., Bourson, J., Kirkiacharian, S., Valeur, B., and Mahuzier, G.,** 4-Bromomethyl-6,7-dimethoxycoumarin as a fluorescent label for carboxylic acids in chromatographic detection, *J. Chromatogr.*, 269, 81, 1983.

196. **Korte, W. D.,** 9-(Chloromethyl)anthracene: a useful derivatizing reagent for enhanced ultraviolet and fluorescence detection of carboxylic acids with liquid chromatography, *J. Chromatogr.*, 243, 153, 1982.

197. **Wendelin, W., Gübitz, G., and Pracher, U.,** Derivatization of carboxylic acids, imides and alcohols with 1-chloromethylbenz[c,d]indol-2(1H)-one (CMBI), *J. Heterocycl. Chem.*, 24, 1381, 1987.

198. **Yamaguchi, M., Takehiro, O., Hara, S., Nakamura, M., and Ohkura, Y.,** 3-Bromomethyl-6,7-methylenedioxy-1-methyl-2(1H)-quinoxalinone as a highly sensitive fluorescence derivatization reagent for carboxylic acids in high-performance liquid chromatography, *Chem. Pharm. Bull.*, 36, 2263, 1988.

199. **Wintersteiger, R.,** Anthracene isocyanate as a new fluorescent label for compounds with an alcoholic group, *J. Liq. Chromatogr.*, 5, 897, 1982.

200. **Takadate, A., Irikura, M., Suehiro, T., Fujino, H., and Goya, S.,** New labeling reagents for alcohols in fluorescence high-performance liquid chromatography, *Chem. Pharm. Bull.*, 33, 1164, 1985.

201. **Goto, J., Goto, N., Shamsa, F., Saito, M., Komatsu, S., Suzaki, K., and Nambara, T.,** New sensitive derivatization of hydroxysteroids for high-performance liquid chromatography with fluorescence detection, *Anal. Chim. Acta*, 147, 397, 1983.

202. **Yamaguchi, M., Iwata, T., and Nakamura, M.,** 3,4-Dihydro-6,7-dimethoxy-4-methyl-3-oxo-quinoxaline-2-carbonyl azide as a highly sensitive fluorescence derivatization reagent for primary, secondary and teriary alcohols in high-performance liquid chromatography, *Anal. Chim. Acta*, 193, 209, 1987.

203. **Karlsson, K.-E., Wiesler, D., Alasandro, M., and Novotny, M.,** 7-[(chlorocarbonyl)methoxy]-4-methylcoumarin: a novel fluorescent reagent for the precolumn derivatization of hydroxy compounds in liquid chromatography, *Anal. Chem.*, 57, 229, 1985.

204. **Wintersteiger, R. and Juan, H.,** Prostaglandin determination with fluorescent reagents, *Prostaglandins Leukotrienes and Med.*, 14, 25, 1984.

205. **Iwata, T., Yamaguchi, M., Hara, S., and Nakamura, M.,** 3,4-Dihydro-6,7-dimethoxy-4-methyl-3-oxo-quinoxaline-2-carbonyl chloride as a highly sensitive fluorescence derivatization reagent for alcohols in high-performance liquid chromatography, *J. Chromatogr.*, 362, 209, 1986.

206. **Horner, L. and Flemming, H.-W.,** Site-specific fluorescence labeling of serine enzymes, *Liebigs Ann. Chem.*, 1985, 1.

207. **Berman, H. A. and Taylor, P.,** Fluorescent phosphonate label for serine hydrolases, pyrenebutyl methylphosphonofluoridate: reaction with acetylcholinesterase, *Biochemistry*, 17, 1704, 1978.

208. **Vaz, W. L. C. and Schoellmann, G.,** Specific fluorescent derivatives of macromolecules. Reaction of dansyl fluoride with serine proteinases, *Biochim. Biophys. Acta*, 439, 194, 1976.

209. **Burnett, T. J., Peebles, H. C., and Hageman, J. H.,** Synthesis of a fluorescent boronic acid which reversibly binds to cell walls and a diboronic acid which agglutinates erythrocytes, *Biochem. Biophys. Res. Commun.*, 96, 157, 1980.

210. **Prakash, C. and Vijay, I. K.,** A new fluorescent tag for labeling of saccharides, *Anal. Biochem.*, 128, 41, 1983.

211. **Anderson, J. M.,** Fluorescent hydrazides for the high-performance liquid chromatographic determination of biological carbonyls, *Anal. Biochem.*, 152, 146, 1986.

212. **Gübitz, G., Wintersteiger, R., and Frei, R. W.,** Fluorogenic labeling of carbonylcompounds with 7-hydrazino-4-nitrobenzo-2-oxa-1,3-diazole, *J. Liq. Chromatogr.*, 7, 839, 1984.

213. **Koenig, P., Reines, S. A., and Cantor, C. R.,** Pyrene derivatives as fluorescent probes of confirmation near the 3' termini of polyribonucleotides, *Biopolymers*, 16, 2231, 1977.

214. **Lee, J. A. and Fortes, P. A. G.,** Labeling of the glycoprotein subunit of (Na,K)-ATPase with fluorescent probes, *Biochemistry*, 24, 322, 1985.

215. **Chen, R. F. and Scott, C. H.,** Atlas of fluorescence spectra and lifetimes of dyes attached to protein, *Anal. Lett.*, 18, 393, 1985.

216. **Fauler G.,** New longwave-excitable coumarin-derived protein labels, *Appl. Fluorescence Technol.*, 1(6), 14, 1989.

217. **Titus, J. A., Haugland, R., Sharrow, S. O., and Segal, D. M.,** Texas Red, a hydrophilic, red-emitting fluorophore for use with fluorescein in dual parameter flow microfluorometric and fluorescence microscopic studies, *J. Immunol. Methods*, 50, 193, 1982.

218. **Hemmilä, I., Dakubu S., Mukkala V., Siitari H. and Lövgren S.,** Europium as a label in time-resolved immunofluorometric assays, *Anal. Biochem.*, 137, 335, 1984.

219. **Reichstein, E., Shami, Y., Ramjeesingh, M., and Diamandis, E. P.,** Laser-excited time-resolved solid-phase fluoroimmunoassays with the new europium chelate 4,7-bis(chlorosulfophenyl)-1,10-phenanthroline-2,9-dicarboxylic acid as label, *Anal. Chem.*, 60, 1069, 1978.

220. **Hijmans, W., Schuit, H. R. E., and Klein, F.,** An immunofluorescence procedure for the detection of intracellular immunoglobulins, *Clin. Exp. Immunol.*, 4, 457, 1969.

221. **Wells, A. F., Miller, C. E., and Nadel, M. K.,** Rapid fluorescein and protein assay method for fluorescent-antibody conjugates, *Appl. Microbiol.*, 14, 271, 1966.

222. **Rinderknecht, H.,** A new technique for the fluorescent labeling of proteins, *Experientia*, 16, 430, 1960.

223. **Ekeke, G. I., Exley, D., and Abuknesha, R.,** Immunofluorimetric assay of oestradiol-17β, *J. Steroid Biochem.*, 11, 1597, 1979.

224. **Burd, J. F., Wong, R. C., Feeney, J. E., Carrico, R. J., and Boguslaski, R. C.,** Homogeneous reactant-labeled fluorescent immunoassay for therapeutic drugs exemplified by gentamicin determination in human serum, *Clin. Chem.*, 23, 1402, 1977.

225. **Khanna, P. L. and Ullman, E. F.,** 4',5'-Dimethoxy-6-carboxyfluorescein: a novel dipole-dipole coupled fluorescence energy transfer acceptor useful for fluorescence immunoassays, *Anal. Biochem.*, 108, 156, 1980.

226. **Khalfan, H., Abuknesha, R., and Robinson, D.,** Fluorigenic method for the assay of proteinase activity with the use of 4-methylumbelliferyl-casein, *Biochem. J.*, 209, 265, 1983.

Chapter 8

# FIBER OPTIC pH SENSORS

### Marc J. P. Leiner and Otto S. Wolfbeis

## TABLE OF CONTENTS

## I. INTRODUCTION

Rapid and continuous monitoring of pH is required in practically all kinds of sciences, including chemical, biomedical, and environmental. The familiar glass pH electrode is the most widely used tool because it is reliable, offers good precision, and measurements can be done rapidly. In medicine, however, where on-line monitoring of blood pH is highly desirable during complex surgical procedures and hospitalization under intensive care, the use of miniaturized glass electrodes mounted on flexible catheters proved not to be practical because of their size, rigid design, and the possible electrical hazard. Moreover, local electrical potentials are known to adversely affect the performance of pH electrodes.

The electrical interference problem of electrodes is not limited to their use in medicine. In any environment where high sensitivity is critical, electrical noise generated by extraneous fields will cause trouble. In addition, pH electrodes do not work in solutions through which an electrical current is flowing, for instance in electrolysis cells and batteries. Other problems inherent to the use of electrodes include the susceptibility of wire leads and couplings to deterioration under corrosive conditions, or conditions of altering temperatures.

Many of these problems can be overcome by optically measuring pH. The technique dates back to the days when litmus was used to indicate the acidity of a solution. Eventually, pH indicator strip tests became popular, the first of which suffered from dye bleeding. This was overcome by covalently immobilizing the dye on a rigid support. However, only during the past decade has optical sensing of pH made some important advances and become precise enough to be of practical utility. Optical sensing techniques are capable of measuring extreme values of pH with great precision. Fiber optic pH sensors have the advantage of very small size and flexibility and offer an attractive possibility for invasive analysis of plant and animal tissues.[1,2] They can be produced easily at very low cost without sacrificing accuracy and are sufficiently simple in design to be considered as disposable. A fascinating feature is the fact that the optical signal can be carried over large distances. This is particularly advantageous when the sample to be measured is unaccessible, for instance in case of remotely[1,3] and invasive sensing.[4-6]

Because of the quite different operating principle as compared to pH electrodes, pH optrodes offer quite new possibilities and at the same time are subject to some limitations that are not encountered with electrodes. The fundamental difference between optical techniques and potentiometric measurements of pH is that the optical technique measures the *concentration* of a dye species (that is related to pH) while pH is defined in terms of activity, which is what potentiometric measurements are based on. Janata[7] has clearly shown the compromises that have to be made in case of optical pH measurements. Difficulties preventing practical applications of pH measurements with fiber optic pH sensors have also been discussed by Seitz,[1] Ratzlaff et al.,[8] and Edmonds et al.[9]

pH optrodes are based upon pH-dependent changes of the optical properties (absorbance, reflectance, fluorescence, refractive index, and the like) of a thin reagent layer, attached to the tip or surface of an optical light guide through which these changes are detected. The reagent layer usually contains a dye which reversibly reacts with the protons of the sample. In the most popular version, the pH-dependent absorption[5,10-12] or fluorescence[4,13-15] is monitored and related to pH.

Because the indicator dye and the sample are in different phases, there is necessarily a mass transfer step required before constant response is reached. This leads to relatively slow response times. Photobleaching and leaching, interferences by ambient light, nonideal optoelectronic equipment, the lack of violet and ultraviolet LEDs and inexpensive blue lasers are further problems encountered in development of fiber optic pH sensors.

A number of optical sensors for pH have been described, each differing in the chemical transducer and the optical principle employed.[6,16-18] Several patents covering both the chemical and instrumental aspect have been issued.[19-30] Most investigations have been made on physiological pHs. In the past, however, the working range has been extended to other pHs as needed in certain industrial applications because it has been recognized that pH optrodes have the potential of becoming useful in special fields of application where potentiometric methods fail or because they can offer considerable economic and sampling advantages.

In these fields, the intrinsic characteristics of small sensor size, electrical isolation, chemical inertness, and corrosion resistance may offer definite advantages over electrochemical sensors. Since dissociation equilibria are strongly affected by the composition of the surrounding medium, the concentration of diluted substances, and temperature, fiber optic pH sensors are likely to perform best under well defined sample conditions. Typical examples are seawater pH, biofermenters, agrifood industries, and other situations where the sample composition is almost invariable.

The measurement of pH is most advanced in the area of biochemical analysis because of the narrow pH range covered. Agrifood (milk), environmental (seawater, geology) and industrial (process control) applications are merely at their beginning. Finally, pH optrodes act as transducers for $CO_2$ and $NH_3$ optrodes, and in certain biosensors in which an enzyme produces protons.

# II. FUNDAMENTALS AND GENERAL ASPECTS

## A. DEFINITION OF pH

According to the definition given by Sörensen and as discussed in the excellent paper by Stavart,[31] pH is the negative logarithm of the hydrogen ion *concentration* [H⁺]:

$$pH = -\log\left[H^+\right] \tag{1}$$

Today, the actual acidity of a solution is related to the *activity* of the hydrogen ions $a_{H^+}$:

$$pH = -\log a_{H^+} \tag{2}$$

Activity and concentration are related by the activity coefficient $f_{H^+}$:

$$a_{H^+} = f_{H^+} \cdot \left[H^+\right] \tag{3}$$

The activity coefficient depends on the ionic strength of the solution and approaches unity for infinitely dilute solutions. In very dilute solution, pH can be related directly to the concentration of the hydrogen ions, i.e., Equation 1 is valid.

The activity coefficient of an ion also varies with temperature and ionic strength according to the limiting Debye-Hückel law (Equation 14). Lewis and Randell have defined ionic strength I by:

$$I = 0.5 \sum_{i=1}^{n} c_i \cdot z_i^2 \tag{4}$$

where $c_i$ is the concentration of the *i*th ion, z is the charge of this ion, and n is the total number of different ions.

## B. OPTICAL SENSING PRINCIPLES

Fluorescence- and absorptivity-based pH sensors rely upon indicator dyes which are weak electrolytes in that they exist in both an acid and base form in the pH range of interest. Acid and conjugate base have different absorption and/or fluorescence properties. According to the Brönsted-Lowry acid/base definition, weak acids and bases are part of an equilibrium according to:

$$HA^z \rightleftharpoons H^+ + A^{z-1} \tag{5}$$

where $A^{z-1}$ denotes the acid form of the dye, the base form, z is the charge number, and H⁺ the hydrogen ion. A mixture of a weak acid and a conjugate base is a so-called buffer. This buffer can be characterized by means of the mass action law:

$$K_c = \frac{\left[A^-\right] \cdot \left[H^+\right]}{[HA]} \tag{6}$$

where $K_c$ is the concentration constant of the acid-base dye, and [HA] and [A⁻] are the *concentrations* of the acid and base form of the dye, respectively. The concentration constant

$K_c$ is related to the thermodynamic constant $K_c$ of a weak acid by the individual activity coefficients $f_x$ of the reaction partners:

$$K_a = K_c \cdot \frac{f_{A^-} \cdot f_{H^+}}{f_{HA}} \tag{7}$$

When expressed in logarithmic form, the activity-based Henderson-Hasselbalch equation is obtained as

$$pH = pK_a + \log \frac{[A^-]}{[HA]} + \log \frac{f_{A^-}}{f_{HA}} \tag{8}$$

where $pK_a$ is $-\log K_a$.

Electrochemical pH determinations are based on a measurement of the electromotive force of a cell having a reversible electrode whose potential is linearly dependent on activity of hydrogen ions, and hence on pH. Optical measurements are a linear function of the solute concentration ($[A^-]$ or $[HA]$), but not the activity. As the pH varies, the relative fractions of the acid and base forms are changed and changes can be detected by means of absorption or fluorescence measurements.

Therefore, many optrode papers use the Henderson-Hasselbalch equation based on the concentration constant $K'_a$ defined in Equation 6 but ignore the activities:

$$pH = pK_c + \log \frac{[A^-]}{[HA]} \tag{9}$$

In Equation 9, pH is defined according to Equation 1 as $-\log [H^+]$. In order to get pH in terms of activity (Equation 2), a mixed constant $K_{ac}$ of the acid-based indicator is defined:

$$K_{ac} = \frac{a_{H^+} \cdot [A^-]}{[HA]} \tag{10}$$

and Equation 9 takes the form:

$$pH = pK_{ac} + \log \frac{[A^-]}{[HA]} \tag{11}$$

where

$$pK_{ac} = pK_c - \log f_{H^+} \tag{12}$$

The mixed pK is only valid for a given ionic strength and temperature (e.g., the ionic strength and temperature of the calibrating buffer solutions). The ionic strength can be kept constant by introducing a background electrolyte (e.g., sodium chloride) into the calibrating buffer solutions. When measuring pH in sample solutions of different ionic strength, the pH error caused by ionic strength effects can be expressed by Equation 12:

$$\Delta pH = \log \frac{f^c_{A^-}}{f^c_{HA}} - \log \frac{f^s_{A^-}}{f^s_{HA}} \tag{13}$$

where subscript c denotes the calibrating solution and s the sample solution.

As Janata[7] has pointed out in his remarkable discussion of the thermodynamic aspects in optical determination of pH, most optrode papers ignore the effect of ionic strength on the dissociation equilibrium (third term in Equation 8) as well solute-solvent and solute-solute interactions. This omission may be tolerated for very dilute solutions, but not in most "real" situations. Particularly when indicators are immobilized on hydrophobic matrices, they have been found to exhibit acid-base properties that are quite different from those in aqueous solution.[17,32,33]

## C. EFFECTS OF IONIC STRENGTH AND TEMPERATURE

The effect of ionic strength on the dissociation equilibrium can be derived from the classical Debye-Hückel theory. This approach works well for univalent electrolytes up to ionic strength $I < 0.1\ M$. For higher ionic strengths and/or multivalent ions the specific interaction theory must be applied. For thermodynamic reasons, individual activity coefficients of anions and cations cannot be measured separately. The activity coefficient $f_x$ of an ion can be calculated with good approximation by the following equation:

$$\log f_x = -\frac{A \cdot z_i^2 \cdot I^{1/2}}{1 + B \cdot d \cdot I^{1/2}} + C \cdot I \tag{14}$$

where A and B are constants that vary with the temperature of the solvent. C is an empirical parameter, and d is the mean ion size parameter ($3$ to $4 \cdot 10^{-8}$ cm). At 25°C, A is 0.509, and B is $0.329 \cdot 10^8$ cm$^{-1}$(Mol/l)$^{-1/2}$ so that the product B·d is approximately one.

The variation of $pK_a$ with the thermodynamic temperature T can be expressed[34,35] by equations of the form:

$$pK_a = \frac{A}{T} + B + C \cdot \log T \tag{15}$$

where A, B and C are constants which may be obtained from the standard changes of enthalpy ($\Delta H°$), entropy ($\Delta S°$) and heat capacity ($\Delta C_p°$) for the dissociation process.

Edmonds et al.[9] have shown that an increase in ionic strength from 0.01 to 3.00 $M$ can shift the $pK_c$ value in Equation 6 by as much as 1.23 units (Table 1). Therefore, the ionic strength of the sample has to be kept constant during pH measurement or calibration.

The general change in $pK_c$ caused by a temperature increase of 5°C (in going from 25 to 30°C) is about +0.2 (Table 2).

In a study[35] on the behavior of phenol red in seawater at temperatures between 5 and 30°C it was found that, within a salinity S of 33 ‰ and 37 ‰, the phenol red equilibrium constant $K_c$ could be described by the expression $K_c = (4834.00/T) - 84.831 + 30.7580 \cdot \log T + 0.004 \cdot (35 ‰ - S)$. In the range of $20.2 - 35.2$ ‰ and at 25°C Monici et al.[36] found a decrease of the pK value of phenol red of 0.044 for a 1 ‰ salinity increase (Figure 1)

In optical pH measurements, the effect of ionic strength on the dissociation constant results in a shift of the pH-dependent dissociation curves. Therefore, ionic strength effects are indistinguishable from signal changes caused by pH. The pH error caused by ionic strength depends on the type of indicator used and on the concentrations of all ionic species in solution. The resulting errors are particularly significant when the composition of the sample solution differs significantly from that of the calibration solution or when it changes during measurements. In electrochemical measurements, pH errors caused by the liquid junction are distinctly smaller than those caused by the effect of ionic strength on the indicators.[7]

Edmonds et al.[9] have pointed out that fiber optic determination of pH in an unknown solution with an accuracy of ±0.1 is unrealistic. To decrease the error caused by ionic strength effects, an appropriate buffer for each type of sample has to be used for calibrating a pH optrode. Although it seems possible to minimize salt effects[37,38] or effects from the sample matrix,[38] pH optrodes therefore cannot be applied to unknown sample solutions to give a reliable pH indication.

**TABLE 1**

**Results Obtained for $pK_c$ with Changing Ionic Strength at 25°C[9]**

| | Strength ($M$) | | | | | | |
|---|---|---|---|---|---|---|---|
| | 0.01 | 0.10 | 0.50 | 1.00 | 1.50 | 2.00 | 3.00 |
| Bromophenol blue | 3.94 | 4.41 | 4.37 | 4.40 | 4.42 | 4.45 | 4.51 |
| Bromocresol green | 4.57 | 4.65 | 4.65 | 4.68 | 4.71 | 4.75 | 4.82 |
| Bromocresol purple | 6.26 | 5.67 | 5.67 | 5.71 | 5.74 | 5.75 | 5.76 |
| Methyl red | 4.90 | 5.04 | 5.08 | 5.13 | 5.31 | 5.07 | 5.01 |
| Phenol red | 7.46 | 6.96 | 6.52 | 6.37 | 6.39 | 6.55 | 6.44 |

**TABLE 2**

**Effect of Temperature on the $pK_c$ of Some Absorbance
Based pH Indicators[9]**

| | $pK_c$ ($\pm$s.d.) | |
|---|---|---|
| | 25°C | 30°C |
| Cresol red | 7.95 ± 0.05 | 8.13 ± 0.04 |
| Phenol red | 7.61 ± 0.05 | 7.79 ± 0.03 |
| Bromocresol green | 4.57 ± 0.03 | 4.71 ± 0.03 |
| Bromothymol blue | 6.82 ± 0.02 | 7.04 ± 0.01 |

However, by making use of two pH sensors, it is possible at least to compensate for ionic strength effects on pH and also to simultaneously measure ionic strength.[37,38] In the method proposed by Opitz and Lübbers,[37] a double pH fluorescence indicator system is used with different ionic strength effects on the respective pK values. Solutions of two differently charged indicators, 8-hydroxypyrene-1,3,6-trisulfonate (charge –4) and 4-methylumbellifer-one (charge –1), have been used. The dependence on the ionic strength of both indicators is sufficiently different to yield two sets of equations to measure both ionic strength and pH. However, both relationships are highly nonlinear, which makes linear regression analysis almost impossible.[7] Using these indicators, this method is suitable for the ionic strength range between 0.1 and 1 $M$ and a pH range between 6.5 and 7.5. The study has been confined to experiments in solution and the method is restricted to a pair of indicators with overlapping pK values and well separated excitation or emission spectra.

Wolfbeis and Offenbacher[31,38] described a method in which a sensor combination is used that measures both pH and ionic strength via two pH determinations. This was achieved by immobilizing 7-hydroxycoumarin-3-carboxylic acid covalently onto porous glass supports with different surface chemistries (Figure 2). In the first sensor, the indicator is embedded in an uncharged microenvironment. Therefore, it is highly sensitive toward changes in ionic strength. In the second sensor, the indicator is embedded in a highly charged environment, and therefore is less sensitive toward changes in ionic strength. In the range I = 100 to 200 m$M$, the second sensor is practically independent of ionic strength, and therefore indicates the true pH of the sample. The difference of the fluorescence signals of the two sensors can be related to ionic strength.

## D. DYE IMMOBILIZATION

Traditionally, pH indicators have been applied by dissolving them in the sample to be tested or by impregnating suitable paper strips. The problem with early pH papers was the bleeding

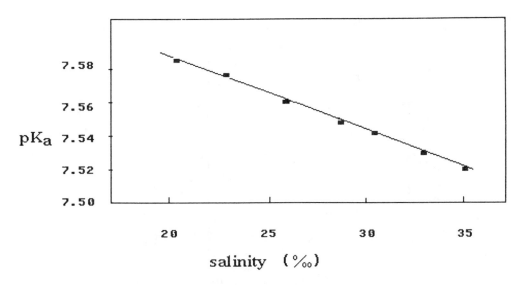

FIGURE 1.    Relationship between pK and salinity for phenol red at 22°C. (Redrawn from Monici, M., Boniforti, R., Buzzigoli, G., DeRossi, D., and Nannini, A., *Proc. SPIE*, 798, 294, 1987. With permission.)

of indicator dye from the pH paper into the sample solution, which later was overcome by covalently immobilizing the dye on the support. There are, however, a few reports on the use of non-immobilized pH-sensitive dyes. Thus, Baldini et al.,[39] tested a liquid dye solution attached to the end of a fiber, thereby avoiding the problem of immobilizing the dye on a polymeric support. The problem is to find a suitable membrane which must be highly selective for hydrogen ions and prevents the dye from leaving the dye cell. A 70-μm cellulose acetate membrane of asymmetric structure with progressive porosity was found to be the best material with respect to proton permeability and dye-retaining capability. The response time was as long as 15 min. Therefore, the use of an acid-base indicator solution at the tip of the optical fiber is not practical.

Recently, Luo and Walt[40] proposed a pH sensor design based on a delivery system capable of delivering fresh reagents for prolonged time periods with controlled release polymers. The feasibility of this technique was proven with a pH sensor based on the delivery of hydroxypy-renetrisulfonic acid (emission 520 nm) and sulforhodamine 640 (emission 610 nm) from an ethylene-vinyl acetate polymer. The ratio of fluorescence emission at 530 and 610 nm, resulting from excitation at 488 nm, was used to quantify pH from 5.5 to 8.0 in phosphate buffer solutions with a precision of ±0.1 pH units. The response time of this sensor is a function of its design, the conditions of the solution being measured, and lies between 0.5 and 3 h. Others[41] introduced HPTS into a low density network structure of a silica gel film. The fluorescence indicator is entrapped by the "cage-like" network while the transport of atoms and small molecules is supported. The results obtained seem to hold promise for biosensor probe application.

All other techniques reported so far rely on dyes immobilized on a polymeric support and attached on the fiber tip. This seems to be a more promising approach for the development of fiber optic pH sensors. Numerous methods of indicator immobilization are known. In the 1970s, the Merck Company marketed nonbleeding pH papers with indicators covalently bound to cellulose. In 1975, Harper[42] linked a series of colorimetric pH indicators to highly porous glass using silane coupling agents. In the past decade, fabrication schemes with indicators immobilized on solid supports such as a variety of hydrophobic polymeric materials,[5,10,15] sintered glass,[14,16,43] and various kinds of cellulose[4,16,44-46] have been proposed. Some of these techniques are described in Chapter 7. The solid support is held in place at the end of the optical light guide by some form of porous membrane. Gel layers can be directly

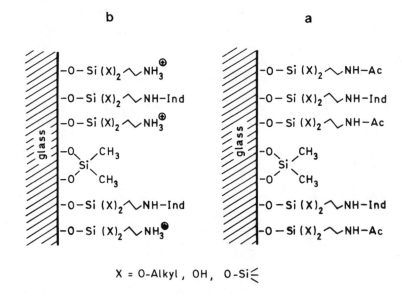

X = O-Alkyl, OH, O-Si$\leqslant$

FIGURE 2.  Scheme of surface chemistry of (a) a pH sensor highly sensitive toward changes in ionic strength and (b) a pH sensor less sensitive toward changes in ionic strength. (From Wolfbeis, O. S. and Offenbacher, H., *Sens. Actuators*, 9, 85, 1986. With permission.)

attached on the fiber by mechanical binding cellulose onto the optical fiber[4] or directly immobilizing acrylamide on surface modified glass fibers.[13]

Basically, there are three ways of dye immobilization, namely physical entrapment, electrostatic binding, and covalent linking via chemical bonds. Sometimes chemical absorption is also used to "immobilize" a dye, but the retainment of a dye is mostly based on a strong hydrophobic interaction between dye and polymer, and a Nernst distribution which favors the presence of the dye in the polymer rather than in the aqueous sample. However, adsorption of the indicator on a hydrophobic polymer can affect the acid-base properties of the dye.[33] The occurrence and the extent of washout and hysteresis are factors that must be accurately examined before a final selection of the dye and the polymeric support is made. Covalently immobilized dyes are not washed out at all, thereby providing a long-term stability of the sensor layer.[14]

When the indicator is confined within a surface layer, buffer capacities of dissolved ionic species will affect the kinetics of pH response.[7] Therefore, the thickness of a surface layer should be very small. In practice, the optimum thickness of the surface layer will be a compromise between signal intensity and signal response time. Proteins, present in biomedical and biological samples, affect the dissociation constant of the acid-base dyes. In order to avoid penetration of proteins inside the sensing layer, the latter may be isolated from the sample by means of a cellulose dialysis membrane. Boisde and co-workers[47] have compared the various methods of immobilization.

## 1. Covalent Linking

Controlled porous glass (CPG) is well suited for covalent immobilization of dyes. Since it is a powder, it requires special techniques for sensor fabrication. Planar sensing areas are obtained by gluing the dyed particles onto optically transparent solid supports.[43] The layer may also be produced by sintering or melting the undyed CPG onto the support[23] which gives a high specific surface onto which the indicators may be bonded. Another approach is to glue a single bead onto the end of a fiber.[18] By heating glass to its glass transition temperature, CPG may be fixed thermally on a plane glass support or on an optical fiber.[14,23,43] A further method of preparing defined CPG layers of 5 to 20 µm thickness is to treat borosilicate glass (with a $B_2O_3$ content of less than 13% and a metal oxide content of 3.5 ± 1.0%) with alkali to give a highly porous material. It is most suitable for dye immobilization but has a poor stability towards alkaline buffers. This can be improved, however, by thermal treatment and silylation.[47]

Thin films of cellulose, in particular the so-called cuprophane foils (as used in blood dialysis) have been used for covalent immobilization of HPTS[24,44] and fluorescein.[16] While cellulose itself has no functional group for direct and stable binding of a reactive dye, it can easily be modified to give, for instance, aminocellulose onto which a variety of pH probes may be bound. In another approach, 3-µm films of cellulose acetate were spread onto planar supports using the spin-coating technique.[45] After hydrolysis of the acetate groups with dilute base, a highly porous cellulose film is obtained onto which direct dyes such as congo red may be bound.

The response time of porous cellulose-based pH sensors is usually very rapid, being in the order of several seconds.[45,48] Cuprophane pH sensors show slower response (in the order of 1 min) and also some hysteresis[33,48] which is particularly expressed when changing from strongly acidic to neutral solution. It is negligible at near neutral pHs.

Very thin gel layers have been fabricated directly on fibers using poly(acrylamide).[13] When surface-activated single fibers were immersed into an aqueous solution containing acrylamide, N,N-methylene-bis(acrylamide), and acryloyl-fluorescein, a very thin polymer/dye layer is obtained after initiation of the polymerization.

## 2. Electrostatic Immobilization

Electrostatic immobilization[15,30] is possible with indicators that are charged in both the acid and conjugate base form. They can be bound onto supports with ion exchange properties. In case of surfaces with a high concentration of ionizable groups (such as ion-exchanging membranes), the surface pH is different from the bulk value due to electrostatic repulsion. Because both fixed and freely diffusing charges contribute to the potential profile, the surface pH value may be affected by any ionic species.[7] If, for instance, HPTS is electrostatically immobilized on quaternized aminoethyl cellulose, its pK is shifted by −1.2 units[48] (from 7.4 to 6.2).

## 3. Chemical Absorption

Narayanaswamy and Sevilla[32] investigated the acid-base equilibria of three indicators on a nonpolar surface. Bromphenol blue, bromthymol blue, and 2,6-dichlorophenol-indophenol were adsorbed on Amberlite XAD-2 resin (a nonionic styrene/divinylbenzene copolymer) as described by Kirkbright et al.[10] In the immobilized phase, the reflectance spectra of the acid form did not level off with increasing pH, as expected. This behavior, different from that in solution, is explained in terms of additional equilibria:

$$HA_{ads} \overset{K_1}{\rightleftharpoons} H^+_{ads} + A^-_{ads}$$

$$\| \quad\quad\quad\quad\quad \|$$
$$\| K_2 \quad\quad\quad\quad \| K_3$$
$$\| \quad\quad\quad\quad\quad \|$$

$$HA_{sol} \overset{K_4}{\rightleftharpoons} H^+_{sol} + A^-_{sol}$$

where $K_1$ and $K_4$ are the dissociation constants of the acid-base dye in the adsorbed and solution phases, respectively, and $K_2$ and $K_3$ are constants governing the adsorption/desorption equilibria. As the pH increases, the concentration of $A^-_{ads}$ on the polymeric surface increases. The charged anionic species ($A^-$) is less strongly adsorbed on the nonpolar surface than the noncharged acidic species (HA), and tends to pass into the solution phase. This causes some of the noncharged acidic species on the surface to dissociate so that the concentration of $A^-$ is almost unchanged, while that of HA decreases.

In a similar study, Bacci et al.[33] pointed out that a desorption of the alkaline species from the hydrophobic polymer should result in an extensive leaching from the resin, particularly at high pH. This indeed has been observed for phenol red, while bleeding for bromthymol blue is negligible. Bleeding could not be related to the structure of the dye. The observed effect was traced back to the absorption mechanism on the polymeric support by assuming that the polymer beads are coated with two or more layers of the dye, and that the inner layers are essentially composed of undissociated molecules which constitute a dye reservoir. The ionic species predominate in the outer layers exposed to the solvent. Therefore, a constant concentration of the undissociated species is present at the surface and a concentration decrease of the undissociated molecules is observed at the surface only when this reservoir is empty. For alizarin, chlorophenol red, and bromthymol blue a hysteresis during a pH cycle has been observed, whereas no hysteresis was found for phenol red.[33]

## E. RESPONSE TIME

Most specifications given in the literature as to response time are incomplete or not reproducible. A 100% response time is defined as the time required for the total signal change ($\Delta$ S) to occur. Because of the mostly exponential response curve, the exact determination of the 100% response time is somewhat difficult. Therefore, most authors prefer to specify response times in terms of 90 or 63% signal change. A 63% signal change has occurred when the signal has dropped to $-\Delta$ S/e or increased to $+\Delta$ S/e. Such a specification implies, however, a single-exponential signal change which, unfortunately, is seldom the case.

pH optrodes with thin gel layers have very fast response (in the order of seconds) when changing from neutral pH to both strongly alkaline and acidic solutions. Much longer response times (in the order of 0.5 to 3 min) are observed when changing from alkaline or acidic solutions to near neutral pH values. Within the physiological pH, the response also depends on the buffer capacity (strong buffers giving faster response).

For these reasons, a useful specification of the response time of a sensor can be given only by exactly specifying the solutions which have been used in the experiment, and the pH range over which the change was observed. We suggest to determine response times by specifying the time required for the signal change to occur from pH = pK to pH = pK + 1 and pH = pK − 1, with exactly specified buffers and temperatures.

An important factor contributing to response time is the rate of diffusion of solvent and ions across the membrane. The response characteristics obtained are a complex nonlinear function of several parameters, such as the concentration of ions in solution, rate and efficiency of diffusion across the sensing layers, and the difference between the initial and the final pH value of the test solution.[17]

Sensors with the indicator being confined within a surface layer have their own buffer capacities and require a mass transfer step before a constant signal is obtained. This limits response time. On the other hand, sensors with smaller surface layers[13,49] attached directly to the distal end of the fiber or indicators immobilized as a monolayer[50] have a fast response, but need more intense radiation sources such as lasers and, in turn, are subject to a greater extend to photodecomposition.

A rapidly responding sensing membrane was obtained[45] by immobilizing a direct azo dye, congo red, onto a 2.9-μm cellulose acetate film that previously had been subject to an exhaustive base hydrolysis. The porous structure of hydrolyzed cellulose acetate minimizes barriers to mass transport, resulting in a rapid response time and a large dynamic range.

## F. SENSOR STABILITY

Photobleaching is likely to be a problem for long-term stability. To some extent, it can be compensated for by multiple-wavelength detection[5,15] or opening a shutter for short time periods during measurements.[49] Most indicators, however, are subject to different rates of photobleaching. In case of most phenol-type indicators, the ratio of emission intensities can be measured under excitation of the phenol and phenolate form. Another approach is to excite at an isosbestic point in the fluorescence excitation spectrum to measure the total amount of indicator present, independent of its form.[51] The ratio method can account for photobleaching, leaching, and light source fluctuations which otherwise cause drift problems. However, multiple wavelength measurements require a slightly more complex optical system and more sophisticated data processing. Another possibility consists in the measurement of pH via fluorescence lifetime which is much less subject to leaching and bleaching because lifetime is independent of fluorophore concentration.[52]

The optical signal must be resolved from interferences by ambient light and sample fluorescence. Therefore, the optrode must be operated either in a dark environment, or the optical signal must be encoded (e.g., by modulation). This, however, does not compensate for interferences by sample fluorescence (e.g., from bilirubin which has the same fluorescence maximum as fluorescein). To overcome this problem, sample phase and pH-sensitive layer may be optically isolated by means of a nontransparent layer;[47] see also Chapter 3. The latter, of course, can cause additional buffer capacity and may substantially prolong the response time.

One possibility to optically isolate sample from sensor is to cover the optrode with a cellophane membrane dyed with azo dyes (these are nonfluorescent dyes) or with finely dispersed silver metal.[69] Such a membrane also protects the sensor from proteins. In comparative studies with electrodes it has been demonstrated that, by using these covers, no measurable Donnan effect is observed and response time are in the same order as without cover.

## III. pH OPTRODES

### A. ABSORBANCE- AND REFLECTIVITY-BASED SENSORS

In case of absorbance-based pH sensors, the Lambert-Beer law can be applied that relates absorption with the concentration ([D]) of the dye species:

$$A = \log \frac{I_o}{I} = \epsilon \cdot [D] \cdot 1 \tag{16}$$

where $I_o$ and $I$ denote the transmitted light intensity in the abscence and presence of the dye at the absorption wavelength, 1 is the effective path length, and $\epsilon$ is the molar absorption coefficient at the analytical wavelength.

In case of reflectance-based pH sensors and only absorption from the alkaline species occurring at the analytical wavelength, the absorbance can be described by the following Equation 17:

$$A = \log \frac{k \cdot I_0^{ref}}{I} = \frac{A_{max}}{10^{(-\Delta+1)}} \tag{17}$$

where $k = I_o/I_o^{ref}$, $\Delta = pH - pK$, and $A_{max} = T \cdot \epsilon \cdot l$. A is the absorbance and $A_{max}$ is the absorbance of the completely dissociated dye. I is the transmitted light intensity at the analytical wavelength, and $I_o^{ref}$ is the transmitted reference light intensity. $I_o^{ref}$ can be measured at any

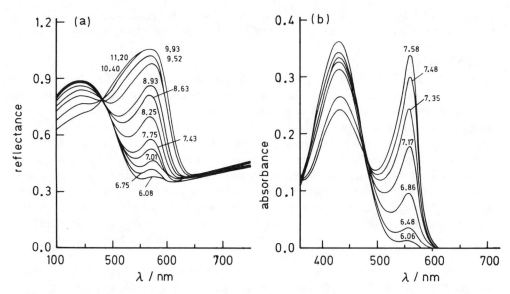

FIGURE 3. Reflectance spectra of phenol red immobilized on XAD-2 (a) and absorption spectra of phenol red in aqueous solution ($1.5 \times 10^{-5}$ $M$) (b) at different pH values. (Redrawn from Bacci, M., Baldini, F., and Scheggi, A. M., *Anal. Chim. Acta*, 207, 343, 1988. With permission.)

wavelength, where the intensity of multiple reflected light is independent of pH. Typically, it is measured at the isosbestic point or at a wavelength at which neither form absorbs. The reference measurement is frequently needed to compensate for optical and instrument variations. $A_{max}$, k, and pK are constants of the sensor.

The first fiber optical pH sensor ever described in detail was specifically developed for the physiological pH range by Peterson et al.[5,56] It is based on monitoring the color of a pH-sensitive dye indicator (phenol red), contained within a proton-permeable implantable envelope. Changes in light absorption of phenol red (Figure 3) with pH are measured by illuminating the dye through a single-strand optic fiber which abuts it, and sensing the back-scattered light through an adjacent optic fiber strand which leads to a remote light detector.

The sensor (Figure 4) was fabricated by inserting a pair of 0.15-mm diameter plastic optical fibers through 3 to 5 cm of a cellulosis dialysis tubing of 300 μm inside diameter and 20 μm wall thickness. Over a 4-mm space, the tubing has been packed with 5 to 10 μm diameter polyacrylamide microspheres containing covalently bound phenol red, and 1 μm diameter polystyrene microspheres for light scattering. UV curing optical adhesive was used to seal off the distal end of the tubing and to seal off and bind the fibers to the proximal end of the tubing.

In solution, phenol red behaves as a weak acid of pK 7.92 ± 0.02 at zero ionic strength[57] and of pK 7.78 ± 0.02 at 0.25 $M$ ionic strength[58] and exists in two tautomeric forms, each having a different absorption spectrum. The dye, when copolymerized with acrylamide, was surprisingly found to be immobilized without previously introducing a chemical link.[5] The polystyrene microspheres provide a very effective light scattering. The dye is illuminated by white light transmitted by one optical fiber. Green light (588 nm) is adsorbed by the base form of the dye to a degree dependent upon pH and by multiple scattering is returned to the other optical fiber for transmission to the detector. Red light (600 nm) is not adsorbed in relation to pH and gives an optical reference signal.

A detailed description of the measuring instrument with respect to mechanical, optical and electronic subsystems has been given.[6] The instrument (Figure 5) is equipped with a 150 W

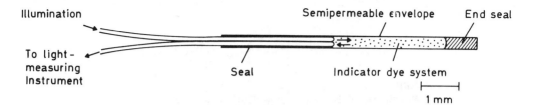

FIGURE 4. Concept of the chemical fiber optic pH sensor of Peterson et al. (Redrawn from Peterson, J. I. and Vurek, G. G., *Science*, 224, 123, 1984. With permission.)

tungsten halogen projector lamp as a light source and a photodiode/operational amplifier combination with a 1 cm$^2$ light sensitive area as light detector. A reference measurement was needed to compensate for optical and instrument variations. A rotating filter wheel, positioned between detector and plastic fiber, allows the measurement of dark, red and green signals. The dark signal is subtracted from the red and green signals to correct for drifts in the photosensor. The corrected green signal is divided by the corrected red signal to produce a normalized signal, which is independent of variations in optical and electrical gain. A green and a red LED serve to determine the position of the filter wheel which is necessary for signal demultiplexing.

The sensor is of flexible construction and hence presents no risk of breakage into sharp bends. The dye is nontoxic, and the sensor is capable of measuring pH of tissue and blood pH over the physiological range. When compared in 0.05 *M* phosphate buffer at 25°C against a standard laboratory pH electrode, the optrode readings had a standard deviation of less than 0.01 pH units. The temperature coefficient, expressed as the change in pH indication per degree Celsius was 0.017 between 20 and 40°C, and a change of 0.01 pH units was observed per 11% change in ionic strength over the range of 0.05 to 0.3 *M*. The response time for the signal to drop to 63% of its initial value is 0.7 min. *In vitro* blood evaluation in heparinized dog blood gave 0.017 pH unit standard deviation.

Markle et al.[59] redesigned this pH sensor for practical use. In the form of a 25-gauge hypodermic needle with an ion-permeable side window and improved instrumentation and signal processing, a 90% response time of 30 s and a precision of ±0.001 pH units was claimed. An evaluation *in vivo* of a fiber optic pH sensor based on the same principle has also been reported by Tait et al.,[60] who confirmed the correlation between pH values obtained from a fiber optic photometric probe and a conventionnal miniature glass pH electrode in response to graded reductions in regional myocardial blood flow.

With a system of four miniature fiber-optic pH probes implanted at different transmural depths, simultaneous measurements of transmural pH gradient in canine myocardial ischemia were performed.[61] Each probe consisted of two optical fibers of 0.075 mm diameter and a semipermeable membrane of 0.12 mm diameter containing a pH sensitive dye, housed in a modified 0.5 mm stainless steel needle. The fiber optic probe system was calibrated with buffers of pH 6.200, 6.840, and 7.384 and showed a 90% response in 60 s. The pH of an additional arterial blood probe agreed with the pH of the arterial blood as determined with standard blood gas analysis (mean difference 0.01 ± 0.005, n = 6). The intramyocardial pH probes agreed closely with that monitoring the pH of arterial blood (mean difference 0.02 ± 0.006, n = 16). The readings were stable over several hours, and the mean pH change over 2 h was 0.03 ± 0.016 (n = 6).

A similar principle was applied by Suidan et al.[62] using sulfo-phenolphthalein as an indicator. Again the ratio of the intensities of green and red light, as returned by the indicator, was measured. The probe was tested at 37°C in the pH range 7.04 to 7.44. With buffer

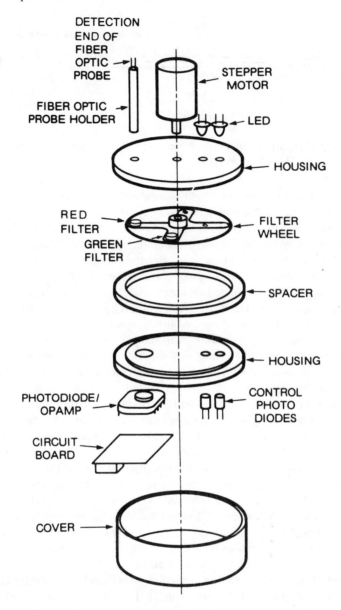

FIGURE 5.    Exploded view of light sensor mechanical assembly (From Ratzlaff, E. H., Harfmann, R. G., and Crouch, S. R., *Anal. Chem.*, 56, 342, 1984. With permission.)

solutions, the mean difference between regular pH and fiber optic pH measurements was 0.010 pH units, with a standard deviation of ± 0.013 pH units. *In vitro* experiments with human blood samples gave a mean difference of –0.005 ± 0.013 pH units.

Other reflectance-based pH sensors[10,17] exploit the spectral changes of bromthymol blue (pK 7.7). A photometric investigation on six indicator dyes absorbed on beads of a copolymer of styrene and divinylbenzene (XAD-2) was performed in order to test the ability of XAD-2 to retain the absorbed indicators.[10] The dyes were adsorbed onto the polymer by simply placing the dry beads in a 0.1% indicator solution in methanol for 4 h.

Using this material, a pH optrode[17] was fabricated from a bifurcated plastic 16-fiber bundle of 1 mm diameter. One half of the bifurcated bundle was employed to transmit visible radiation to the sensitive tip, and the other was used to conduct the radiation reflected from the sensitive tip to the detector. The dyed polymer beads are retained by a porous membrane of poly(tetrafluoroethylene) (PTFE). The greatest sensitivity in pH measurement was obtained at a wavelength of 593 nm. No significant ionic strength effect was observed in a series of buffer solutions containing 0.1, 0.5, and 1 $M$ sodium chloride, respectively.[10] Temperature coefficients of immobilized bromothymol blue and thymolphthaleine, respectively, between 25 and 45°C were said to be $0.013 \pm 0.003$ and $0.015 \pm 0.003$ per degree Celsius. The response time for 63% of the total signal change to occur was 65 s. It was concluded that the rate-determining step that governs the response time is the rate of diffusion of solvent and ions across the hydrophobic PTFE membrane.

In a consecutive paper,[63] the development of a small portable solid state instrument for use with the pH optrode[17] has been reported. The instrument employs two light sources: a measuring LED with an emission wavelength of 635 nm, and a reference LED with an emission wavelength of 820 nm. The latter is not adsorbed by the indicator. The measuring LED was modified for connection to an optical fiber by removal of part of its body, so that there was a flat, polished surface perpendicular to the direction of light emission and very close to the emitting junction. A spherical glass bead (approximately 1 mm diameter), acting as a focus lens, was cemented in an indentation made over the emitting area. The light from the two sources was combined by using a bifurcated light conduit consisting of 16 fibers, with 8 fibers going to source and detector. Signal demultiplexing was achieved by using the time-division method. The pH sensor was tested in a buffer solution by varying the pH of this solution. In the pH range 7 to 9 the response is linear, with a resolution of ±0.01 pH units (Figure 6). A similar two-wavelength time division multiplexed system was used in an inexpensive fiber optic pH sensor with reportedly rapid response.[64] It measures the changes in the absorption of an indicator dye and provides a reference channnel to compensate for losses in the lightpath, with solid-state LED sources and detector being used.

In order to develop a fiber optic pH sensor for seawater monitoring, Monici et al.[36] adsorbed phenol red on nonionic spheres of polystyrene-divinylbenzene (amberlite XAD-2 resin). The design of the sensor typ (Figure 7) has been improved with respect to robustness and sensitivity by making use of a microlens to collimate and to expand the light beam. Further reduction of the time constant (e.g., by proper choice of the proton-permeable membrane, smaller resin sphere dimensions, and appropriate packing) are necessary to fulfill the requirements of ocean depth profiling.

An alternative design for a miniature fiber optic pH sensor for chemical process control and medical applications has been proposed in studies performed by Boisdé and others.[11,22,65] Thymol blue and bromophenol blue have been adsorbed on microspheres of cross-linked polystyrene-divinylbenzene copolymer (XAD-4) of about 1 mm diameter (see Chapter 14). The probe has been constructed according to Figure 8 using silica fibers with low losses over large distances. The apparent pK values (2.05 and 12) of thymol blue on the support differ significantly from those in solution (1.65 and 9.2) and seem to be related to the concentration of the dye on the support. The apparent pK value shifts toward the real pK value upon aging, and response time decreases after 3 months of testing. The excellent reversibility, as observed at highly acidic and basic pH, is decreased at neutral pH and shows some hysteresis.

Similarly, cresol red was adsorbed on an anion exchange resin.[66] The sensor was then constructed as described before.[17] In accordance with the two pK values of the dye, the response in pH change of the normalized reflected intensity was linear from 0.025 to 0.60 $M$ HCl and in the pH range 6.1 to 7.2. The response time was found to be shorter for the acidic pH range than for the neutral pH range. During a test period of 3 months, the sensor proved to be stable when kept in distilled water. The sensor has been applied for pH measurements in pasteurized dairy milk. In order to avoid matrix effects, milk solutions of different pH were used for calibration.

Goldfinch and Lowe[46] have covalently immobilized bromcresol green and bromthymol

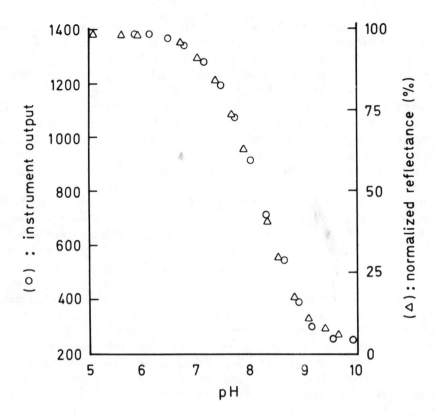

FIGURE 6.    Probe response vs. pH. (O) measured with optoelectronic instrument; $\Delta$, measured with conventional instrumentation. (Redrawn from Guthrie, A. J., Narayanaswamy, R., and Welti, N. A., *Talanta*, 35, 157, 1988. With permission.)

blue on cellulose strips to obtain a pH sensitive layer whose absorbance changes with pH. A LED served as the light source, and a photodiode as the light detector. The bromthymol blue-dyed layer changes its color from yellow to blue over the pH 6 to 10 range and exhibits a pK of around 7.5. The bromcresol green membrane responds to pH over the 3.3 to 6.8 range with a pK of approximately 5.1. This optoelectronic sensor provides the basis for the enzyme sensors developed by the same group.

An optosensing technique for determining pH in solutions of low buffering capacity and low ionic strength has also been reported.[67] Three commercially available pH strips with immobilized indicators of different pK values (1.80 ± 0.22, 3.82 ± 0.17, 5.28 ± 0.01) were tested. The fibrous indicator pad was attached in a modified flow cell. Over a 3-d time period, more than 2365 injections were made on a single pH path, and no deterioration of the indicator could be observed. A small decrease in signal height was attributed to the displacement of the pad in the cell. An increase in response time was observed with decreasing phosphate buffer concentration.

## B. FLUORESCENCE-BASED SENSORS

Fluorescence is particularly well suited for optical sensing because emitted light returning from the sensor easily can be distinguished from the exciting light. Owing to the sensitivity of fluorescence, even low fluorophore levels can be detected. This is of particular importance when inner filter effects due to high analyte or indicator concentrations are to be avoided. For weakly absorbing species, i.e., when A < 0.05, the intensity $I_f$ of fluorescence light returning from the sensor is proportional to the intensity of the exciting radiation, $I_o$, and the concentration ([D]) of the fluorescent dye in the sensor:

Phenol Red adsorbed on Amberlite XAD-2 resin

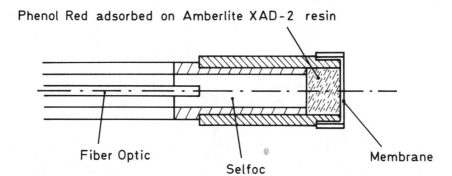

FIGURE 7.   Schematic representation of the tip of a pK sensor based on reflectance. (Redrawn from Monici, M., Boniforti, R., Buzzigoli, G., DeRossi, D., and Nannini, A., *Proc. SPIE*, 798, 294, 1987. With permission.)

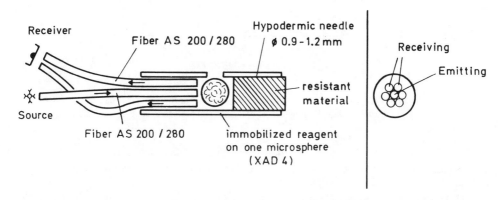

FIGURE 8.   Schematic diagram of a pH optrode (Redrawn from Boisdé, G. and Pérez, J. J., *Proc. SPIE*, 798, 238, 1987. With permission.)

$$I_f = k' \cdot I_o \cdot \theta \cdot \varepsilon \cdot l \cdot [D] \tag{18}$$

where l is the length of the light path in the sensing layer, $\varepsilon$ is the molar absorptivity, $\theta$ is the is the quantum efficiency of fluorescence, and k′ is the fraction of emission that can be measured. At constant $I_o$, Equation 18 can be simplified to

$$I_f = k \cdot [D] \tag{19}$$

where $k = k' I_o \theta \varepsilon l$.

A variety of fluorescent indicators is known (Chapter 7), but only a few meet the requirements of an excitation maximum beyond 400 nm to allow the use of inexpensive plastic fiber optics as light guide as well as inexpensive light sources such as halogen lamps or light emitting diodes. Large Stokes shifts are required in order to conveniently separate scattered exciting light from fluorescence light with inexpensive light filters. Further requirements are the lack of toxicity, photostability, and the presence of functional chemical groups suitable for covalent immobilization.[68]

In a study on fluorescent indicators for physiological pH values,[68] certain coumarins and the trisodium salt of 8-hydroxy-1,3,6-pyrenetrisulfonate (HPTS) were found to be most useful indicators for measuring near-neutral pH values. At physiological ionic strengths, the pK of HPTS is approximately 7.3. HPTS may be excited at either 470 or 405 nm to give a

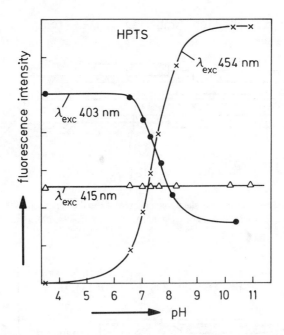

FIGURE 9.    Plot of HPTS fluorescence intensity at 520 nm as a function of pH value at various excitation wavelengths $\lambda_{exc}$. (From Wolfbeis, O. S., Fürlinger E., Kroneis, H., and Marsoner, H., *Fresenius Z. Anal. Chem.*, 314, 119, 1983. With permission.)

fluorescence at 520 nm (Figure 9). It fluoresces from the phenolate form between pH 1 and 11, irrespective of whether the phenol form (at 405 nm) or the phenolate form (at 470 nm) is excited.

Probably the first pH fluorosensor ever described was obtained[44] by covalent immobilization of HPTS (as its trisulfochloride) onto amino groups of chemically modified cellulose dialysis membranes (cuprophane) of 8 to 19 µm thickness, and silica beads (µ-Bondapak-NH$_2$) that were fixed on solid supports. The membranes were attached onto the walls of a flow-through cell and solutions of various pH were passed by. By measuring the fluorescence at 530 nm under 460-nm excitation, a 90% response of about 1.8 min was found, using 0.17 *M* phosphate buffers (Figure 10).

Because HPTS has three sulfonato groups, it can be immobilized conveniently and essentially irreversibly on strong anion exchange materials. Thus, Zhujun and Seitz[15] obtained a pH sensor by electrostatic immobilization of HPTS on an anion-exchange membrane. It allowed the measurement of pH in the range 6 to 8 by relating the ratio of fluorescence intensities emitted at 510 nm and excited at 405 nm (selective for acid form) and 470 nm (specific for base form). The ratio was not affected by source fluctuations, drift, and slow loss of reagent, all of which can affect the single intensity measurement. The sensor showed an approximately 10% loss in intensity after 4 h of continuous illumination. HPTS was found to be chemically decomposed in the presence of even very small amounts of free radicals.[48] When irradiated with day light, radicals are formed in some anion exchange membranes containing halogenated polymeric compounds.

An important observation is the effect of indicator loading on the response curve of a sensor (Figure 11). With increasing indicator amounts being immobilized, the relative signal change becomes smaller and the pK is shifted to lower values. The authors have given an equation[15] that allows calculation of the curves.

The preparation and performance of two optical sensing materials for measurement of near-neutral pH values were also described.[14] HPTS and 7-hydroxycoumarin-3-carboxylic acid (HCC), respectively, were covalently immobilized on surface-modified controlled porous glass (CPG) platelets. Analytical excitation and emission wavelenghts were, respectively, 410 and 455 nm for the HCC-based sensor, and 465 and 520 nm for the HPTS-based sensor. On CPG, the pK values (6.8 for HPTS and 6.9 for HCC) were lower than those determined in solution (7.30 for HPTS and 7.01 for HCC).

The decrease in pK and shifts in the emission maxima of immobilized HPTS were attributed to changes of chemical structure after immobilization. Since HPTS has three sulfonic acid groups, up to three covalent sulfonamide links may be formed by the immobilization chemistry. The pK shift was estimated to be –0.53 units per covalent link. The effect of ionic strength could be reduced (but not eliminated) by chemical silanization of the glass surface. The fluorescence spectra of immobilized HPTS (Figure 12) and HCC (Figure 13) are similar to those in solution, except for a slight shift to longer wavelengths for HPTS. Too heavy loading of CPG with indicator resulted in a background emission from the neutral species. These observations coincide with those of Narayanaswamy and Sevilla[32] and Bacci et al.[33] observed with absorbance-based acid-base dyes immobilized on hydrophobic polymer supports.

A miniaturized fluorescence-based pH sensor designed for *in vivo* applications (along with an $O_2$ and $CO_2$) has been described by Gehrich et al.[4] HPTS was covalently bound to hydrophilic cellulose matrix attached to the distal end of a single optical fiber. The cellulose matrix was coated with an opaque cellulose overcoat which provides both mechanical integrity and optical isolation from environmental optical interferences (Figure 14). The device employs a xenon arc lamp with a mechanical filter wheel and a bulk optical lens system. Fluorescence intensity is measured at 520 nm with 460 nm excitation and with 410 nm excitation. These intensities are dependent upon the concentrations of the basic and acid form of the dye. The ratio of the two signals can be related to pH and is relatively insensitive to optical throughput.

Another type of fluorescence-based pH sensor[18] consists of an individual porous glass bead fixed with an epoxy resin to a single optical fiber. The probe relies on changes of the fluorescence characteristics of immobilized aminofluorescein with pH. The instrumental configuration consisted of an argon ion laser, a double monochromator, a photomultiplier tube, and a photometer. Considerable variations in the absolute response to pH were found due to the irregular shapes and the range of sizes of the beads.

In order to avoid the cumbersome procedures preparing miniaturized sensors by attaching preformed sensing materials onto single fibers, it has been proposed[13] to coat the reagent layer directly on the end face of a fiber via thermal or photopolymerization. In a typical experiment, a fluoresceinamine was copolymerized with acrylamide and N,N,′methylenebis(acrylamide) onto the distal end of surface-modified single glass/glass optical fibers of 100/140 μm diameter. The polymer-modified single fiber sensors have a response time of 9 s or less, when measured in McIlvaine's standard buffers. One disadvantage is the poor sensor reproducibility due to difficulty in controlling the polymerization process, the need for an argon ion laser, and a rather sophisticated detection techniques.

To further decrease the response time and the reproducibility of single fiber optic pH sensors, Kawabata et al.[50] immobilized a monolayer of fluorescein directly on the end surface of a surface modified optical fiber. Again, an argon ion laser was used as an exciting source. Fluorescence was detected by a photomultiplier after passing a monochromator. Photodecomposition of the indicator occurred with an output power exceeding 30 mW. The transient response time of the sensor was less than 10 ms. However, since fluorescence from the monolayer was very weak, it was necessary to improve the signal-to-noise ratio by electronic smoothing. As a result, the observed response time was governed finally by the time constant of the amplifier and found to be 9 s for 63% response. Because of high fluctuations in the background signal, the precision in pH measurement was ±0.1 pH units. Obviously, in this case there will always be a trade-off between response time and precision.

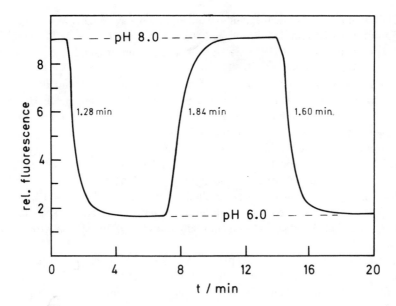

FIGURE 10.   Response of HPTS, immobilized on cellulose, towards changes in pH. The times given are the 99% response times. Excitation at 465 nm, emission taken at 520 nm. (Redrawn from Fürlinger, E., Ph.D. thesis, University of Graz, 1982. With permission.)

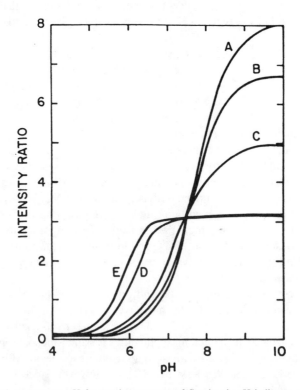

FIGURE 11.   Calculated response vs. pH for varying amounts of fluorigenic pH indicator on a membrane. (A) Corresponds to a low concentration of indicator which does not significantly absorb excitation radiation; (B) through (E) correspond to increasing levels of indicator. (From Zhujun, Z. and Seitz, W. R., *Anal. Chim. Acta*, 160, 47, 1984. With permission.)

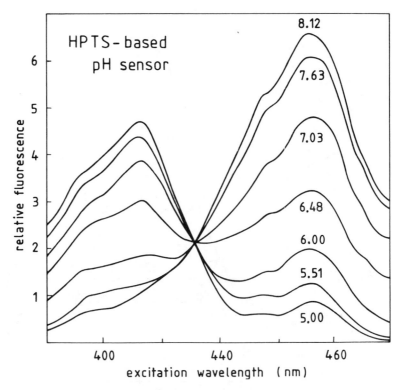

FIGURE 12. Excitation spectra of glass-immobilized HPTS at various pH values (phosphate buffer, 23°C, emission taken at 520 nm). (From Offenbacher, H., Wolfbeis, O. S., and Fürlinger E., *Sens. Actuators*, 9, 73, 1986. With permission.)

Czolk[70] has used Celgard™ membranes as a solid support onto which tetr(sulfophenyl)porphyrin was immobilized. When placed at a fiber end, fluorescence intensity responds in the pH 4.0 to 5.5 range. The dye can be excited with the 633-nm line of the helium-neon laser. Because of the presence of two emission bands, an internal referencing is possible by ratioing the intensities as measured at two wavelengths.

## C. ENERGY-TRANSFER-BASED SENSORS

Jordan and Walt[49] developed a single-fiber optic pH sensor based on energy transfer from a pH insensitive fluorophore, eosin (the donor), to a pH sensitive absorber, phenol red (the acceptor). The dyes were co-immobilized with acrylamide on the distal end of a surface-silanized single optical fiber. The pH-sensitive layer had a thickness of approximately 10 μm. An argon ion laser was used to excite eosin with light of 488 nm. The fluorescence spectrum of eosin overlaps with the absorption spectrum of the base form of phenol red. As the pH increases, the concentration of the base form of phenol red increases, resulting in an increased energy transfer from eosin to phenol red and in a diminished fluorescence intensity of eosin. Thus, changes in the absorption of phenol red as a function of pH are detected as changes in the fluorescence signal of eosin. The intensity of the fluorescence signal was observed with a photon-counting detector at the emission maximum of eosin (546 nm) after passing a dichroic mirror, a longpass filter, and a monochromator. The precision was ±0.008 pH units when measured with McIlvaine buffer solutions. The time required for 63% response is 0.07 min for a pH change from 7.1 to 6.5. Some photobleaching of the base form of phenol red has been observed when continuously exciting the sensor over a 10-min period. This problem could be solved by opening a shutter for short time periods during measurements.

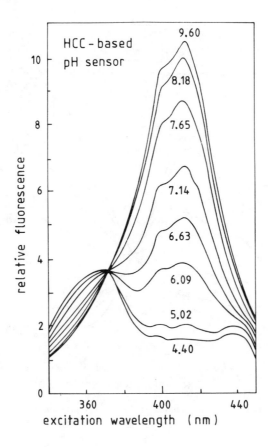

FIGURE 13. Excitation spectra of glass-immobilized HCC at various pH values (phosphate buffer, 23°C, emission taken at 520 nm). (From Offenbacher, H., Wolfbeis, O. S., and Fürlinger E., *Sens. Actuators*, 9, 73, 1986. With permission.)

Wolfbeis and Schulman[52] have demonstrated the feasibility of measuring pH via fluorescence lifetime. A strongly fluorescent primary absorber was used as the pH-independent fluorescer whose lifetime is measured. A pH-dependent dye, whose absorption occurs at the wavelength of the fluorescence of the primary absorber, quenches the fluorescence as a function of pH and thereby reduces its lifetime. The advantage of this system relies on the independence of fluorescence lifetime on fluorophor concentration and light source and detector fluctuations.

A sensor system for pH or $pCO_2$ measurements was described that is based on the energy transfer from a fluorescent donor to a fluorescent and pH-sensitive acceptor.[53] By measuring the ratio of the fluorescence intensities of donor and acceptor, an internally referenced pH-sensitive signal is obtained.

## D. OTHER pH SENSOR TYPES

Recently, Attridge et al.[54] discussed a novel design of a fiber optic pH sensor which exploits the variation in refractive index of the cladding with pH. A coaxial directional coupler was applied by sensing changes in the refractive index of a polymer in which was bound electrostatically a phthalein or sulfophthalein dye. The coupler consists of two monomode waveguides placed a few micrometers apart over a well-defined interaction length. The feasibility of making a pH sensor has been demonstrated and further work has to be done for designing an experimental sensor.

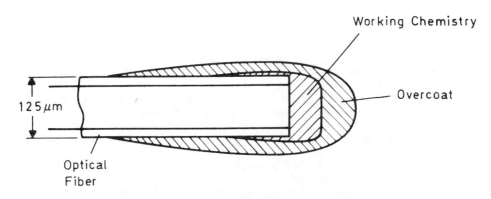

FIGURE 14.    Diagram of an intravascular pH-optrode. The pH sensitive dye (HPTS) is bound to a cellulose matrix. (Redrawn from Gehrich, J. L., Lübbers, D. W., Opitz, N., Hansmann, D. R., Miller, W. W., Tusa, J. K., and Yafuso, M., *IEEE Trans. BioMed. End.*, 33, 117, 1986. With permission.)

A variety of chemical materials is known to undergo a strongly pH-dependent swelling. This could be exploited in an interferometric pH sensor if interferences by other ions can be eliminated. Similarly, one of the numerous other effects of pH on the optical and physical properties of certain materials may be used in pH sensors yet to be developed.

It has been shown[55] that, rather than immobilizing a fluorescent probe on the fiber tip, pH values can be measured in fermentation broths by simply adding a fluorescent pH indicator to the solution. By measuring the fluorescence with plain fibers at two wavelengths, an optical pH determination is possible which very favorably compares with electrode measurements.

# REFERENCES

1. **Seitz, W. R.,** Chemical sensors based on fiber optics, *Anal. Chem.*, 56, 16A, 1984.
2. **Peterson, J. I. and Vurek, G. G.,** Fiber-optic sensors for biomedical applications, *Science*, 224, 123, 1984.
3. **Borman, S.,** Optrodes, *Anal. Chem.*, 53, 1616A, 1981.
4. **Gehrich, J. L., Lübbers, D. W., Opitz, N., Hansmann, D. R., Miller, W. W., Tusa, J. K., and Yafuso, M.,** Optical fluorescence and its application to an intravascular blood gas monitoring system, *IEEE Trans. BioMed. Eng.*, 33, 117, 1986.
5. **Peterson, J. I., Goldstein, S. R., Fitzgerald, R. V., and Buckhold, D. K.,** A fiber optic pH probe for physiological use, *Anal. Chem.*, 52, 864, 1980.
6. **Goldstein, S. R., Peterson, J. I., and Fitzgerald, R. V.,** A miniature fiber optic pH sensor for physiological use, *J. Biomech. Eng.*, 102, 141, 1980.
7. **Janata, J.,** Do optical sensors really measure pH?, *Anal. Chem.*, 59, 1351, 1987.
8. **Ratzlaff, E. H., Harfmann, R. G., and Crouch, S. R.,** Absorption-corrected fiber optic fluorometer, *Anal. Chem.*, 56, 342, 1984.
9. **Edmonds, T. E., Flatters, N. J., Jones, C. F., and Miller, J. N.,** Determination of acid-base indicators: implications for optical fibre probes, *Talanta*, 35, 103, 1988.
10. **Kirkbright, G. F., Narayanaswamy, R., and Welti, N. A.,** Studies with immobilised chemical reagents using a flow cell for the development of chemically sensitive fibre-optic devices, *Analyst*, 109, 15, 1984.
11. **Boisdé, G. and Pérez, J. J.,** Miniature chemical optical fiber sensors for pH measurements, *Proc. SPIE*, 798, 238, 1987.
12. **Nylander, C.,** Chemical and biological Sensors, *J. Phys. E.*, 18, 736, 1985.
13. **Munkholm, C., Walt, D. R., Milanovich, F. P., and Klainer, S. M.,** Polymer modification of fiber optic chemical sensors as a method of enhancing fluorescence signal for pH measurement, *Anal. Chem.*, 58, 1427, 1986.
14. **Offenbacher, H., Wolfbeis, O. S., and Fürlinger, E.,** Fluorescence optical sensors for continuous determination of near-neutral pH values, *Sens. Actuators*, 9, 73, 1986.
15. **Zhujun, Z. and Seitz, W. R.,** A fluorescence sensor for quantifying pH in the range from 6. 5 to 8. 5, *Anal. Chim. Acta*, 160, 47, 1984.
16. **Saari, L. A., and Seitz, W. R.,** pH sensor based on immobilized fluoresceinamine, *Anal. Chem.*, 54, 821, 1982.
17. **Kirkbright, G. F., Narayanaswamy, R., and Welti, N. A.,** Fibre-optic pH probe based on the use of an immobilised colorimetric indicator, *Analyst*, 109, 1025, 1984.
18. **Fuh, M. R. S., Burgess, L. W., Hirschfeld, T., and Christian, G. D.,** Single fibre optic fluorescence pH probe, *Analyst*, 112, 1159, 1987.
19. **Peterson, J. I. and Goldstein, S. R.,** U. S. Patent 4,200,110, 1980.
20. **Cramp, J. H., Ferguson, R. F., and Brian, D. B.,** European Patent Appl. 61,884, 1982.
21. **Kirkbright, G. F. and Welti, N. A.,** European Patent Appl. 126,600, 1984.
22. **Boisdé, G. and Perez, J. J.,** European Patent Appl. 88,400, 719-6, 1988.
23. **Wolfbeis, O. S., Kroneis, H., and Offenbacher, H.,** German Off. 3,343,636, 1987
24. **Wolfbeis, O. S., Kroneis, H., and Offenbacher, H.,** U.S. Patent 4,568,519, 1986.
25. **Peterson, J. I. and Goldstein, R. S.,** U.S. Patent Appl. 855,384, 1978; through Chem. Abstr. 90:99677, 1979.
26. **Peterson, J. I.,** U.S. Patent Appl. 855,397, 1978.
27. **Peterson, J. I.,** U.S. Patent 4,194,877, 1980.
28. **Offenbacher, H. and Kroneis, H.,** U.S. Patent 4,776,869, 1988.
29. **Hirschfeld, T.,** European Patent Appl. 304,141, 1986.
30. **Seitz, W. R. and Zhujun, Z.,** U.S. Patent 4,548,907, 1985
31. **Stewart, P. A.,** Modern quantitative acid-base chemistry, *J. Physiol. Pharmacol.*, 61, 1444, 1983.
32. **Narayanaswamy, R. and Sevilla, F. S.,** Reflectometric study of the acid-base equilibria of indicators immobilised on a styrene/ divinylbenzene copolymer, *Anal. Chim. Acta*, 189, 365, 1986.
33. **Bacci, M., Baldini, F., and Scheggi, A. M.,** Spectrophotometric investigations on immobilized acid-base indicators, *Anal. Chim. Acta*, 207, 343, 1988.
34. **Ramette, R., Culberson, C. H., and Bates, R. G.,** Acid-base properties of tris(hydroxymethyl) aminomethane, (Tris) buffers in seawater from 5 to 40°C, *Anal. Chem.*, 49, 867, 1977.
35. **Robert-Baldo, G. L., Morris, M. J., and Byrne, R. H.,** Spectrophotometric determination of seawater pH using phenol red, *Anal. Chem.*, 57, 2564, 1985.
36. **Monici, M., Boniforti, R., Buzzigoli, G., DeRossi, D., and Nannini, A.,** Fibre-optic pH sensor for seawater monitoring, *Proc. SPIE*, 798, 294, 1987.

37. **Opitz, N. and Lübbers, D. W.,** New fluorescence photometrical techniques for simultaneous and continuous measurements of ionic strength and hydrogen ion activities, *Sens. Actuators*, 4, 473, 1983.

38. **Wolfbeis, O. S. and Offenbacher, H.,** Fluorescence sensor for monitoring ionic strength and physiological pH values, *Sens. Actuators*, 9, 85, 1986.

39. **Baldini, F., Brenci, M., Conforti, G., Falciai, R., and Mignani, A. G.,** Model for an optical fiber pH sensor, in *Proc. NATO ASI on Optical Fiber Sensors*, Erice, Italy, May 2 to 10, 1986, Martinus Nijhoff, The Hauge, 1986, 437.

40. **Luo, S. and Walt, D. R.,** Fiber optic sensors based on reagent delivery with controlled release polymers, *Anal. Chem.*, submitted.

41. **Knobbe, E. T., Dunn, B., and Gold, M.,** Organic molecules entrapped in a silica host for use as biosensor probe materials, *Proc. SPIE*, 906, 39, 1988.

42. **Harper, B. G.,** Re-usable glass-bound pH indicators, *Anal. Chem.*, 47, 348, 1975.

43. **Offenbacher H.,** Fluorescence Optical Sensors for Clinical Analysis, Ph.D. thesis, University of Graz, 1984.

44. **Fürlinger E.,** A Spectroscopic Study on Fluorescent pH Indicators and their Immobilization on Polymeric Supports, Ph.D. thesis, University of Graz, 1982.

45. **Jones, T. P. and Porter, M. D.,** Optical pH sensor based on the chemical modification of a porous polymer film, *Anal. Chem.*, 60, 404, 1988.

46. **Goldfinch, M. J. and Lowe, C. R.,** Solid-phase optoelectronic sensors for biochemical analysis, *Anal. Biochem.*, 138, 430, 1984.

47. **Boisde, G., Biatry, B., Magny, B., Dureault, B., Blanc, F., and Sebille, B.,** Comparison between two dye immobilization techniques on optodes for the pH measurement by absorption and reflectance, *Proc. SPIE*, 1172, 239, 1989.

48. **Leiner, M. J. P. and Wolfbeis, O. S.,** Unpublished results.

49. **Jordan, D. M. and Walt, D. R.,** Physiological pH fiber-optic chemical sensor based on energy transfer, *Anal. Chem.*, 59, 437, 1987.

50. **Kawabata, Y., Tsuchida, K., Imasaka, T., and Ishibashi, N.,** Fiber-optic pH sensor with monolayer indicator, *Anal. Sci.*, 3, 7, 1987.

51. **Lübbers, D. W. and Opitz, N.,** Blood gas analysis with fluorescence dyes as an example of their usefulness as quantitative chemical sensors, *Anal. Chem. Symp. Ser.*, 17, 609, 1983.

52. **Wolfbeis, O. S. and Schulman, S. G.,** Optical measurement of pH via fluorescence lifetime, unpublished results, 1990.

53. **Leader, M. J. and Kamiya, T.,** European Patent Appl. 283,116, 1988.

54. **Attridge, J. W., Leaver, K. D., and Cozens, J. R.,** Design of a fibre-optic pH sensor with rapid response, *J. Phys. E*, 20, 548, 1987.

55. **Junker, B. H., Wang, D. I. C., and Hatton, T. A.,** Fluorescence sensing of fermentation parameters using fiber optics, *Biotechnol. Bioeng.*, 32, 55, 1988.

56. **Goldstein, S. R. and Peterson, J. I.,** A miniature fiber optic pH sensor suitable for in vivo application, *Adv. Bioeng.*, 1977, 81, 1977.

57. **Popa, G., Luca, C., and Iosif, E.,** Spectrophotometric determination of the dissociation constants of pH indicators from the sulphonphthalein class, *Omagin Rahuta Ripan*, 1966, 457, 1966, *Chem. Abstr.*, 67: 94493, 1967.

58. **Kadin, H.,** Second derivative spectophotometric determination of the dissociation constant of covalently bound phenol red/ polyacrylamide, *Anal. Lett.*, 17, 1245, 1984.

59. **Markle, D. R., McGuire, D. A., Goldstein, S. R., Patterson, R. E., and Watson, R. M.,** A pH measurement system for use in tissue and blood, employing miniature fiber optic probes, in *1981 Adv. Bioeng.*, Viano, D. C., Ed., Am. Soc. Mech. Eng., New York, 1981, 123.

60. **Tait, G. A., Young, R. B., Wilson, G. J., Steward, D. J., and MacGregor, D. C.,** Myocardial pH during regional ischemia: Evaluation of a fiber optic probe, *Am. J. Physiol. Heart Circ. Physiol.*, 15, H232, 1984.

61. **Watson, R. M., Markle, D. R., Ko, Y. M., Goldstein, S. R., McGuire, D. A., Peterson, J. I., and Patterson, R. E.,** Transmural pH gradient in canine myocardial ischemia, *Am. J. Physiol. Heart Circ. Physiol.*, 15, H232, 1984.

62. **Suidan, J. S., Young, B. K., Hetzel, F. W., and Seal, H. R.,** pH measurement with a fiber-optic tissue pH monitor and a standard blood pH meter, *Clin. Chem.*, 29, 1566, 1983.

63. **Guthrie, A. J., Narayanaswamy, R., and Welti, N. A.,** Solid-state instrumentation for use with optical-fibre chemical sensors, *Talanta*, 35, 157, 1988.

64. **Grattan, K. T. V., Mouaziz, Z., and Palmer, A. W.,** Dual wavelength optical fibre sensor for pH measurements, *Biosensors*, 3, 17, 1987/88.

65. **Boisdé, G., Blanc, F., and Pérez, J. J.,** Chemical measurements with optical fibers for process control, *Talanta*, 35, 75, 1988.

66. **Moreno, M. C., Martinez, A., Millan, P., and Camara, C.,** Study of pH sensitive optical fibre sensor based on the use of cresol red, *J. Mol. Struct.*, 143, 553, 1986.
67. **Woods, B. A., Ruzicka, J., and Christian, G. D.,** Measurement of pH in solutions of low buffering capacity and low ionic strength by optosensing flow injection analysis, *Anal. Chem.*, 58, 2496, 1986.
68. **Wolfbeis, O. S., Fürlinger, E., Kroneis, H., and Marsoner, H.,** A study of fluorescent indicators for measuring near neutral ("physiological") pH-values, *Fresenius Z. Anal. Chem.*, 314, 119, 1983.
69. **Offenbacher, H. and Schwarzenegger, E.,** European Patent Appl. 300,990, 1989.
70. **Czolk, R.,** Studies on the fluorimetric application of light guides as exemplified by the pH dependence of the fluorescence of tetra-(p-sulfophenylporphyrin; Diplom-Arbeit, Universität und Kernforschungs-Zentrum Karlsruhe, 1986; in German.

# INDEX

## A